HAZARDOUS WASTE SITE SOIL REMEDIATION

Environmental Science and Pollution Control Series

1. Toxic Metal Chemistry in Marine Environments, *Muhammad Sadiq*
2. Handbook of Polymer Degradation, *edited by S. Halim Hamid, Mohamed B. Amin, and Ali G. Maadhah*
3. Unit Processes in Drinking Water Treatment, *Willy J. Masschelein*
4. Groundwater Contamination and Analysis at Hazardous Waste Sites, *edited by Suzanne Lesage and Richard E. Jackson*
5. Plastics Waste Management: Disposal, Recycling, and Reuse, *edited by Nabil Mustafa*
6. Hazardous Waste Site Soil Remediation: Theory and Application of Innovative Technologies, *edited by David J. Wilson and Ann N. Clarke*
7. Process Engineering for Pollution Control and Waste Minimization, *edited by Donald L. Wise and Debra J. Trantolo*
8. Remediation of Hazardous Waste Contaminated Soils, *edited by Donald L. Wise and Debra J. Trantolo*
9. Water Contamination and Health: Integration of Exposure Assessment, Toxicology, and Risk Assessment, *edited by Rhoda G. M. Wang*
10. Pollution Control in Fertilizer Production, *edited by Charles A. Hodge and Neculai N. Popovici*
11. Groundwater Contamination and Control, *edited by Uri Zoller*
12. Toxic Properties of Pesticides, *Nicholas P. Cheremisinoff and John A. King*
13. Combustion and Incineration Processes: Applications in Environmental Engineering, Second Edition, Revised and Expanded, *Walter R. Niessen*
14. Hazardous Chemicals in the Polymer Industry, *Nicholas P. Cheremisinoff*
15. Handbook of Highly Toxic Materials Handling and Management, *edited by Stanley S. Grossel and Daniel A. Crowl*

Additional Volumes in Preparation

HAZARDOUS WASTE SITE SOIL REMEDIATION

Theory and Application of Innovative Technologies

edited by

David J. Wilson
Vanderbilt University
Nashville, Tennessee

Ann N. Clarke
ECKENFELDER INC.
Nashville, Tennessee

CRC Press is an imprint of the
Taylor & Francis Group, an **informa** business

CRC Press
Taylor & Francis Group
6000 Broken Sound Parkway NW, Suite 300
Boca Raton, FL 33487-2742

First issued in paperback 2019

ISBN-13: 978-0-8247-9107-0 (hbk)
ISBN-13: 978-0-367-40222-8 (pbk)

Visit the Taylor & Francis Web site at
http://www.taylorandfrancis.com

and the CRC Press Web site at
http://www.crcpress.com

Library of Congress Cataloging-in-Publication Data

Hazardous waste site soil remediation : theory and application of
innovative technologies / edited by David J. Wilson, Ann N. Clarke.
p. cm. -- (Environmental science and pollution control series ; 6)
Includes bibliographical references and index.
ISBN 0-8247-9107-X (alk. paper)
1. Hazardous wastes--Environmental aspects. 2. Soil remediation.
I. Wilson, David J. II. Clarke, Ann N. III. Series:
Environmental science and pollution control ; 6.
TD879.H38H39 1994
628.5'5--dc20 93-30052
CIP

To friends and adversaries around the world
whose work in the lab and in the field made this book possible,
with hopes for better, cheaper cleanups in the future.

Preface

The cost of remediating the tens of thousands of Superfund and Resource Conservation and Recovery Act (RCRA) hazardous waste sites in the U.S. alone has been estimated at $750 billion. These estimates are likely to increase in the future. The financial burden this presents to corporations, consumers, and taxpayers is formidable. Obviously, the development of more effective, less costly remediation technologies would be of great benefit to all concerned.

In the past, reliance has been placed on a relatively limited number of technologies that may no longer be acceptable to regulatory agencies (capping, excavation, and reburial) or are extremely expensive (incineration). Recently, a number of innovative technologies have surfaced that provide the environmental engineer with additional options. In this book we have attempted to provide introductions to these alternatives. We have not included the older technologies noted above because extensive literature on them already exists.

Our book is directed to environmental managers and regulators and to engineers who are not necessarily experts in remediation technologies. These people must read proposals, reports, and recommendations involving the various technologies with understanding and insight. They must be able to communicate with the experts on the technologies, and they must be able to justify their recommendations to their management. It is our hope that this book will be of help in all these tasks.

We note that the treatment of the various technologies in this book shows considerable latitude. This is due to variations in the nature and complexity of the technologies and to their respective degrees of development at the time the book was written. We have included a rather substantial amount of theory because most of these technologies are sufficiently new that there is no large body of practical experience on which one can rely. Under such circumstances, theory may provide a helpful guide in evaluating and selecting techniques to deal with a given menu of contaminants at a specific site.

The first chapter, by James H. Clarke, Danny D. Reible, and Robert D. Mutch, Jr., addresses fundamental aspects of groundwater hydrology and mass transport that provide the stage on which any remediation scheme is played. Much of this chapter is oriented toward the behavior of dense nonaqueous phase liquids (DNAPLs), which present the environmental engineer with some challenging problems in remediation. The second chapter, by Robert D. Mutch, Jr., and Joanna I. Scott, addresses some difficulties that arise in connection with pump-and-treat operations in such diffusion-limited systems as fractured porous bedrock. In these systems, diffusion limitations often lead to frustratingly long remediation times and groundwater contaminant concentrations that exhibit rebound once the pump-and-treat operation ceases.

The third chapter, on the chemical stabilization of contaminated soils, is by Jesse R. Conner and provides a discussion of one of the better established of the innovative technologies by a leader in the field. The chapter addresses such questions as what compounds can be immobilized, what additives are most effective under various conditions, which mixing and materials handling techniques are useful in particular situations, and what costs can be expected.

The fourth chapter, by David J. Wilson and Ann N. Clarke, deals with soil vapor extraction, which in recent years has become widely accepted for the removal of volatile organics from the vadose zone. Mathematical modeling techniques are of some interest in connection with this technology, for assessment and design work; these are discussed in some detail. Wilson and Clarke are joined by Paul R. dePercin in the following chapter. This chapter addresses thermally enhanced soil vapor stripping. Thermal enhancement permits vapor extraction of compounds having vapor pressures that are too low to permit efficient conventional vapor extraction at ambient temperatures.

The ex situ technique of thermal desorption of soil contaminants is discussed in Chapter 6, by Richard J. Ayen, Carl P. Swanstrom, and

Carl R. Palmer. This method provides a lower-cost alternative to incineration for the removal of organics, including those which are virtually nonvolatile at ambient temperatures.

The very broad, rapidly developing subject of enhanced biodegradation for on-site remediation of soils and groundwater is discussed by Ronald E. Hoeppel and Robert E. Hinchee in Chapter 7. Their presentation includes the necessary microbiological background to give the nonspecialist access to current concepts and developments in bioremediation. This is then followed by discussions of the biodegradation of several environmentally important classes of organic compounds as well as of the biotransformations of several environmentally significant metals. The chapter also addresses some of the regulatory hurdles that have tended to hold back the development of bioremediation techniques.

In Chapter 8, Ann N. Clarke, Robert D. Norris, and David J. Wilson discuss the technique of air sparging in the zone of saturation, a rather new method that has been demonstrated to be effective in Europe and may provide a means for increased rates of DNAPL removal from the zone of saturation.

Kenton H. Oma describes in situ vitrification in Chapter 9. This technology has the advantage of remediating soil contaminated with both non-volatile metals and organic compounds, but it is limited to use above the water table.

The book closes with Chapter 10, by David J. Wilson and Ann N. Clarke, on soil surfactant washing and (in situ) flushing. This technique has been under study for some time but has not yet been used in field-scale work. The reasons for this, as well as the promise that these surfactant methods hold, are discussed.

In a field that is developing as fast as the remediation of hazardous waste sites, it is difficult to predict future trends with any confidence. In looking into the future, one can certainly expect methods that are energy-intensive to labor under a progressively increasing handicap in the years to come as energy costs continue to rise. One expects that techniques for site characterization (permeabilities, distribution and concentrations of contaminants, etc.) will become more accurate and less costly as new instruments and methods are developed. This will be augmented by the use of improved methods of data processing and interpretation, such as kriging. One can hope that applied microbiologists will be successful in their attempts to find organisms that are able to break down refractory organics (particularly chlorinated compounds) under a range of common environmental conditions. One might expect mathematical modeling to improve and to become more widely used in selection of technologies on a site-specific basis and in the cost-effective design of remediation schemes employing, perhaps, several technologies at once or in sequence. And one may hope that methods will be forthcoming for the rapid removal of DNAPLs from media in which diffusion rates are so frustratingly slow.

We would like to emphasize the need for more research support from both government and industry for the development of improved technologies. This is especially critical in the development of technically improved and more cost-effective treatment of the vapor phase emissions common to many of the innovative technologies. Obviously any improvements in the effectiveness of remediation techniques themselves can be expected to result in more timely cleanups and in quite substantial savings for industry, consumers, and taxpayers.

David J. Wilson
Ann N. Clarke

Contents

HAZARDOUS WASTE SITE SOIL REMEDIATION

1

Contaminant Transport and Behavior in the Subsurface

James H. Clarke
ECKENFELDER INC.
Nashville, Tennessee

Danny D. Reible
Louisiana State University
Baton Rouge, Louisiana
and University of Sydney
Sydney, Australia

Robert D. Mutch, Jr.
ECKENFELDER INC.
Mahwah, New Jersey

INTRODUCTION

An understanding not only of the nature of soil contamination for a given site, but also of the equilibrium and mass transport processes that govern the potential migration of particular chemicals, is a critical first step in any successful soil contamination evaluation and remediation effort. In the following sections of this chapter, information is presented concerning the important physical and chemical properties of the contaminants and the subsurface environment, together with a practical and applied discussion of potential migration pathways, upper limits of contamination based on equilibrium assumptions, and fundamental mass transport processes that govern the fate of a chemical in the subsurface.

Initially, the focus is on the development of a conceptual and predictive model based on the nature of contamination, the actual geological complexity of the subsurface, and the mass transport processes themselves, which is unquestionably oversimplified. Nevertheless, this approach results in an increased understanding of the equilibrium limits and the transport processes, which is most useful as a basis for further evaluation. Often, the assumption of uniform and homogeneous site conditions can enable the use of mathematical models that yield analytical, closed-form solutions that may be reliable within an order of magnitude. This is, in many cases, sufficient for further evaluations. Should more precise estimates be needed, the additional complexity, concerning both the contamination and the subsurface conditions, will typically dictate the need for the use of dynamic numerical models that require an extensive amount of data for reliable calibration and verification. Our objective is not to take the analysis to this level in this treatment, but rather, to focus on providing a good understanding of the contaminant migration pathways that can occur, the definition of pertinent equilibrium states and potential upper limits of contamination, and those steady-state mass transport analyses that can be used to generate predictive estimates most easily. The major focus is on the transport and behavior of organic compounds of low water solubility.

NATURE OF THE SUBSURFACE ENVIRONMENT

Although a comprehensive presentation of groundwater hydrogeology is beyond the scope of this book, there is merit in reviewing some of the fundamental principles of subsurface geology and groundwater movement. These principles are then expanded in subsequent sections of the chapter concerning contaminant migration. Several excellent books are available as resources in fundamental quantitative hydrogeology (Davis and DeWiest, 1966; Freeze and Cherry, 1979; Cedergren, 1977).

It is useful to think in terms of a hierarchy of evaluations concerning the movement of chemicals in the subsurface. Initially, one needs a good understanding of subsurface geology and the movement of water within the subsurface. This can be obtained through the performance of fairly fundamental hydrogeological investigations designed to map the subsurface geology in the vicinity of the chemical release. Typically, soil borings are performed to needed depths, and geological strata are delineated through examination of geologic samples from those borings. These investigations can then provide cross sections of the geology in the area of the chemical release. Invariably, this type of subsurface investigation will delineate the presence of different geological zones in which water movement varies. In addition to soil borings, monitoring wells are located strategically in the vicinity of the site so that information about the direction and magnitude of groundwater movement in the vicinity of the release can be determined. In a majority of hydrogeologic settings, the basic three-dimensional patterns and approximate velocities of groundwater can be developed through a competently designed and executed hydrogeologic investigation. At the same time, there are many complex hydrogeologic regimes where even fundamental questions like "What is the direction of groundwater flow?" may be frustratingly difficult to answer.

Figure 1 illustrates a typical groundwater flow system. The flow of groundwater within this system is governed largely by two parameters: hydraulic conductivity and hydraulic gradient, the latter being the more easily quantifiable parameter. It represents the change in hydraulic (or potentiometric) head per unit distance found at various locations within the groundwater system. It is these potentiometric differentials that keep groundwaters in constant motion, always striving to resolve the energy inequities within the groundwater reservoir. With rare exceptions, resolution is never achieved and the cycle of groundwater recharge, migration, and ultimate discharge goes on as it has for billion of years.

Perhaps the most fundamental and useful relationship, involving the most important parameters of the groundwater flow system, is Darcy's law. Darcy's law states that the flow of groundwater through a unit cross-sectional area is directly proportional to the hydraulic gradient and the hydraulic conductivity:

$$Q = KiA$$

where Q = flow
K = permeability or hydraulic conductivity
i = hydraulic gradient
A = cross-sectional area

For porous media, Darcy's law can be rewritten to yield an expression for the average interstitial groundwater velocity as follows:

$$\overline{V} = \frac{Ki}{\eta}$$

where $\overline{V}$ is the average interstitial groundwater velocity and η is the porosity. (Note that $\overline{V}$ is *not* the molecular velocity, which is variable in space and always higher than $\overline{V}$ due to the tortuosity effects.)

The most elusive parameter in Darcy's law is the hydraulic conductivity, K. Both the quantity and the velocity of groundwater flow are functions of the hydraulic conductivity of the soil or rock through which the groundwater is moving. It is important to appreciate that hydraulic conductivity is not a direct measure of groundwater velocity, even though it is often expressed in the units of velocity (i.e., cm/s). Hydraulic conductivity is a property of all natural materials that describes the quantity of water conducted through a unit area of the material under a hydraulic gradient of unity. Hydraulic conductivity is one of the most

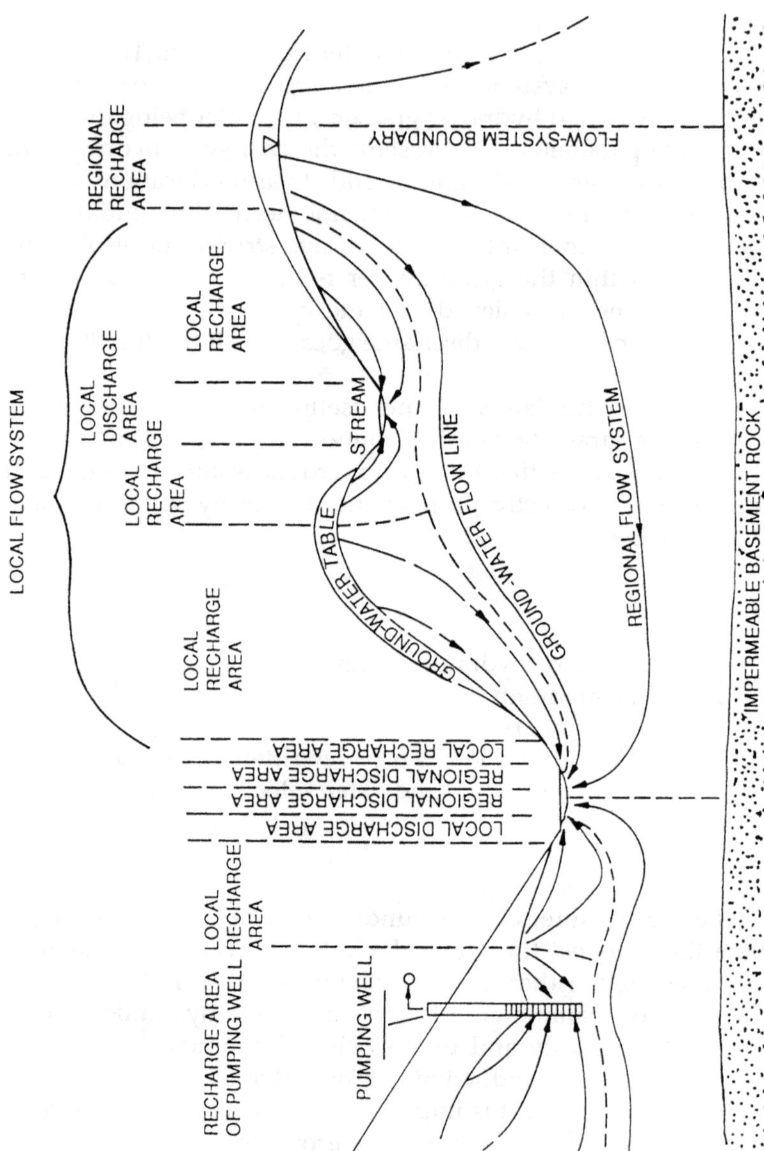

Figure 1 Typical groundwater flow system.

variable units of measure in nature. It can vary by a factor of 10 billion within commonly encountered geologic materials. For example, gravel can have a permeability as high as 10 cm/s, while the permeability of a clay can often be as low as 1×10^{-9} cm/s.

Once monitoring wells have been installed and developed in the vicinity of a chemical release, water-level measurements can then be used to determine the hydraulic heads and hydraulic gradients that exist within the groundwater system of interest. Typically, information is needed concerning both horizontal and vertical gradients, although the latter are often overlooked, sometimes with disastrous consequences. Figure 2 shows a water-level contour map that can be used to delineate horizontal hydraulic gradients in the area of interest. Locations having the same water level are linked together as an equipotential line, and groundwater flow occurs perpendicular to the equipotential lines (in isotropic systems) and in the direction of decreasing potential.

Once we have a sufficient understanding of the occurrence and movement of water in the groundwater system of interest, it then makes sense to look at the behavior and migration of chemicals within the soil/rock and groundwater system. Depending on the state of the chemical, the nature of its release, the time that has elapsed since the release, and the properties of the system, the chemical can occur in a bulk liquid phase, in a dissolved or sorbed form, or in all three forms. In many cases, initial investigations in the vicinity of chemical releases, such as those that have occurred in landfills and waste disposal sites, generated data primarily at locations distant from the source of the release. At these locations, in many cases, the chemicals were present in a dissolved or sorbed state and predictive tools using mass transport processes for materials in these states were appropriate. We now know, however, that in many cases, as the source of the release is approached, some chemicals are present within the subsurface in an undissolved/unsorbed state. *Nonaqueous-phase liquids* (NAPLs), for instance, can occur in the subsurface as a myriad of entrapped droplets or globules of pure chemicals or mixtures.

NONAQUEOUS-PHASE LIQUIDS

Nonaqueous-phase liquids (NAPLs) are organic liquids that are relatively insoluble in water. Nonaqueous-phase liquids are typically differentiated into two classifications. Nonaqueous-phase liquids with densities less than that of water are referred to as *light nonaqueous-phase liquids* (LNAPLs). Examples include gasoline, kerosene, jet fuel, and nonchlorinated industrial solvents such as benzene and toluene.

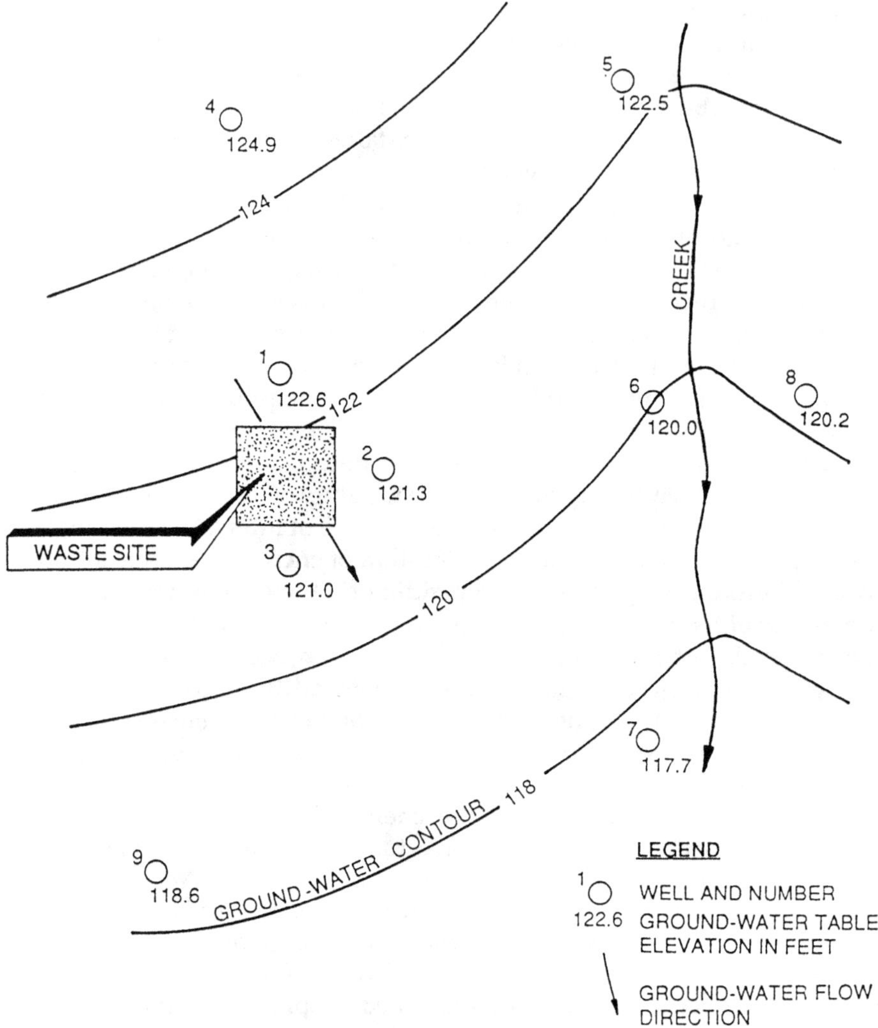

Figure 2 Groundwater contour map.

Nonaqueous-phase liquids with densities greater than that of water are referred to as *dense nonaqueous-phase liquids* (DNAPLs). Examples include chlorinated solvents such as trichloroethylene, methylene chloride, trichloroethane, and dichlorobenzene. LNAPL and DNAPL materials behave very differently in the subsurface environment.

Table 1 Solubilities, Densities, and Common Cleanup Standards of Selected DNAPL Chemicals

Chemical	Solubility in water (μg/L)	Density (g/cm^3)	NJDEPE draft groundwater cleanup standards (ppb)
Trichloroethylene	1,100,000	1.47	1
Tetrachloroethylene	200,000	1.63	0.4
1,2-Dichloroethane	8,700,000	1.26	0.3
1,1-Dichloroethylene	400,000	1.20	1
1,1,1-Trichloroethane	4,400,000	1.35	30
Methylene chloride	20,000,000	1.32	2
trans-1,2-Dichloroethylene	600,000	1.26	100
cis-1,2-Dichloroethylene	800,000	1.28	10

Since NAPLs have relatively low solubilities in water, they are capable of moving as a separate phase through groundwater systems and are very slowly solubilized. Having densities greater than that of water, DNAPLs tend to sink vertically through aquifers. These factors, coupled with the fact that many of the DNAPL chemicals are considered toxic at even low parts per billion levels, dictate that even relatively small amounts of DNAPL entering the subsurface can contaminate large portions of an aquifer.

Table 1 lists a number of common DNAPL chemicals along with their densities, solubilities, and one set of typical groundwater cleanup standards, in this case, New Jersey's proposed standards. Of particular note are these compounds' solubilities. From one perspective, these solubilities are sufficiently low that these chemicals will, in fact, behave as a separate phase in groundwater before ultimately being solubilized. However, from another perspective, it can be seen that their solubilities are several orders of magnitude higher than New Jersey's proposed cleanup standards. Consequently, despite their relatively low solubilities compared to other chemicals, their solubilities are sufficiently high to render groundwater nonpotable even when concentrations are only a minute fraction of the solubility limits.

DNAPL chemicals are used extensively by a wide variety of industries. In addition to the chemical industry, many industries, such as electronic and instrument manufacturers and aerospace industries, use large quantities of DNAPL chemicals. Furthermore, DNAPL usage is not restricted to large industries. DNAPL chemicals are used by dry cleaning

Table 2 U.S. Production of Dense Organic Solvents, 1986

Solvent	10^6 lb	10^6 kg
1,2-Dichloroethane[a]	12,940	5,871
1,1,1-Trichloroethane[b]	648	294
Carbon tetrachloride[a]	627	284
Methylene chloride[b]	561	255
Chloroform[a]	422	191
Tetrachloroethylene[b]	405	184
Trichloroethylene[b]	165	75

Source: Feenstra et al. (1991).
[a]American Chemical Society (1986).
[b]Halogenated Solvents Industry Alliance (personal communication).

establishments, machine shops, photographic processing shops, and printing shops. The total quantity of DNAPL chemicals produced in the United States is very high, as indicated in Table 2 (Feenstra et al., 1991).

Over the last decade, the full implications of LNAPL and DNAPL contamination of groundwater have become increasingly clear in the United States. The presence of DNAPLs in the subsurface is now widely regarded as a major impediment to aquifer restoration. After some initial ambivalence, the U.S. Environmental Protection Agency (USEPA) has recently acknowledged the fact that DNAPLs are "likely to be a significant limiting factor in site remediation" (Huling and Weaver, 1991). The behavior of LNAPLs and DNAPLs in relatively homogeneous porous media is complex and difficult to predict. In fractured media DNAPL behavior is even more complex. In this chapter we place more emphasis on the behavior of DNAPL than LNAPL, due primarily to the greater propensity of DNAPLs to contaminate groundwater systems.

CAPILLARY PRESSURE RELATIONSHIPS

An understanding of DNAPL transport in the subsurface begins with an understanding of the capillary phenomena associated with multiphase fluid flow. Although relatively new to the environmental field, capillary behavior has been studied in considerable detail by petroleum engineers since the early twentieth century. Leverett (1940) described the basic capillary behavior of groundwater and of groundwater/oil systems in porous media. Rose and Bruce (1949) described laboratory apparatus and techniques for determining the capillary pressure–saturation relationships of petroleum reservoir rocks. The work of Leverett and Rose and Bruce was expanded upon by Russell and Dickey (1950) in describing the factors controlling the porosity, permeability, and capillary properties of petroleum reservoirs. In 1956, William Lamb of the Massachusetts Institute of Technology studied dual-phase fluid flow in regard to the potential of storing oil in an earthen reservoir. This study looked at the capillary aspects of dual-phase fluid flow through compacted clay as well as the geotechnical aspects of designing a large, open petroleum reservoir (Lamb, 1956).

Capillary phenomena arise from the fact that when two immiscible fluids are in contact with each other in the subsurface, the molecular attractions between similar molecules in each fluid are greater than the attractions between the different molecules of the two fluids. Molecular attraction is greater within the more dense fluid, and as a result, the surface of the contact is drawn into a curvature that is concave toward the more dense fluid. An interfacial tension develops across the plane of the concave surface. The result of the interfacial tension is a pressure difference between the two immiscible fluids. This pressure difference, called *capillary pressure*, is determined as

$$P_c = P_D - P_w$$

where P_D is the pressure of DNAPL and P_w is the pressure of water. Berg (1975) derived an expression for capillary pressure in terms of energy change or work when a drop of immiscible oil is immersed in water. The oil assumes a spherical shape in the water, and in the process, work is performed.

The work performed by the interfacial tension results in a change in surface area of the oil drop (dA_o):

$$W = \sigma \, dA_o \qquad (1)$$

where W = work
σ = interfacial tension
dA_o = change in surface area of the oil drop

Fluid pressures both inside and outside the drop also contribute to the total work performed. Both the pressure of the water, P_w, and the pressure of the oil, P_o, act normally to the interface. The fluid pressure–related work is

$$W = P_o \, dV_o - P_w \, dV_o \qquad (2)$$

where dV_o is the change in the volume of the oil drop. Berg (1975) points out that at equilibrium the work done by interfacial tension is equal to the work done by fluid pressure. Therefore, equations (1) and (2) can be equated.

$$\sigma\, dA_o = P_o\, dV_o - P_w\, dV_o \tag{3}$$

Defining capillary pressure P_c as

$$P_c = P_o - P_w$$

and rearranging terms, we can rewrite equation (3) as follows:

$$P_c = \frac{\sigma\, dA_o}{dV_o} \tag{4}$$

Since the oil drop is spherical, its area, A_o, is $4\pi r^2$, where r equals its radius. Differentiating this expression to obtain dA_o, we obtain

$$dA_o = 8\pi r\, dr$$

Similarly, the volume, V_o, of the spherical drop equals $\frac{4}{3}\pi r^3$, which when differentiated yields $4\pi r^2\, dr$. Substituting these two differentials into equation (4) yields

$$P_c = \frac{8\pi r\, dr\sigma}{4\pi r^2\, dr}$$

This equation reduces to

$$P_c = \frac{2\sigma}{r} \tag{5}$$

In the subsurface, where the effect of the solid media must also be taken into account, the capillary pressure has been shown to equal

$$P_c = \frac{2\sigma \cos\theta}{r} \tag{6}$$

where θ is the contact angle between the fluid boundary and the solid surfaces of the medium. However, in most hydrogeologic systems, the aquifer is water wet before introduction of the second-phase DNAPL and θ can generally be assumed to equal zero. Consequently, $\cos\theta$ becomes unity and it can be dropped from equation (6). The expression therefore reduces to equation (5).

In fractured rock, the fractures behave more as two parallel plates than as the nearly circular interstices of a porous media. The expression for the displacement pressure of a fracture of aperture b can be determined through a force balance as illustrated in Figure 3, assuming once again that in an initially water-wet media, $\theta = 0$.

Referring to Figure 3, it can be seen that the force due to interfacial tension (F_σ) equals

$$F_\sigma = 2l\sigma$$

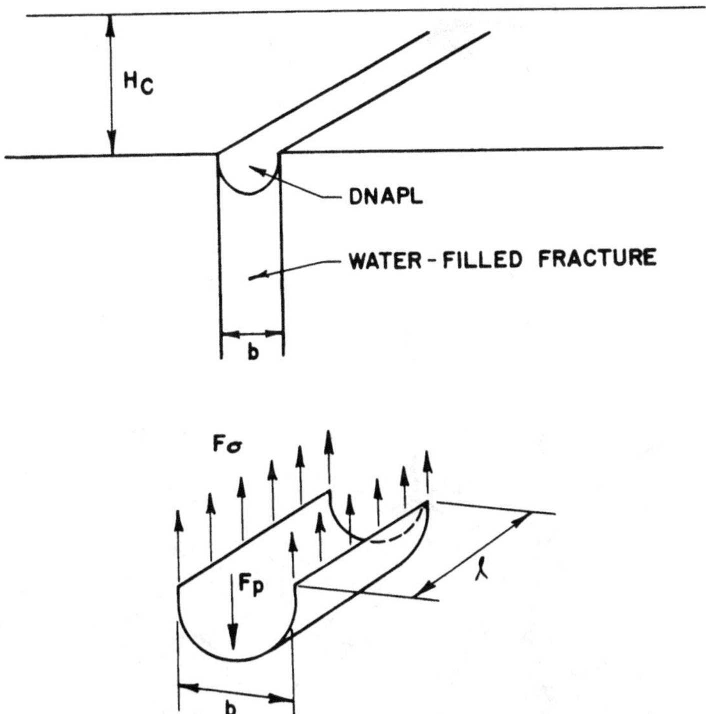

Figure 3 Force balance on DNAPL invading a fracture.

In addition, the force due to the pressure difference across the interface, F_p, equals

$$F_p = P_c lb$$

At equilibrium, $F_\sigma = F_p$

$$2l\sigma = P_c lb$$

Solving for P_c, we derive

$$P_c = \frac{2l\sigma}{lb} = \frac{2\sigma}{b} \tag{7}$$

It can therefore be seen that capillary pressure is proportional to the interfacial tension between the two fluids and inversely proportional to the radius of curvature of the DNAPL globule [Equation (5)].

For a globule of DNAPL to enter and move through a saturated soil, the pressure of the DNAPL above the formation must exceed the

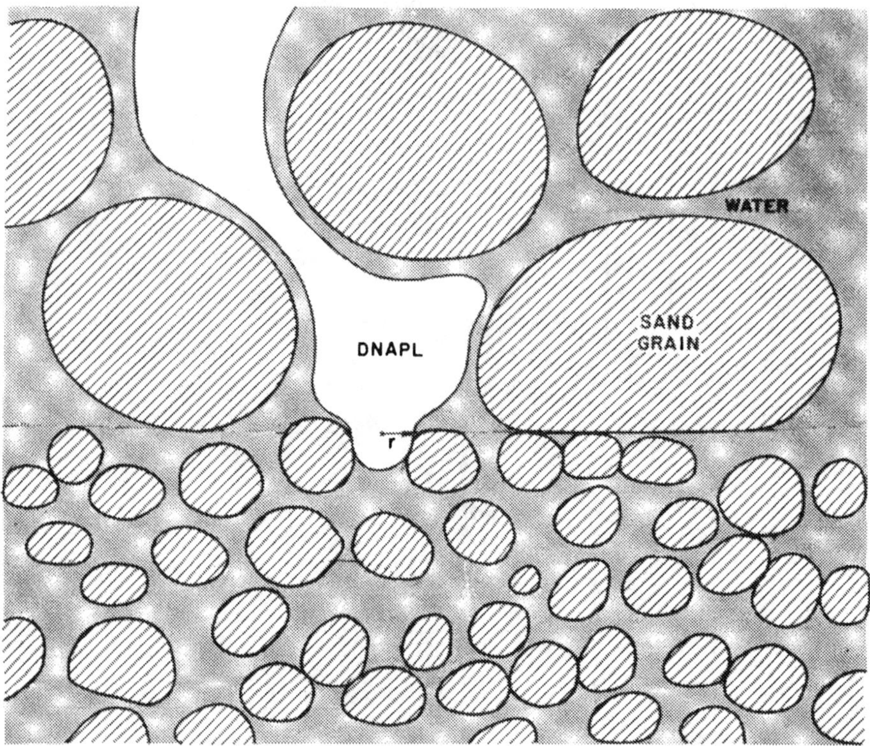

Figure 4 DNAPL invading a finer-grained sand.

capillary pressure of the stratum it is seeking to invade. In this context, capillary pressure is often referred to as *entry pressure* or *displacement pressure*. As illustrated in Figure 4, in order for an accumulation of DNAPL to enter a saturated porous medium, the radius of curvature of the DNAPL globule must conform to the dimensions of the pore space it is entering. Consequently, the displacement pressure will be the capillary pressure existing at such time as the radius of curvature equals the pore radius. The radius of curvature of DNAPL can achieve the necessary reduced magnitude only if the fluid pressure within the DNAPL accumulation, P_D, becomes sufficiently high. It can reach the necessary pressure by accumulating to a great enough thickness above the stratum being invaded. The greater pressure within the DNAPL arises as a result of the higher density of the DNAPL compared to the surrounding water as illustrated in Figure 5.

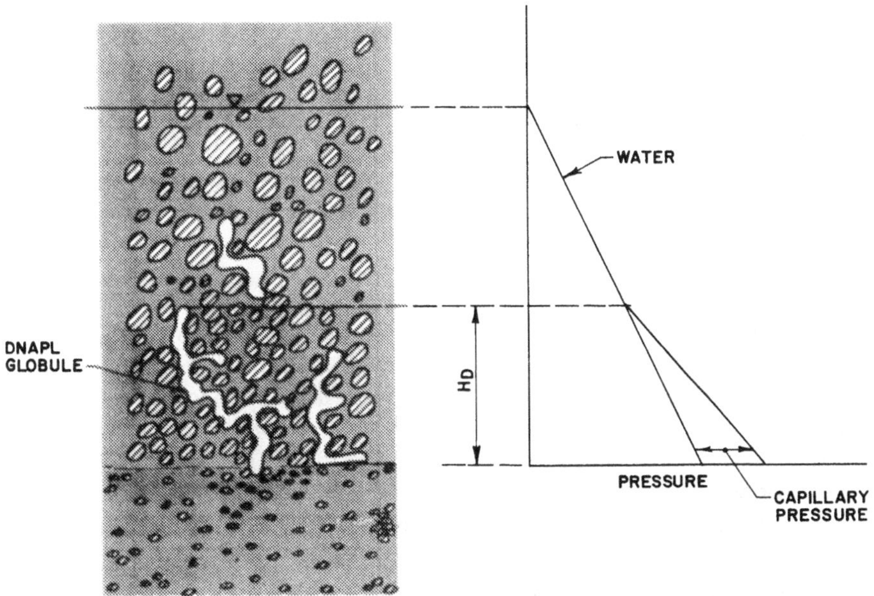

Figure 5 Pressure diagram of DNAPL accumulation above a finer-grained sand.

The capillary pressure at the base of a DNAPL accumulation, P_D, equals the height of the DNAPL accumulation above the base, H_D, times the density differential between DNAPL and water, $\Delta\rho g$, or

$$P_D = \Delta\rho g\, H_D \tag{8}$$

Equating this expression with equation (5) and solving for H_D yields

$$H_D = \frac{2\sigma}{\Delta\rho g\, r} \tag{9}$$

Equation (9) indicates that the height to which a DNAPL must accumulate above a finer-grained stratum before intruding into the pores of that stratum is a function of the interfacial tension between the DNAPL and the groundwater, the density differential, $\Delta\rho g$, and the pore throat radius of the finer-grained stratum. The determination of the necessary height of DNAPL accumulation above a stratum is a critical determination in judging whether, and to what extent, DNAPL will intrude into various formations.

Similarly, equating equations (7) and (8) yields an expression for the necessary height of DNAPL accumulation above a fracture of aperture, *b*:

$$H_D = \frac{2\sigma}{\Delta\rho g\, b} \tag{10}$$

RESIDUAL SATURATION

Once a DNAPL intrudes into and passes through a porous medium, it leaves behind a trail of residual DNAPL contamination, as illustrated in Figure 6. The spatial distribution of DNAPL within the subsurface is often quite complex, owing to the DNAPL's tendency to follow preferentially pathways of least capillary resistance. In a porous medium the residual contamination takes the form of isolated droplets and/or globules of pure-phase DNAPL held in the soil interstices by capillary forces. These droplets and globules are themselves immobile, due to capillary pressure, but represent long-lived sources of groundwater contamination as they dissolve slowly into the surrounding mobile groundwater. Schwille (1988) conducted numerous laboratory experiments illustrating the behavior of dense chlorinated solvents in porous media. These experiments involved sand tank models and glass bead simulated porous media. He photographed the behavior of the chlorinated solvents and their tendency to form residual saturations in the porous media, and found that the residual DNAPL consisted of either globules or droplets, depending on the characteristics of the medium.

Schwille (1988) also conducted parallel glass plate experiments to explore the behavior of dense chlorinated solvents in fractured media. His work indicated that a residual saturation in the fractures will remain in the wake of DNAPL migration through a fracture flow system. Notwithstanding these experiments, much less is known about the accumulation of DNAPL in fractured rock. As illustrated in Figure 7, it is believed that in a fractured rock medium, DNAPL will accumulate in several forms: (1) as droplets held within fractures as a result of capillary pressures [as illustrated by Schwille (1988)], (2) as small pools and puddles lying on relatively flat-lying fractures, or (3) as accumulations within dead-end fractures.

Two terms are used to quantitatively describe the degree of DNAPL residual saturation: residual saturation and retention factor. *Residual saturation* equals the volume of DNAPL divided by the volume of media pore space. It can therefore be thought of as the fraction of the interstices filled with DNAPL. *Retention factor* is defined as the volume of

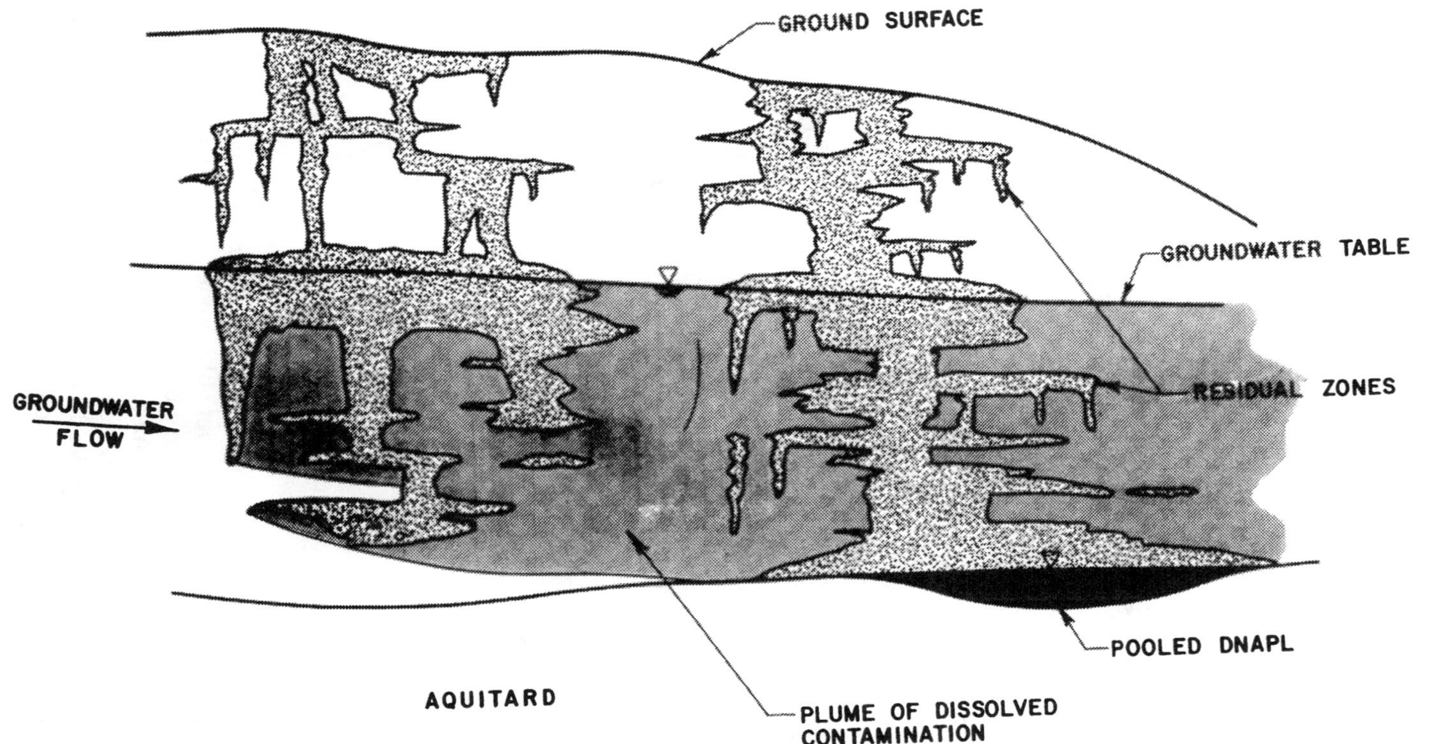

Figure 6 Typical DNAPL behavior in porous media.

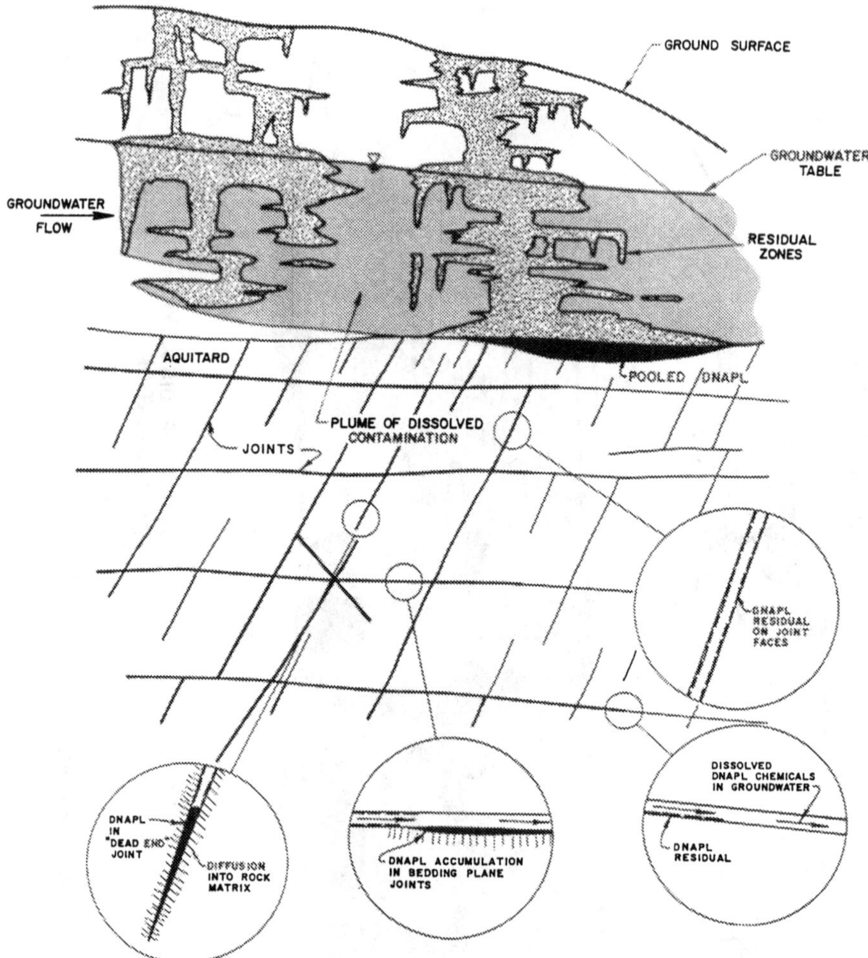

Figure 7 DNAPL occurrence in fracture-controlled groundwater systems.

DNAPL per unit volume of formation. In the metric system it is most often expressed as liters of DNAPL per cubic meter of formation. Mercer and Cohen (1990) provide a summary listing of values of residual saturation and retention factor from numerous laboratory studies of LNAPL and DNAPL behavior. Unsaturated porous media laboratory studies indicate that residual saturation typically ranges from 0.10 to 0.20. In saturated media, residual saturation tends to be higher, ranging from 0.15 to 0.50. Schwille (1988) found that the retention factor for DNAPL chemicals in sandy soils ranged from 3 to 30 L/m^3 in the unsaturated zone and 5 to 50 L/m^3 in the saturated zone.

Virtually no information exists regarding DNAPL residual saturation in fractured rock, owing to the virtual impossibility of simulating fractured rock behavior in the laboratory and the difficulty of conducting field studies in fractured rock. Based on the results of his parallel-plate experiments, Schwille estimated that the DNAPL residual saturation remaining on parallel glass plates equaled 0.5 L/m^3 or 0.025 L/m^2 of joint surface area. Using this figure together with a knowledge of joint spacing allows one to estimate the retention factor associated solely with residual DNAPL left on joint faces (not included are puddles on nearly flat-lying joints or accumulations in dead-end fractures). For example, a cubic meter of rock with a uniform fracture spacing of 0.1 m in all directions would have a total joint surface area of 60 m^2/m^3. Applying Schwille's laboratory-observed retention factor on joint surfaces of 0.025 L/m^2, a retention factor of 1.5 L/m^3 of rock can be calculated. Similarly, if the joints in a rock are uniformly spaced at distances of 1 m, then 1 m^3 of rock would have a surface area of 6 m^2 and the DNAPL retention factor would be 0.15 L/m^3.

While calculations such as these provide a general indication of the retention factor of fractured rock, in practice the residual saturation of DNAPL in fractured rock can be expected to vary widely. Experience in the field has shown that DNAPL tends to migrate through fractured rock with a high degree of heterogeneity. DNAPL will preferentially migrate through the portion of the fracture system having the widest apertures and therefore the lowest displacement pressures. Once DNAPL-wetted pathways are established, DNAPL tends to follow such pathways preferentially, resulting in considerable variability of contaminant distribution in the rock. Often, significant portions of the rock are completely free of DNAPL residuals, while other areas can have very high concentrations. As an upper bound, one can consider the case where an entire portion of a fracture network is fully saturated with DNAPL. In such a case, the retention factor can be readily calculated from the fracture porosity of the rock. For example, a fracture porosity of 0.001 translates to a retention factor of 1 L/m^3. Similarly, a fracture porosity of 0.01 (a relatively high number for fractured rock) results in a retention factor of 10 L/m^3. The latter figure probably represents an upper bound for the DNAPL retention factor of fractured media.

In actuality, DNAPL is never likely to fully occupy a rock's fracture network or a soil's interstices. There will always be some residual water remaining. The degree of DNAPL saturation will be a function of capillary pressure. As capillary pressure increases, the DNAPL will intrude into progressively finer pore spaces or fracture apertures, forcing water out. Ultimately, the DNAPL saturation will reach a maximum at a point where water reaches the point of irreducible saturation. A typical capillary pressure curve is illustrated in Figure 8. Of particular note is the fact that water saturation at any particular capillary pressure is less during the wetting process (water saturation increasing) than during drying (DNAPL saturation increasing). This situation is referred to as hysteresis and its effect is illustrated in Figure 8. Once DNAPL has intruded into a formation, there will be some minimum residual saturation of DNAPL below which the DNAPL saturation will not fall even after drainage of the DNAPL.

The residual DNAPL is comprised of isolated blobs and ganglia of DNAPL that have been cut off from the continuous DNAPL body. Because of the disconnected nature of the DNAPL at residual saturation, the DNAPL is incapable of flow. For flow to occur, the globules must become continuous and a pressure gradient must exist. The globules become continuous as DNAPL saturation increases. The relative mobility of DNAPL in the subsurface is typically illustrated with a relative permeability curve, as shown in Figure 9. Referring to this figure, it can be

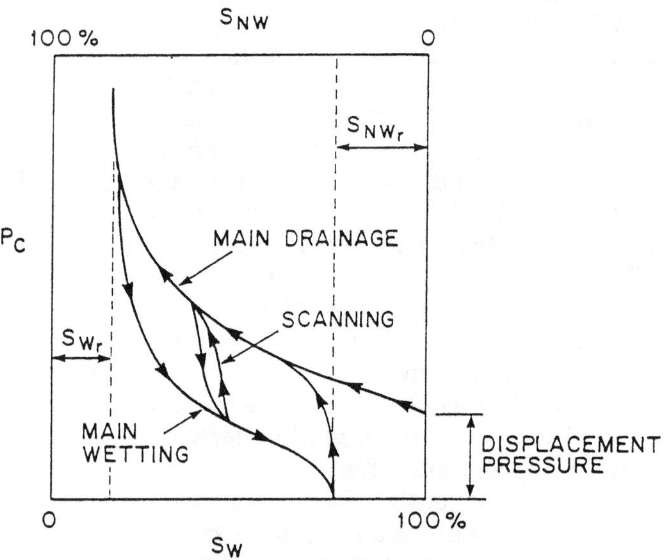

Figure 8 Typical capillary pressure curve, S_W, wetting-phase saturation; S_{NW}, nonwetting-phase saturation; S_{Wr}, residual wetting phase; S_{NWr}, residual nonwetting phase.

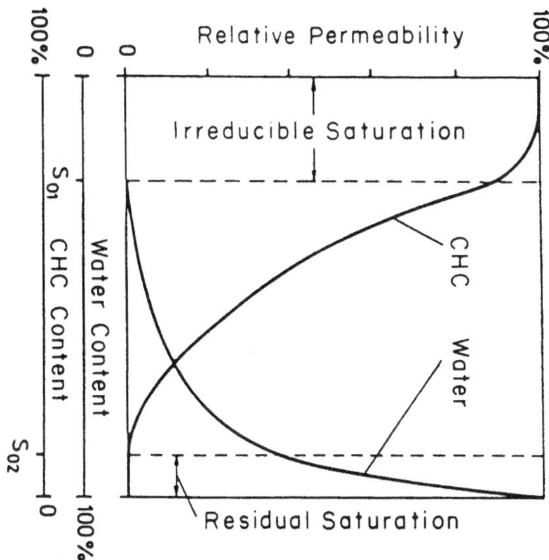

Figure 9 Relative permeability curve. (From Schwille, 1988.)

seen that at DNAPL chlorinated hydrocarbons (CHC) residual saturation, the relative permeability of the DNAPL is zero and therefore the DNAPL is immobile and the water is mobile. The water is thus free to flow through the zone of residual DNAPL globules. At the opposite end of the diagram, DNAPL can reach sufficiently high degrees of saturation that the continuity of water in the pores is disrupted, reducing the relative water permeability to zero. This can occur in "pools" of DNAPL lying above aquitards. In such pools, the DNAPL is mobile, but the water is not. Therefore, no groundwater can circulate through these zones, severely limiting the rate of dissolution and removal of such pool-like accumulations.

Schwille (1988) also conducted experiments to determine the rate of dissolution of simulated pools of chlorinated hydrocarbons sitting at the base of aquifers using sand tank models. He conducted his experiments using 1,1,1-trichloroethane and trichloroethylene. His experiments employed a 1.5 × 0.5 m trough filled with a uniform sand with a hydraulic conductivity of 3.5×10^{-3} m/s (3.5×10^{-1} cm/s). The temperature was held between 20 and 22°C. A flow of groundwater was established, above the pool-like accumulation, at flow rates ranging from 0.45 to 6.7 m/day. The results of Schwille's solubilization experiments are presented in Table 3. Schwille found that the rate of DNAPL removal was

Table 3 Results of Solubilization Experiments in the Flat Trough

Run number	Volume flow rate (L/h)	CHC concentration (mg/L)	Water flow velocity (m/day)	Removal rate for total trough area (g/day)	Theoretical removal time (days)	Removal rate for 1 m^2 (g/day)
1. 1,1,1-TCA[a]						
3	1.5	170	0.7	6.1	447	8.1
1	4.5	89	2.0	9.6	284	12.8
2	14.8	85	6.7	20.1	91	40.1
2. TCE[b]						
1	1	90	0.45	2.2	1690	2.9
6	2	67	0.9	3.2	1137	4.3
2	2	87	0.9	4.2	873	5.6
3,4	4	73	1.8	7.0	521	9.3
5	6	77	2.7	11.1	329	14.8

Source: Schwille (1988).
[a]Total amount 1,1,1-TCA added = 2728 g; total amount removed = 214 g = 7.8%.
[b]Total amount of TCE added = 3650 g; total amount removed = 190 g = 5.2%.

related to the groundwater flow velocity. It should be noted, however, that he used velocities that are higher than generally encountered in actual field situations.

Schwille's work on dissolution of pooled DNAPL suggests that these pool-like accumulations will persist in the subsurface for exceedingly long periods of time. For example, using his measured TCE removal rate of 2.9 g/day per square meter at a groundwater flow velocity of 0.45 m/day, one can calculate that a relatively small pool of TCE containing 250 L and having a surface area of three m^2 would remain in the subsurface for a period of nearly 125 years. It can be reasonably extrapolated that in the absence of any other removal mechanisms, larger pools or pools of lower-solubility DNAPLs such as dichlorobenzene could persist in the subsurface for periods of many centuries. In the case of lower-solubility materials such as PCBs, residence times estimated in millennia are not unusual.

MIGRATION OF LNAPLS

The migration of LNAPL chemicals in the unsaturated zone is similar in many respects to the migration of DNAPL. The LNAPL will move generally downward to the unsaturated zone following pathways of least capillary resistance and leaving behind a trail of residual contamination.

Field studies have shown that the migration of LNAPLs in the unsaturated zone can be spatially complex, similar to the pattern illustrated in Figure 6. If the LNAPL reaches the top of the groundwater table, it can accumulate as a "pool" of free product. A relatively rapid and large release of LNAPL will often depress the water table. The depression in the water table will ultimately subside as the layer of free product spreads laterally, and as a result, thins. Fluctuations in the groundwater table cause the pool of free product to ride up and down with the groundwater table, producing what is known as a *smear zone* of residual soil contamination, as illustrated in Figure 10. LNAPL materials that have sufficient volatility also migrate through the unsaturated zone as a result of volatilization from the free product layer and the zones of residual contamination. The gas-phase migration of LNAPL material can be substantial when the stratum has a relatively high permeability or where utility trenches backfilled with permeable materials intersect the contaminated zone. In fact, it is often due to soil gas migration that underground storage tank leakage is first discovered.

Groundwater monitoring wells constructed through layers of LNAPL floating on the groundwater table will usually exhibit substantially greater thicknesses of LNAPL in the well than actually exist in the aquifer. This phenomenon occurs because the layer of LNAPL floats on the capillary fringe some distance above the groundwater table. Consequently, it enters the well at an elevation above the water table and accumulates on the water surface in the well. As it accumulates, it depresses the water surface and continues to accumulate until such time as its top level in the well is coincident with the top of the floating layer in the aquifer. Because of the difficulties in estimating the distance of the floating LNAPL in the aquifer above the groundwater table, it is difficult to estimate the relationship between the observed thickness of LNAPL in the well and the actual thickness of LNAPL in the aquifer. Numerous examples have shown that the thickness in the well is usually four or more times greater than the actual thickness in the aquifer.

Zones of Contamination

The potential impact of nonaqueous-phase liquids on soils and groundwater in the vicinity of a chemical release, together with the ultimate transport and migration of constituent chemicals in their dissolved or sorbed states, makes it prudent to distinguish at least three types of zones of contamination. Figure 11 shows the impact of a release of DNAPL and LNAPL on the subsurface, for both porous media and an underlying fractured rock system. Zone 1 represents the unsaturated zone which could be contaminated to retention capacity by both

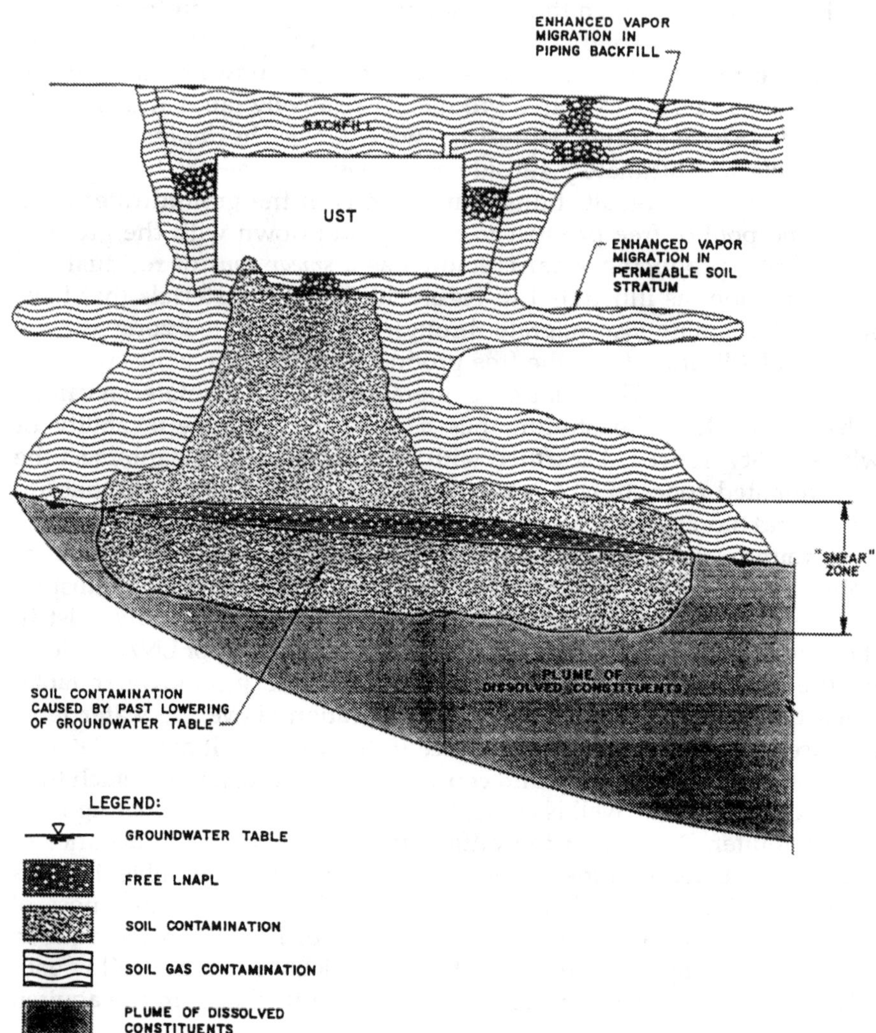

Figure 10 Typical LNAPL occurrence in soil.

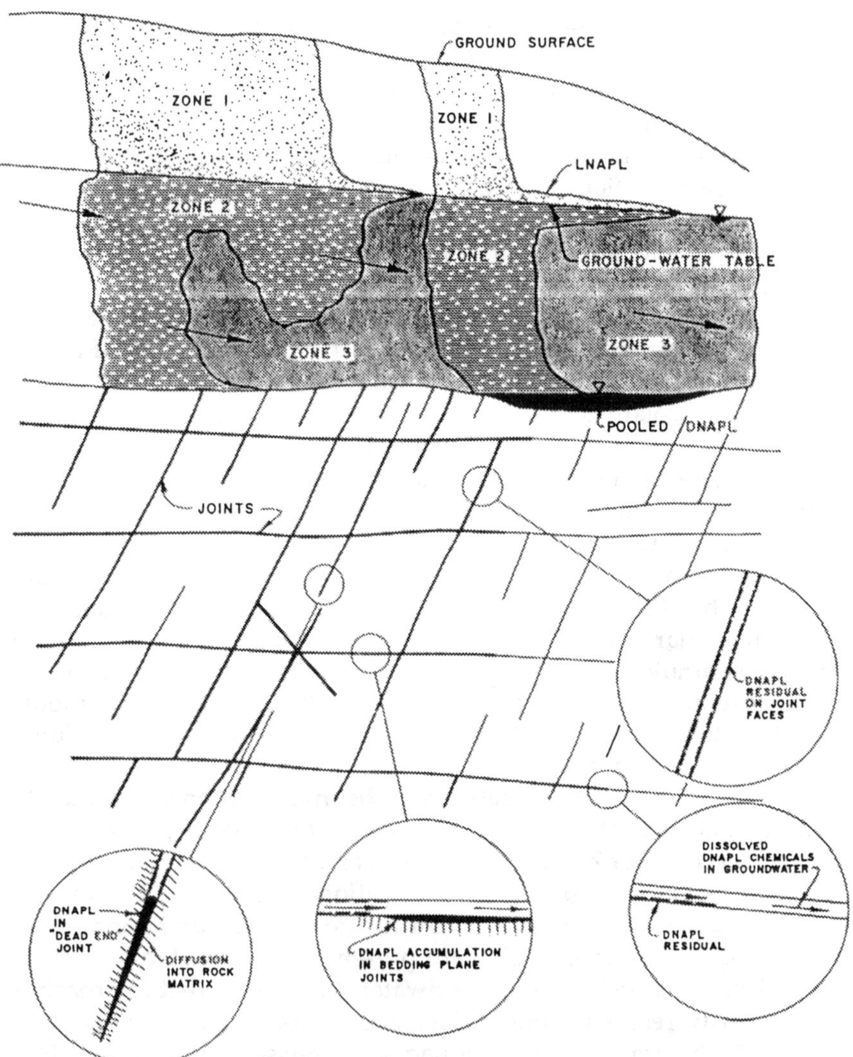

Figure 11 DNAPL migration in subsurface.

DNAPLs and LNAPLs. The chemicals distinguish themselves, however, as they move through the unsaturated zone into the capillary fringe. Nonaqueous-phase liquids with densities greater than water continue to migrate through the capillary zone into the saturated zone, and if they are present in sufficient quantity, continue their migration vertically downward until they reach an aquitard or a system where the capillary pressure is sufficient to bring their vertical migration to a halt. Light non-aqueous-phase liquids will float on top of the water in the capillary fringe and move as free product along the top of the water table in a direction of decreasing hydraulic gradient. Zone 2 represents residual DNAPL or LNAPL contamination found below the groundwater table in the saturated zone. Ultimately, the LNAPL or DNAPL will be present in an environment where they are either dissolved in the aqueous phase, volatilized in the unsaturated zone, or sorbed to either saturated or unsaturated soils. This situation has been distinguished as Zone 3 in the saturated zone—a zone of dissolved-sorbed contamination typically identified as the *plume*.

As illustrated in Figure 11, DNAPL materials in fractured bedrock systems tend to accumulate in dead-end joints, to settle into small depressions within the bedding plane joints, and to form thin discontinuous and irregular films or small droplets on joint faces. Once in a state of residual saturation, the DNAPL is largely immobile. However, since the joints are also occupied by groundwater, the DNAPL constituents are free to solubilize into the flowing groundwater, producing a plume of dissolved contamination.

Inspection of Figure 11 reveals that if the investigation is confined to Zone 3, with no indication of the presence of nonaqueous-phase liquids in either Zone 1 and/or Zone 2, the choice of remediation technology can be seriously in error. In particular, excavation of contaminated soils in the unsaturated zone will not prevent future groundwater contamination if nonaqueous-phase liquids are present below the depth of excavation. Also, the selection of groundwater pump and treat approaches will necessarily result in long delays and contaminant concentration tailing due to the presence of the nonaqueous-phase materials as well as other interactions with the soil skeleton. The sorption process, which is critical to contaminant migration in Zone 3, will be discussed further in a following section, on the soil/water interface.

CHEMICAL PARTITIONING IN THE SUBSURFACE

Although it is rarely the case that any environmental system of interest is at equilibrium, utilization of equilibrium principles, concepts, and models can often provide two types of useful and important information: conservative estimates of upper limits of contamination, and the direction in which transport is occurring across an environmental interface if the system is not at equilibrium. Employment of equilibrium concepts and models requires the availability of a few important chemical and physical parameters for both the constituent of interest and the environment in which it is present. Typically, one can generate a great deal of information and fairly reliable predictive estimates if the water solubilities, vapor pressures, Henry's constants, and other kinds of partition coefficients are known for the chemical of interest. The temperature and organic content of the soil are major soil properties of interest in equilibrium assessments involving partitioning from soil to other media.

We can immediately distinguish several types of important interfaces:

1. Fluid/fluid partitioning
   a. Air/water
   b. Air/NAPL
   c. Water/NAPL
2. Fluid/solid partitioning
   a. Water/soil
   b. Air/water/soil
   c. Air/water/soil/NAPL

In the sections that follow, equilibrium conditions and quantitative predictive models will be evaluated for several selected interfaces of interest: air/water, air/NAPL, water/NAPL, and water/soil. Figure 12 is a schematic depiction of a chemodynamic equilibrium state for transport of a dissolved constituent across the air/water interface. Several good books are available for reference concerning equilibrium concepts and models (Thibodeaux, 1979; MacKay, 1991).

Chemical Partitioning Across the Air/Water Interface

The partitioning of a chemical of interest across the air/water interface is described by its Henry's constant for the equilibrium state. The *Henry's constant* is simply the concentration of the chemical in the vapor phase divided by its concentration in the aqueous phase at equilibrium. Given the many different ways of expressing concentrations in both the vapor and aqueous phases, there are several ways of expressing the Henry's constant with corresponding differing systems of units. The reader is encouraged to use extreme caution in selecting a Henry's constant to ensure that the system of units represented is consistent with the

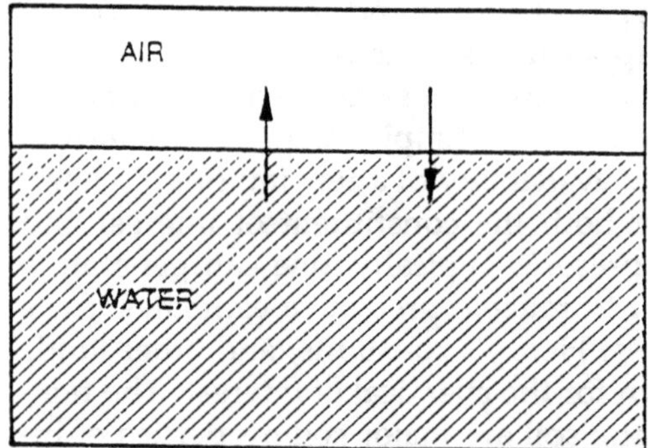

Figure 12 For a closed system, dynamic equilibrium occurs when the desorption rate equals the absorption rate.

application of the Henry's constant that is needed. Table 4 presents several of the more common expressions for the Henry's constant.

With respect to the information presented in Table 4, several clarifying comments are needed. The notation used to distinguish the Henry's constant is an attempt to be consistent with several forms that exist in the literature. Following the convention in Thibodeaux's text, A refers to the constituent, and 1 and 2 refer to the vapor and aqueous phases, respectively. Note that three of the Henry's constants are dimensionless: H_{AX}, H', and H_A. Finally, there are a variety of useful relationships

Table 4 Forms of Henry's Constant

(1) $$H_{A1} = \frac{P_{A1}}{X_A} = \frac{\text{partial pressure of } A \text{ in air (atm)}}{\text{mole fraction of } A \text{ in water}}$$

(2) $$H_{AX} = \frac{Y_A}{X_A} = \frac{\text{mole fraction of } A \text{ in air}}{\text{mole fraction of } A \text{ in water}}$$

(3) $$H = \frac{P_{A1}}{C_{A2}} = \frac{\text{partial pressure of } A \text{ in air (atm)}}{\text{concentration of } A \text{ in water (mol/m}^3\text{)}}$$

(4) $$H' = \frac{C'_{A1}}{C'_{A2}} = \frac{\text{concentration of } A \text{ in air (g/cm}^3\text{)}}{\text{concentration of } A \text{ in water (g/cm}^3\text{)}}$$

(5) $$H_A = \frac{\gamma_{A2} P^\circ_A}{P_T} = \frac{\text{(activity coefficient for } A \text{ in water)} \times \text{[pure component vapor pressure of } A \text{ (atm)]}}{\text{total pressure (atm)}}$$

Table 5 Additional Relationships for Henry's Constants

(1) $H' = \dfrac{H}{RT}$

where $R = 8.2 \times 10^{-5}$ atm · m^3/mol · K
T = temperature (K)

(2) $H' = \dfrac{16.04(P^\circ)(M)}{(S)(T)}$

where P° = pure component vapor pressure of A (mm Hg)
M = molecular weight of A (g)
S = water solubility of A (ppm)
T = temperature (K)

(3) $H = \dfrac{P^\circ M}{760S}$

where P° = pure component vapor pressure of A (torr)
M = molecular weight of A (g)
S = water solubility of A (ppm)

among Henry's constants presented in Table 4. One of these is shown in Table 5 along with other ways of getting two of the Henry's constants from fundamental chemical properties for the constituent of interest.

For organic compounds that are only sparingly soluble in water, perhaps the most popular form of the Henry's constant is H'. Note that this can be derived from a knowledge of the pure component vapor pressure, the molecular weight, the water solubility, and the temperature. Also note that the water solubility for the compound appears in the denominator, and consequently, for chemicals that are miscible in water, this relationship is not valid. With these types of compounds, H_A should be used. This requires knowledge of the activity coefficient of the compound in water which can be obtained from the published literature or estimated from thermodynamic principles (e.g., use of Scatchard–Hildebrand theory (Prausnitz, 1969)).

Table 6 presents information concerning the values of the concentration-based dimensionless Henry's constant, along with the water solubility, pure component vapor pressure, and molecular weight for several organic chemicals of interest. As we have seen, the Henry's constant for immiscible organic compounds is directly proportional to the pure component vapor pressure and inversely proportional to the water solubility. Consequently, compounds having low pure compo-

Table 6 Henry's Constants and Chemical Properties for Selected Organic Chemicals

Chemical	H'	S (ppm)	$P°$ (mmHg)	M (g)
Dieldrin	8.9×10^{-4}	0.25	1×10^{-7}	381
Lindane	2.2×10^{-5}	7.3	9.4×10^{-1}	291
DDT	1.7×10^{-3}	1.2×10^{-1}	1×10^{-7}	354
Aroclor 1242	2.4×10^{-2}	0.24	4.1×10^{-4}	254
Aroclor 1254	1.2×10^{-1}	1.2×10^{-2}	7.7×10^{-5}	326
Benzene	2.4×10^{-1}	1780	95.2	78
Toluene	2.8×10^{-1}	1100	60	92
Trichloroethylene	4.2×10^{-1}	1100	60	131
Tetrachloroethylene	3.4×10^{-1}	400	2×10^{-2}	166

Source: Lyman et al. (1990).

nent vapor pressures, but also sufficiently low water solubilities can have relatively high values for the Henry's constant. Inspection of the information in Table 6 reveals that, for example, we would expect comparable equilibrium ratios not only for benzene, toluene, trichloroethylene, and tetrachloroethylene, but also for Aroclor 1254 and to some extent, Aroclor 1242, irrespective of the fact that the pure component vapor pressures for these six compounds range over seven orders of magnitude. It is important to realize that equilibrium analyses across the air/water interface, which rely only on a knowledge of the pure component vapor pressure, can be greatly in error. In fact, when one looks at transport across the interface in a nonequilibrium situation, it becomes clear that the pure component vapor pressure does not even enter into the analysis, since all of these hydrophobic organic compounds have sufficiently high Henry's constants to be liquid-phase controlled (Thibodeaux, 1979).

The discussion above refers to the concept of equilibrium across the air/water interface and provides a definition of the equilibrium state in terms of the Henry's constant. To use this equilibrium concept in a quantitative predictive model, we look at the partitioning of a chemical across the air/water interface as follows (McCall et al., 1983):

Redefine the Henry's constant as

$$H' = \frac{\%\ M_{\text{air}}/\%\ V_{\text{air}}}{\%\ M_{\text{water}}/\%\ V_{\text{water}}}$$

where $\%\ M_{\text{air}}$ = % of the total mass of the chemical in the air compartment at equilibrium

$\% \, M_{\text{water}}$ = % of the total mass of the chemical in the water compartment at equilibrium

$\% \, V_{\text{air}}$ = % of the total system volume that is air

$\% \, V_{\text{water}}$ = % of the total system volume that is water

Also, from conservation of mass,

$$\%M_{air} + \% \, M_{water} = 100$$

Let's assume that we know H', $\% \, V_{\text{air}}$, and $\% \, V_{\text{water}}$. We can solve the foregoing equations for $\% \, M_{\text{air}}$ and $\% \, M_{\text{water}}$, to yield

$$\% \, M_{\text{air}} = \frac{100}{1 + \% \, V_{\text{water}}/(H' \cdot \% \, V_{\text{air}})}$$

$$\% \, M_{\text{water}} = 100 - \% \, M_{\text{air}}$$

Example

Look at trichloroethylene and Aroclor 1254.

$H'(\text{TCE}) = 0.42$

$H'(1254) = 0.12$

Case 1. $\% \, V_{\text{air}} = 50$, $\% \, V_{\text{water}} = 50$.
At equilibrium:

$\% \, M_{\text{air}}(\text{TCE}) = 30$

$\% \, M_{\text{water}}(\text{TCE}) = 70$

$\% \, M_{\text{air}}(1254) = 11$

$\% \, M_{\text{water}}(1254) = 89$

Case 2. $\% \, V_{\text{air}} = 90$, $\% \, V_{\text{water}} = 10$.

$\% \, M_{\text{air}}(\text{TCE}) = 79$

$\% \, M_{\text{water}}(\text{TCE}) = 21$

$\% \, M_{\text{air}}(1254) = 52$

$\% \, M_{\text{water}}(1254) = 48$

Chemical Partitioning Across the Air/NAPL Interface

An evaluation of the chemical partitioning across the air/NAPL interface is especially important when nonaqueous-phase liquids are present in the unsaturated zone. The nonaqueous-phase liquid could either be a "pure" compound such as trichloroethylene, or a mixture of several volatile constituents, such as gasoline.

With a pure component we can generate an upper limit for component concentration in the vapor phase of the pore space simply by using the compound's pure component vapor pressure. This is the highest concentration that we could hope to achieve. If more precise estimates of the concentration in the vapor phase in the pore space are needed, we will need to generate a site-specific partition coefficient or air/NAPL Henry's constant. An approach has been presented that uses a combination of experimentally determined removal rates in laboratory air stripping soil columns, together with a mathematical model that incorporates the partitioning (Wilson et al., 1988).

If we are interested in the volatilization of a constituent of a nonaqueous-phase liquid into the vapor phase of the pore space, we can use Raoult's law as an approximation. This relates the equilibrium concentration in the pore space to the mole fraction of the chemical present in the mixture. If we assume an activity coefficient for the constituent in the nonaqueous phase equal to 1, this approach gives the partial pressure of the constituent in the pore space, P_{A1}, as the product of the mole fraction of the constituent in the NAPL, X_{A1}, times its activity coefficient in the NAPL, γ_A (which we take to be 1), times its pure component vapor pressure, $\mathring{P}_A$ ($P_{A1} = X_A \gamma_A \mathring{P}_{A1}$). If more precise predictions are needed, we could use Scatchard–Hildebrand theory to provide a better estimate of the activity coefficient of the constituent in the NAPL. For example, Scatchard–Hildebrand theory would yield an activity coefficient for benzene in octane (gasoline surrogate) of 1.5 (Reible et al., 1989).

Chemical Partitioning Across the Soil/Water Interface

Equilibrium partitioning between soil and water is typically described by an overall soil/water distribution coefficient K_D. These soil water distribution coefficients can be measured in the laboratory using actual samples of soil and aqueous solutions containing the chemical of interest. For organic constituents of low water solubility and low reactivity, research has demonstrated that the single most important property of the soil is its naturally occurring organic content. For these "hydrophobic" organic compounds, it is possible to define a new soil/water distribution coefficient K_{oc} such that $K_D = f_{oc} K_{oc}$, where f_{oc} is the naturally occurring organic content fraction of the soil. This approach yields the soil/water coefficient K_{oc}, which is relatively independent of soil type and which can be correlated to the water solubility or the octanol/water partition coefficient, K_{ow}, for the organic compound of interest. That is, for relatively inert organic compounds of low water solubility, to estimate K_D all we need to know is the organic content of the soil and the water solubility of the compound.

Table 7 Laboratory Studies: Batch K_D Determinations

Soil	BET surface area (m^2/g solids)	f_{oc} (g_{oc}/g_{solids})	K_D (cm^3/g)			
			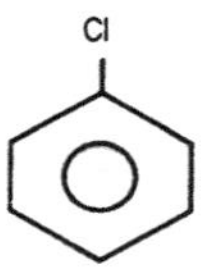 Cl	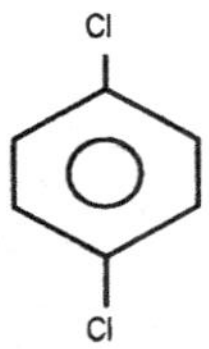 Cl, Cl	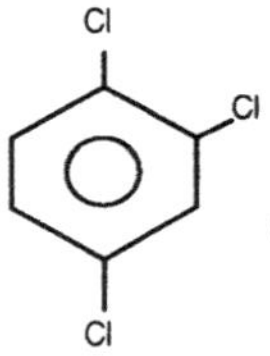 Cl, Cl, Cl	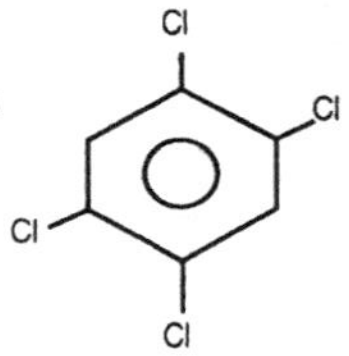 Cl, Cl, Cl, Cl
Aqueous material *C*	4.4	0.0073	1.2	4.4	14.5	37.9
Aqueous material *E*	3.2	0.0008	0.4	1.1	2.5	6.2
Kaolin	12	0.0006	0.6	1.1	2.4	4.9
α-Al_2O_3	120	0.0001	0.6	0.9	1.5	2.2
SiO_2	500	0.0001	4.2	6.0	7.6	12.1

Source: Schwarzenbach and Giger (1985).

For the approach above to have validity, the naturally occurring organic content of the soil should be at least one-tenth of 1%. The correlations that are available are well demonstrated at this level of naturally occurring organic content. In fact, recent investigations have shown that when the organic content of the soil is less than 0.1%, the overall soil/water distribution coefficient, K_D, is not simply related to the soil/water distribution coefficient referenced to organic content, K_{oc} (Stauffer and MacIntyre, 1986).

Table 7 provides information from laboratory studies involving determinations of K_D values for a homologous series of chlorinated benzenes. Information is provided concerning both the surface area and the naturally occurring organic fraction for each soil type. Analysis of the data presented in the table reveals that when the organic fraction drops below that present in aquifer material C, the corresponding decrease in K_D values is substantial and occurs even as the surface area for a given soil increases by two orders of magnitude for aquifer material E through α-Al_2O_3. In fact, the K_D values do not increase until the surface area increases to 500 m^2/g for SiO_2. Also note that the range in K_D values is substantially greater for aquifer material C than for SiO_2, which might imply more of a chemical partitioning in the case of aquifer material C and more of a physical adsorption in the case of SiO_2. For reference purposes, the surface area of activated carbon is typically within the range 1200 to 1400 m^2/g. Finally, it can be seen that the relationship $K_D = f_{oc}K_{oc}$ simply does not hold for any of the soil types having an organic content below that of aquifer material C.

Using an approach similar to the one taken for the air/water interface, we can develop quantitative equilibrium models for the soil/water interface as follows. Look at the equilibrium partitioning of a chemical between the soil and water compartments. Let

$$K_D = \left.\frac{C_{\text{soil}}}{C_{\text{water}}}\right|_{\text{eq}} = \frac{\text{g/g}}{\text{g/mL}}$$

If the solids density is given by ρ (mL/g), then

$$K_D = \frac{\%\ M_{\text{soil}}/(\rho \cdot \%\ V_{\text{soil}})}{\%\ M_{\text{water}}/\%\ V_{\text{water}}}$$

where $\%\ M_{\text{soil}}$ = % of the total mass of the chemical on the soil at equilibrium

$\%\ M_{\text{water}}$ = % of the total mass of the chemical in the water at equilibrium

$\%\ V_{\text{soil}}$ = % of the total system volume that is soil

$\%\ V_{\text{water}}$ = % of the total system volume that is water

Also, $\% \, M_{\text{soil}} + \% \, M_{\text{water}} = 100$.

Assume that we know K_D, ρ, $\% \, V_{\text{soil}}$, and $\% \, V_{\text{water}}$; then

$$\% \, M_{\text{water}} = \frac{100}{1 + \rho K_D \cdot \% V_{\text{soil}} / \% V_{\text{water}}}$$

and $\% \, M_{\text{soil}} = 100 - \% \, M_{\text{water}}$.

The sorption coefficient referenced to organic content, K_{oc}, has been correlated to several other properties of various chemicals, in particular, water solubility, S, and octanol/water partition coefficient, K_{ow}. Three examples are as follows:

1. Kenaga and Goring (1978)

   $\log K_{oc} = -0.55 \log S + 3.64$ (S in mg/L)

   $r^2 = 0.71$

   106 chemicals used—wide variety

   $\log K_{oc} = 0.544 \log K_{ow} + 1.37$

   $r^2 = 0.74$

   45 chemicals used—wide variety

2. Karickhoff et al. (1979)

   $\log K_{oc} = -0.54 \log S + 0.44$ (S in mole fraction)

   $r^2 = 0.94$

   10 chemicals used—mostly aromatic or polynuclear aromatic compounds

   $\log K_{oc} = 1.00 \log K_{ow} - 0.21$

   (*Note*: This can also be written as $K_{oc} = 0.63 K_{ow}$)

   $r^2 = 1.00$

   10 chemicals used—mostly aromatic or polynuclear aromatic compounds

3. Chiou et al. (1979)

   $\log K_{oc} = -0.557 \log S + 4.277$ (S in μ mol/L)

   $r^2 = 0.99$

   15 chemicals used—chlorinated hydrocarbons

(The *Handbook of Chemical Property Estimation Methods* by Warner J. Lyman, William F. Reehl, and David H. Rosenblatt is an excellent resource with respect to use of the above and similar correlations.)

Note that

$$S(\text{mole fraction}) = \frac{S(\text{mg/L}) \times 10^{-3}}{\text{molecular weight} \times 55.51}$$

and

$$S(\mu\text{mol/L}) = \frac{S(\text{mg/L}) \times 10^{+3}}{\text{molecular weight}}$$

Caution must be exercised in use of the correlations above. In particular, one should choose a correlation derived from properties of chemicals similar to those of interest, if possible.

Example

Benzene has a water solubility of 1780 mg/L at 20°C and a molecular weight of 78.11 g. Estimate K_{oc} using correlations I through III.

Correlation I: $K_{oc} = 71$ mL/g
Correlation II: $K_{oc} = 186$ mL/g
Correlation III: $K_{oc} = 71$ mL/g
(variability: factor of 2)
Geometric mean: $K_{oc} = 98$ mL/g

Note: Benzene was used in the derivation of correlation I.

(experimental result $K_{oc} = 83$ mL/g).

Table 8 presents information concerning the partitioning of various organic chemicals of interest between soil and water at equilibrium. The analysis uses the correlation of Chiou et al. (1979) to determine a K_D for a soil having an organic content of 1%. The amounts (expressed as percentages of the total mass) of the constituent in each compartment are then estimated from the equilibrium model presented above, with the assumption that the percent of the total volume that is water is 20 and the percent of the total volume that is soil is 80. This is an example of the utility of knowledge of the equilibrium state and estimates derived from an equilibrium model. The reader should exercise caution, however, in using experimentally determined data from contaminated soils in the field with an equilibrium model. The analysis of the equilibrium state in a two-compartment soil/water model incorporates the assumption that the concentration of the organic constituent in the aqueous phase is be-

Table 8 Two-Compartment Equilibrium Model: Soil/Water Partitioning in the Saturated Zone

Chemical	Molecular Weight (g)	S (ppm)	S (μmol/L)	K_{oc}[a] (mL/g)	K_D[b] (mL/g)	$\%M_{water}$[c]	$\%M_{soil}$[c]
Benzene	78	1,800	23,000	70	0.7	12.5	87.5
Trichloroethylene	130	1,100	8,500	120	1.2	7.7	92.3
Chlorobenzene	110	500	4,500	170	1.7	5.6	94.4
Napthalene	130	30	230	920	9.2	1.1	98.9
Aroclor 1242	260	0.10	0.38	32,000	320	0.031	99.969
DDT	350	0.0031	0.0089	260,000	2,600	0.0038	99.9962

[a]Used correlation of Chiou et al. (1979): $\log K_{oc} = -0.557 \log S + 4.277$, where S is the water solubility in μmol/L.

[b]$K_d = f_{oc}K_{oc}$, where f_{oc} is the fractional organic content (assumed to be 0.01 in these calculations).

[c]Corresponding to $\%\ V_{water} = 20$, $\%\ V_{soil} = 80$.

low its solubility limit. Often, equilibrium calculations using measured field data will generate aqueous-phase concentrations of target constituents that are above the solubility limit. This indicates the presence of a nonaqueous phase, which further complicates the analysis. Equilibrium estimates of aqueous-phase concentrations that are above the solubility limit should obviously not be used in exposure analysis components of a risk assessment or as mobile concentrations in any transport and fate predictions.

Chemical Partitioning Across the Water/NAPL Interface

The presence of nonaqueous phases in the subsurface can easily be seen to greatly complicate any attempts at contaminated groundwater collection and treatment, since before being collected and treated, the constituents of the NAPL must dissolve into the moving groundwater. This is usually due to the fact that NAPL residuals are immobile and "pooled" accumulations in the subsurface are often widely disseminated, difficult to locate, and even more difficult to pump directly. Consequently, actual constituent removal is necessarily indirect, resulting from dissolution to the moving aqueous phase and subsequent collection and treatment.

The water solubilities of many organic chemicals of interest range from very low parts per billion to typically less than 20,000 to 30,000 parts per million. In the ideal best case, the constituents in the NAPL exert their water solubilities at the NAPL/water interface. The actual transport of constituents from the NAPL would necessarily result in concentrations less than those achieved at equilibrium. Using this best-case assumption, one can estimate the number of pore volumes that would have to be flushed through the system to dissolve the constituents of interest from the NAPL and the corresponding time required using an average groundwater velocity from Darcy's law. The time required can easily run into the hundreds and thousands of years if low water solubility NAPLs are present in any substantial quantities. This situation is further complicated if nonaqueous phases are present in fractured rock systems and the major transport mechanism is diffusion from the relatively immobile rock matrix water into the groundwater flowing in the fractures (Mutch et al., 1992).

DISSOLVED CONSTITUENT TRANSPORT IN THE SATURATED ZONE

At a sufficient distance from any nonaqueous phase, NAPL constituents will be present at concentrations well below their solubility limits (zone 3). The theory for mass transport in Zone 3 is very well developed and several excellent references are available (Bear, 1988; Freeze and Cherry, 1979). Resources concerning the actual numerical modeling of the transport of dissolved constituents in the saturated zone are available as well (Walton, 1989; Wang and Anderson, 1982). Table 9 presents information concerning the major transport and fate processes occurring in the subsurface and groups them according to their impact on contaminant velocity and concentration. We look at advection, diffusion/dispersion, and sorption/retardation in this chapter. Biodegradation is discussed elsewhere in the book. These processes are evaluated further in Table 10 with respect to their impact on the time of travel of a dissolved constituent from the point of release to a point of concern and their impact on the concentrations at any point in the system, assuming continuous release of a constant concentration.

Advection is simply mass transport due to bulk groundwater movement. If this were the only mass transport process occurring, the aver-

Table 9 Major Transport and Fate Processes Affecting Velocity and Concentration of Dissolved Constituents in the Zone of Saturation

Advection, diffusion/dispersion: processes that act to transport and spread/mix the dissolved constituents

Sorption/retardation: process that acts to reduce the velocity/mixing of the dissolved constituents

Dispersion, sorption, degradation: processes that act to reduce the concentrations of the dissolved constituents in the mobile aqueous place

Table 10 Major Mass Transport Processes in the Saturated Zone: Additional Information

Process	Impact on tot[a]	Impact on C (continuous release)
Advection	$\overline{V}_{\text{chem}} = \overline{V}_{\text{aq}} = \frac{iK}{h}$	$C = C_0$
	$\text{tot}_a = \text{distance}/\overline{V}_{\text{aq}}$	
Dispersion	leading/tailing	$C \neq C_0$
	$\text{tot}_d < \text{tot}_a$	$C < C_0$
Retardation	$\overline{V}_{\text{chem}} < \overline{V}_{\text{aq}}$	$C \neq C_0$
	$\text{tot}_r > \text{tot}_d$	$C < C_0$

[a]tot, time of travel; $\overline{V}_{\text{aq}}$, average groundwater velocity; C, concentration of the dissolved constituent; C_0, initial concentration at point of continuous release; tot_a, time of travel if only advection is present; tot_d, time of travel with dispersion in addition to advection; tot_r, time of travel with sorption present as well.

age velocity of the chemical would be equal to the average velocity of the groundwater, which could be determined from Darcy's law through knowledge of the hydraulic gradient, the hydraulic conductivity, and the porosity of the system. Under most conditions, advection is accompanied by dispersion. *Dispersion* is the term given to the overall mass transport process resulting from both molecular diffusion, which always occurs if there is a concentration gradient in the system, and the mixing of the constituent due to turbulence and velocity gradients within the system. In the subsurface, turbulence is normally low. However, there are substantial velocity gradients due to geological heterogeneities, and those gradients turn out to be the major cause of dispersion in the saturated zone. *Retardation* is the decrease in chemical velocity compared to the average groundwater velocity resulting from the interaction of the dissolved constituent with the soil/rock skeleton (sorption).

With this brief background on the major transport processes occurring in the subsurface, let's look at the transport and dispersion of a single dissolved, nonreacting solute in a uniformly homogeneous and isotropic porous medium. These assumptions will enable a rigorous mathematical analysis that yields closed-form analytical solutions. These solutions can provide a good vehicle to increase our understanding of the fate and transport of a dissolved constituent in the simplest case for analysis. This will form a basis into which additional complexities can be incorporated as needed. We start with a graphical depiction of plume development and behavior in a thin, vertically well-mixed aquifer.

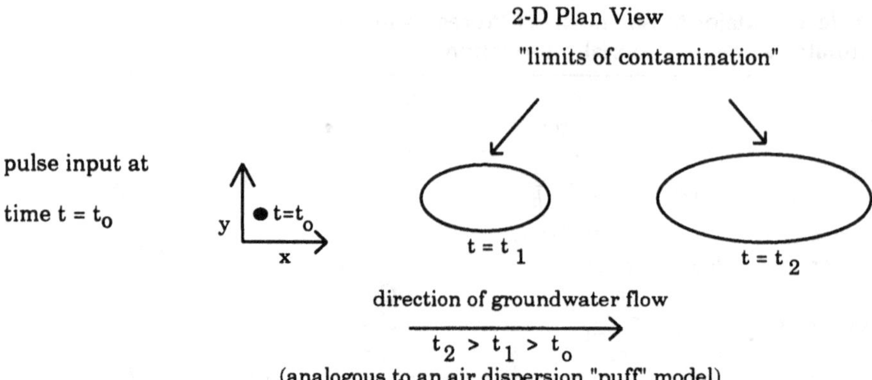

Note:

1. The chemical moves downgradient as a unit or *puff* in the direction of groundwater flow.
2. The puff disperses along both the longitudinal axis, x, and the transverse axis, y.
3. The longitudinal dispersion (D_x) is greater than the transverse dispersion (D_y).

In the case of continuous input, the limits of contamination always emanate from the source and the plume continues to move downgradient and continues to disperse along both the longitudinal and transverse directions until a *steady-state* situation is reached at some time t_{ss}.

At steady state, the flux of mass of the chemical coming into a system is balanced by the mass flux leaving the system and concentrations become constant (time independent) at all points within the system.

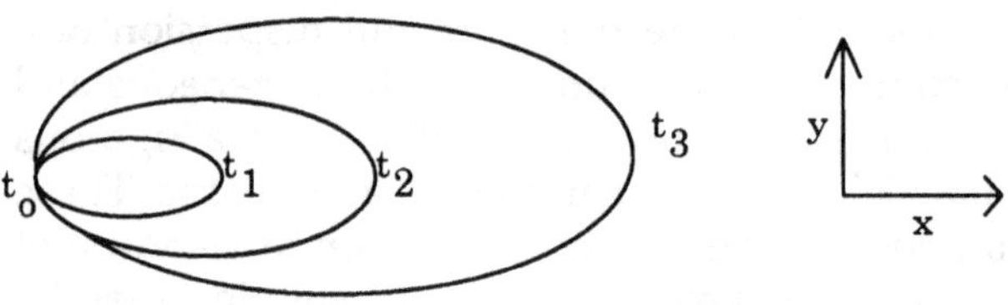

$t_3 > t_2 > t_1 > t_o$

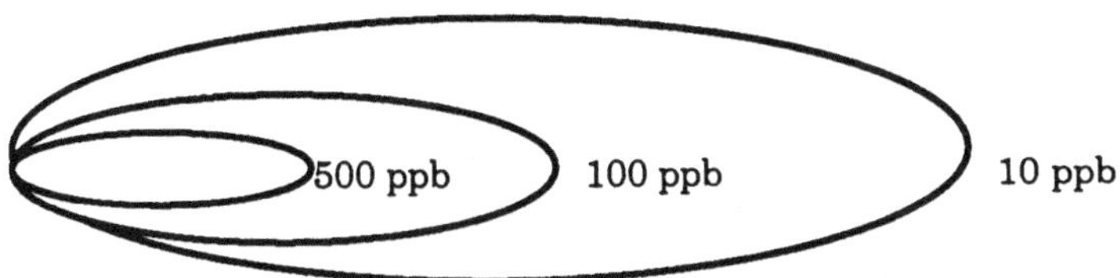

At the steady-state conditions, we can map true *isocons* or contours of constant concentration.

So far we have looked only at nonreacting solutes, such as Cl^-, which neither interact with the soil skeleton nor degrade either chemically or biologically. In most cases the dissolved constituent will sorb to the soil matrix and a partitioning of the chemical between the mobile groundwater phase and the immobile solid soil phase will occur. The extent to which this sorption process occurs depends on both the properties of the chemical and the properties of the soil. Sorption reduces both the velocity and the dispersion of the chemical until steady-state conditions are reached, after which time there is no net sorption, and retardation ceases to be an active process.

Now let's look at chemical transport in the subsurface in terms of a one-dimensional soil column experiment with continuous input of a constant concentration of a nonreactive tracer, so that we can begin to develop a more quantitative approach. t_{opv} is the time needed for one pore volume displacement.

$$t_{opv} = \frac{l}{\overline{V}}$$

We could also look at the concentrations for a given fixed time, at varying distances, x, from the source:

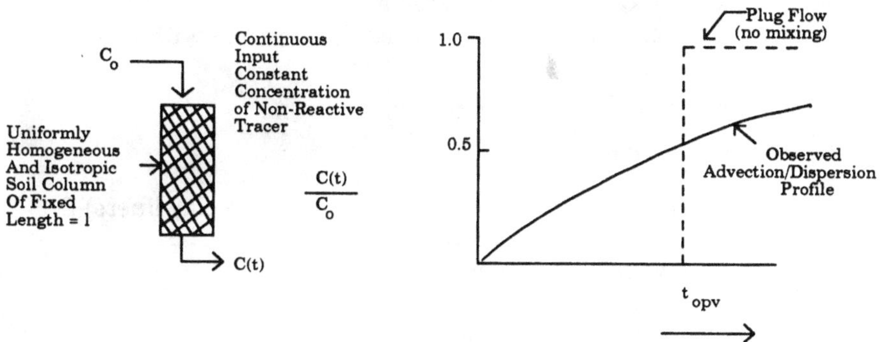

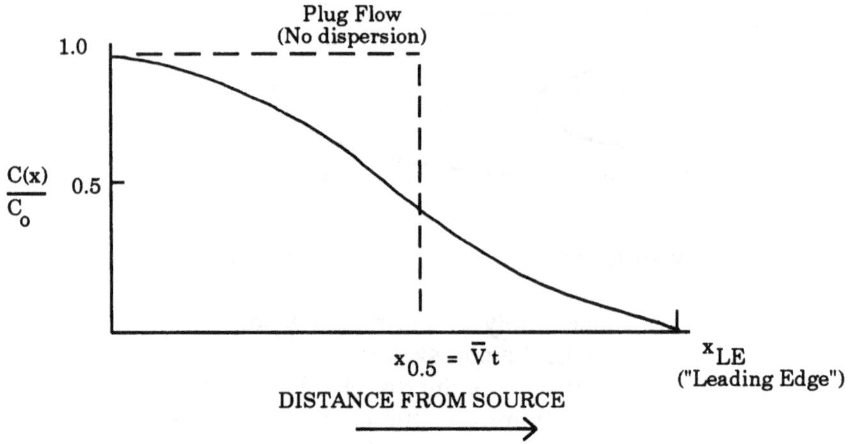

Important observations:

1. There is a leading edge corresponding to a first appearance of the chemical at time $t < t_{opv}$ (in the fixed-x case) or to nonzero concentrations at distances x greater than the average velocity location $\overline{V}t$.
2. For times $t << t_{ss}$, dispersion acts to reduce the concentrations at all locations from C_o, the input concentration. In a one-dimensional view, however, $C(x) \to C_o$ as $t \to t_{ss}$ for all x.

The dispersion process is handled mathematically through dispersion coefficients which are typically approximated as follows:

$$D = \overline{V}\alpha + D_m \text{ (molecular diffusion coefficient)}$$

where D has units of length^2/time, $\overline{V}$ is the average groundwater velocity (length/time), and α is the dispersivity (length), a characteristic of the given subsurface environment (α typically increases with increasing scale due to heterogeneities).

Note: For small ions and molecules

$$D_m \sim 1 \times 10^{-5} \text{ cm}^2/\text{s}$$

Consequently, unless $\overline{V}$ is very low (e.g., in compacted clay liners) $D \sim \overline{V}\alpha$ is usually a good approximation.

Question: How can we measure D?

Answer: Fit predicted concentrations to those measured using a nonreacting tracer. Actually, a very simple way is to determine the concentration profile with distance, x, from the source (for fixed t) and use

$$D = \frac{(x_{0.16} - x_{0.84})^2}{8t}$$

where $x_{0.16}$ and $x_{0.84}$ are the distances corresponding to 16% and 84% of the concentration distribution, respectively.

Perhaps the best way at this time (and the simplest) is to assume that the dispersivity is 5 to 10% of the flow path length (travel distance). This assumption is well founded statistically and appears to be valid based on experience.

With this background we look at a mathematical formulation in one dimension as follows:

One-dimensional advection–dispersion model:

$$D\frac{\partial^2 C(x,t)}{\partial x^2} - \overline{V}\frac{\partial C(x,t)}{\partial x} = \frac{\partial C(x,t)}{\partial t}$$

There are two closed-form analytical solutions, as follows:

1. Hunt (1978)

$$C(x,t) = \frac{C_0}{2}\exp\left(\frac{x\overline{V}}{2D}\right)\left\{\exp\left(\frac{-|x|\overline{V}}{2D}\right)\text{erfc}\left[\frac{|x| - \overline{V}t}{2(Dt)^{1/2}}\right] - \exp\left(\frac{|x|\overline{V}}{2D}\right)\text{erfc}\left[\frac{|x| + \overline{V}t}{2(Dt)^{1/2}P}\right]\right\}$$

for $-\infty < x < +\infty$ with a continuous input C_o at $x = 0$.

2. Ogata and Banks (1961)

$$C(x,t) = \frac{C_0}{2}\left\{\text{erfc}\left[\frac{x - \overline{V}t}{2(Dt)^{1/2}}\right] - \exp\left(\frac{x\overline{V}}{D}\right)\text{erfc}\left[\frac{x + \overline{V}t}{2(Dt)^{1/2}}\right]\right\}$$

for $0 \le x < \infty$ and a continuous input C_0 at $x = 0$.

Note: In the range $0 \le x < \infty$, the two solutions above are identical.

Ogata and Banks make a good case for neglecting the second term at sufficiently large distances from the source, so that

$$C(x,t) \simeq \frac{C_0}{2}\,\text{erfc}\left[\frac{x - \overline{V}t}{2(Dt)^{1/2}}\right] \qquad 0 \leq x < \infty$$

Note: Above, erfc (z) = 1 − erf(z) (erfc; complementary error function; erf, error function) and erf(−z) = −erf(z), erf(0) = 0, and erf(z) → 1.0 as z → ∞ [actually, erf(2.0) = 0.995 and erf(3) = 0.999978].

Example

Show that at steady state ($t \to \infty$) $\overline{C}(x,t) = C_0$ for $0 \leq x < \infty$.

$$\overline{C}(x,t) = C_0 \exp\left(\frac{x\overline{V}}{D}\right) \qquad -\infty < x < 0$$

Example

Consider using the closed-form analytical solution to the one-dimensional advection–dispersion model to estimate the location of the leading edge of the plume at any time, t, given values for $\overline{V}$ and D. Using the Ogata–Banks solution yields

$$\text{erfc}\left[\frac{x - \overline{V}t}{2(Dt)^{1/2}}\right] = \frac{2C(x,t)}{C_0}$$

As indicated above, erf (z) → 0 when z → ∞. In fact, erfc (3.0) = 0.000022, so that when

$$\frac{x - \overline{V}t}{2(Dt)^{1/2}} = 3.0$$

then

$$\frac{C(x,t)}{C_0} = 0.000011$$

or

$$C(x,t) = 0.0011\% \text{ of } C_0$$

(which may or may not be measurable).

Example:

Let $\overline{V}$ = 1.0 ft/day and, D = 5.0 ft^2/day. Compute $x_{0.5}$ and $x_{0.000011}$ for t = 300, 500, 1000, and 5000 days.

$$x_{0.5} = \overline{V}t$$

$$x_{0.000011} = \overline{V}t + 6(Dt)^{1/2}$$

t (days)	$x_{0.5}$ (ft)	$x_{0.000011}$ (ft)
300	300	530
500	500	800
1000	1000	1420
5000	5000	5950

Question: For a given t, what is the greatest downgradient distance, $x_{1.0}$, at which $C(x,t) = C_0$?

$$C(x,t) = C_0 \qquad \text{when} \quad \operatorname{erfc} \frac{x - \overline{V}t}{2(Dt)^{1/2}} = 2$$

Approximate using $\operatorname{erfc}(-3.0) = 1 + \operatorname{erf}(3) = 1.999978$ so that $x_{1.0} = \overline{V}t - 6(Dt)^{1/2}$.

t (days)	$x_{0.5}$ (ft)	$x_{1.0}$ (ft)
300	300	70
500	500	200
1000	1000	580
5000	5000	4000

Next look at the solution to the advection–dispersion equation in two dimensions.

$$D_x \frac{\partial^2 C(x,y,t)}{\partial x^2} + D_y \frac{\partial^2 C(x,y,t)}{\partial y^2} - \overline{V}_x \frac{\partial C(x,y,t)}{\partial x} = \frac{\partial C(x,y,t)}{\partial t}$$

From Wilson and Miller (1978),

$$C(x,y,t) = \frac{f'_m B^{1/2} \exp\,(|x/B|)\exp(-r/B)}{4_\eta(2\pi r D_x D_y)^{1/2}} \operatorname{erfc}\left\{\frac{-[(r/B) - 2u]}{2u^{1/2}}\right\}$$

$$\text{for ``large''}\ \frac{r}{B}\left\{\begin{matrix}10\%\ \text{error when}\ \frac{r}{B} > 1 \\ 1\%\ \text{error when}\ \frac{r}{B} > 10\end{matrix}\right\}$$

where f'_m = mass injected per unit length
η = porosity

D_x, D_y = longitudinal and lateral dispersion coefficients, respectively

$\overline{V}_x$ = average groundwater velocity in the x direction

$$B = \frac{2D_x}{V_x}$$

$$r = \left(x^2 + \frac{D_x}{D_y}y^2\right)^{1/2}$$

$$u = \frac{r^2}{4D_x t} \qquad (\textit{note}\text{: all the time dependence is in } u)$$

At a steady state, $t \to \infty$ and $u \to 0$ and, $\text{erfc}(-\infty) = 1 + \text{erf}(\infty) = 2$, so that

$$\overline{C}(x,y) = \frac{f'_m B^{1/2} \exp(|x/B|) \exp(-r/B)}{2_\eta (2\pi r D_x D_y)^{1/2}}$$

Also, concentration along the plume centerline at steady state is given by

$$\overline{C}(x,0) = \frac{f'_m}{2\eta(\pi x \overline{V}_x D_y)^{1/2}}$$

What about sorption–retardation?

If the dissolved solute interacts with the soil skeleton through a sorption process, we need to distinguish two concentrations:

C_{aq} = concentration in the mobile aqueous phase
C'_{soil} = concentration in the *immobile* solid soil phase

(The differential equations describe the spatial and temporal behavior of C_{aq}.) Look at the one-dimensional model in general:

$$D\frac{\partial^2 C_{\text{aq}}}{\partial x^2} - \overline{V}\frac{\partial C_{\text{aq}}}{\partial x} + \sum_i S_i = \frac{\partial C_{\text{aq}}}{\partial t}$$

where S_i is a source or sink. Treat sorption as a sink as follows:

$$S_{\text{sorption}} = -\frac{\rho}{\eta}\frac{\partial C_{\text{soil}}}{\partial t}$$

where ρ is the bulk solids density (g/mL) and η is the porosity.

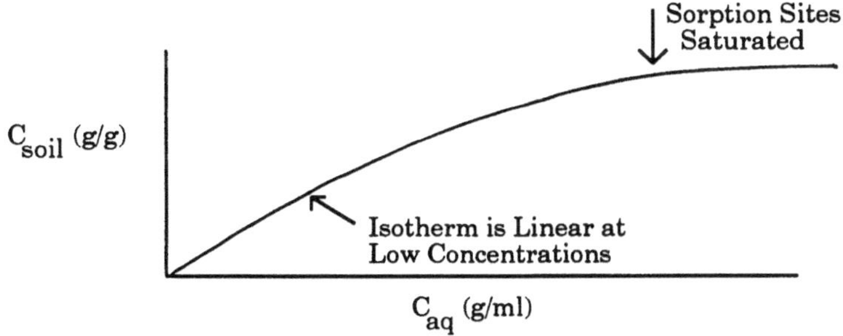

If we assume local equilibrium in the linear portion of the isotherm (low concentrations), we can work with a constant equilibrium distribution coefficient.

$$K_D = \frac{C_{soil}}{C_{aq}}$$

or

$$C_{soil} = K_D C_{aq}$$

$$\frac{\partial C_{soil}}{\partial t} = \frac{\partial C_{soil}}{\partial C_{aq}} \frac{\partial C_{aq}}{\partial t} = K_D \frac{\partial C_{aq}}{\partial t}$$

so that the one-dimensional equation becomes

$$\frac{\partial C}{\partial t}\left(1 + \frac{\rho}{\eta} K_D\right) = D\frac{\partial^2 C}{\partial x^2} - \overline{V}\frac{\partial C}{\partial x}$$

Define

$$R = 1 + \frac{\rho}{\eta} K_D \qquad (R = \text{retardation factor})$$

so that

$$\frac{\partial C}{\partial t} = \frac{D}{R}\frac{\partial^2 C}{\partial x^2} - \frac{\overline{V}}{R}\frac{\partial C}{\partial x}$$

Thus the average velocity of the chemical has been retarded by a factor R. Consequently, we can think of the sorption process as acting effectively to retard the transport of the chemical. (However, the process actually results in a time scaling, and eventually, if steady-state conditions are reached, retardation will cease to be an active process.)

Figure 13 depicts the effect of retardation on the mobilities of different constituents emanating from a source. Retardation results in a "chromatographic" effect, where those constituents having the highest

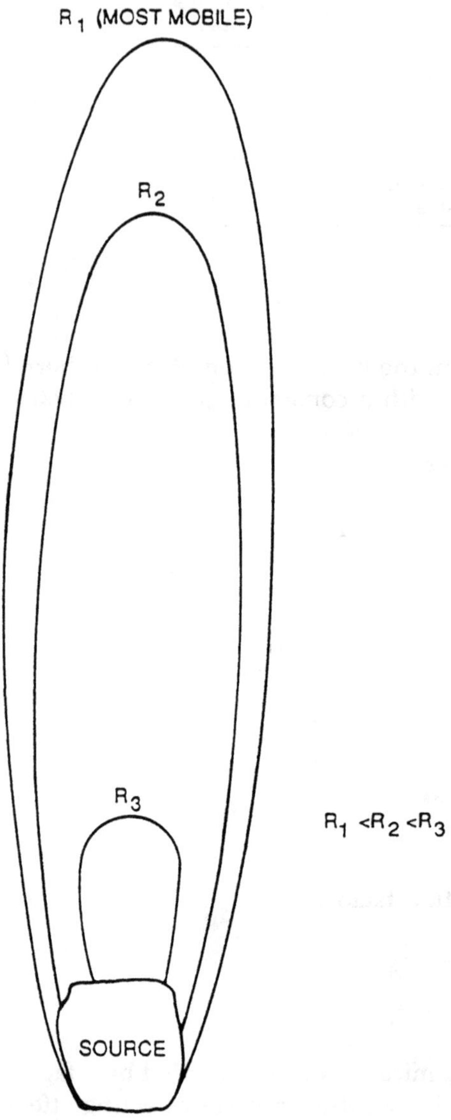

Figure 13 Plan view of contaminant transport retardation.

R values (greatest sorption/lowest water solubilities) have the lowest velocities and those having the lowest R values (least sorption/highest water solubilities) have the highest velocities.

TRANSPORT OF METALS IN THE SUBSURFACE

To this point the discussion has focused on the behavior and transport of hydrophobic organics within the saturated zone. Techniques have been presented for developing a quantitative description of the equilibrium condition concerning the partitioning of hydrophobic organics between water and soil. This has enabled incorporation of the sorption/retardation process for hydrophobic organics into a linear advection–dispersion–sorption model.

Transport in the subsurface is much more complex for heavy metals, due to the much more complex chemistry of heavy metals in aqueous systems. The actual form in which a metal may be present (ionization state, hydration state, etc.) is very much dependent on environmental conditions, and these conditions often vary substantially over the scales of interest. Consequently, to date, attempts at describing the sorption process for heavy metals using a linear partition coefficient have typically failed. The complex chemistry of heavy metals in the subsurface may require departure from the linear model. Furthermore, a more basic concern is whether metals achieve an equilibrium state in saturated zones of interest in the first place. The reader is referred to publications by Theis (1988), for example, and others for additional information on the contaminant transport of heavy metals in the subsurface.

NONEQUILIBRIUM EFFECTS

To this point, contaminant transport in the subsurface has been discussed from the standpoint of a uniform and homogeneous aquifer and sorption occurring under local equilibrium conditions. The aquifer is rarely uniform and homogeneous, and the actual geological complexity can often have major impacts on contaminant velocities within the subsurface. Furthermore, the local equilibrium assumption may in some cases be unrealistic as well, depending on groundwater velocities and sorption kinetics.

The major impact of geological complexities upon contaminant velocities within a saturated zone results from the presence of zones of immobile water or zones where the hydraulic conductivities are very low. These could be interbedded clay and silt lenses in a porous medium or the rock matrix itself in a fractured rock aquifer. Depending on the antecedent contamination times, there may have been a substantial flux of contaminants into these zones of low hydraulic conductivity as a result of diffusion. Attempts to purge the system of these contaminants by pumping water and collecting it at the surface will be frustrated by the presence of the contaminants in these zones of immobile water. The constituents of concern will have to diffuse into a zone of higher hydraulic conductivity before they can be collected. In fact, depending on the antecedent contamination times, several years of groundwater pumping may be necessary to reverse concentration gradients within these zones so that diffusion out of these zones into the zone of higher hydraulic conductivity can occur.

REFERENCES

Bear, J. 1988. *Dynamics of Fluids in Porous Media*. Dover, New York; originally published in 1972.

Berg, R. R. 1975. Capillary pressures in stratigraphic traps. *Am. Assoc. Pet. Geol. Bull.* 59(6):939–956.

Cedergren, H. R. 1977. *Seepage, Drainage, and Flow Nets*, 2nd ed. Wiley, New York.

Chiou, C. T., L. J. Peters, and V. H. Freed. 1979. A physical concept of soil–water equilibrium for non-ionic organic compounds. *Science* 206:831–832.

Davis, S. N., and R. J. M. DeWiest. 1966. *Hydrogeology*. Wiley, New York.

Feenstra, S. M., et al. 1991. *Removal of Dense Non-aqueous Phase Liquids (DNAPLs) from Fractured Bedrock Formations: Principles and Potential Remedial Technologies*. Final Report, May 24.

Freeze, R. A., and J. A. Cherry. 1979. *Groundwater*. Prentice Hall, Englewood Cliffs, NJ.

Huling, S. G., and J. W. Weaver. 1991. Dense nonaqueous phase liquids. *Groundwater Issue*. EPA/540/4-91-002. U.S. Environmental Protection Agency, Office of Research and Development, Office of Solid Waste and Emergency Response, March.

Hunt, B. 1978. Dispersive source in uniform groundwater flow. *J. Hydraul. Div. ASCE* 104:75–85, January.

Karickhoff, S. W., D. S. Brown, and T. A. Scott. 1979. Sorption of hydrophobic pollutants on natural sediments. *Water Res.* 13:241–248.

Kenaga, E. E., and C. A. I. Goring. 1978. Relationship between water solubility, soil sorption, octanol water partitioning, and bioconcentration of chemicals in biota. 3rd *Aquatic Toxicology Symposium Proceedings*. ASTM STP 707.

Kueper, B. H., and D. B. McWhorter. 1991 "The behavior of dense, nonaqueous phase liquids in fractured clay and rock, *Ground Water* 29(5):716–728.

Lamb, T. 1956. The storage of oil in an earth reservoir. *J. Boston Soc. Civil Eng.*, July.

Leverett, M. D. C. 1940. Capillary behavior in porous soils. *Pet. Technol.*, August.

Lyman, W. J., W. F. Reehl, and D. H. Rosenblatt. 1990. *Handbook of Chemical Property Estimation Methods*. American Chemical Society, Washington, DC.

MacKay, D. 1991. *Multimedia Environmental Models*. Lewis Publishers, Chelsea, MI.

Mackay, D. M., and J. A. Cherry. 1989. Groundwater contamination: pump-and-treat remediation. *Environ. Sci. Technol.* 23(6):630–636.

McCall, P. J., R. L. Swann, and D. A. Laskowski. 1983. Partition models for equilibrium distribution of chemicals in environmental compartments. In: R. L. Schwann and A. A. Eschenroeder (eds.), *Fate of Chemicals in the Environment*. ACS SS225. American Chemical Society, Washington, DC.

Mercer, J. W., and R. M. Cohen. 1990. A review of immiscible fluids in the subsurface. 120–121.

Mutch, R. D., Jr., Scott, J. I., and Wilson, D. J. 1992. Cleanup of fractured rock aquifers: implications of matrix diffusion. *Environ. Monit. Assess.*, in press.

Ogata, A., and R. B. Banks. 1961. *A Solution of the Differential Equation of Longitudinal Dispersion in Porous Media*. Geological Survey Professional Paper 411-A.

Prausnitz, J. M. 1969. *Molecular Thermodynamics of Fluid-Phase Equilibria*, Prentice Hall, Englewood Cliffs, NJ.

Reible, D. D., M. E. Malhiet, and T. H. Illangasekare. 1989. Modelling gasoline fate and transport in the unsaturated zone. *J. Hazard. Mater.*

Rose, W. and W. A. Bruce. 1949. Evaluation of capillary character in petroleum reservoir rock. *Pet. Trans. AIME*, May, pp. 127–142.

Russell, L. D., and P. A. Dickey. 1950. Porosity, permeability, and capillary properties of petroleum reservoirs. In: P. D. Trask (ed.), *Applied Sedimentation*. Wiley, New York.

Schwarzenback, R. P., and W. Giger. 1985. Behavior and fate of halogenated hydrocarbons in groundwater. In: C. H. Ward, W. Giger, and P. L. McCarty (eds.), *Groundwater Quality*. Wiley-Interscience, New York.

Schwille, F. 1988. *Dense Chlorinated Solvents in Porous and Fractured Media: Model Experiments*. Lewis Publishers, Chelsea, MI.

Stauffer, T. B., and W. G. MacIntyre. 1986. Sorption of low polarity organic compounds on oxide minerals and aquifer material. *Environ. Toxicol. Chem.* 5:949–955.

Theis, T. L. 1988. Reactions and transport of trace medals in groundwater. In: J. R. Kramer and H. E. Allen (eds.), *Metal Speciation: Theory, Analysis, and Application*. Lewis Publishers, Chelsea, MI.

Thibodeaux, L. J. 1979. *Chemodynamics: Environmental Movement of Chemicals in Air, Water and Soil*. Wiley-Interscience, New York.

Walton, W. C. 1989. *Analytical Groundwater Modelling: Flow and Contaminant Migration*. Lewis Publishers, Chelsea, MI.

Wang, H. F., and M. P. Anderson. 1982. *Introduction to Groundwater Modelling*. W. H. Freeman, New York.

Wilson, J. L., and P. J. Miller. 1978. Two dimensional plume in uniform groundwater flow. *J. Hydraul. Div. ASCE*, April, pp. 503–513.

Wilson, D. J., A. N. Clarke, and J. H. Clarke. 1988. Soil clean up by in situ aeration. I. Mathematical modeling. *Sepr. Sci. Technol.* 23:991.

2

PROBLEMS WITH THE REMEDIATION OF DIFFUSION-LIMITED FRACTURED-ROCK SYSTEMS

Robert D. Mutch, Jr.
and
Joanna I. Scott
ECKENFELDER INC.
Mahwah, New Jersey

INTRODUCTION

A great many groundwater-based public water supplies in the United States and tens of thousands of private homes rely on fractured bedrock aquifers for their water. Several bedrock aquifers, such as the Passaic Formation aquifer in northern New Jersey, have been granted sole source status and thus have been accorded special federal protection. At the same time, there has been an alarming increase in the incidence of contamination of these important aquifers as a result of improperly controlled waste disposal, leaking underground storage tanks, industrial spills, and other sources. Numerous groundwater extraction and treatment programs have been initiated and still more are planned to remediate fractured rock aquifers.

Groundwater extraction and treatment systems often have two principal purposes: to control the further spread of contamination, and to clean up the contaminated section of the aquifer. Controlling the further spread of contamination plumes, while more complicated in fractured rock aquifers than in simple porous media, is not especially problematic. However, restoring aquifer water quality once the aquifer has become contaminated is considerably more difficult. Not only are dense nonaqueous-phase liquids (DNAPL) more difficult to purge from fractured rock (Mackay and Cherry, 1989), but the phenomenon of *matrix diffusion* (the diffusion of contaminants from the mobile water in the fractures to the virtually stagnant water within the rock matrix) can make aquifer restoration frustratingly slow at best, and, at worst, an exercise in futility. This phenomenon has become especially critical in recent years as mandated post-cleanup contaminant concentration levels have gravitated to the low parts per billion range. Since the initial concentrations of some of the more soluble contaminants may exceed 100 or even 1000 parts per million, the required improvement in water quality may be 99.99% or more.

In this chapter some of the problems inherent in attempting to control and remediate contaminated fractured rock aquifers are discussed. In the first section we describe a typical fractured rock aquifer, including groundwater flow characteristics and the basic mechanisms of contaminant transport. The second section includes various strategies for arresting plume migration in fractured rock aquifers. The more difficult problem of actually removing contaminants from a fractured rock aquifer and restoring it to usable condition is treated in the third section, with special emphasis on matrix diffusivity.

FRACTURED ROCK AQUIFERS

Fractured rocks generally consist of blocks of relatively intact rock (the rock matrix) separated by a network of fractures. The mean spacing of fractures in rock can vary from a few centimeters in highly fractured rock to 10 m or more in granite and some other crystalline rocks. The mean fracture width (fracture aperture) is typically a few millimeters or less. In some extensively fractured rocks, the fractures are so numerous and close together that, at any scale of practical interest, the rock essentially behaves as an equivalent porous medium. When fractures are farther apart, groundwater flow patterns can be much more complex than in porous media; in the most extreme cases each fracture is essentially a separate flow zone with its own properties. Complex fractured rock aquifers that cannot be treated as equivalent porous media are the concern of this chapter.

The bulk of groundwater flow in most fractured rock follows the line of least resistance through the fracture system with little, if any, flow through the matrix blocks. The water within the rock is therefore of two types: mobile water, which is predominantly in the fractures, and essentially immobile water within the pores of the rock matrix. A useful way to describe this situation is to define two porosities: a matrix porosity of the rock matrix (exclusive of the fractures) and a fracture porosity. Some literature values for matrix and fracture porosities are given in Table 1. The matrix porosity can be quite high, as much as 25% in some sedimentary rocks, but the fracture porosity is much lower, 0.5% or less. This means that the bulk of the water in fractured rocks is contained within the pores of the rock matrix. For example, if a rock has a matrix porosity of 10% and a fracture porosity of 0.5%, only roughly 5% of the water in the rock is mobile; approximately 95% of the water is in pores within the rock matrix and is essentially immobile.

In some fractured rock aquifers with low-to-moderate fracturing, the flow field may be discontinuous, with groups of fractures forming semiautonomous flow zones. The poor communication between the various flow zones in such an aquifer complicates control and remediation efforts. It is also difficult to measure and characterize such complex flow fields.

The groundwater flow rate in a fractured rock aquifer depends on the overall hydraulic conductivity of the rock and the hydraulic gradient across the rock. The overall hydraulic conductivity, which is essentially that of the fracture flow system in most cases, can be estimated from pumping tests or other in situ measurements of hydraulic conductivity. The hydraulic conductivity of the intact rock matrix, as measured in laboratory tests, is typically considerably lower. For example, a hydraulic gradient of 0.01 across a section of rock with an overall hydraulic conductivity of 1×10^{-4} cm/s (typical of a moderately productive fractured rock aquifer) will produce a flow rate per unit area of rock of 8.5×10^{-4} m^3/m^2 per day. The groundwater flow rate varies widely for different rock types and locations and frequently for different locations within the same fractured rock aquifer.

Since flow is restricted to a small fraction of the total rock volume, the actual groundwater flow velocity (seepage velocity) in fractured rock is generally much higher than in a porous medium with the same overall flow rate. Seepage velocities in fractured rock can be as low as a few millimeters per day in low-permeability rock or as high as several tens of meters per day in prolific fractured rock aquifers. For example, a flow rate of 8.6×10^{-4} m^3/m^2 per day in a fractured rock with a fracture porosity of 0.0025 corresponds to an average seepage velocity of 0.34 m/day.

Dissolved contaminants that enter a fractured rock will be carried through the fractures along with the flowing water (advection). Some contaminant, however, will be transported faster (or slower) than the average groundwater velocity by the process of longitudinal dispersion.

Table 1 Typical Parameter Values for Fractured Rock

Parameter	Type of rock	Reported range	Reference
Matrix porosity	Sandstone	14.8% average	Barrell (1914)
	Shale	8.2% average	Barrell (1914)
	Limestone	5.3% average	Barrell (1914)
	Sandstone, siltstone, quartzite, chert, and conglomerate	<1 to >25%	Manger (1963)
	Limestone, dolomite, chalk, and marble	0.2 to 20%	Manger (1963)
	Shale, claystone, and slate	0.5 to 15%	Manger (1963)
	Granite and gneiss	0.06 to 0.4%	Neretnieks (1985)
Fracture porosity	Sandy shale petroleum reservoirs	1×10^{-3} to 2×10^{-3}	Streltsova (1976a)
	"Fissured reservoirs"	1×10^{-3} to 5×10^{-3}	Streltsova (1976b)
	Conasauga Shale (Tennessee)	2×10^{-3}	Walter and Thompson (1982)
	"Siliceous reservoirs"	1×10^{-3} to 5×10^{-3}	Yurochko (1982)
	Nolichuky Shale (Tennessee)	1×10^{-3}	Smith and Vaughan (1985)
	Culebra Dolomite (New Mexico)	2×10^{-3}	Kelley et al. (1987)
	Igneous and metamorphic	$\sim 1 \times 10^{-4}$	Freeze and Cherry (1979)
Matrix diffusivity	Low-molecular weight (<500 g/mol) solutes in water	1.5×10^{-9} m²/s	Rasmuson and Neretnieks (1981)
	Clayey soil	$4.6–7.5 \times 10^{-10}$ m²/s	Barone et al. (1988)
	Queenstone Shale (Ontario)	$1.4–1.6 \times 10^{-10}$ m²/s	Barone et al. (1989)
	Queenstone Shale (Ontario)	3.7×10^{-12} m²/s	Pankow et al. (1986)
	Granite	2.3×10^{-14} to 1.6×10^{-12} m²/s	Bradbury and Green (1985)
	Granite and gneiss	1×10^{-14} to 7×10^{-13} m²/s	Neretnieks (1985)

Longitudinal dispersion is caused by local variations in fracture aperture and groundwater velocity. The longitudinal dispersivity, which has units of length, has been shown to have a high degree of scale dependence; that is, the larger the scale of the transport situation, the larger the measured longitudinal dispersivity. Longitudinal dispersivity varies from as little as a few millimeters in laboratory columns to more than 100 m in large scale plumes (Anderson, 1979). Longitudinal dispersion is important in determining the rate of expansion at the edges of a contaminant plume but has little effect on concentrations in the main part of a plume.

Contaminants in fractured rock may sorb to the walls of the fractures and therefore be transported more slowly than the groundwater. Sorption can usually be described by a *retardation factor*, which is simply the groundwater velocity divided by the average velocity of the contaminant. A contaminant with a retardation factor of 3 would be transported through the fractures at a rate one-third that of the water or of an unsorbed contaminant. Metals, in particular, can have large retardation factors, as much as 1000 or more.

As contaminated groundwater moves through the fractures of a bedrock aquifer, the contaminants will diffuse into the essentially stagnant matrix pore water of the rock, as illustrated in Figure 1. The rate at which a contaminant diffuses depends on the difference between the concentration in the fracture and in the rock matrix, the area over which diffusion takes place, and the diffusivity of the contaminant. The diffusivity depends on the properties of both the rock matrix and the contaminant. Some literature values for diffusivities in various rock types are given in Table 1.

From one perspective, the diffusion of contaminants into the rock matrix is beneficial in that it retards the advance of the contaminant plume through the fractured rock. Lever and Bradbury (1985) reported that matrix diffusion leads to effective retardation factors in excess of 100 and can reduce peak concentrations by three to four orders of magnitude, provided that the groundwater velocity is relatively small. However, when the objective is to purge contamination from an aquifer, the contaminants must diffuse back out of the rock matrix before they can be flushed from the rock. This diffusion-controlled release of contaminant from the rock matrix can greatly prolong aquifer cleanup efforts over what would be possible in a simple porous medium of equivalent hydraulic conductivity.

Diffusion into and out of the rock matrix will be even slower if the contaminant sorbs to the rock matrix. A contaminant with a retardation factor of 3 will diffuse at a rate three times slower than an unsorbed

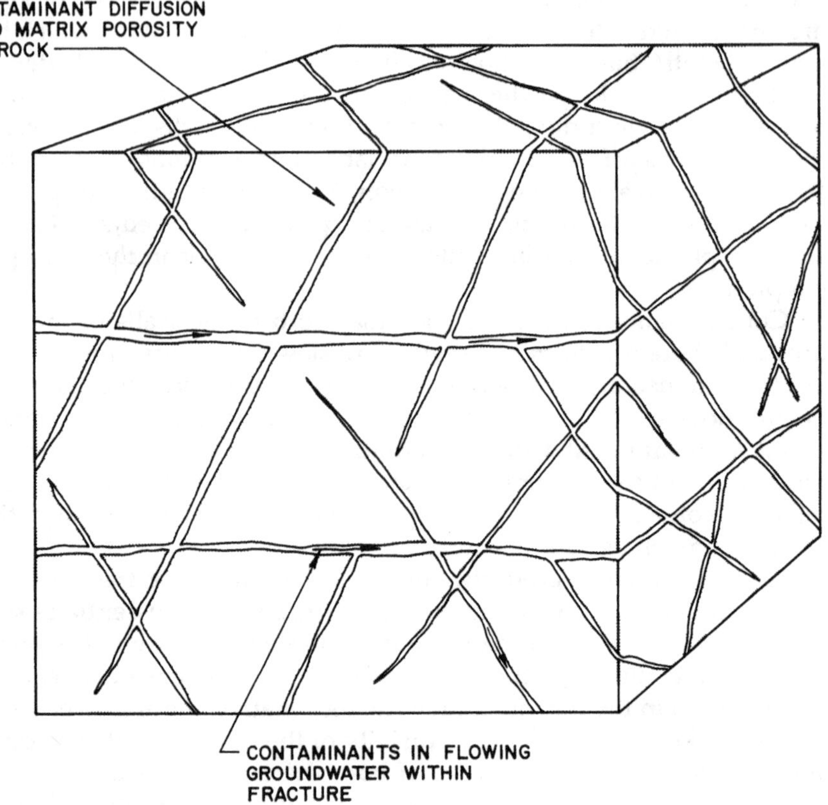

Figure 1 Matrix diffusion in fractured rock aquifers

contaminant. Less of the contaminant will diffuse into the rock, but what does diffuse in will take longer to come out again.

There are other phenomena that may affect the rate of contaminant removal from fractured rock. Contaminants may be removed from the system through biological, chemical, or radioactive decay. Undissolved chemicals, especially dense nonaqueous-phase liquids (DNAPL), may exist in the fractures. The rate at which DNAPL is removed will depend on the rate at which it dissolves into the groundwater, which in turn depends on such factors as the solubility of the DNAPL, the concentration already in the groundwater, and the area of contact between DNAPL and groundwater. If the DNAPL is in a narrow, dead-end fracture, the area of contact between the DNAPL and the mobile water will be small, and the rate of dissolution may be very slow.

STRATEGIES FOR PLUME CONTAINMENT IN FRACTURED ROCK

Arresting the migration of a plume of contaminated groundwater in a fractured rock aquifer can be more difficult than in a porous medium for several reasons. First, the fracture flow system of some aquifers can be highly variable and complex, with a number of semiautonomous flow zones in poor hydraulic communication with one another. A system of groundwater extraction wells must intersect each important flow zone to assure complete capture of the plume. Second, well yields in fractured rock are frequently highly variable. Wells that happen to intersect a major part of the fracture flow system will have higher yields than wells that intersect only minor fractures. It is not uncommon for wells drilled into fractured rock to have little or no yield, even when the formation as a whole is quite transmissive. Third, groundwater extraction systems designed for fractured rock aquifers with low hydraulic conductivities frequently consist of many, closely spaced, low-pumping-rate recovery wells. Such systems often have high construction costs and high operation and maintenance costs. Moreover, such systems are often difficult to operate because of the low yields of individual wells and the tendency of wells to dewater during dry periods of the year.

The complexity of groundwater flow in fractured rock generally varies in inverse proportion to the extent of fracturing. Rocks with widely spaced fractures, such as typical granites, usually have complex groundwater flow systems. In such rocks, groundwater flow is, to a large extent, controlled by the orientation and character of individual fractures. At the other extreme, extensively fractured rock may behave essentially as a porous medium. In such cases, arresting a plume's migration is achieved following conventional approaches of aquifer testing and capture zone development (Walton, 1970; Javendel and Tsang, 1986).

For fractured rock aquifers with complex flow systems, there are a number of techniques available to enhance plume capture. These include hydrofracturing and explosive blasting. There is also a philosophical approach, the *observational method*, which can improve the effectiveness of groundwater extraction systems in any medium, but which may be particularly valuable in complex systems such as fractured rock aquifers.

Hydrofracturing

Hydrofracturing is a traditional approach for increasing well yields in low-permeability fractured rock. In hydrofracturing, a segment of open rock is isolated with inflatable packers and water is injected into the segment at high pressure. The pressure creates artificial fractures that radiate outward from the borehole. The increased fracturing in the vicinity of the well increases the probability that the well will communicate with the natural fracture flow system.

The following example illustrates the type of yield enhancement typically achievable by hydrofracturing. As part of the construction of a groundwater extraction and treatment system at a large eastern manufacturing facility, four extraction wells were constructed in a fractured sandstone of moderate hydraulic conductivity. It was estimated, based on numerical modeling of the aquifer system, that to establish the required capture zone, the four wells would have to have a total yield of at least 16 gal/min. Past experience at the site indicated that wells constructed in this fractured rock aquifer had widely varying yields, and that despite the overall moderate hydraulic conductivity of the aquifer, some wells had little or no yield. The four new wells proved to be consistent with this prior experience: two of the wells initially had moderate yields and two had essentially no yield (Table 2). The total yield of the four extraction wells was 9.5 gal/min, insufficient to ensure plume capture.

In an attempt to increase well yields, all four wells were hydrofractured by injecting 1800 gallons of water into each well at pressures up to 500 psi. The resulting increased yields are also given in Table 2. The two wells with original moderate yields experienced modest increases in yield of around 50 to 100%. More important, however, the two wells, which originally had essentially zero yields, now had usable yields. The

Table 2 Example of Results of Hydrofracturing

Well	Pre-hydrofracturing yield (gpm)	Post-hydrofracturing yield (gpm)
1	6.5	9.5
2	3.0	6.5
3	<0.1	2.0
4	<0.1	2.7
Total yield	9.5	20.7

total yield of all four extraction wells increased to about 20 gallons per minute, sufficient to establish the required capture zone. As shown by this example, hydrofracturing can frequently salvage a previously unproductive well. It may also reduce the number of wells required to capture a plume.

Explosive Blasting

A more novel approach to capturing a plume in fractured rock is the use of explosive blasting. In this technique, a line of closely placed shot holes is drilled in front of and perpendicular to an advancing plume, as illustrated in Figure 2. Explosive blasting is conducted in each shot hole to create a high-permeability corridor of pulverized rock across the path of the plume. An extraction well drilled into the pulverized rock will often have a yield many times that of a well constructed in the native rock. One, or at most a few, of these higher-yield wells can then be used to lower the potentiometric surface in the pulverized rock corridor to create a highly efficient collection drain in the rock.

One of the particular advantages of this technique is that one can substitute a small number of high-yield extraction wells, and possibly even a single well, for the large number of low-yield extraction wells typical of conventional extraction systems in low-permeability fractured rock. Fewer wells generally means lower construction costs and lower operational and maintenance costs. While the blasted rock approach may have higher capital costs, these costs can often be substantially offset by reductions in long-term operational and maintenance costs. The blasted rock approach can also help allay concerns that portions of the plume may be missed by all the wells and thus circumvent the groundwater extraction system.

Observational Method

The observational method is an approach for dealing with the inevitable uncertainties that arise in geotechnical or geoenvironmental projects. It was originally developed by Carl Terzaghi and is described by Peck (1969). Every geoenvironmental project will have uncertainty associated with it, whether it is the extent of subsurface contamination, the transmissivity of a particular aquifer, or the vertical hydraulic conductivity of a critical aquitard. The observational method is a particularly attractive way of managing the inherent uncertainties associated with the design of groundwater extraction systems, as well as other subsurface remedial projects (Smyth and Quinn, 1991). The complex flow systems found in typical fractured rock aquifers can lead to especially large uncertainties in the design of groundwater extraction systems; therefore, it may be

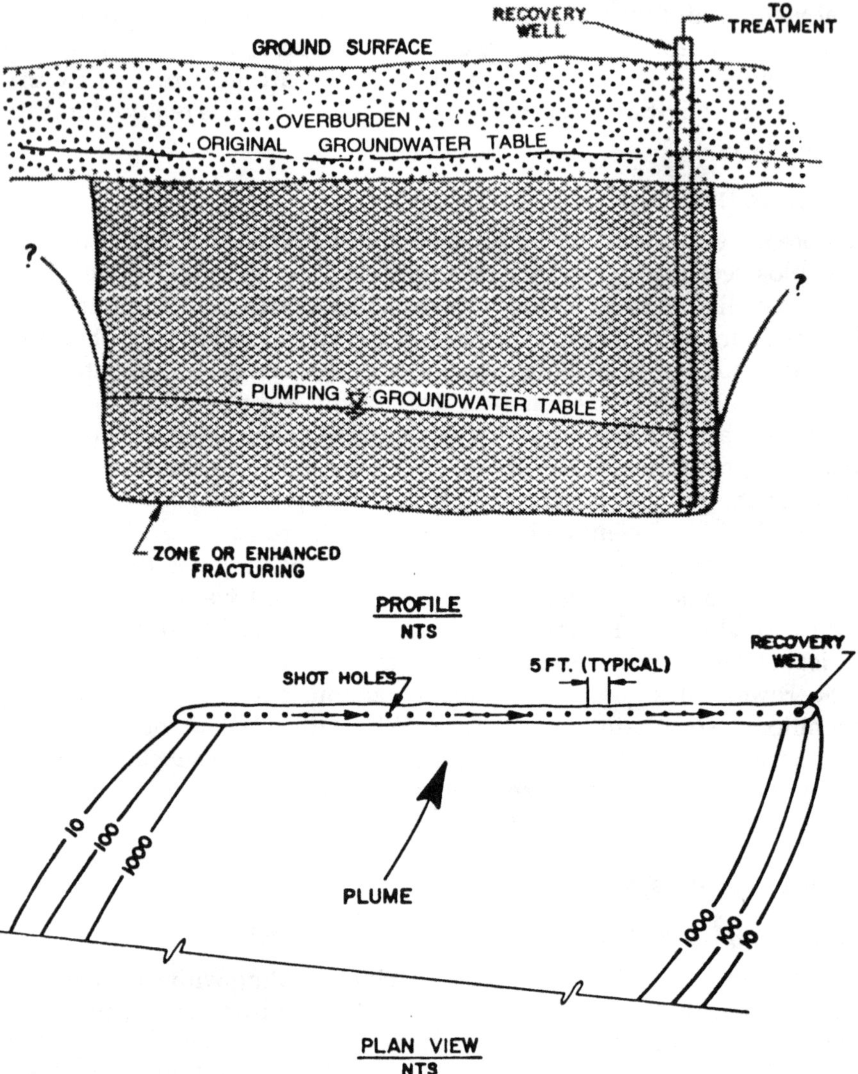

Figure 2 Groundwater collection in fractured bedrock by means of a high-permeability blasted rock zone.

particularly advantageous to use the observational method with fractured rock aquifers.

The essential characteristics of the observational method are as follows:

1. The design is based on a hydrogeologic/geotechnical investigation of sufficient scope to determine the general conditions and the probable range of key parameters. However, the investigation does not seek to define precisely all key hydrogeologic parameters through a protracted investigation.
2. The design is based on the most probable subsurface conditions rather than on worst-case conditions.
3. Sufficient flexibility is included in the design, or specific contingency plans are developed, to cover all reasonable deviations from the most probable conditions on which the design is based.
4. The performance of the particular remedial measure is monitored during and following construction and the design is modified, or a particular contingency plan is implemented, as necessary to achieve the particular remedial objectives.

Examples of contingency plans for groundwater extraction systems would include hydrofracturing wells with insufficient yields, increasing the pumping rates over what was originally planned (with a concurrent increase in treatment capacity for the extracted groundwater), adding additional pumping wells, or if explosive blasting was employed, additional blasting to enlarge the pulverized rock zone.

The observational method has several advantages. First, the design is based not on worst-case conditions but on the most probable conditions; this often results in considerable cost savings to the owner or operator of the facility. Second, if actual conditions do deviate from the most probable conditions, a predesigned and preapproved contingency plan can be rapidly implemented. Finally, the observational method explicitly recognizes the uncertainties that are always present, even in a worst-case design. The owner, consultant, and involved regulatory agencies can all be aware of the uncertainties in advance and have more reasonable expectations for the system.

A limitation of the observational method is that it is not applicable to situations where a reasonable contingency plan cannot be developed for all reasonable deviations from the most probable conditions, a "fatal flaw." A potential disadvantage of the observational method is that the costs for designing the multiple contingency plans can, in some circumstances, be unreasonable. However, this is rarely the case for groundwater extraction systems, where the usual options, such as more wells or additional blasting, are relatively easy to design.

REMEDIATION OF FRACTURED ROCK AQUIFERS

Removing contaminants from fractured rock aquifers can be very difficult. In particular, contaminants that have diffused into the rock matrix will generally have to be removed by diffusion, a slow process. The extent of the problem will be illustrated in this section using the results of a simple numerical model. The model, which was developed by Mutch et al. (1993), is qualitatively accurate and can be used to determine semi-quantitatively the types of behavior to be expected of contaminated fractured rock aquifers. A more sophisticated numerical model is generally not warranted given the sparse and uncertain data available for most sites. The model is for a single fracture in the zone of saturation surrounded by intact rock matrix. The water in the fracture is mobile, but the pore water in the rock matrix is assumed to be stagnant. Contaminants are carried through the fractures by advection and hydrodynamic dispersion but move into and through the rock matrix only by diffusion.

Each model simulation starts with a *contamination phase,* in which contaminated water enters the rock. During the following *remediation phase,* the source of contaminant is assumed to have been removed so that only clean water enters the rock. The model calculates the contaminant concentration in the fracture and rock matrix during the course of contamination and remediation. The calculated concentration in the fracture is most representative of what would be measured in a monitoring well.

The model uses average parameters of the rock. The parameters needed include:

Groundwater flow rate (Q)
Mean fracture spacing (d_f)
Matrix diffusivity (D_m)
Matrix porosity (n_m)
Fracture porosity (n_f)
Longitudinal dispersivity (α_L)
Length of fracture (L)
Length of the contamination period (t_A)
Concentration of contaminant during contamination period (C_0)

A complete description of the model can be found in Mutch et al. (1993).

In this section the numerical model is used to investigate four aspects of the remediation of fractured rock aquifers; (1) natural remediation of a typical fractured sedimentary rock aquifer, (2) the effect of the various aquifer parameters on aquifer cleanup rate, (3) natural remediation of a typical fractured igneous or metamorphic rock aquifer, and (4) the effectiveness of groundwater extraction systems in the remediation of fractured rock aquifers. Some approaches that may help improve cleanup rates in fractured rock conclude the discussion.

Natural Remediation of a Typical Fractured Sedimentary Rock Aquifer

In this section we assume that any source of contaminants to a fractured rock aquifer, such as contaminated soil overlying the aquifer, has been removed. The clean water which then flows through the fractured rock aquifer will carry away contaminants; the numerical model was used to simulate this natural remediation of a fractured sedimentary rock aquifer. The model parameters chosen to represent a typical sedimentary rock aquifer are given in Table 3. The literature values in Table 1 were used to select values for the matrix and fracture porosities and the matrix diffusivity; reasonable values were selected for the remaining parameters. During the contamination phase an arbitrary contaminant concentration of 10,000 units was assumed in the water flowing into the rock.

Table 3 Parameter Values Used in the Model

Parameter	Representative value: sedimentary rocks	Representative value: Igneous and metamorphic rocks	Range of values investigated (sedimentary rocks only)
Contamination period, t_A (yr)	20	20	1 to 100
Length of fracture, L (m)	100	100	20 to 500[a]
Groundwater flow rate, Q (m/day)	8.64×10^{-4}	8.64×10^{-4}	8.64×10^{-5} to 8.64×10^{-3}[a]
Fracture spacing, d_f (m)	1	5	0.2 to 3
Fracture porosity, n_f	0.0025	0.0005	0.0005 to 0.01[a]
Matrix porosity, n_m	0.075	0.001	0.005 to 0.25
Matrix diffusivity, D_m (m^2/s)	3.1×10^{-11}	1.0×10^{-13}	3.1×10^{-12} to 3.1×10^{-10}
Longitudinal dispersivity, α_L (m)	5	5	0.5 to 100[a]

[a]Results not shown.

The results of the simulation for these representative parameters are shown in Figure 3. The logarithm of the concentration in the fracture is plotted against the time since remediation began (the contamination period is not shown). Initially, the concentration in the fracture drops quickly as clean water flushes contaminants from the fracture; the concentration reaches 1000 units (90% improvement) in about 12 years. After this, however, progress is much slower, requiring, for example, about 120 years to reach 100 units (99% improvement), and about 380 years to reach 10 units (99.9% improvement). In fact, after about 80 years, the graph is essentially linear, meaning that each tenfold improvement in concentration requires the same amount of time (about 260 years for this case.)

The reason for this behavior becomes clear if the concentrations in the rock matrix are examined. Figure 4 presents the contaminant levels in the fracture water and the rock matrix at various times during the course of cleanup. Initially (i.e., at the end of the 20-year period of contamination) the concentration is nearly 10,000 units in the fracture, the same as in the incoming groundwater. Concentrations are lower in the rock matrix and reach a minimum of about 100 units at the midpoint of the block.

After the source is remediated, clean water begins to flow through the rock and to flush contaminant out of the fractures. The concentration in the fracture initially drops quickly and soon becomes lower than in the adjacent rock. Contaminant begins to diffuse out of the rock and into the fracture, where it is carried away by advection. Farther into the rock, however, contaminant continues to diffuse away from the fracture toward the lower concentration in the center of the block (e.g., the concentration profile after 10 years). This two-way diffusion continues for some time until the concentration reaches a maximum at the midpoint of the block, after which diffusion is entirely toward the fracture. In the present case, this occurs after about 50 years and the concentration at the midpoint of the block reaches a maximum of about 1100 units, 10 times higher than when remediation began. Once diffusion is entirely toward the fracture the concentration profiles in the rock retain the same shape and concentrations decrease at the same rate everywhere. This occurs after about 80 years in the present case and corresponds to the straight-line portion of Figure 3.

Another reason for the long cleanup times can be seen by calculating the total mass of contaminants in the fracture and the rock matrix. These calculations indicate that at all times during remediation, the mass of contamination in the rock matrix is much greater than that in the fracture. For the representative case at the beginning of remediation

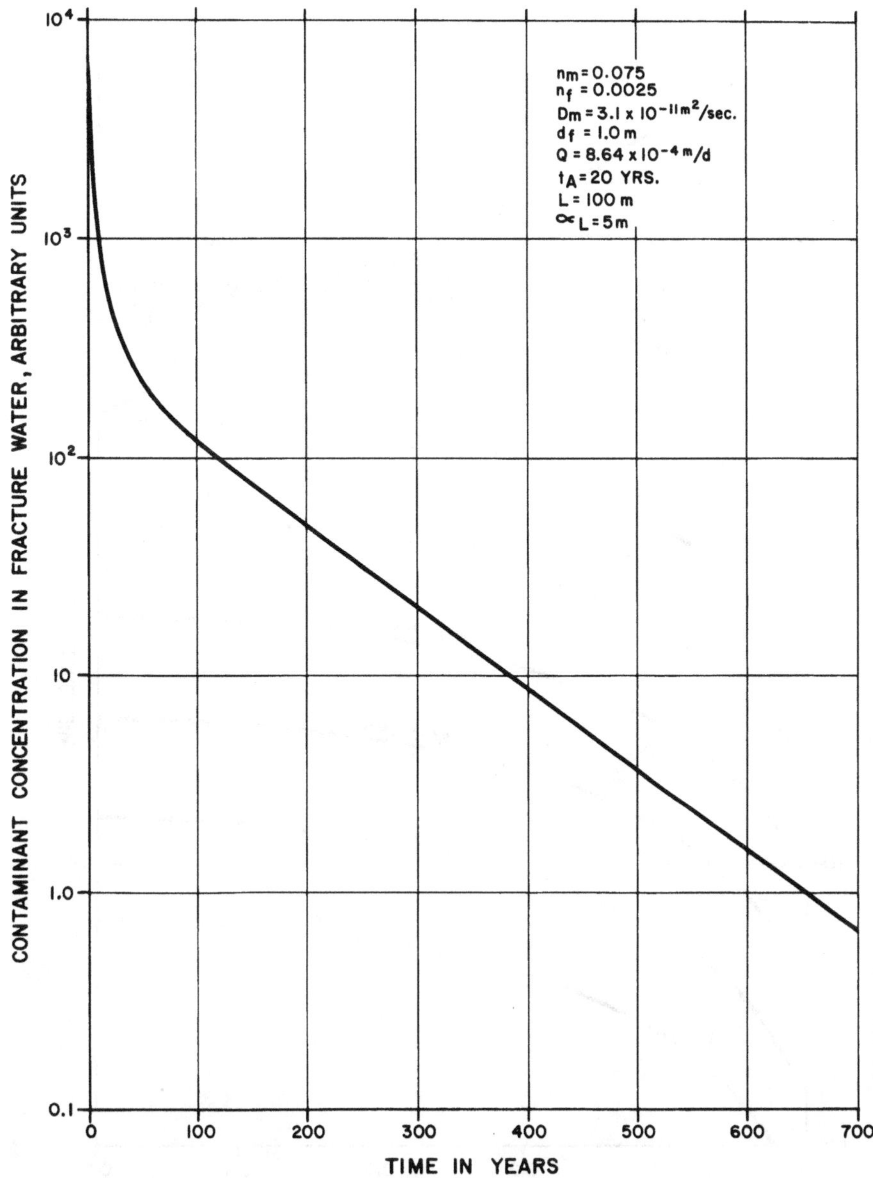

Figure 3 Contaminant concentration in the fracture water versus time after remediation began—representative fractured sedimentary rock aquifer.

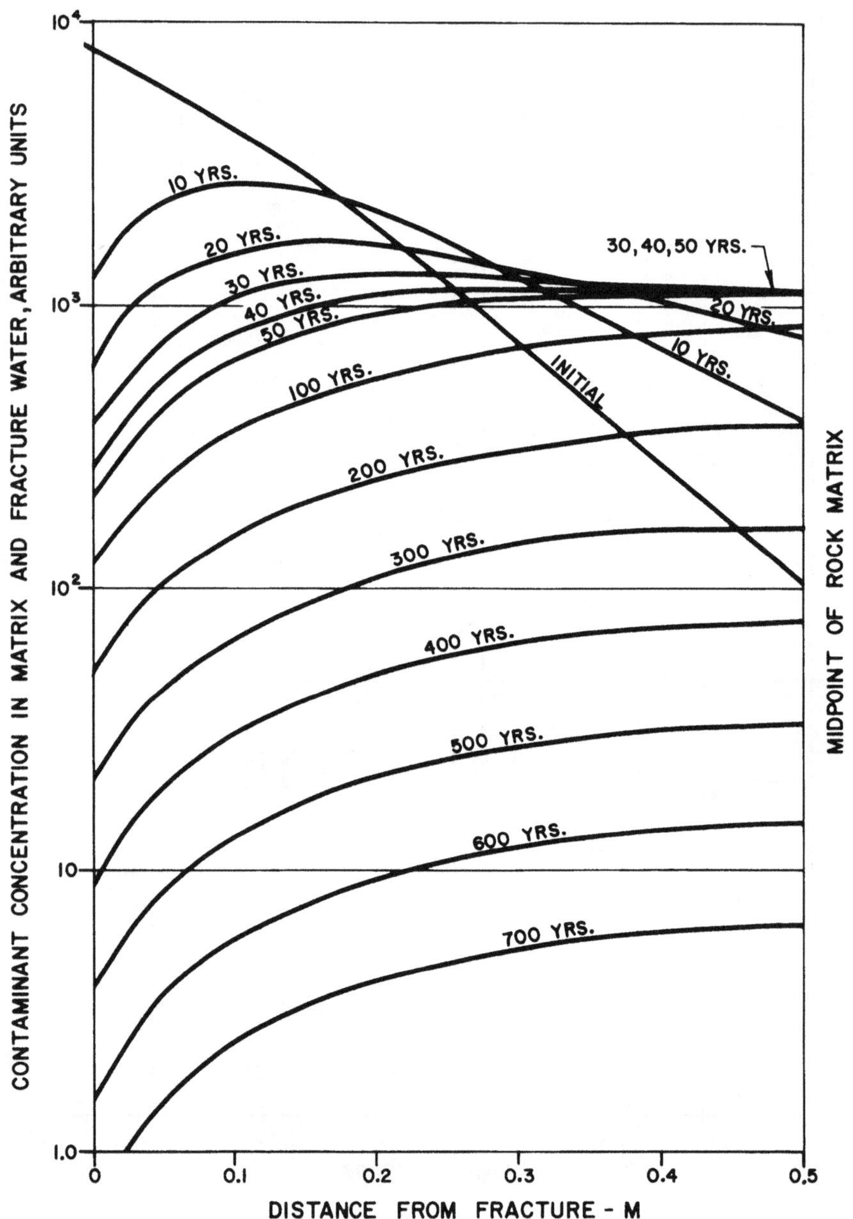

Figure 4 Contaminant concentrations in the rock matrix during remediation for a representative fractured sedimentary rock aquifer.

some 90% of the contaminant mass is held within the relatively immobile pore water, with only about 10% in the flowing fracture water (the concentration is higher in the fracture water, but the volume of mobile water is much less than the volume of immobile pore water). As the concentration in the fracture drops during remediation, the percentage of contaminant actually in the fracture becomes even lower, about 0.3%. This means that only a small portion of the contaminant is available to be removed by the flowing groundwater; the remainder is contained in the rock matrix and is removed more slowly by diffusion.

The general behavior seen in Figure 3, a fast drop in concentration in the fracture followed by a long period in which the concentration decreases more slowly, will be typical of any fractured rock aquifer for which matrix diffusion is an important contaminant transport mechanism. The actual rate of remediation, however, is strongly dependent on the properties of the rock. In the next section we explore the effects of the various model parameters on the cleanup rate.

Effects of Aquifer Parameters on Aquifer Cleanup Rate

The relative importance of the various model parameters in determining aquifer cleanup time can be seen by varying each one over a range of potential values while keeping the other parameters fixed (a sensitivity analysis). This was done for the typical fractured sedimentary rock aquifer discussed earlier. The representative parameter values and the range over which the parameters were varied are given in Table 3; the ranges of variation for the matrix porosity and diffusivity and the fracture porosity are based on the literature values in Table 1.

The effect of varying the matrix diffusivity can be seen in Figure 5. Both the initial rate of decrease and the final linear rate are strongly affected by the diffusivity. Higher diffusivities are characterized by a slower rate of initial cleanup but a much faster ultimate cleanup. For example, when the diffusivity is 3.1×10^{-10} m^2/s (10 times higher than the representative value) cleanup is 90% complete after about 45 years and 99.99% complete after about 160 years (compared to 10 years and 640 years for the representative case). In the final linear region, each additional order of magnitude takes about 34 years (compared to 260 years in the representative case). The higher matrix diffusivity means that contaminant diffuses back into the fracture at a much higher rate than in the representative case that leads not only to higher initial concentrations in the fracture (for the same flow rate) but also to faster ultimate removal from the rock matrix. At the other extreme, when the diffusivity is 3.1×10^{-12} m^2/s (10 times lower than the representative value), cleanup is 90% complete in less than 5 years and 99% complete after

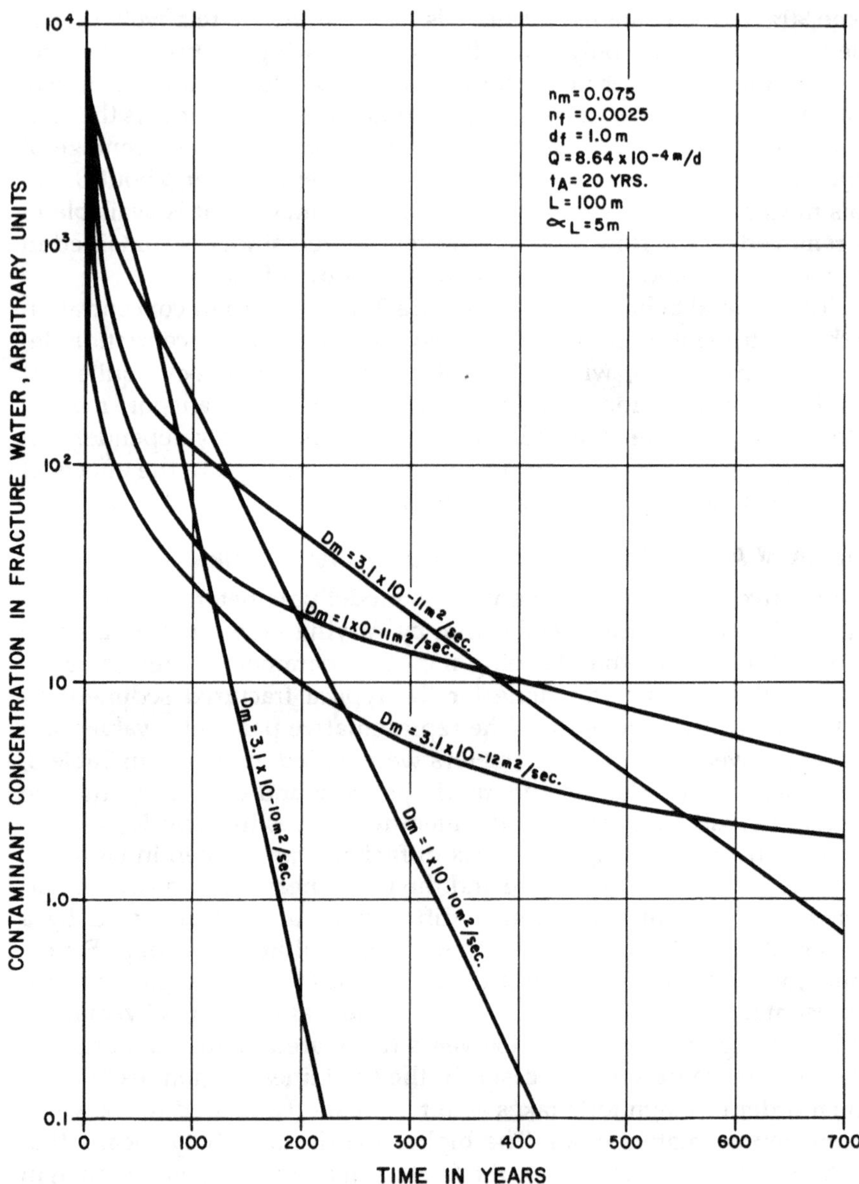

Figure 5 Effect of matrix diffusivity on the remediation of a fractured sedimentary rock aquifer.

about 35 year, but cleanup will not be 99.99% complete until after about 1300 years, and each additional order-of-magnitude improvement will require about 2400 years.

The results of varying the fracture spacing are shown in Figure 6. The overall behavior is similar to that for matrix diffusivity (Figure 5). For example, increasing the fracture spacing from 1 to 3 m is roughly equivalent to reducing the diffusivity by a factor of 10 and leads to a very slow final cleanup rate. On the other hand, a fracture spacing of 0.2 m has a relatively rapid final cleanup rate with each order-of-magnitude improvement requiring only about 25 years.

The strong dependence of cleanup rates on the matrix diffusivity and fracture spacing, as seen in Figures 5 and 6, indicates that the cleanup times predicted by the model should be taken as only a rough indication of the time frame to be expected. Matrix diffusivities and fracture spacings are rarely known accurately enough to make firm predictions. Furthermore, the model uses average parameter values and fractured rock is usually quite heterogeneous, making predictions of actual cleanup times even more difficult.

The other model parameters have less effect on cleanup times than do the matrix diffusivity and fracture spacing. For example, the effect of varying the matrix porosity is shown in Figure 7. The initial drop in concentration is dependent on the matrix porosity, but the plotted curves then become nearly parallel. Eventually, each order-of-magnitude change in concentration requires about the same amount of time for all values of the matrix porosity. The initial drop is steeper and the ultimate cleanup rate is faster for lower matrix porosities. This occurs because low matrix porosity means less immobile water for contaminant to diffuse into or out of, and therefore a higher percentage of contaminant is in the fracture water, where advection can remove it.

Despite the basic similarity in behavior for the different matrix porosities, the times required for a particular level of cleanup are quite different. For example, the time required to reduce the concentration in the fracture to 10 units (99.9% removal) is about 80 years for a matrix porosity of 0.005 but is about 600 years for a matrix porosity of 0.25. The effects of changing the groundwater flow rate and the length of the fracture are not shown here but are quite similar to the effect of varying the matrix porosity. These parameters have no direct effect on the rate of diffusion through the rock matrix, and changing them has little effect on the final diffusion-controlled phase.

The length of the contamination period also affects the cleanup time, as shown in Figure 8. Contamination periods ranging from 1 year to 100 years were simulated as well as an "infinite" period (long enough

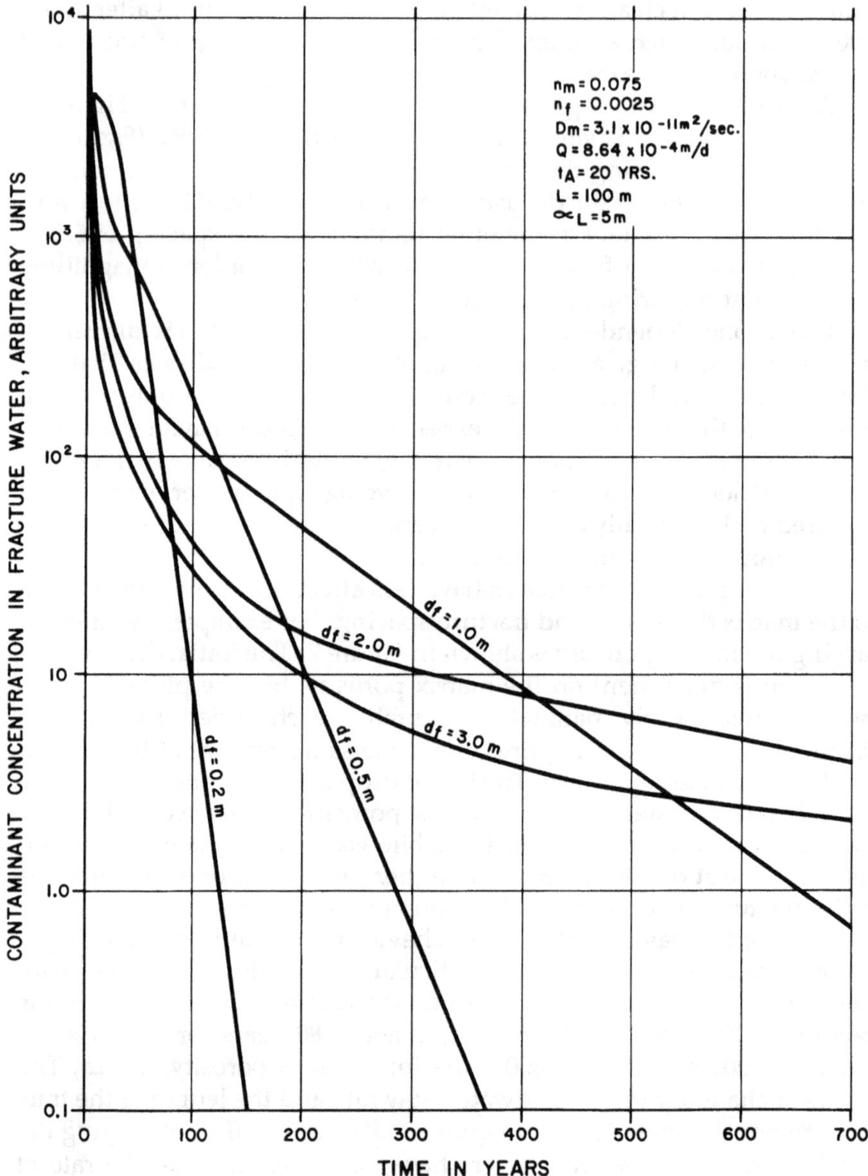

Figure 6 Effect of mean fracture spacing on the remediation of a fractured sedimentary rock aquifer.

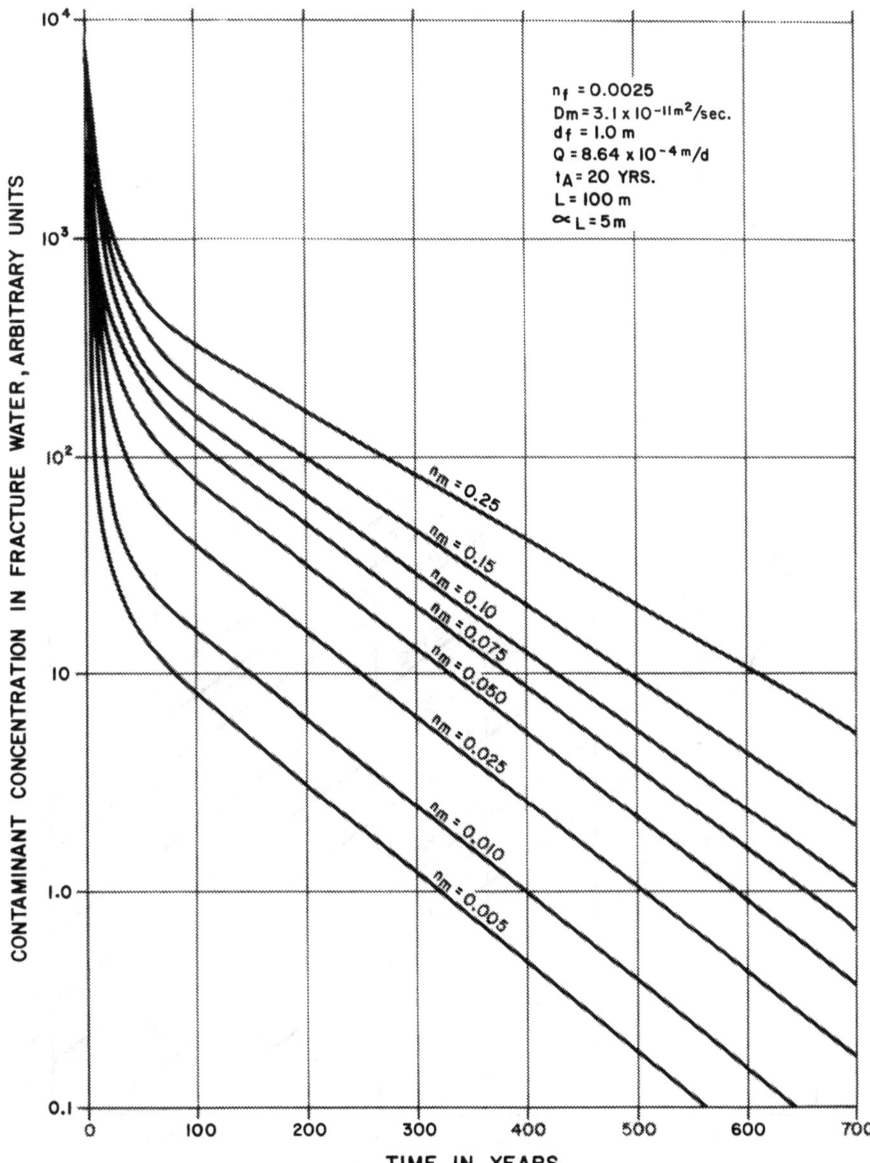

Figure 7 Effect of matrix porosity on the remediation of a fractured sedimentary rock aquifer.

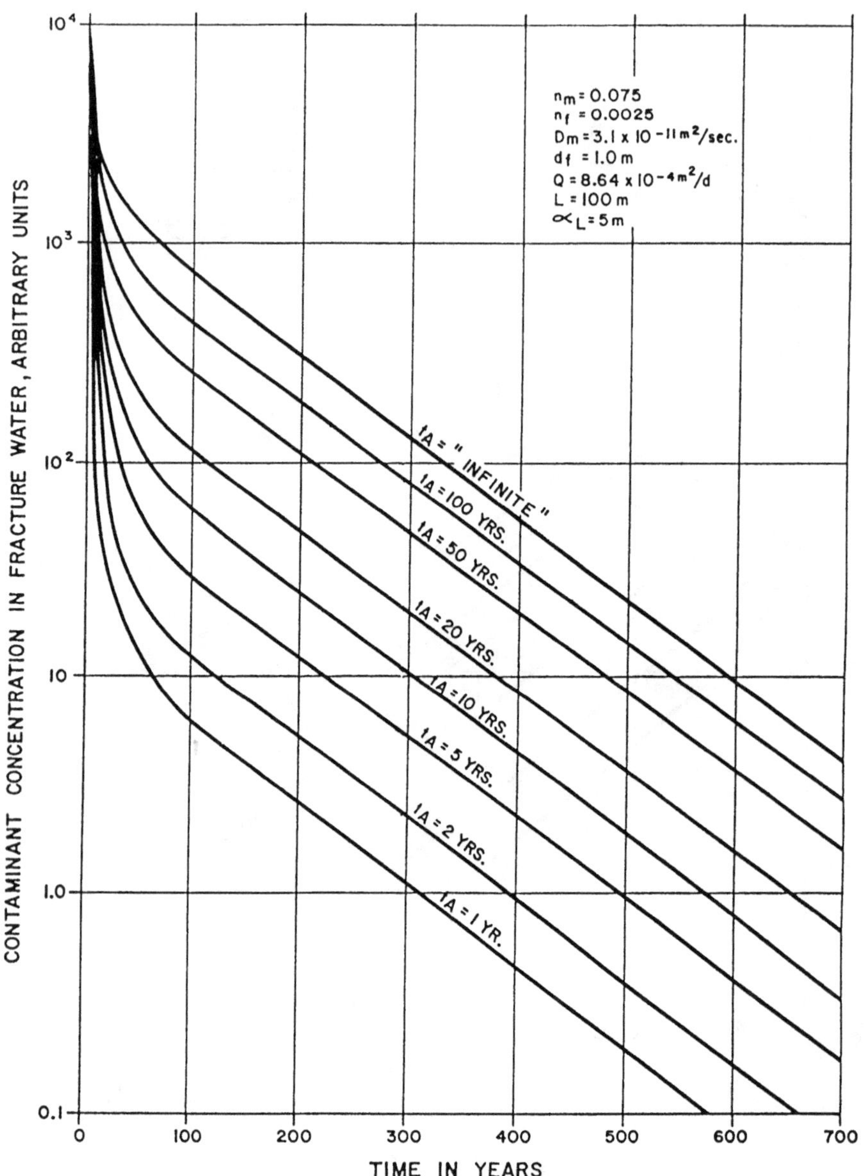

Figure 8 Effect of the length of the contamination period on the remediation of a fractured sedimentary rock aquifer.

to completely saturate the rock matrix with contaminant). Again, as with the matrix porosity and groundwater flow rate, the main effect is on the initial flushing period, with the shorter contamination periods having faster initial cleanup rates. After the first 50 to 100 years, however, all the cases remediate at the same rate and require about 260 years for each order-of-magnitude improvement. Note that even a short period of contamination can lead to a long cleanup time if a high degree of water quality improvement is required. For example, a contamination period of only one year necessitates more than 300 years to attain a 99.99% improvement in water quality.

The final two parameters that were varied, the fracture porosity and the longitudinal dispersivity, were found to have essentially no effect on concentrations in the fracture (not shown). The fracture porosity has no direct effect on either the rate of diffusion out of the rock matrix or the groundwater flow rate in the fracture. The longitudinal dispersivity has little effect on concentrations in the fracture once the contaminant has spread throughout the fracture system, which occurs rapidly at all but the slowest flow rates.

Therefore, the model parameters can be divided into three groups, based on how strongly they affect the rate of cleanup of the fracture. The first group consists of the matrix diffusivity and the mean fracture spacing. These parameters have a very strong effect on the diffusion rate out of the fracture and, therefore, on the overall rate of cleanup. Both the initial drop in concentration and the ultimate cleanup rate are strongly affected by changes in the matrix diffusivity or fracture spacing. The second group of parameters have a strong effect on the initial drop in concentration but not on the ultimate cleanup rate. This group consists of the matrix porosity, the recharge rate, the length of the fracture, and the length of the contamination period. These parameters affect primarily the rate at which contaminant is transported through the fracture and only indirectly affect the rate of diffusion out of the rock matrix. The third group includes the fracture porosity and the longitudinal dispersivity; these parameters have essentially no effect on the concentrations in the fractures or the cleanup rate.

Natural Remediation of a Typical Fractured Igneous or Metamorphic Rock Aquifer

Igneous and metamorphic rock aquifers often have low matrix diffusivities, low matrix and fracture porosities, and wide fracture spacings. Extrapolating from Figures 5 and 6, one can predict that the initial cleanup rate of such an aquifer will be very fast, but that final cleanup rates will be extremely slow. These predictions are borne out by Figure 9, which

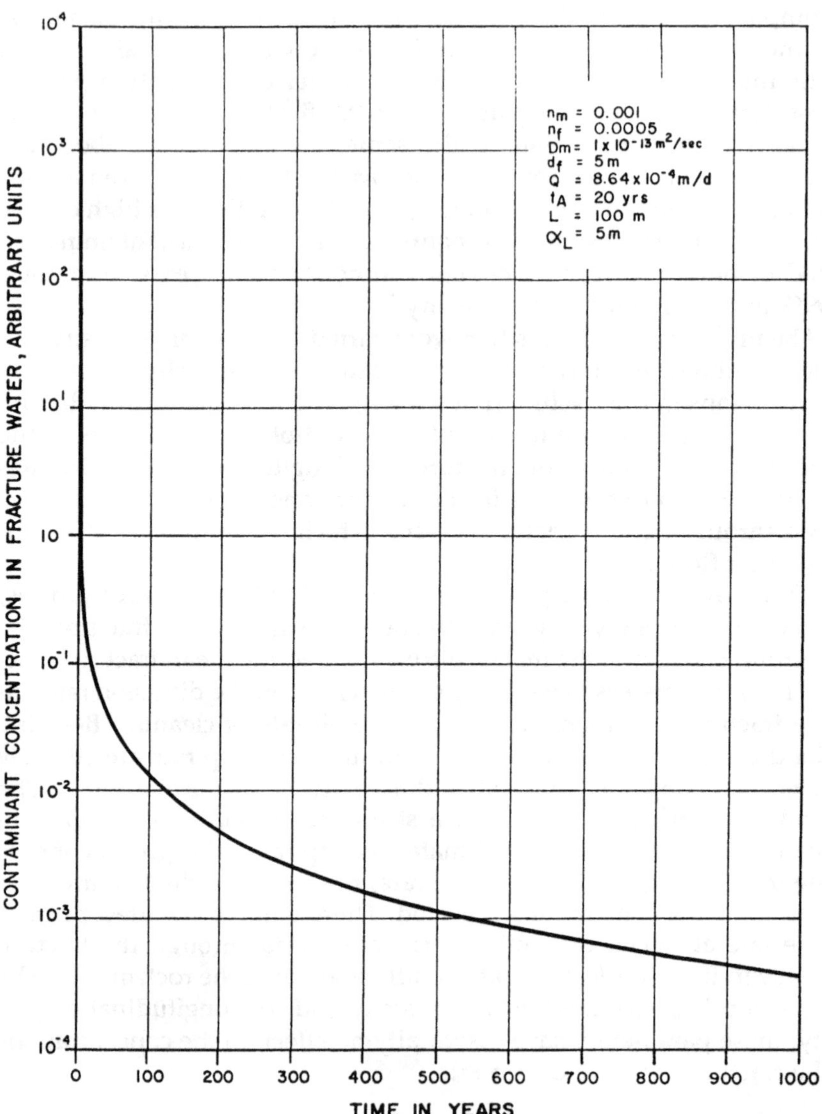

Figure 9 Contaminant concentration in the fracture water versus time after remediation began—representative fractured igneous or metamorphic rock aquifer.

gives the results of simulating natural remediation of a representative fractured igneous or metamorphic rock parameter. The parameters used are given in Table 3. The initial cleanup is very rapid; a 99.99% improvement in water quality occurs in the first 5 years, 99.999% within 20 years, and 99.9999% within 120 years. If a higher degree of aquifer quality is required, however, it may be nearly impossible to achieve because of the pronounced "tailing" effect shown in Figure 9.

Effectiveness of Groundwater Extraction Systems in Restoring Fractured Rock Aquifers

Groundwater extraction and treatment systems are generally designed with two purposes in mind. The first, and more easily achievable, goal is to prevent migration of contaminants farther downgradient. The second goal is to flush contaminants from the aquifer and ultimately to meet water quality criteria. As discussed earlier, a carefully designed pumping system should to be able to stop further migration of a plume in a fractured rock aquifer. However, in situations where cleanup rates are controlled by matrix diffusion, pump and treat systems may not be very effective in speeding up the cleanup of fractured rock.

The mathematical model was used to investigate the effectiveness of groundwater extraction and treatment systems in fractured rock aquifers. Pumping was simulated by increasing the groundwater flow rate. A typical example is shown in Figure 10, which is for the representative sedimentary rock aquifer discussed earlier. The parameters used were the same as in Table 3 except that the groundwater flow rate was increased by a factor of 5 during the pumping period. This increase in flow is generally representative of what is achievable by a typical groundwater recovery system. After the pumping period ended, the groundwater flow rate was returned to the natural flow rate. The solid curve in Figure 10 is the concentration achieved by natural flushing (no pumping); it is the same as in Figure 3. The dashed curve shows the results of pumping for a 22-year period, which results in a 99% improvement in water quality in the fracture. However, when pumping ceases, the contaminant concentration in the fracture quickly rises to levels nearly as high as would have occurred if no pumping had been done. This rebound effect negates most of the improvement in water quality achieved during the pumping period. In fact, the same degree of remediation would be achieved without pumping in 37 years. A second simulation, represented by the dashed line in Figure 10, indicates that in order to keep a 99% improvement in water quality after remediation stops, pumping must continue for nearly 100 years. The same degree of

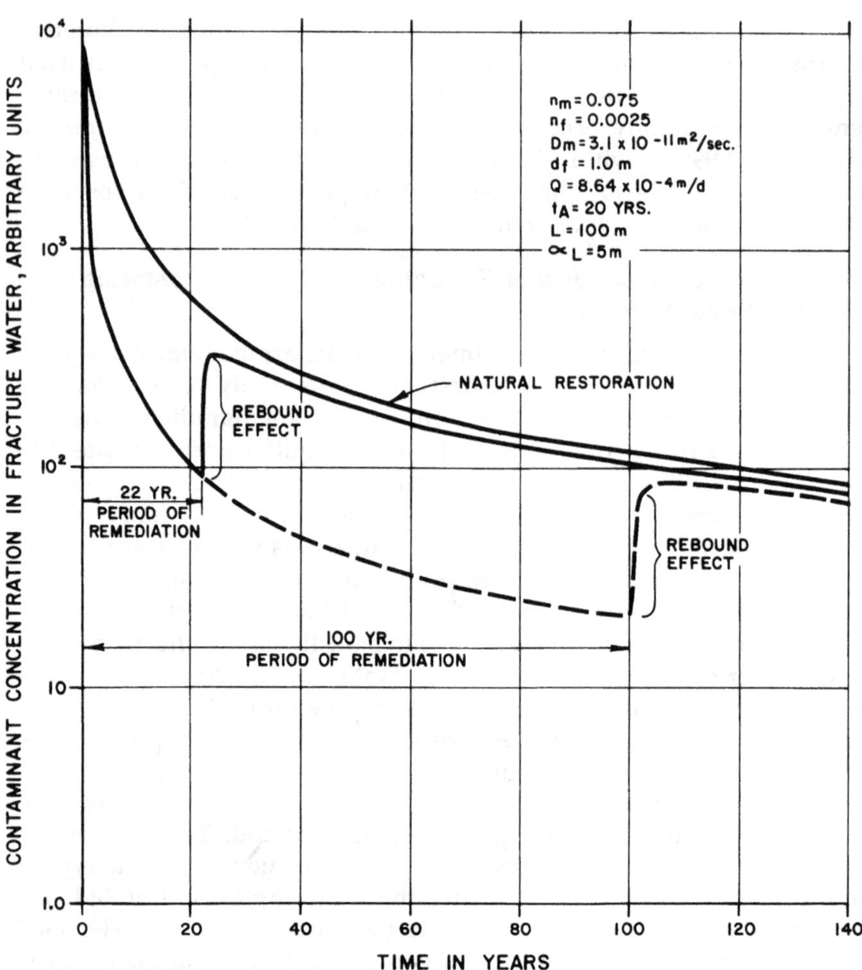

Figure 10 Effectiveness of groundwater extraction and treatment systems in accelerating fractured rock aquifer remediation.

water quality improvement in the fracture would be achieved by natural flushing in about 140 years.

The rebound effect seen in Figure 10 occurs because the increased flow rate is assumed to affect only the fractures—transport of contaminant out of the rock matrix is still by diffusion only. The higher flow rate flushes contaminants out of the fracture more quickly and reduces the concentration in the fracture. Pumping, however, has no effect on the diffusivity of the rock matrix, and only marginally increases the concentration gradient between the fracture and the rock. Since the bulk of the contaminant is in the pore water, pumping has little effect on the ultimate rate of cleanup. This indicates that groundwater extraction and treatment programs in fractured rock aquifers are likely to meet with only limited success in restoring water quality.

Potential Strategies to Increase Cleanup Rates in Diffusion-Controlled Fractured Rock Aquifers

Our earlier discussion presents a bleak picture: If remediation rates are controlled by matrix diffusion and high levels of water quality improvement are required, remediation of fractured rock aquifers may take hundreds or even thousands of years. In addition, the remediation rate is only marginally increased by groundwater pumping. Are there any ways to speed up remediation of such aquifers? In this section we present some techniques that may work in certain cases.

First, the model results are valid only if there is no advective transport through the rock matrix. In some fractured rock aquifers, especially sedimentary rocks with high porosity (approaching 25%), the hydraulic conductivity of the rock matrix may be high enough for there to be some flow through the rock matrix. Even a low flow rate through the rock matrix could increase cleanup rates dramatically.

Even if there is no groundwater flow through the rock matrix under natural conditions, it may be possible to induce such flow. A carefully designed groundwater extraction system may be able to produce substantial pressure gradients across the porous matrix separating adjacent fractures. Then, if the hydraulic conductivity of the rock matrix is not too small, there may be enough advection through the matrix blocks to speed up markedly transport of contaminant out of the rock matrix.

If advection through the rock matrix cannot be achieved, the only hope is to speed up diffusion rates out of the rock. The rock properties that most strongly affect diffusion rates are the diffusivity and the fracture spacing. Of these, it may be possible to change the mean fracture spacing. For example, pumping might be able to force flow through small fractures, which, under natural conditions, are not part of the fracture flow system. This would effectively decrease the mean fracture spacing and significantly speed up cleanup rates as shown in Figure 6. Further fracturing of the rock by such means as hydrofracturing or blasting would also decrease the mean fracture spacing. Note, however, that such increased fracturing must be achieved over the entire volume of the plume, not just near a few wells or in a downgradient trench as suffices to enhance plume capture. On the other hand, there is no need to pulverize the rock. A small decrease in mean fracture spacing, for example, from one meter to 0.2 m, would have a large effect on cleanup rates (see Figure 6).

A more speculative remediation technique is dewatering. Most organic and some inorganic contaminants would be transported out of dry rock by the process of gaseous diffusion, which is a much faster process than is aqueous diffusion in saturated rock (gaseous diffusion constants are typically about four orders of magnitude larger than aqueous diffusion constants). To be effective, dewatering would have to include the rock matrix, not just the fractures. Dewatering the rock matrix may be quite difficult since not only would the fracture flow system have to be dewatered, but also, the humidity in the fractures would have to be kept low enough for water to evaporate from the rock matrix. The process, even if feasible, would probably be quite slow but would still probably be faster than natural remediation.

Whether any of these potential remediation techniques would be feasible or cost-effective will depend on the properties of the particular fractured rock aquifer and the site.

SUMMARY

Techniques such as hydrofracturing and explosive blasting can improve the efficiency of groundwater extraction systems and help with plume containment. Use of the observational method can help handle the large uncertainties inherent in the design of groundwater extraction and treatment systems in fractured rock aquifers.

Cleanup of fractured rock aquifers, particularly of sedimentary rocks, can be exceedingly slow if the release of contaminants from the porous rock matrix is diffusion controlled. Tens, hundreds, or even thousands of years may be required. This is especially true where an improvement in water quality greater than 99% is necessary. The time required to achieve a particular level of aquifer water quality restoration is particularly sensitive to the matrix diffusivity and the mean fracture spacing of the rock. The initial rate of cleanup is also sensitive to the matrix porosity, the recharge rate, the fracture length, and the length of the contamination period. Even relatively short periods of contamination such as might result from a quickly detected spill can lead to long periods of cleanup if one is dealing with a porous medium from which contaminant removal is diffusion controlled.

Because of their typically low matrix diffusivities and large fracture spacings, igneous and metamorphic rocks are capable of a more rapid initial restoration in water quality than are most sedimentary rocks. A 99.999% or more improvement in water quality may be achievable in relatively short periods of time. Further improvements in water quality, however, may be extremely time consuming because of the slow but persistent diffusion of contaminant out of the rock matrix into the mobile fracture water.

Although effective in preventing the further spread of a contaminant plume in the aquifer, conventional groundwater extraction and treatment programs are not likely to result in significantly more rapid rates of cleanup in aquifers where matrix diffusion controls contaminant transport. Pumping does improve aquifer quality as long as it is in operation, but contaminant concentrations rebound relatively rapidly to nearly no-action levels upon cessation of pumping. It may be possible to increase the rate of cleanup of pumping if it results in substantial advection through the rock matrix, or if it causes the opening and flushing of additional fractures. In some cases, techniques that increase the degree of fracturing of the rock or dewater the rock may also be effective in increasing the rate of remediation.

REFERENCES

Anderson, M. P. 1979. Using models to simulate the movement of contaminants through groundwater flow systems. *CRC Crit. Rev. Environ. Control* 9:97–156.

Barone, F. S., E. K. Yanful, R. M. Quigley, and R. K. Rowe. 1988. Effect of multiple contaminant migration on diffusion and adsorption of some domestic waste contaminants in a natural clayey soil. *Can. Geotech. J.* 26:189–198.

Barone, F. S., R. K. Rowe, and R. M. Quigly. 1989. Laboratory determination of chloride diffusion coefficient in an intact shale. *Proceedings of the 42nd Canadian Geotechnical Conference*, pp. 241–248.

Barrell, J. 1914. The strength of the earth's crust. *J. Geol.* 22.

Bradbury, M. H. and A. Green. 1985. Measurement of important parameters determining aqueous phase diffusion rates through crystalline rock matrices. *J. Hydrol.* 82:39–55.

Freeze, R. A., and J. A. Cherry. 1979. *Groundwater*. Prentice Hall, Englewood Cliffs, NJ.

Javendel, I., and C. F. Tsang. 1986. Capture zone type curves: a tool for aquifer cleanup. *Ground Water*, 24.

Kelley, V. A., J. F. Pickens, M. Reeves, and R. L. Beauheim. 1987. Double-porosity tracer-test analysis for interpretation of the fracture characteristics of a dolomite formation. *Proceedings of the Solving Ground Water Problems with Models Conference*, NWWA, pp. 147–169.

Lever, D. A., and M. H. Bradbury. 1985. Rock-matrix diffusion and its implications for radionuclide migration. *Miner. Mag.* 49:245–254.

Mackay, D. M., and J. A. Cherry. 1989. Groundwater contamination: pump-and-treat remediation. *Environ. Sci. Technol.* 23:630–636.

Manger, G. E. 1963. *Porosity and Bulk Density of Sedimentary Rocks*. U.S. Geological Survey Bulletin 1144-E.

Mutch, R. D., Jr., J. I. Scott, and D. J. Wilson. 1993. Cleanup of fractured rock aquifers: implications of matrix diffusion. *Environ. Monit. Assess.*

Neretnieks, I. 1985. Transport in fractured rocks. *International Association of Hydrogeologists Memoirs*, Vol. XVII, Part 1, *Proceedings, Hydrogeology of Rocks of Low Permeability*, pp. 301–318.

Pankow, J. F., et al. 1986. An evaluation of contaminant migration patterns at two waste disposal sites on fractured porous media in terms of the equivalent porous medium (EPM) model. *J. Contam. Hydrol.* 1:65–76.

Peck, R. B. 1969. Ninth Rankine lecture, advantages and limitations of the observational method in applied mechanics. *Geotechnique* 19:171–187.

Rasmuson, A., and I. Neretnieks. 1981. Migration of radionuclides in fissured rock: the influence of micropore diffusion and longitudinal dispersion. *J. Geophys. Res.* 86(B5):3749–3758.

Smith, E. D., and N. D. Vaughan. 1985. Experiences with aquifer testing and analysis in fractured low-permeability sedimentary rocks exhibiting nonradial pumping response. *Hydrology of Rocks of Low Permeability*. International Association of Hydrogeologists, pp. 137–139.

Smyth, J. D., and R. D. Quinn. 1991. The observational approach in environmental restoration. *Proceedings of the 1991 Specialty Conference, Environmental Engineering Division of ASCE*, Reno, NV.

Streltsova, T. D. 1976a. Hydrodynamics of groundwater flow in a fractured formation. *Water Resour. Res.* 12:405–414.

Streltsova, T. D. 1976b. Advances and uncertainties in the study of groundwater flow in fissured rocks. *Adv. Groundwater Hydrol.*, pp. 48–56.

Walter, G. R., and G. M. Thompson. 1982. A repeated pulse technique for determining the hydraulic properties of tight formations. *Ground Water* 20:186–193.

Walton, W. 1970. *Groundwater Resource Evaluation*. McGraw-Hill, New York.

Yurochko, A. I. 1982. In: North, F. K. 1985. *Petroleum Geology*. Allen & Unwin, Boston, p. 607.

3

Chemical Stabilization of Contaminated Soils

Jesse R. Conner
RUST Remedial Services Inc.
Clemson Technical Center, Inc.
Anderson, South Carolina

SHORT HISTORY OF STABILIZATION TECHNOLOGY

The field of stabilization and solidification (S/S) has lately begun to mature into an accepted environmental technology. Because it is driven by regulation that essentially mandates its use for many waste streams, it is becoming a standard unit process in hazardous waste treatment and disposal. Until recently, few reference sources existed [1], but this situation has been rectified in the last few years by publications of the U.S. Environmental Protection Agency (USEPA) [2], numerous papers, and a definitive book in 1990 by the author [3].

Terminology

Most of the terminology in the S/S field includes terms borrowed from other environmental areas and from other technologies, but these familiar words and phrases are often given new and specific meanings. There is no "official" set of definitions yet, although the USEPA has unofficially promulgated some of the definitions given here. In S/S there has been a tendency to use the terms *chemical fixation, immobilization, stabilization,* and *solidification* interchangeably. Recently, USEPA has defined stabilization as follows:

1. *Stabilization* refers to those techniques that reduce the hazard potential of a waste by converting the contaminants into their least soluble, mobile, or toxic form. The physical nature and handling characteristics of the waste are not necessarily changed by stabilization.

2. *Solidification* refers to techniques that encapsulate the waste in a monolithic solid of high structural integrity. The encapsulation may be of fine waste particles (microencapsulation) or of a large block or container of wastes (macroencapsulation). Solidification does not necessarily involve a chemical interaction between the wastes and the solidifying reagents, but may mechanically bind the waste into the monolith. Contaminant migration is restricted by vastly decreasing the surface area exposed to leaching, and/or by isolating the wastes within an impervious capsule.

Common terms, especially acronyms, with which the novice should quickly become familiar are listed below and discussed further as we proceed.

Regulatory Processes and Treatment Standards

Best demonstrated available technology (BDAT)
Comprehensive Environmental Response, Compensation and Liability Act (CERCLA), or Superfund
Delisting
Land disposal restrictions (LDR), or landbans
Publically owned treatment works (POTW)
Resource Conservation and Recovery Act (RCRA)
VHS model

Facility or Treatment Types

CERCLA/RCRA
Ex situ
In situ
Minimum technology unit/requirement (MTU/MTR)
Monofill
Off site
On site
Remedial action
Sanitary landfill
Secure landfill
Superfund
Treatment, storage, and disposal facility (TSDF)

RCRA/CERCLA Waste Categories

California list
Characteristic

First third
F-waste
Landban
Listed
Second third
Third third

Leaching and Other Tests

Extraction procedure toxicity test (EPT)
Toxicity characteristic leaching procedure (TCLP)
Multiple extraction procedure (MEP)

Origins

S/S has its roots in other processes and equipment developed for different purposes. Most of the history of the development of S/S systems for general use on waste residues dates only from about 1970. However, the beginnings of most present-day commercial S/S systems go back to four primary areas of technology that were practiced long before 1970: (1) radioactive waste solidification and disposal, (2) mine backfilling, (3) soil stabilization and grouting, and (4) production of stabilized base courses for road construction. Of these, only radioactive waste treatment is a S/S process in the present sense. The other three applications had other utilitarian purposes, although they frequently used wastes such as fly ash in the process. Portland cement or fly ash or both were used in mine backfilling, sodium silicate plus setting agents, cement or organic polymerizing systems for grouting and soil stabilization, and lime/fly ash for road base construction. There are many isolated instances where waste residue generators, especially waste disposal site operators, used cement, fly ash, lime, soil, and various combinations of these materials to solidify liquids for disposal in landfills where some stability was required in the fill material. Nearly all of this early work involved a need for solidification only, and rarely, if ever, were leaching or other performance tests conducted or required.

Early Practices

In the 1950s, the nuclear industry recognized the need for solidification of radioactive waste in drums and other containers before such waste could be shipped or buried at government-controlled disposal sites in the United States. At first, much of the liquid waste containing low-level radioactivity was simply absorbed into various mineral sorbents such as vermiculite, or solidified by making a concrete mixture with very large quantities of portland cement. In Europe, radioactive wastes were typically solidified in concrete and buried at sea in drums. The large amount of cement used in solidification of these wastes, especially when the water content is high, results in low volume efficiency—the percentage of the final container occupied by the waste, not the cement. This makes the container more difficult to handle because of its high bulk density, and greatly increases the disposal cost, already very high. Since the major cost in disposal of containerized radioactive wastes is in the cost of the container, its transportation, and disposal, the obvious way to cut costs was by getting more waste and less S/S agent into the container. Also, the solidification process with cement at that time was not well controlled and was somewhat unpredictable, particularly when constituents that retard the setting of portland cement were present in the waste streams, as they often were. Because of these problems, urea-formaldehyde and asphalt systems came into some use to provide more consistency, lower weight, and better space efficiency. By the late 1950s, it was realized that the addition of sodium silicate to the portland cement process often provided better results overall than did other processes [4]. Later, organic polymer processes were also used for radioactive waste solidification [5]. The nuclear industry also experimented with, and used, deep underground disposal of intermediate level wastes using cement/fly ash/clay compositions, pumping the fluid mixtures into fractured shale zones, where they solidified and became immobilized [6].

Other than internal studies at various nuclear installations and government nuclear regulatory agencies [7], there was no published work done on leachability, environmental degradation, or any other performance characteristics of solidified waste until the early 1970s. At that time, several companies began applying scientific principles to solidification of various waste materials. One used a lime/fly ash process that incorporated sulfates [8] from the waste to contribute to strength development in the final solid. Another used a (later) patented process [9] based on the combination of soluble silicates and silicate setting agents, usually sodium silicate solution and portland cement. The former activities were aimed primarily at the sludges expected to be produced by lime/limestone scrubbing of flue gases to remove sulfur oxides. The latter, on the other hand, aimed at a wider variety of sludges, especially those emanating from manufacturing, metal finishing, and metal producing operations, and predicated its business on a mobile service whereby treatment units could be brought to the remedial location and the waste treated on site. Another process in England began commercial operations using its cement/fly ash process [10] for solidifying industrial inorganic waste streams at central treatment sites. From a business standpoint, this period was not the most propitious for anyone entering the waste residuals treatment field because there were few laws or regulations in the United States or elsewhere concerning the disposal of waste residues, and therefore little driving force for this kind of treatment.

Beginning about 1975, in anticipation of the coming of RCRA, S/S systems began to receive attention from governmental agencies, waste generators, and engineering firms. After the passage of RCRA, solid waste disposal companies began to take an interest in hazardous waste treatment technology, and a number of vendors appeared on the scene offering processes, chemicals, and services.

Regulatory Background

Without regulation, S/S technology would scarcely exist, much less flourish, in a world of harsh economic realities; and the characteristics of the technology itself are defined by these regulations. The impetus for S/S of hazardous wastes was provided by the Resource Conservation and Recovery Act (RCRA) of 1976 [11], including the subsequent 1984 amendments (HSWA), and the Comprehensive Environmental Response, Compensation and Liability Act (CERCLA) [12], otherwise known as Superfund. RCRA includes provisions for developing criteria to determine which wastes are hazardous, establishing standards for siting, design, and operation of disposal facilities, encouraging the states to develop their own regulatory programs, and other specifics. The HSWA reauthorized RCRA and made major changes including the establishment of the land disposal restrictions (LDRs), also known as landbans. Regulations promulgated under both RCRA and HSWA direct in detail the generation, handling, treatment, and disposal of wastes. CERCLA, and its reauthorization in 1986—known as SARA—established a massive remedial program for the cleanup of existing sites that threaten the environment. In addition to the landbans, SARA "requires that remedial actions meet all applicable, relevant, and appropriate public health and environmental standards. Therefore, the Superfund program must be consistent with the BDAT approach when disposing of contaminated soils and debris from Superfund sites" [13]. Undoubtedly, this will also apply eventually to non-Superfund remedial actions.

In the promulgation of the various landbans, specific technologies are specified as "best demonstrated available technology" (BDAT). S/S treatment is one of the most important BDATs now and will continue to be in the future. While land disposal of hazardous wastes is not the method of choice from an environmental standpoint, it is developing an increasingly important place in the overall waste management scheme of the future for several reasons. There are many hazardous wastes that are simply not amenable to techniques such as thermal, chemical, or biological destruction. When these wastes cannot be feasibly reused in any beneficial way, there is no other recourse but land disposal. Even when other techniques are used, they usually generate residues which are themselves hazardous. In the cleanup of abandoned sites under the Superfund program and remediation of other old disposal practices by private entities, on-site or in situ treatment and disposal often remain the safest and least expensive alternatives, and one of the primary techniques here is S/S. Even before the landbans, S/S had already been accepted as a basis for *delisting* in a number of petitions under RCRA. S/S, termed *solidification* or *encapsulation*, is specifically cited under CERCLA (40 CFR 300) as a method to be considered for remedying releases from contaminated soils and sediments.

Other than in site-specific remedial action work under CERCLA/SARA or in private cleanup operations, the two major regulatory systems that affect S/S technology are the landban system and the delisting system. On May 31, 1985, the USEPA published a proposed schedule [14] for the statutory landbans, including the identification of three groups of listed and unlisted hazardous wastes that were to go into effect at different times. These three groups, collectively called the *landban*, subsequently became known as the *first third*, *second third*, and *third third*, and they went into effect on August 8, 1988, June 8, 1989, and May 8, 1990, respectively. Several of these profoundly affect S/S technology. In addition to the landbans, SARA "requires that remedial actions meet all applicable, relevant, and appropriate public health and environmental standards. Therefore, the Superfund program must be consistent with the BDAT approach when disposing of contaminated soils and debris from Superfund sites." Undoubtedly, this will also eventually apply to non-Superfund remedial actions. As other wastes not presently listed as hazardous, or specifically excluded under RCRA, become regulated, they will also become subject to landban provisions and regulations.

Delisting is an amendment to the lists of hazardous wastes, granted by the USEPA when it is shown that a specific waste stream no longer has the hazardous characteristics for which it was originally listed. This may be accomplished by S/S treatment. The treatment methods and leaching levels to be achieved are not established by statute or by rule but are determined by the USEPA on a case-by-case basis. The de facto allowable leaching levels are a multiple of the drinking water standards and are established by the use of groundwater attenuation models. The delisting approach takes into account the toxicity, persistence, and mobility of constituents, as well as waste quantity and disposal location.

Delisting is expensive, complex, and time consuming; it is important for S/S technology because many of the delistings granted to date have been based on the use of this technology.

The USEPA land disposal restrictions (LDRs) of 1985–1990 under the 1984 reauthorization of the Resource Conservation and Recovery Act [15] (RCRA), otherwise known as the landbans, coupled with finalization of the toxicity characteristic [16] (TC) rule, have had the greatest impact on hazardous waste stabilization technology in the past 20 years or so. Very stringent metal leaching requirements for certain USEPA-listed wastes have necessitated entirely new approaches to formulation, and the advent of leaching limits for some organics has opened a new area of research in their immobilization.

The present TC regulatory levels are shown in Table 1. These are the cutoff levels that define a waste as being hazardous under RCRA. Using the toxicity characteristic leaching procedure (TCLP), a concentration of any listed constituent in the leachate at or above these levels designates the waste as hazardous. It remains hazardous until treated, by stabilization or other means, to reduce its leachability below the TC levels. There are actually three subsets within the TC: metals, pesticides, and other organics. The levels for metals and pesticides have existed for some years; the other 26 organic levels are new. The metal levels apply not only to the definition of a hazardous waste—the original purpose—but to the LDR maximum leaching levels for disposal of "characteristic"[†] waste at a RCRA treatment, storage, and disposal facility (TSDF), otherwise known as a secure landfill. In addition, the TC metal and pesticide levels have been applied to remedial work under the Comprehensive Environmental Response, Compensation and Liability Act [17] (CERCLA) and its later reauthorization, otherwise known as Superfund, as well as to state-lead and private remediations.

For now, the other 26 organic levels apply generally only to the definition of a hazardous waste and to some remedial work on a project-by-project basis. Eventually, it is expected that they, or some variation thereof, will be applied at TSDFs and in most remedial work. This will have another substantial impact on stabilization technology, and the USEPA will doubtless continue to add new organic compounds, and perhaps some metals, to the list. Potentially, any of the 231 compounds on the USEPA's hazardous constituent list could be incorporated into

---

[†]Characteristic wastes are those hazardous wastes that are not specifically listed by USEPA and assigned a hazardous waste number but which are found to be hazardous by one of four characteristics: corrosivity, reactivity, ignitability, or toxicity.

Table 1 Toxicity Characteristic Regulatory Levels

USEPA Hazardous waste no.	Constituent	Regulatory level (mg/L)
D004	Arsenic	5.0
D005	Barium	100.0
D018	Benzene	0.5
D006	Cadmium	1.0
D019	Carbon tetrachloride	0.5
D020	Chlordane	0.03
D021	Chlorobenzene	100.0
D022	Chloroform	6.0
D007	Chromium	5.0
D026	Cresol (total or individual)	200.0
D016	2,4-D	10.0
D027	1,4-Dichlorobenzene	7.5
D028	1,2-Dichloroethane	0.5
D029	1,1-Dichloroethylene	0.7
D030	2,4-Dinitrotoluene	0.13
D012	Endrin	0.02
D031	Heptachlor and its hydroxide	0.008
D032	Hexachlorobenzene	0.13
D033	Hexachloro-1,3-butadiene	0.5
D034	Hexachloroethane	3.0
D008	Lead	5.0
D013	Lindane	0.4
D009	Mercury	0.2
D014	Methoxychlor	10.0
D035	Methyl ethyl ketone	200.0
D036	Nitrobenzene	2.0
D037	Pentachlorophenol	100.0
D038	Pyridine	5.0
D010	Selenium	1.0
D011	Silver	5.0
D039	Tetrachloroethylene	0.7
D015	Toxaphene	0.5
D017	2,4,5-TP (Silvex)	1.0
D040	Trichloroethylene	0.5
D041	2,4,5-Trichlorophenol	400.0
D042	2,4,6-Trichlorophenol	2.0
D043	Vinyl chloride	0.2

Table 2 Some BDAT Regulatory Levels for Metals

	Cutoff level of metal in TCLP leachate (mg/L)			
Waste code	Cadmium	Chromium	Lead	Nickel
F006-12	0.66	5.20	0.510	0.320
F024		0.073	0.021	0.088
K002-6		0.094	0.370	
K048-52		1.70		0.20
K061	0.140	5.20	0.240	0.320
P073-4				0.320
U144-6			0.510	
Lowest BDAT level	0.066	0.073	0.021	0.088
TC level	1.0	5.0	5.0	

the TC. The only other organic regulatory leaching levels in effect concern a specific group of *F-wastes*, F001 to F005, established under the solvent ban in 1986. It is likely that EPA will make some changes in this category, setting some total constituent analysis (TCA) requirements and perhaps eliminating the TCLP requirements. For the moment, however, F001 to F005 constitute the only organic leaching requirements under the LDR.

For wastes specifically listed by the USEPA as hazardous and assigned F, K, P, or Q numbers, the regulatory situation vis-à-vis stabilization is much more complex. Allowable leaching levels for metals (and some organics in specific wastes) using the TCLP were determined by the USEPA using the best demonstrated available technology (BDAT) method for each waste number or code. The BDAT method did not require the leaching level to be consistent for each metal from waste code to waste code, and it is not. Table 2 shows the various levels for several metals in different listed waste codes, the lowest BDAT level for any listed waste, and the TC levels for these metals.

Examination of Table 2 quickly reveals that the overall regulatory system is not scientifically consistent for a given metal. This is especially true in view of the fact that wastes with different regulatory leaching requirements are frequently disposed of side by side in the same landfill. Nevertheless, the requirements are legal standards that must be met. These levels apply for certain residues that contain combinations of wastes with different USEPA-listed waste codes. Although the individual leachability of a given metal in a given waste was determined by the USEPA to be achievable, this is not necessarily true for the mixtures of wastes that are often unavoidable in the real world. An example of this situation is hazardous waste incinerator ash, which by the USEPA's rules carries with it the codes of all wastes burned in creating that ash.† This poses some rather difficult problems in chemistry [18], especially in meeting the lowest landban levels illustrated in Table 2. Some of these levels are actually below the drinking water standards for the metal, as is the case for lead.

The USEPA is now considering a special set of rules for the treatment and disposal of contaminated soil and debris (CS&D). Advanced notices of proposed rulemaking (ANPRMs) were published for debris [19] and soil [20] in 1991. It is likely that these rules will be the primary ones to affect the treatment of contaminated soils in the future, probably starting in late 1993.

CHARACTERISTICS OF CONTAMINATED SOIL AND DEBRIS

A good knowledge of the nature of waste streams is vital to an understanding of S/S technology. The treatment of contaminated soils is a special case. Such wastes are becoming increasingly more important in the S/S field as remedial programs become more active. While the discussion in this book is aimed primarily at contaminated soils, the technology applies to a much wider variety of waste streams, and in fact, much of it comes from work on industrial process and wastewater treatment residues, which are mostly sludges, dusts, filter cakes, and liquids.

Soils usually do not require solidification, although some may be processed to suppress dust and provide greater stability in a landfill. Also, they require the *addition* of water in treatment, unlike other waste streams, where water removal is practiced to reduce volume. Many waste types are not amenable to S/S treatment or do not require it. In general, solid, nonhazardous wastes require no treatment. Liquid nonhazardous wastes may require solidification only to prepare them for land disposal. Nonaqueous hazardous wastes, such as spent solvents, are best treated by other means, such as recovery or incineration. Aqueous wastes that are designated as hazardous by the USEPA, especially those which are liquid, are among those most commonly considered for S/S treatment. The recent landbans [21] have brought many new streams into the S/S area, including solid hazardous wastes that had been going untreated to permitted RCRA treatment, storage, and disposal (TSD) facilities for direct landfill in secure cells.

---

†This is known as the "derived-from" rule.

Unique Characteristics of Soils

Natural soils contain clay, rock, silt, sand, and many natural organic substances. When they become contaminated, it is usually by infiltration of metal species (and other hazardous components) in solution or by mixture with sludges, filter cakes, and process residues in old landfills. Such contaminants may originate in accidental spills, from deliberate dumping, or from leaching of older landfills. The interaction of metals with soils is very complex. Adsorption and ion exchange by clay minerals, reaction with insolubilizing anions present in the soil, and complexation by humic substances in the organic fraction of the soil all occur. Complexation may lead to either increased or decreased solubility of the metal, depending on the ligands present. So-called "contaminated soils" are frequently the result of improper disposal of these residues and chemically often resemble the residue more than a natural soil. Therefore, here we will discuss a wide variety of nonsoil wastes where they are pertinent.

Soluble metal salts are often distributed throughout large, hard pieces of clay and porous rock. This may have occurred over a period of many years, and outward diffusion of the species so that it can react with agents can be very slow, yet rapid enough to leach at unacceptable rates. This is especially true where the species must be oxidized, reduced, or otherwise treated in a multistep process such as Cr^{6+} reduction. Here, excess reducing agent is destroyed in the final process steps and is not available to reduce chromium diffusing from the particle interior at a later time. In some situations the only solution may be to grind the soil or debris to a very fine powder so that the reactions may be carried to completion within a reasonable time. In others, long-lived agents may be incorporated in the S/S matrix, remaining there to react with the metal as it diffuses out of the soil particles. There are even situations in which the agent may be generated within the matrix as it is used.

Special Problems Associated with Debris

Some of the major classes of debris that is commonly found in contaminated soils are:

Metal slags
Glassified slags
Glass
Concrete (excluding stabilized hazardous waste)
Masonry and refractory bricks
Metals cans, containers, drums, and tanks

Metal nuts, bolts, pipes, pumps, valves, appliances, or industrial equipment
Scrap metal [as defined in 40 CFR 261.1(c)(6)]
Paper, cloth, and so on

Ferrous metal can be removed during the treatment operation by magnetic means. Other large pieces of debris can be screened out from the soil matrix and handled separately. Small pieces—less than 2 in. in maximum dimension—can usually be treated along with the soil itself if the debris is nonporous or does not pose a leaching problem in the long term. The problems with debris are different depending on whether the S/S treatment is to be done in situ or ex situ. When doing in situ treatment, large pieces of debris can interfere with proper mixing or even damage equipment. With ex situ treatment, large debris must be removed to allow the soil to travel through the process equipment. There are two classes of debris that result in different handling and treatment problems, whether in situ or ex situ. Nonporous debris, if small, can be included in the treatment mixture since the contamination clinging to the surface will usually be accessed by the treatment chemicals. If large, the debris can be removed and processed separately by encapsulation in treated soil poured or placed around it, by grinding to smaller particle size and including in the soil treatment, or by surface washing followed by disposal as nonhazardous waste or recovery in the case of metals.

In the case of porous debris, the problem is similar to that discussed above with naturally occurring "debris" such as porous rock. Wood, some slags and refractories, paper, cloth, even concrete, can absorb soluble contaminants that could later leach from the debris. In most cases this may not pose an environmental problem if sufficient excess reagent is present in the mixture to react with the species as they diffuse out. However, as discussed previously, species such as Cr^{6+} will not be immobilized because the immobilizing agent is no longer active in the treated soil. Therefore, this sort of debris must be separated from the soil. It may then be ground to sufficiently fine particle size to allow proper treatment, or handled by some other means, such as a central waste treatment facility.

Pretreatment/Preprocessing of Soil and Debris

Most pretreatment operations involve either particle size reduction to allow proper treatment, or chemical treatment that must be done separately from the S/S operation per se. Particle size reduction is done with standard industrial equipment used for this purpose (see the section "Delivery Systems"). Chemical treatments include chromium reduction, cyanide destruction, and oxidation of certain organics and metal complexes. These will be discussed later in this chapter.

PRINCIPLES OF STABILIZATION

General Categories of Stabilization Processes

S/S technology can be broken down into two basic classes of systems: (1) solidification only and (2) stabilization or immobilization. Solidification processes utilize chemically reactive formulations which, together with the water and other components in sludges and other aqueous wastes, form stable solids. In this sense, *stable* means that the solids are physically stable under normal or expected environmental conditions and will not revert to the original liquid, semiliquid, or unstable solid state. The compositions resulting from these processes may or may not be hazardous as defined by waste characteristic tests such as leaching. Chemically solidified wastes normally have load-bearing strengths (compacted) in excess of 1 ton per square foot and may have other specified physical characteristics.

Stabilization systems not only solidify the waste by chemical means, but also insolublize, immobilize, encapsulate, destroy, sorb, or otherwise interact with selected waste components. The purpose of these systems is to produce solids that are nonhazardous, or less hazardous than the original waste. The degree of hazard for these kinds of materials and systems is usually defined by leaching tests.

Most present commercial S/S processes are quite simple conceptually [22] and utilize standard mechanical equipment in their operation. As a matter of fact, an inorganic S/S mechanical system is truly an assembly of mixers, chemical storage and feeding devices, pumps, conveyors, and ancillary equipment. This is shown schematically in Figure 1. The waste to be treated is conveyed by pump, mechanical conveyor, or other means into a surge tank or feed hopper, which in turn feeds the waste into the mixer, where it is mixed with the S/S reagents. Depending on the process used, one or more dry and/or liquid components may be added to the waste in the mixer. The mixing process normally takes 1 to 15 min, depending on the mechanical system used, the size of the batch, the type of waste, and the amounts and types of reagents being used. After mixing is complete, the waste, still in liquid or semisolid form (in many cases), is removed from the mixer by pumping, in the case of liquids, or by screw conveying or dumping in the case of semisolids. The treated waste is then moved by pump or conveyance device to an area where it can harden and develop its final physical and chemical properties. If the waste has been pretreated, the hazardous

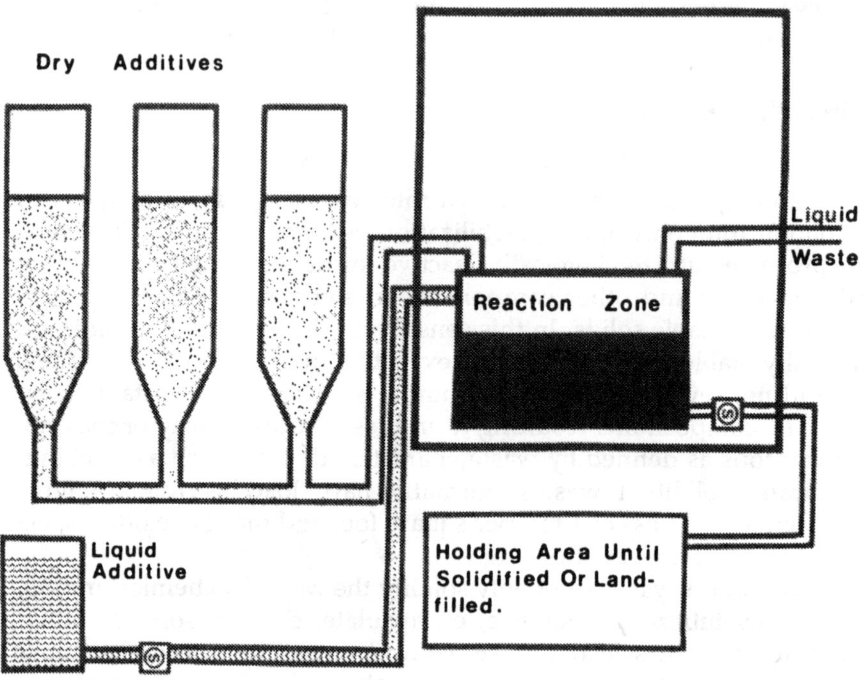

Figure 1 Schematic of S/S system.

components, usually heavy metals, have already been converted into a relatively insoluble form. If it has not been pretreated, the metals are usually immobilized during the mixing process. By the time the waste exits the mixer, it has largely developed its final chemical properties unless these properties depend on a physical, monolithic form.

Hardening, or curing, often takes place in a temporary container or impoundment near the S/S plant, with the solid subsequently being conveyed to the disposal site. Alternatively, the treated waste can be conveyed directly into the disposal site and solidified in place in its final location, regulations permitting. In principle, the latter technique is preferred because it involves fewer steps in handling of the waste and thus is less expensive. Also, a waste that is still liquid or semisolid can be poured into place in the landfill with minimum void space, yielding minimal permeability and maximum landfill space utilization in the final product. When waste is solidified directly in its final resting place, it becomes, sooner or later, essentially a monolithic form which exhibits the minimal permeability associated with the process [23–25]. It is unwise, however, to assume that the chemical properties as measured on such a monolith are necessarily those that ultimately will be associated with the waste over the centuries. Natural processes change the physical character of the material in the landfill just as they do natural soils and rocks. For this reason, chemical stability tests such as leaching tests are often conducted in the worst-state condition; that is, the waste is ground up on the assumption that a variety of factors may cause physical breakdown of the monolith. Conducting tests in this way attempts to assure that the chemical properties which have been developed are the best that can be achieved with the S/S process, and that the leaching rate from the solidified material will probably be considerably less, thereby providing a substantial safety factor.

Comparison of Processes

S/S systems are of two basic types: inorganic or organic, according to the nature of the solidification chemicals used, not the waste composition. Organic systems have been little used for industrial wastes except in the area of radioactive waste solidification and will not be discussed further. Inorganic systems are those using inorganic reagents that react with certain waste components; they also react among themselves to form chemically and mechanically stable solids. These systems are based on reactions between binders, catalysts, and setting agents which occur in a controlled manner to produce a solid matrix. The matrix itself often is, or becomes, a pseudomineral. This type of structure displays properties of stability and a rigid, friable structure similar to that of many soils and rocks.

Different processes exhibit different setting and curing reactions. Most commercial inorganic S/S systems, however, solidify by very similar reactions, which have been thoroughly studied in connection with portland cement technology used in concrete making. While the pozzolanic reactions of the processes using fly ash and kiln dusts are not identical to those of portland cement, the general reactions are alike. One reason for this is presented in an interesting way by Cote [26]. The compositions of most of the primary reagents used in inorganic S/S systems were plotted on a ternary diagram using three oxide combinations: SiO_2, $CaO + MgO$, and $Al_2O_3 + Fe_2O_3$. All of these reagents have the same active ingredients as far as solidification reactions are concerned; this is shown in Figure 2. The combinations of these five oxides express the essential composition of any of these materials, even though the actual compounds are not simple oxides but more complex silicates and aluminates in many cases. It is interesting that all of the reagents, with the exception of power plant fly ash, have their

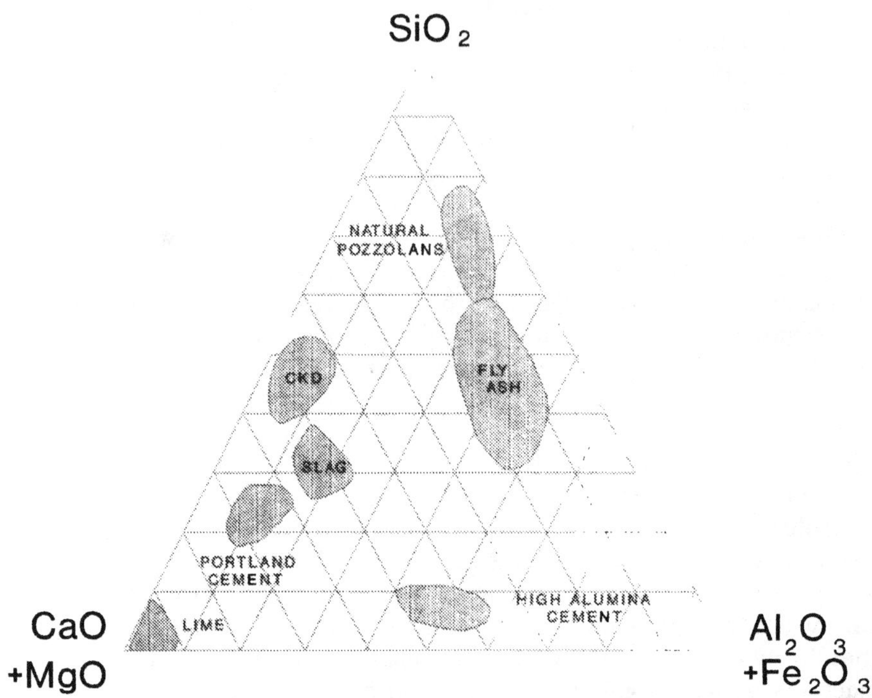

Figure 2 Ternary diagram showing common S/S reagent compositions.

origin in natural limestone and clay formations, and all are inexpensive as industrial chemicals go.

The cementitious reactions that occur in these processes require the pH to be above 10. This is initiated by the dissolution of free lime from the solid and continues throughout the setting and curing stages of the mixture. A *false set* can happen when enough free lime is present initially to start the reactions, but availability decreases as the process proceeds and the reactions stop or slow down. Therefore, any reaction that competes successfully for the calcium ion may inhibit setting. The cementitious or pozzolanic reactions also require sufficient free water if they are to go to completion. Water is used up in the hydration reaction of these chemical systems, so sufficient water must be provided. The water content as measured by total solids determination is not necessarily all available for reaction.

The inorganic processes can most meaningfully and conveniently be grouped into two categories: those that use bulking agents, such as class F fly ash, and those that do not. In this context, we mean by a *bulking agent* an addition that primarily adds to the total solids and viscosity of the waste, thus preventing settling out of the suspended waste components before solidification can occur; it may also help produce a solid with better physical properties. Examples of these two groups of processes are (1) cement-based or cement/soluble silicate systems (no bulking agents), and (2) cement/fly ash, lime/fly ash, cement/clay, or lime/clay (systems with bulking agents).

There are two types of bulking agents: those that are essentially inert in the system and act as described previously, and those that also have reactive capacity or pozzolanic activity. A pozzolan is defined as a material that does not exhibit cementing ability when used by itself but when used in combination with other materials, such as portland cement, lime, and so on, will interact with these agents, resulting in a cementitious reaction.

The most noticeable effect of the difference between systems with and without bulking agents is that systems with bulking agents often have lower chemical costs by virtue of replacing some of the more expensive cementing materials with the less expensive waste products such as fly ash. Systems without bulking agents sometimes have lower overall costs because of the lower weight and volume increase associated with them. Generic inorganic chemical and solidification processes that have been either used or proposed for use are shown below. The most important systems today are marked with an asterisk:

Portland cement-based (major ingredient is cement)*
Portland cement/lime
Portland cement/clay
Portland cement/fly ash*
Portland cement/soluble silicate*
Lime/fly ash*
Cement or lime kiln dust*

Many of the various permutations and combinations of these systems, including various types of additives, are patented or are covered with patents applied for, but most of the generic system types are believed to be in the public domain, at least in the United States and Canada. All of these processes have been used commercially for solidification of water-based waste liquids and sludges and filter cakes. A large body of technical information on leachability, physical properties, and general stability is available.

In volume of waste treated, the lime–fly ash process has probably been the most used in the United States, although it has been very narrowly applied, primarily to flue gas desulfurization sludges. For other types of wastes and industrial sludges, the kiln dust and portland cement–based processes are the most widely used at the present time. In the United States, one of the most flexible techniques has been the portland cement–sodium silicate process. This process has been applied to a variety of wastes, and a good technical base, including leaching test information, is available. Another technique, the portland cement–fly ash process, has been used in Europe and Canada but has not been applied very broadly in the United States to date. This process works well with certain types of waste; like the lime–fly ash process, it involves large additions of the solidifying agents and therefore large volume increases. In summary, the inorganic systems are characterized by:

Relatively low cost
Good long-term stability, both physically and chemically
Documented use on a variety of industrial wastes over a period of at least 10 years
Widespread availability of the chemical ingredients
Nontoxicity of the chemical ingredients
Ease of use in processing (processes normally operated at ambient temperature and pressure and without unique or very special equipment)
Wide range of volume increase
Inertness to ultraviolet radiation
High resistance to biodegradation
Low water solubility
Relatively low water permeability
Good mechanical and structural characteristics

Process Equipment

An S/S mechanical system is, as stated previously, truly an assembly of mixers, chemical storage and feeding devices, pumps, conveyors, and ancillary equipment. More detailed schematics of mechanical sequence and layout are shown in Figures 3 and 4 for fixed and mobile systems, respectively. A more detailed discussion is given in Conner [3]. This aspect of stabilization is also covered later in the section "Delivery Systems."

Principles of Immobilization

Immobilization, often called stabilization, is a separate subset of S/S technology even though it is usually spoken of and used in conjunction with solidification. It can be practiced where there is no need to change a waste's physical properties, as in the detoxification of contaminated

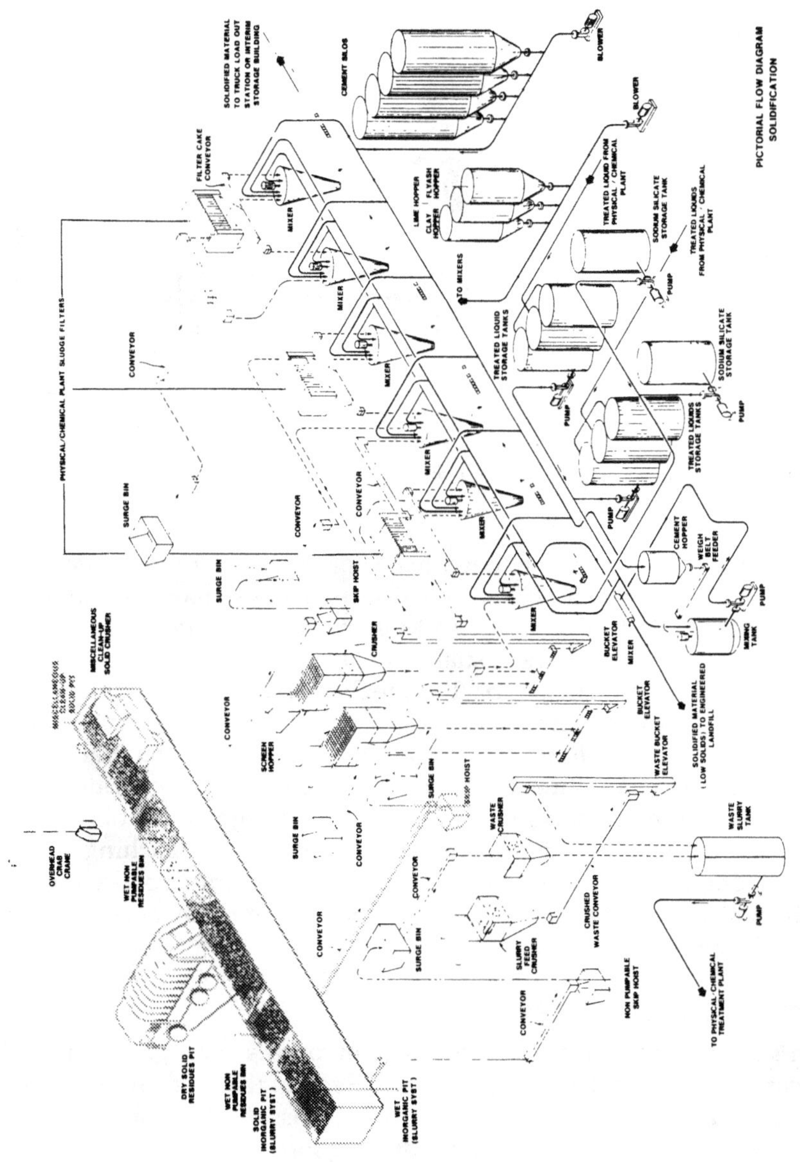

Figure 3 Schematic of fixed S/S system.

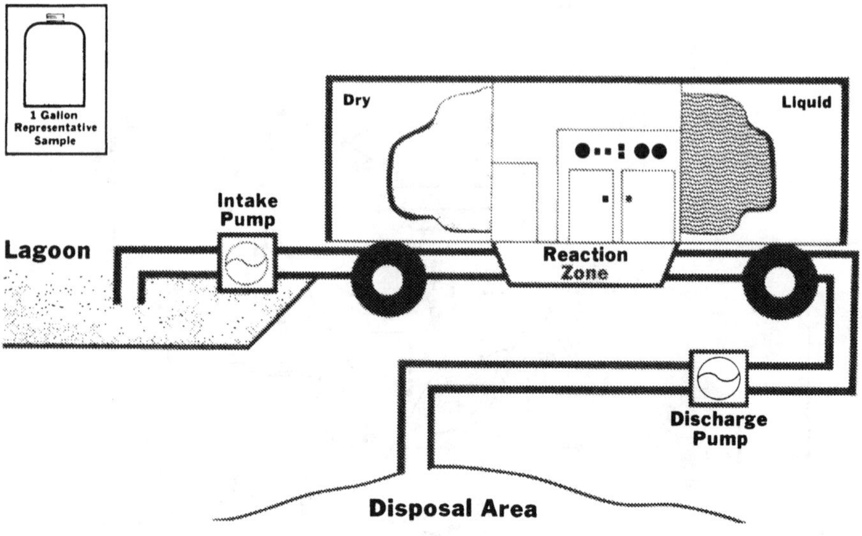

Figure 4 Schematic of mobile S/S system.

soil. In another sense, it has been used for many years in wastewater treatment. In fact, much of what we do understand about immobilization is attributable to that field, and where specific processes or additives are used, they have usually been derived from water and wastewater treatment chemistry. The mechanisms of immobilization are different for the three primary groups of pollutants: metals, other inorganics, and organics. Here we explore immobilization in general and as applied to metals, as an introduction to the subject and to provide a context for the later, more detailed discussions. Since immobilization is generally defined as a treatment product's resistance to leaching in the environment, it is essential to understand the leaching process and the tests by which it is measured. The reader requiring more in-depth treatment of leach testing should consult Conner [3].

Leaching

When ground or surface water contacts or passes through a material, each constituent dissolves at a finite rate. Even in the most impermeable solidified waste—or clay, concrete, brick, or glass, for the matter—water will eventually permeate if there is a driving force. Wherever water penetrates, some of the waste dissolves; there is no such thing as a completely insoluble material. Therefore, when a waste, treated or not, is exposed to water, a *rate* of dissolution can be measured. We call this process *leaching*; the water with which we start, the *leachant*; and the contaminated water that has passed through the waste, the *leachate*. The capacity of the waste material to leach is called its *leachability*.

Leaching is a rate phenomenon and our interest, environmentally, is in the *rate* at which hazardous or other undesirable constituents are removed from the waste and into the environment via the leachate. This rate is usually measured and expressed, however, in terms of *concentration* of the constituent in the leachate. This is because concentration determines the constituent's effects on living organisms, especially human beings.[†] Concentration is the primary basis for water quality standards and water quality standards, especially drinking water standards [27], are normally the basis for leaching standards. Thus we speak of the leaching rate of a constituent but usually measure it as concentration in the leachate.

When evaluating a material for leachability, we usually compare the concentration of the hazardous constituent in the leachate to that in the original waste. This tells us what proportion of the constituent dissolved out during the test, which becomes a measure of the leachability of the material. If the test conditions can be expressed in terms of a time-related number, such as equivalent years of field exposure to rainfall, we can state leachability in true rate terms: for example, pounds of constituent per year or percent of the original content per year. Multiplying the hazardous constituent's concentration in the original waste by the total quantity of the waste gives us the total amount of the constituent, and thus the hazard potential posed by that quantity of waste. Interestingly, as the hazardous constituents leach, the hazard potential gradually diminishes. Thus if the leaching rate is controlled so as not to exceed the allowable environmental standards in the ground or surface water, leaching should really be a beneficial process in the long term. This assumes, of course, that the leaching rate remains constant at an acceptable level, or decreases with time. As we shall see later, a number of scenarios are possible, depending on the disposal conditions.

Measurement of Leachability

For obvious reasons, actual tests cannot be run on the solidified waste in the specific disposal site for thousands of years—the only absolutely certain way—before choosing a S/S technology. As with the testing of

---

[†]There are also cumulative effects determined on the basis of total exposure over a long period of time.

any other product requiring long-term stability, we must make judgments and reach decisions based on information that is available or can be obtained rather quickly. In this case the information consists of actual field data, about 20 years' worth at present, laboratory test results, some understanding of the chemistry of the system, and models that tie everything else together. A large body of information has been built up in this fashion, primarily for the metal constituents specified by the USEPA [28]: arsenic, barium, cadmium, chromium, lead, mercury, selenium, and silver.

It is not possible here to explore the details of leaching test methods and results, but methods are listed later, and a good summary is given by Darcel [24]. The environmental acceptability of a hazardous waste for land disposal in the United States is now based primarily on the USEPA's toxicity characteristic leaching procedure (TCLP) [29], which has replaced the older extraction procedure toxicity (EPT) test for most purposes. In Canada (Ontario Province), the standard leachate extraction procedure is very similar to the EPT. In addition, certain states (e.g., California) have their own tests which must be used in addition to the EPT or TCLP. The USEPA has also used two other test methods in delistings: the multiple extraction procedure (MEP) [30] and the oily waste extraction procedure (OWEP) [31].

All of these tests used for regulatory purposes are batch procedures in which the waste is contacted with a leachant for a specific period of time, agitating the mixture to achieve continuous mixing. Chemical equilibrium is often obtained [32], especially when the solidified waste is crushed before extraction. After extraction and separation of the fluid from the solids, the leachate is analyzed for specific constituents. It should be noted that most of these tests use a leachant/waste ratio of 20:1, so the maximum concentration of constituent that can theoretically be attained in the leachate is 5% of that in the original solid. The leachant in most cases is dilute acetic acid, buffered in some procedures. The total amount of acid added varies with the test and/or with the alkalinity of the waste. The final pH of the leachate at the end of the test is controlled by the alkalinity of the waste in most cases where the leachant is deionized water or dilute acid. As we will see, final pH is one of the prime controlling factors in metal leaching. The TCLP test is designed to simulate the leaching potential of a waste in a "mismanagement" scenario, where it is disposed in a landfill designed for municipal refuse. Such landfills are known to generate organic acids during decomposition of organic matter in the refuse; the purpose of acetic acid in the leachant is to simulate those acids.

Leaching Tests Versus Actual Leaching

Neither the TCLP nor its predecessor, the ETP, actually simulate any real-world set of conditions. Arguably, however, they do perhaps create a worst-case environment for leaching. Attempts to correlate these laboratory leaching tests with field data have not been successful. Natural minerals do not exhibit leach rates even approaching those measured for waste forms in the laboratory. If they did, the world's land masses would be very different from what they actually are. The same processes that inhibit leaching of natural substances may also apply to buried waste forms in the real world. Nevertheless, these procedures are so widely used and required as specification tests that arguments over their validity are largely academic in practice. It is not unusual, however, to use several tests or even a whole test battery for specific disposal situations.

Leaching Criteria

The other aspect of leachability testing involves the criteria that are applied in judging whether the concentrations of constituents in the leachate are acceptable (i.e., the impact on the environment will be negligible). The basis for these criteria in the United States has been the national interim primary drinking water standards [33]. For the characterization of a nonlisted waste as nonhazardous, the concentration of any listed constituent may not exceed 100 times the drinking water standard. As it is applied to the definition of a nonlisted, or characteristic, waste as hazardous, this set of criteria covers the eight metals, six pesticides/herbicides, and 26 other organics listed in Table 1. For listed wastes, the regulatory situation is much more complex, as discussed previously.

Factors Affecting Leachability

Two sets of factors, or variables, affect the leachability of a treated waste: (1) those that originate with the material itself, and (2) those that are a function of the leaching test. The combination of the two sets determines the leachability of the material. Detailed discussion of these factors is not possible here, but some understanding of them is important because in a very basic way they determine whether or not a material meets disposal criteria.

Test Method Factors. The major variables are normally specified for given test protocol, but latitude as to the specification and controllability of the parameter can, and does, cause significant problems with reproducibility. The following list is derived primarily from the EPT and

TCLP, but most of the comments would apply to other batch testing protocols and, to a lesser extent, to column and dynamic leach testing.

1. *Surface area of the waste*. Ideally, the waste should be tested in the same physical condition as that expected in the landfill. In practice, this is not possible; in fact, in many test protocols, the surface area is increased deliberately by grinding or crushing the waste to achieve a worst-case condition. S/S processes may produce the waste form as a monolith, as a soil-like material, and even in granular or powder form. From the standpoint of immobilization, the ideal S/S process should contain the constituents of concern even if the physical structure is destroyed. In practice, most processes rely on a combination of chemistry and micro- or macro-encapsulation. The EPT test allowed the use of either of two compromises. The waste could be crushed to moderate particle size—9.5 mm—to represent the effects of handling and other environmental conditions. Alternatively, a monolithic form could be subjected to an impact procedure (the structural integrity option) and tested in whatever condition results from that procedure. Presumably, any specimen surviving the impact as a monolith would remain that way in the disposal environment, an assumption that is not justified by supporting data. In the TCLP test, the sample must be reduced to −9.5-mm particle size before testing.

Under the EPT procedure, neither preparation method was sufficiently controlled to allow good replication of results, let alone comparison of various S/S products that fracture in different ways. As a result, the sample preparation is dependent on the technician doing the preparation and the equipment used. For example, the crushing procedure specified the maximum particle size but not the minimum size or the size distribution. Unfortunately, the newer TCLP procedure does little to correct this fault in the methodology, although it does eliminate the variable of the structural integrity option. The variability of test results between laboratories and waste forms is therefore not decreased greatly by the TCLP.

2. *Extraction vessel*. None of the batch tests adequately spell out the geometry of the extraction vessel, except when the zero-headspace extractor (ZHE) is used for leaching of organic constituents in the TCLP. This geometry affects the degree of abrasion to which the waste form is subjected during the tests and therefore the ultimate particle size of the waste being tested.

3. *Agitation techniques and equipment*. All batch methods use some form of agitation to permit more rapid attainment of equilibrium be-

tween the waste specimen and the leachant. The TCLP requires a rotary extraction device and gives suitable specific examples. Most workers now use a rotary device of the National Bureau of Standards (NBS) design [34]. There is no analogy for such agitation techniques in a modern waste landfill, where the leachant is normally stationary or flowing very slowly around the waste particles and the leaching process is diffusion limited [35]. The major objection to this sort of agitation technique is that it changes the particle size distribution of the waste, increasing the effective surface area exposed to the leachant. However, with S/S processes that produce soil-like products, the physical state of the waste is usually of relatively little importance.

4. *Nature of the leachant.* In principle, the leachant should be that which is actually in contact with the waste in the landfill. However, this condition is usually difficult to reproduce in a practical leaching test. Leaching water pH, oxidation-reduction potential (E_h), even composition, all change with time and are usually not known with any degree of accuracy in any case. The trend in present testing protocols is to use an aggressive leaching solution with moderately low pH—thus the EPT and TCLP leachants.

The leachant must also be usable in everyday laboratory situations: ordinary laboratory reagents, easy and reproducible preparation of solutions, atmospheric pressure, in air. The choice of mildly acidic solutions using carbonic acid (CO_2-saturated water) or acetic acid meets these requirements and also has some basis in natural co-disposal situations. Acetate buffer systems have been used in estimating the availability of trace metals in agriculture, and natural precipitation and some soil waters contain CO_2. Thus the choice of these systems is a compromise.

5. *Ratio of leachant to waste.* It is obvious that no ratio can be selected that would represent real conditions in any given landfill at all times, let alone in a variety of landfill scenarios. Wastes, especially after S/S treatment, contain large amounts of soluble, nontoxic components that generate common ion and total ionic strength effects that can reduce the solubility of certain constituents of concern when the ratio is low. Investigators have generally determined that higher ratios are more appropriate [36].

6. *Number of elutions used.* In some test protocols, such as the MEP [5], the waste is extracted with successive batches (elutions) of leachant. This is designed to approach the natural, continuous replacement of fresh leachant that occurs in many landfills, and it has been used to simulate column leaching test procedures, which are more difficult to

set up and to reproduce. The MEP has been used by EPA to simulate long-term leaching for up to 1000 years, and by other investigators in sequential batch methods [37]. It has generally been assumed that the initial elutions contain the maximum concentrations of constituents, because they are exposed to the highest concentrations present on the fresh waste surfaces. However, this is not necessarily true.

7. *Time of contact*. Early in the development of standard leaching test procedures, the effect of time was studied in some detail [38]. The goal was to achieve or approach equilibrium at reasonable test times. It was found that 24 h was a good compromise in most situations, and most protocols have used this time as a standard. Recently, EPA modified this for the TCLP, reducing test time to 18 hours for practical scheduling reasons in a laboratory. It is believed that compared to other uncontrolled variables this change does not substantially affect the results.

8. *Temperature*. Solubility of constituents is a function of temperature, and leaching test results are, at least partially, functions of solubility of the species being investigated. Temperature in the disposal site varies as a function of time, depth, location, and chemical reactions occurring in the landfill. Obviously, all of these conditions cannot be used; in fact, they are not even known in most cases. Therefore, the laboratory standard temperature of 20 to 25°C is normally used. The TCLP standard, for example, is 22 ± 3°C.

9. *pH adjustment*. Probably the greatest operating variable in the EPT was the pH adjustment procedure. The TCLP methodology improves on this by eliminating the periodic pH adjustment procedure and using one of two solutions, depending on the alkalinity of the waste. pH level and control are extremely important factors in evaluating leachability, especially for metals. The final pH of the leachant should always be recorded, since it is this value that often determines whether a waste meets regulatory standards for metals. Unfortunately, measurement of final TCLP leachate pH was not required in the original test methodology, and so is often neglected by testing personnel.

10. *Separation of extract*. This seemingly straightforward test element is quite important. Many metals can exist in waste forms and in the waste–leachant mixture in colloidal state. The filtration procedures are designed to remove these colloids but do not always do so. Any cloudy filtrate should be considered suspect. Also, filtration takes time, and the longer the time, the more contact with the leachant. Although not always controllable, filtration time should always be recorded.

11. *Analysis.* Analytical procedures are spelled out in detail in SW-846 [39]. Laboratories experienced in water analysis but new to analysis of hazardous waste residuals and leachates often have difficulty in both accuracy and reproducibility in the latter. Matrix effects and interferences are the rule rather than the exception, and are often very strong in the high-ionic-strength leachates encountered in this field. In the past, regulatory levels were typically in the milligram per liter range and were expressed only to 0.1 mg/L accuracy. Present levels are falling rapidly, sometimes requiring accuracy to several micrograms per liter (parts per billion) or below. This has placed greater demands on the analyst and the equipment.

Waste Form Factors and Immobilization Mechanisms. A number of factors, characteristics, mechanisms, or containment systems may affect the degree of immobilization of constituents in the waste. Often, several of these are in operation at the same time. It is beyond the scope of this chapter to discuss all of these in detail, but some are covered in the next section under immobilization of metal species, and those interested in more depth are referred to the various references, especially Conner [3]. Figure 5 demonstrates schematically how they work, and they are listed below:

- pH control
- Redox potential control
- Chemical reaction
  - Carbonate precipitation
  - Sulfide precipitation
  - Silicate precipitation
  - Ion-specific precipitation
  - Complexation
- Adsorption
- Chemisorption
- Passivation
- Ion exchange
- Diadochy
- Reprecipitation
- Encapsulation
  - Microencapsulation
  - Macroencapsulation
  - Embedment
- Alteration of waste properties
  - Particle size

Adsorption

Example: Interaction of single layer of clay mineral with polar groups

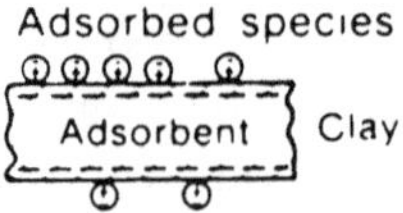

Chemsorption

Example: Organic compounds on clays

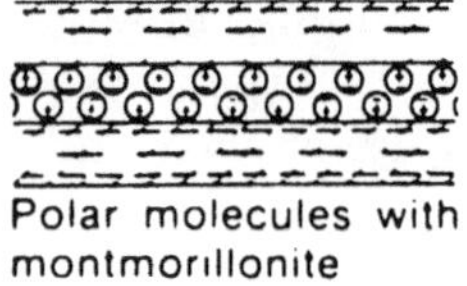

Polar molecules with montmorillonite

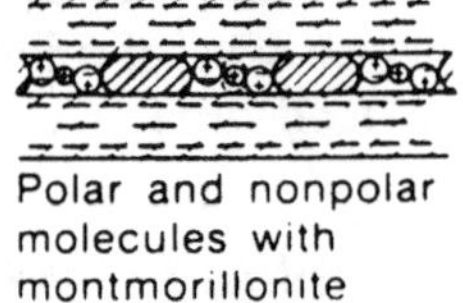

Polar and nonpolar molecules with montmorillonite

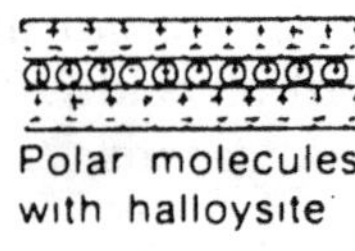

Polar molecules with halloysite

Passivation

Example: Reaction of gypsum with barium solution

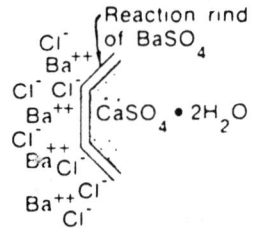

Reaction rind of $BaSO_4$

Cl^- Ba^{++} Cl^- Cl^- Ba^{++} Cl^- Ba^{++} Cl^- Ba^{++} Cl^- Cl^-

$CaSO_4 \bullet 2H_2O$

Diadochy

Example. Substitution in the calcite crystal lattice

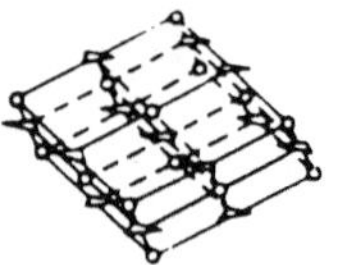

• Ca^{2+} or Mg^{2+} Fe^{2+} Mn^{2+} etc

▱ CO_3^{2-}

Reprecipitation

Example: Production of sideronatrite from sodium sulfate

$$Fe_2(SO_4)_3 + 2Na_2SO_4 + 8H_2O$$
$$\longrightarrow 2Na_2Fe(SO_4)_2 \bullet 3H_2O + H_2O + H_2SO_4$$

Comparative Solubilities at 25°C

$Na_2SO_4 \bullet 7H_2O$	524 g/ℓ
$Na_2Fe(SO_4)_2 \bullet 3H_2O$	0.16g/ℓ

Figure 5 Hazardous constituent containment systems. (From Ref. 40.)

The leaching mechanisms that control and are controlled by these containment systems are listed below. Detailed discussion of these mechanisms is beyond the scope of this book, but certain aspects will be covered later in the chapter as we discuss immobilization of various species.

Solubilization
Transport through the solid
  Convective transport
  Molecular diffusion
    Solid state
    Pore solution
Transport through the leached layer
Transport through the solid–liquid boundary layer
Bulk diffusion in the liquid leachant
Chemical reactions in the leachant
Biological attack

Equilibrium Constants

Before discussing immobilization of specific constituents, it will be worthwhile to review the equilibrium constant concept since so many of the reactions that affect leaching occur between relatively simple molecules and ions, and are reversible. Although this may seem elementary, it is frequently misunderstood. For any reaction

$$aA + bB \rightleftharpoons cC + dD$$

the equilibrium constant is expressed as

$$K_e = \frac{[C]^c \times [D]^d}{[A]^a \times [B]^b} \quad (1)$$

where [A], [B], [C], and [D] are the activities of the respective species. In the case of an ionizable species, such as most metal compounds in aqueous media,

$$AB \rightleftharpoons A + B$$

the dissociation constant is expressed as

$$K_d = \frac{[A] \times [B]}{[AB]} \quad (2)$$

The reactions of interest in leaching occur as heterogeneous reactions; that is, they involve both a solid and a liquid. In equation (2) the solid is AB, and its concentration is a constant. Multiplying both sides of equation (2) by [AB] results in a different value for K_d, which we call K_{sp}, the solubility product:

$$K_{sp} = [A] \times [B] \tag{3}$$

From this comes the important conclusion that for dissociation of solids such as metal compounds, the concentration of the solid metal compound in the waste does not affect the concentration of metal ion in the leaching solution. This remains true as long as any solid metal species exists in contact with the leaching solution. Furthermore, since K_{sp} is a constant for a given system at a given temperature, changing the concentration of either A or B automatically changes the concentration of the other, giving rise to the common-ion effect. This is quite important in metal immobilization for systems at equilibrium, since the metal concentration in the leachate can be controlled by the addition or removal of its associated anion from another species. It is rarely possible to calculate the expected effect with any degree of accuracy in complex waste–reagent systems; however, the degree of departure of measured values from the theoretical concentrate indicates the extent of other factors operating in the system.

IMMOBILIZATION OF METALS IN SOIL MATRICES

More is known about metal immobilization than about that of any other hazardous constituent group encountered in S/S technology. Metals are the only really hazardous constituents that cannot be destroyed or altered by chemical or thermal methods, and so must be converted into the most insoluble form possible to prevent their reentry into the environment. Although this can be done with all metals, the difficulty and cost of such treatment varies greatly with the form of the metal in the waste—its speciation—as well as with the amount present.

The discussion of metal immobilization can be approached systematically from several different viewpoints: by metal, by species in the raw waste, by immobilization mechanism or process, or by waste type. The following lists contain the various parameters under the first two headings; mechanisms and waste types were discussed earlier. Column a is a list of metals that are regulated by federal or state agencies in the United States.† These are the metals of primary concern in S/S technol-

†This discussion excludes metals and other constituents regulated by the Nuclear Regulatory Commission. Radioactive nuclides are normally not present in industrial and domestic wastes. They have different chemistries and are regulated at different levels and by different test methods.

ogy and are commonly referred to as the *toxic metals* or *RCRA metals*. Column b lists additional metals of possible concern in certain situations. Asterisks denote RCRA metals [41].

a. Regulated metals	b. Other metals
Antimony	Aluminum
Arsenic*	Magnesium
Barium*	Manganese
Beryllium	Potassium
Cadmium*	Sodium
Chromium*	Tin
Cobalt	
Copper	
Lead*	
Mercury*	
Molybdenum	
Nickel*	
Selenium*	
Silver*	
Thallium	
Vanadium	
Zinc	

Next are listed the types of metal compounds that may be found in waste streams. The asterisk denotes the more common species. This listing does not include anions containing the toxic metals themselves, or individual organic complexes.

Acetates
Bromides
Carbonates*
Chlorides*
Citrates
Cyanides*
Ferricyanides
Ferrocyanides
Fluorides
Hydroxides*
Iodides
Nitrates*
Oxalates

Oxides*
Phosphates*
Silicates*
Sulfates*
Sulfides*
Organic complexes*

Solubility Principles in Metal Leaching

The theoretical solubilities of metals, as determined from equilibrium concepts, or those measured in pure water or other simple systems are often much different than actual solubilities in complex waste–leachant systems. Table 3 compares published value ranges for individual species with those obtained in leaching tests on treated wastes. In considering Table 3 it is important to understand that such comparisons assume a certain speciation for the metal, which in most cases has not been confirmed. However, the comparison serves a useful function—to point out the limitations of technology in real applications, or in some cases, to show how the complex interactions of several mechanisms can achieve results better than those expected from theory.

Nevertheless, as a starting point in choosing possible chemical reactions for various purposes, it is useful to know the basic water/acid/base solubilities of various species. Figure 6 shows the solubilities of pure metal hydroxides in water. In general, pH values in the alkaline

Table 3 Comparison of Published Metal Solubility Values with Actual Leaching Results on Stabilized Wastes

Metal[a]	Published value (mg/L)	Leaching test result (mg/L)
Antimony (S)	1.2	<0.05
Arsenic (S)	0.3	<0.01
Barium (SO_4)	1.5	<0.02
Cadmium (S)	3.0×10^{-9}	<0.01
Chromium (OH)	0.2	<0.01
Copper (S)	0.12	<0.002
Lead (OH)	0.001	0.003
Mercury (S)	2.0×10^{-21}	<0.001
Nickel (OH)	6.1	<0.005
Selenium	—	<0.05
Silver (S)	2.0×10^{-12}	<0.003
Zinc (S)	1.0×10^{-4}	<0.005

[a]Metal species: (S), sulfide; (OH), hydroxide; (SO_4), sulfate.

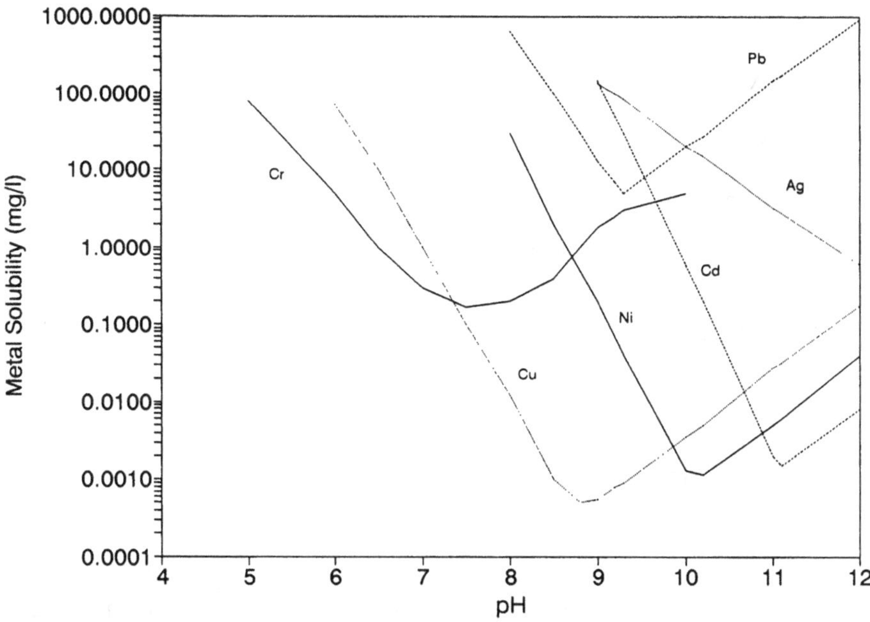

Figure 6 Metal hydroxide solubility as a function of leachate pH.

range are required to minimize metal hydroxide solubility in water, and the hydroxide form is the most common speciation of metals in waste treatment technologies. If all metals exhibited a roughly log-linear relationship between solubility and pH, as does silver, the solution would be simple: maintain a high pH. However, most metals of environmental interest show amphoteric behavior, with a minimum at some pH value and higher solubility at both lower and higher values of pH. The minimum solubilities occur at different pH values. Thus, from Figure 6, achieving a minimum solubility of about 0.1 mg/L for chromium (as the hydroxide) by adjustment to pH 7.5 would increase the solubilities of the other metals to unacceptable values. Conversely, if the leachability of lead, nickel, and copper were optimized, chromium leaching would be far above the lowest landban requirement. Although the actual values and exact minima in complex systems are not the same as with the simple system shown here, and vary considerably with the system under consideration, the principle is the same.

The other problem is that frequently, even at optimum pH, the metal hydroxide solubility is not sufficiently low to meet standards. This is the case for chromium when its minimum solubility of about 0.1 mg/L

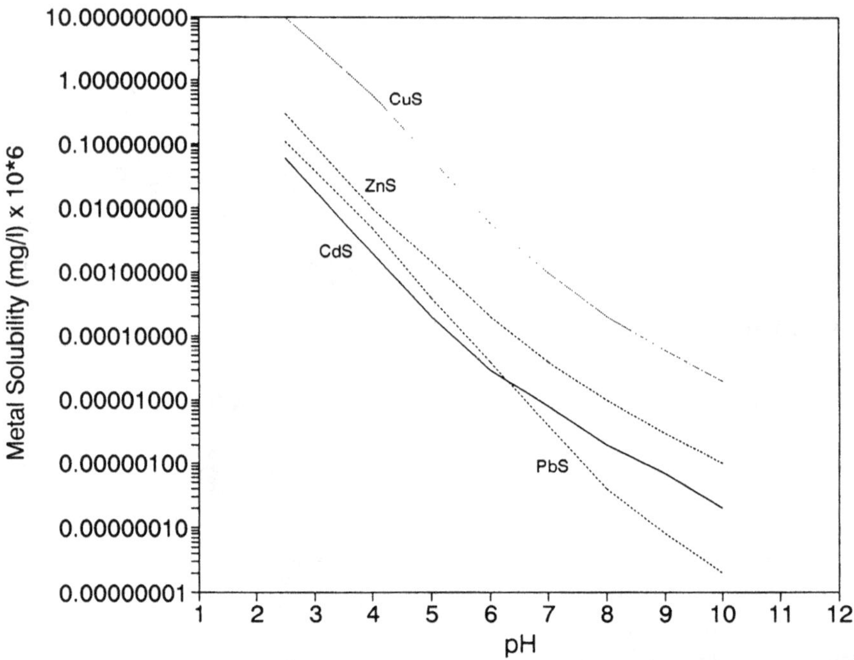

Figure 7 Metal sulfide solubility as a function of leachate pH.

is compared with the lowest landban requirement of 0.073 mg/L. It is apparent that speciation of metals in forms other the hydroxide is necessary in many instances. For example, metal sulfides exhibit better behavior with regard to pH dependency, as shown in Figure 7. The solubility decreases in an approximately log-linear relationship as pH increases. Unfortunately, chromium does not form a sulfide and so is not susceptible to treatment with sulfide ion. However, chromium is incorporated into certain stable silicate matrices, and can also be coprecipitated with other metal species into low-solubility forms. Metals can also be speciated as carbonates, basic carbonates, silicates, phosphates, and with various ligands to achieve lower effective solubilities. Also, it is important to realize that the published metal solubility values are not the results that will be achieved in real systems. This is demonstrated clearly by the data given in Table 3.

To exemplify the points made above, it is useful to look again at the relationship between final leachant pH (at the end of a leaching test) and metal leaching. Posing that pH at the optimum point on the pH curve that represents lowest leachability of the metal in the system under consideration can be done by testing a series of formulations, all using the same reagent system, on the same waste or on a relatively homogeneous group of wastes. One such study was conducted at the author's laboratory on electric arc furnace dust, a USEPA-listed (K061) waste of considerable environmental concern. Samples from four different sources were mixed with several different cementitious reagents at various mix ratios. The leachability of the RCRA metals of interest for each of these mixtures was determined by the TCLP test. These simple systems were found to immobilize all metals except lead to RCRA characteristic (TC) and K061 listed levels; lead exhibited widely varying results that seemed to be related to final leachate pH, so lead leaching was plotted against final leachate pH. The results are shown in Figure 8; the curved line delineates the zone where all the leaching data fall. The horizontal lines define various regulatory levels: RCRA toxicity characteristic (TC) and two land disposal restriction (LDR) levels. Thus a region bounded by the intersection of a horizontal line and diagonal lines to the left and right defines the probable pH range in which a particular regulatory leaching requirement can be achieved. It is evident that any of these reagents will stabilize lead to the TC level (5.0 mg/L) as long as the pH is between about 8 and 11. To meet the LDR level for K061,

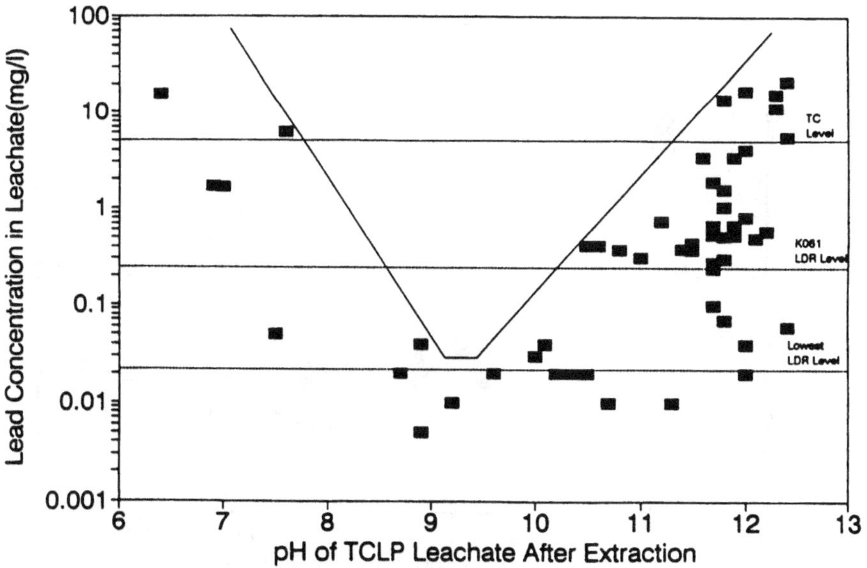

Figure 8 Leachability of stabilized K061 dust for cement-based reagents without additives.

however, the pH must lie in a narrower range—about 9 to 10. It would be impossible to ensure that the lowest LDR level could be met with this formulation system at any pH if, for example, the waste contained other RCRA codes which necessitated that more stringent level.

To meet the K061 LDR level consistently would necessitate formulating to keep the final leachate pH in the range of 9 to 10, not an easy task when a variety of properties must be achieved in the waste form. Another approach would be to use a reagent formulation that broadens the allowable range (i.e., shifts the curve downward and/or changes its shape). This was done through the use of a proprietary additive in a cementitious system, and the results for 11 different such formulations on the four different K061 wastes are shown in Figure 9. Now the K061 level can be met at any pH between about 8 and 11, and the lowest LDR level can be met between approximately pH 9+ and 10+.

Metal Speciation in Soils

The stabilization mechanism exemplified by Figure 9 is not simply a pH adjustment or buffering effect. Even a precise pH adjustment would only have placed the final leachate pH in the range 9 to 10, and buffering, by definition, would simply have kept it there. Even if that were

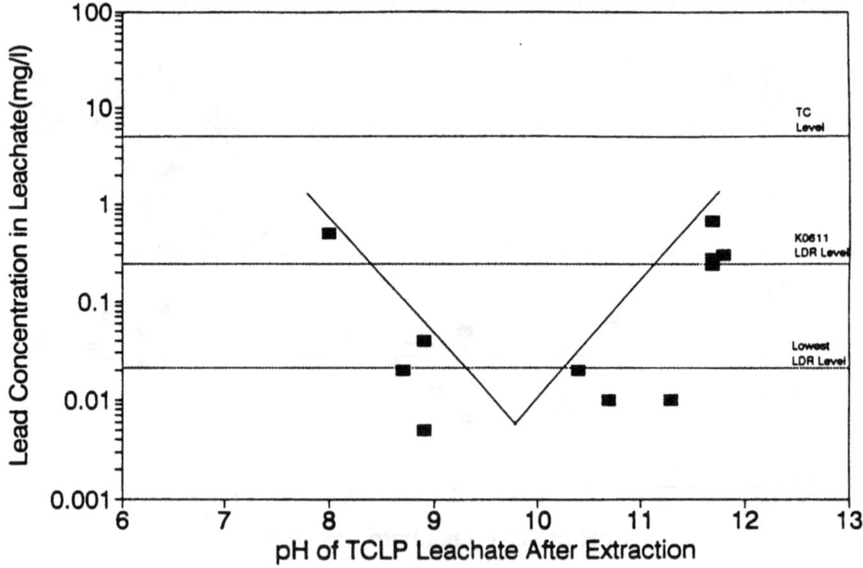

Figure 9 Leachability of stabilized K061 dust for cement-based reagents with additives.

satisfactory, in this system it still would not have ensured meeting the lowest LDR level. And such precise pH manipulations are often impractical in actual systems and disposal scenarios. The effect of the additive is to convert solubilized lead to a less mobile, less pH-sensitive form. This may also be done for many other waste constituents by a number of the mechanisms listed earlier. Having already reviewed the effects and limitations of pH control, we will, however, look at some ways that metal constituents can be speciated in less mobile forms.

In the case of metal constituents, especially in soils, much of the confusion about solubility in systems stems from the fact that there may be several species of a given metal in a waste or stabilized waste form. For example, it is not unusual to find lead existing simultaneously as $Pb(OH)_2$, $PbSO_4$, $PbCl_2$, and perhaps as carbonates and silicates, all in the same waste form. Each species has its own solubility product (K_{sp}) for the medium under consideration—the leachant used in the study or existing in the environment, modified by the dissolved salts from the waste form. Furthermore, the amount of metal in solution is determined by the interrelationship between its various species by the common-ion effect, as well as by common anions in solution and, as we have seen, by pH of the leaching medium. Some of the waste reactants and products commonly found in soils, and reagents that may be used in S/S to treat soils, are listed in Table 4.

Immobilization of Soluble Metal Species

The problem of speciation is minimal when the metal species to be fixed is in solution. Unless it is complexed in soluble, stable form, the metal can usually be precipitated from solution as a known species that exhibits minimum solubility under the expected disposal conditions (or in meeting the required leaching test). This is especially important for the heavy metals of environment interest that exhibit amphoterism, such as arsenic, cadmium, chromium, lead, and zinc. Because most systems are quite alkaline, usually above pH 11 (at least initially), the solubility of the metal hydroxide in the treated waste may actually be higher than that in the original, untreated sludge. This is often the case with Cr^{3+}.

The redox potential, E_h, establishes the ratio of oxidants and reductants existing within the waste–environment system, and may affect the valence state of a metal in that system. The effects of redox processes are commonly expressed as E_h–pH diagrams. The scale ranges from oxidizing (positive) to reducing (negative) potentials, and the resulting diagram shows the domains, or stability fields, in which different species can exist. Such diagrams are useful for conceptual purposes in systems.

Table 4 Reactants and Products in Stabilization

Reactants		
Waste	Reagents	Products
Metal hydroxides	Metals	Metal hydroxides
Metal silicates	Soluble silicates	Metal oxides
Cl^-	Reactive silicates	Metal carbonates
SO_4^{2-}	Reactive silica	Metal silicates
S^{2-}	Fe^{2+}	Metal phosphates
CN^-	Fe^{3+}	Inorganic complexes
NO_3	Al^{3+}	Organic complexes
Organics	S^{2-}	
Silt	Organic sulfur compounds	
Water	Carbonates	
Bioorganisms	Phosphates	
Surfactants	Xanthates	
Organometals	NaOH	
	CaO	
	Surfactants	

In more specific terms, chromium and arsenic are the most common examples of the influence of valence state on solubility. It is usually necessary to speciate chromium as Cr^{3+}, and arsenic as As^{5+}, to meet even the TC levels, to say nothing of the lower LDR levels. The presence of strong oxidants or reductants can change the valence state of a number of the metals, affecting their chemical speciation, and therefore their mobilities, by orders of magnitude. For some metals (e.g., arsenic) both valence and speciation as either cation or anion can change easily with redox potential [42]. Of the metals of interest in non-nuclear use, seven have more than one possible valence state in aqueous systems (As, Cr, Fe, Hg, Mn, Ni, and Se). Also, nitrogen and sulfur have multiple valence states which affect speciation of the metals in a given system. Ag, Cu, Cd, and Zn can be strongly influenced by redox processes even though they have only one valence state in aqueous systems [43].

Immobilization of Low-Solubility Metal Species

The confusion about speciation is exacerbated by the fact that contaminated soils that are treated by S/S are not solutions; they are mixtures of soils with sludges, filter cakes, waste solutions, and other residues from industrial processes or wastewater treatment systems.

The metals may have been precipitated with lime or other alkali, or with agents such as sulfide, to produce metal hydroxides, sulfides, or other compounds that have low solubility under the conditions of precipitation. In this state there is little immediate reaction between the metal species and the reagents. If the metal species in the waste is more soluble than the species that would be formed by anions or ligands introduced by the system, there may be gradual respeciation, beginning at the particle surface, but total respeciation would be expected to occur only over a long period of time. Furthermore, the soluble anion or ligand may be quickly destroyed by reaction with other components of the system. On the other hand, dissolution of the original species by the leaching medium followed by reprecipitation may hasten the speed of respeciation. The ultimate result, then, is usually a mixture of metal species dispersed in a cementitious matrix. Simplistic models of waste systems based on a single speciation of a metal in a waste thus are usually invalid. While it is desirable to know the speciation of the metals in the raw waste, in these complex systems complete analysis is very time consuming, expensive, and in many cases, uncertain.

We have already discussed redox potential effects and one chemical reaction—hydroxide precipitation; from Table 4 we saw something of the chemical complexity common in stabilization systems. To achieve LDR requirements, specific immobilization reactions involving one or more of the mechanisms described earlier may be required. The most important classes of metal compounds in stabilized wastes, other than hydroxides, are sulfides, silicates, and carbonates. Others—phosphates, inorganic and organic complexes, and various reactions that bond metals to insoluble substrates—are also used occasionally.

Sulfide precipitation has been one of the most widely used methods to immobilize metals stabilization treatment. The low solubilities required for highly toxic metals such as mercury are often achievable only by speciation as sulfides, since the metal sulfides have solubilities several orders of magnitude lower than the hydroxides throughout the pH range. Also, their solubilities are not as sensitive to changes in pH. However, metal sulfides can resolubilize in an oxidizing environment, and there is currently disagreement over the acceptability of metal sulfide sludges in uncontrolled landfills. It is necessary to maintain pH 8 or above to completely prevent evolution of H_2S. While excess sulfide ion is necessary for the precipitation reaction, the excess must be kept to a minimum so that free sulfide removal treatment is not required before the waste can be landfilled. Precipitation is normally done with Na_2S or NaHS. One interesting exception in sulfide precipitation is that of chromium, which does not precipitate as the sulfide but as the hydroxide. In some situations, organosulfur compounds may have certain advantages over the inorganics.

Another important method of immobilizing metals by chemical reaction is silicate precipitation by the use of soluble silicates. The reactions of polyvalent metal salts in solution with soluble silicates have been studied extensively over many years. The best known process uses a combination of portland cement and sodium silicate. The "insoluble" precipitates that result from such interactions are usually not well characterized, especially in the complex systems representative of most wastes. Metal silicates are nonstoichiometric compounds in which the metal is coordinated to silanol groups, SiOH, in an amorphous silica matrix. The reactions of soluble silicates in solution was best summarized by Vail [44], who states: "The precipitates formed by the reaction of the salts of heavy metals with alkaline silicates in dilute solution are not the result of the neat stoichiometric reactions describing the formation of crystalline silicates, but are the product of an interplay of forces which yield hydrous mixtures of varying composition and water content." These reaction products are usually noncrystalline and therefore very difficult to characterize structurally. They are most often described as hydrated metal ions associated with silica or silica gel. Iler [45] mentions "that many ions are held irreversibly on silica surfaces by forces still poorly understood in addition to ionic attraction." The composition and form of the metal "silicates" formed from metal ions and soluble silicates are functions of the conditions under which they are formed: temperature, concentration, addition rate, metal ion speciation, presence of other species, and so on. In addition to the use of soluble silicates, metal silicates can be formed from the cementitious reactions in typical soluble silicate systems.

In certain cases, metal carbonates are less soluble than are their corresponding hydroxides. In cement chemistry, the natural formation of carbonates from carbon dioxide from the air is termed *carbonation*. Carbonate ion concentration in a system depends on both CO_2 partial pressure and pH. The carbonate species, CO_3, dominates at pH values larger than 10.3. The carbonation process at alkaline pH is

$$Me(OH)_2(solid) + H_2CO_3 \rightarrow MeCO_3(solid) + 2H_2O$$

The pH at which carbonation occurs depends on the solubility products of the carbonate and hydroxide species, and the CO_2 concentration. Patterson et al. [46] found that the formation of hydroxide precipitates controlled the solubility of zinc and nickel over a wide range of pH values but that cadmium and lead solubilities were controlled by carbonate precipitates in a narrower range. This is in keeping with the results of treatability tests conducted by the author on lead-bearing wastes using soluble and "insoluble" carbonates.

Some recent work has been done on the use of phosphates for metal immobilization in municipal incinerator ash [47], and phosphates have been used in water treatment for many years. Phosphate chemistry is complex and varied. Compounds containing monomeric PO_4^{3-} are called orthophosphates or simply phosphates. The simple phosphate salts of the toxic metals have low water solubility, although they are soluble in acids. However, other phosphates have the potential to sequester the metals as water-soluble species. This property is the basis for detergent and water treatment applications of phosphates. Therefore, the presence of phosphates in the waste or the system may be harmful or beneficial to immobilization, depending on the phosphate species.

One of the most intriguing prospects for metal immobilization is co-precipitation with other metal species. The removal of toxic metals from wastewater with systems that co-precipitate and/or flocculate them with iron and aluminum salts is well known and widely used [48–50]. More recently, it has been used to reduce the solubility of various toxic metals in hazardous waste treatment [51]. The ratio of Fe^{2+} to Fe^{3+} is important, with ratios of 1:1 to 1:2 reported as yielding optimum results. Soils of hydrous metal oxides are stabilized by the presence of excess ferric ion, but acquire a negative charge and destabilize and flocculate under alkaline conditions. As the system becomes alkaline, the ferrous ion is also easily oxidized to ferric, and precipitates as the hydroxide. These reactions remove other metal ions from solution, reducing their concentrations to levels below those obtained with simple hydroxide precipitation. Other complexes can also produce insoluble metal species which may have lower solubilities than the simple metal compounds.

Ordinarily, investigators think in terms of immobilization of metals with inorganic species. However, many organic materials also form low-solubility species with certain metals. Among those described in the literature are humic acids [52], treated leather waste [53], and casein [54]. The most widely publicized insoluble substrate for heavy metal immobilization has been insoluble starch xanthate (ISX) [57,56]. ISX is produced by treatment of starch with cross-linking agents, then xanthating it with CS_2 in the presence of an alkali metal base such as sodium hydroxide. The resulting product is a particulate solid with the structure

$$-\!\!\left[\text{starch}-\text{O}-\overset{\displaystyle \text{S}}{\overset{\|}{\text{C}}}-\text{S}\right]\!\!-$$

In contact with metal ions, the metal links to the sulfur group much as it would with the S^{2-} in inorganic sulfides. Cellulose xanthates are reported to operate in much the same way.

The existence of metal species in a waste, or their precipitation during treatment or during leaching of the waste form, is not always as large particles easily removed during laboratory filtration. Formation of colloids—very small nonsettleable particles not in true solution—is a common phenomenon in chemistry. We have found that it is also more common in stabilization systems than previously believed, and have often observed the presence or formation of colloidal lead species which are not removed by the filter used in the TCLP (0.7 μm) but are removed to a much greater degree by the 0.45-μm filter used in the EPT. This is vividly illustrated by the following data from a study in which three filtration methods were used in a modified TCLP test:

Filter	Lead concentration in filtrate (mg/L)
Unfiltered	29.10
Regular TCLP filter	15.30
Whatman No. 42	0.04

The high-retention Whatman paper removes virtually all colloidal-size particles, but the TCLP glass fiber filter does not. The Whatman paper was checked previously with dilute, true solutions of lead compounds to verify that it does not preferentially sorb lead from solution.

The question raised by these data is this: What filter is appropriate for use in leaching tests? The filters used in the TCLP and EPT were chosen for good retention and fast filtration, the latter property being improved by the change of filters from the EPT to the TCLP. At present, EPA's reaction to these facts is that any metal not removed by the filter is presumed to be mobile in the soil environment. However, this stance is not necessarily supported by scientific data. Are colloids mobile in "soil," especially in view of the fact that soils are very variable materials in themselves? If so, at what size range, since the colloidal spectrum in these systems may be a continuum from true solution to settleable particles in a given system? This phenomenon is probably a prime factor in the large differences sometimes seen between EPT and TCLP test results, especially for lead.

Workers in the field have found methods to minimize the formation of colloidal lead by the use of additives or by aging of the waste form. However, the fundamental question of mobility raised by the presence of colloidal species has not been addressed. It may be very important when working with groundwater models where assumptions are made about the mobility of species in the geo structure. Perhaps more care needs to be exercised in specifying the filtration conditions used in leaching tests if they are intended to simulate any real set of conditions in the environment.

Problems with Complexed Metals

Metals exist in solution in forms other than the simple ions or molecules that we have discussed. Actually, metal ions in solution are usually solvated, that is, they are associated with water molecules in a definite arrangement. The maximum number of water molecules associated with the ion depends on the metal's coordination number. When the water molecules are replaced by other ions or molecules, the result is termed a metal complex. The chemical bonds involved are covalent rather than ionic, that is, electrons are shared by each of the bonded atoms. A common example of such a complex often encountered in environmental chemistry is the cupriammonium complex.

A metal complex may be either inorganic, such as the copper complex ion, or organic. It may be an ion like the cupriammonium complex or a neutral molecule. Finally, it may be either soluble or insoluble. In this section our concern is with soluble complexes; the insoluble type provides a means of immobilizing some metals and other species and is discussed later. Many complexes, such as the cupriammonium complex, are relatively easy to break up by simple means such as pH adjustment. Others, like the ferricyanide complex, are more stable. Even more stable is a special type of complex formation known as a *chelate*. In chelates, the metal is bound chemically to at least two sites on the ligand. This results in a ring formation that is inherently more stable than the simple structures formed by ammonia and cyanide. Most chelating agents are organic, but some are inorganic ligands such as phosphate.

Citrate, gluconate, glycine, EDTA (ethylenediaminetetraacetic acid), and nitrilotriacetate are all chelating agents that form stable, water-soluble metal complexes. A relatively simple structure, calcium gluconate, illustrates the ring structure:

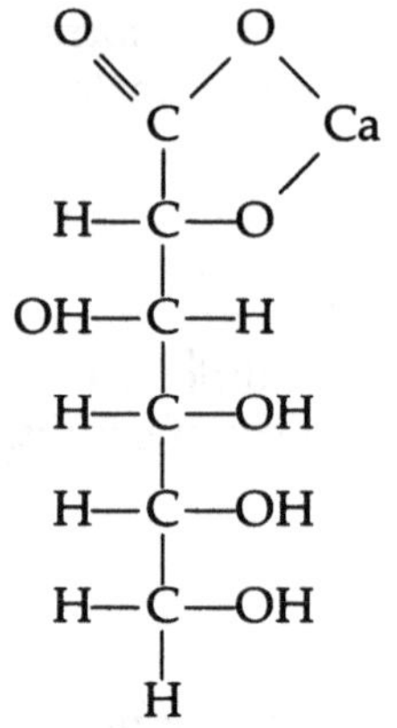

```
 O      O
  \\  /   \
    C      Ca
    |     /
 H—C—O
    |
OH—C—H
    |
 H—C—OH
    |
 H—C—OH
    |
 H—C—OH
    |
    H
```

EDTA forms a more complex, multiple-ring structure with metals:

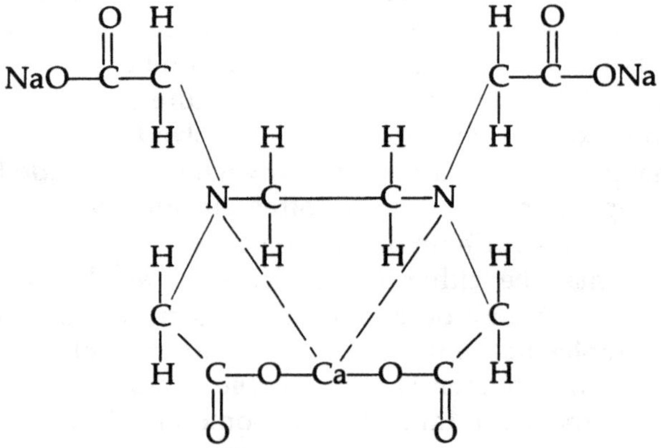

```
        O  H                           H  O
        ||  |                           |  ||
 NaO—C—C                            C—C—ONa
           |  \                        / |
           H   \    H          H     /   H
                \   |          |    /
                 N—C————————C—N
                / \ |          | / \
           H  /   \H         H/    \  H
           | /      \        /      \ |
           C         \      /        C
           | \        \    /       / |
           H  C—O—Ca—O—C   H
              ||            ||
              O             O
```

EDTA is the prototypical chelating agent, but many others, including natural organic acids, are also found in wastes. When a ligand forms a water-soluble complex, it is called a sequestering agent. In this form, the metallic ion is inactivated and no longer participates in its usual chemical reactions. However, since it remains in solution, it is readily leached from wastes, treated or not, and shows up in the subsequent analysis.

Since the sequestered metal species is not chemically reactive, it may not be precipitated by the usual methods, including pH adjustment. In fact, most S/S reagents at ambient temperature are ineffective in precipitating chelated metals. Strong oxidizing agents, often at elevated temperature, are required to destroy the complex and release the metal ion within a reasonable time frame. Like other chemical species, complexing agents vary in their ability to solubilize metals. Just as with other compounds, the chelate structures exist in solution in an equilibrium mixture with chelating agent (L) and metal ion (M):

$$M^{n+} + L^{n-} \rightleftharpoons ML$$

The measure of effectiveness of the chelating agent is the stability constant, K, which is expressed as

$$K_{ML} = \frac{[ML]}{[M^{n+}][L^{n-}]} \tag{4}$$

The larger the positive value of K, the greater is the tendency of the chelate to form.

Chelating agents are used commercially not only to keep a metal in solution, as in electroplating, but also to dissolve insoluble metal species in operations such as boiler tube cleaning in the power industry [57]. In this case an "exchange" reaction takes place:

$$ML + B \rightleftharpoons MB + L$$

where B is the precipitating anion and MB is the insoluble species. This reaction can be described in terms of the stability constant, K_{ML}, and the solubility product for MB [58]:

$$K_{exchange} = \frac{1}{K_{ML} \times K_{MB}} \tag{5}$$

If $K_{exchange}$ is less than 1, chelation will take precedence over precipitation and the insoluble phase will dissolve. This assumes the absence of interfering factors such as very low pH or competition of other cations, which may be complexed.

In addition to strong oxidizing conditions, some chelates may be broken up by very high pH, by sodium sulfide, and by proprietary organosulfur precipitants [59]. However, all of these methods involve more complex or expensive treatment than is common in the S/S field: multiple reaction steps, elevated temperatures, and corrosive or toxic reagents. Often the best method for treatment of metal complexes is pretreatment by the generator before mixing with other waste streams, which may complicate the process. For example, a small amount of complexed nickel from an electroless plating bath may necessitate the treatment of a much larger volume of wastewater treatment sludge. The cost would be much less if the plating bath were managed separately.

Many of the metals of environmental interest form very stable chelates: cadmium, cobalt, copper, lead, mercury, nickel, and zinc. We discuss their treatment further under the individual metal sections later in this chapter.

Long-Term Stability of Immobilized Metals

Much concern has been expressed by environmentalists and regulators about the long-term stability of fixed species. Certain leaching test procedures purport to be equivalent to tens of hundreds of years of natural leaching action in the environment [60]. However, no actual long-term data are available because the technology has only been practiced for about 20 years at this time. Nevertheless, there is no alternative to S/S technology for management of hazardous metals. Therefore, it is especially important to understand the fundamental chemistry of these systems so that we can make intelligent estimates of long-term effects. As pointed out earlier, slow leaching of metals at a controlled rate is not detrimental to human health and the environment.

The real concern is the sudden release of contaminants due to breakdown of the matrix. In principle, this could occur, for example, when the mechanism is pH control and the buffering action of the alkali is finally used up in an acid environment. Knowledge of the structures and species formed in a given S/S system can allay these fears by allowing us to simulate the sudden release conditions that could occur and measure the resulting effects. Work of this sort is being done and the results are encouraging.

Immobilization of Individual Metals

It is not possible here to discuss in detail the environmentally related chemistry of all of the metals of environmental concern; the interested reader is referred to Conner [3] for more information. Three metals that are especially difficult to immobilize under certain circumstances, and that have interesting chemistries applicable to other metals, are reviewed briefly below.

Arsenic

Arsenic is actually classified chemically as a nonmetal or metalloid, although it is grouped with the metals for most environmental purposes. Its principal valence states are +3, +5, and −3. Most arsenic compounds are highly toxic, causing dermatitis, acute and chronic poisoning, and possibly cancer. As little as 0.1 g of arsenic trioxide, As_2O_3, the most important commercial form, can be fatal when ingested. It is also highly toxic to other forms of animal life, and damage to plants has been observed [61]. The other arsenic species are also poisonous, although generally less so.

The compounds of arsenic commonly found in waste materials are:

Arsenic trioxide: As_2O_3
Cacodylic acid (dimethylarsinic acid): $(CH_3)_2AsO_2H$
Metal arsenates: calcium, lead, copper salts
Metal arsenites: sodium arsenite and copper acetoarsenite
Arsenic sulfides: As_2S_3

Systems containing the simple As^{3-}, As^{3+}, and As^{5+} ions in solution are unknown. The oxide is amphoteric and thus is soluble in both acids and bases. Arsenites are often present as complexes such as Scheele's green ($CuHAsO_3$) and copper acetoarsenite, described previously. Arsenates derived from the arsenic acids are oxidizing agents. The organic arsenicals are of special interest in S/S because they are frequently found in wastes, especially in remediation work at old lagoons and other contaminated disposal sites.

Arsenic chemistry is complex, involving a variety of valence states, anionic and cationic species, and both inorganic and organic compounds. These are all commonly found in arsenic wastes, often at the same time. To complicate the issue further, the valence state changes easily and reversibly with redox potential. It is not unusual to find cycling between one state and another in the same waste system, depending on, for example, the degree of aeration resulting from different depths in a waste lagoon.

A number of S/S processes specifically for arsenic have been described in the literature. These are:

Portland cement
Portland cement, sand, and Ca or Mg salt
Sulfate, ferrous/ferric ions, $Ca(OH)_2$
H_2O_2 and ferric ions
$Fe(OH)_3$ at $pH > 6.5$
$Fe_2(SO_4)_3$ and CaO or $CaCO_3$
Portland cement, sodium silicate, and EDTA
Portland cement, sodium silicate, and polyethylene imine
$CaCO_3$ at pH 6
Red mud

In addition to these specific references, information on arsenic in the literature is available in a number of general leaching studies [62–69] as well as in reference works such as Conner [3].

The problem with organic arsenicals is more serious. It may be necessary to alter or destroy the organic species to fix arsenic adequately in this form, particularly if the concentration is high. In many ways this situation is similar to the complexation problem we find with some nickel-bearing wastes—the organic "complexes" are soluble and stable, and considerable chemical energy may be required to "break" them.

Chromium

Chromium belongs to group VIB of the periodic table. It has three valence states, +2, +3, and +6, but the latter two are the most common. Cr^{6+} is acidic, forming chromates $(CrO_4)^{2-}$ and dichromates $(Cr_2O_7)^{2-}$, while the other valence states are basic. The primary environmental problems associated with chromium are with Cr^{6+} compounds. While the trivalent state is also poisonous [89], these compounds have no established toxicity [8], and due to their low solubility, create little environmental risk.

The major chromium compounds that may be encountered in S/S work are composed of Cr^{3+} and Cr^{6+} valence states. The chromate compounds discussed previously are usually reduced to the trivalent state and precipitated with bases in wastewater treatment. The resultant $Cr(OH)_3$ is the form in which the majority of chromium is encountered in S/S treatment. Because of its low solubility, chromium as this species has not been a problem in the past, except for a few specific listed waste streams which have very low LDR requirements—0.073 mg/L.

In view of the status of chromium immobilization work, most technical emphasis on S/S treatment of chromium-containing wastes has been and probably will continue to be placed on more efficient and cost-effective reduction of Cr^{6+}, or in the direct stabilization of this species. The hexavalent form is encountered in a number of wastes either by mixing of untreated chromate solutions with wastewater treatment sludge or because it is intrinsic to the waste, as in some incinerator residues and in wood-treating wastes. It is also a major problem in the remediation of sites where soil has been contaminated with chromate or chromic acid solutions, a frequent situation with old plating operations. The classical way to manage the environmental problem with Cr^{6+} has been a two-step process: reduce the chromium to the trivalent state and then precipitate it as the hydroxide. Reducing agents for this purpose are shown in Table 5. Table 6 shows the amounts of reagents used, and residue produced, for the most common reducing agents. The reactions for bisulfite and ferrous sulfate reduction are

Table 5 Useful Reducing Agents for Chromium Reduction

Agent	Optimum pH range	Comments
$FeSO_4 \cdot 7H_2O$	<3	Reduction occurs over a wide pH range but at lower rates; inexpensive
$Na_2S_2O_5$ $NaHSO_3$	2–3	Reduction rate depends on pH; evolution of SO_2 possible; inexpensive
$Na_2S_2O_4$	>7	Operates at high pH; relatively expensive
Soluble sulfides	7–10	Less effective at high pH, evolution of H_2S at low pH
Reductive resins		Very expensive; used with precious metals
Hydrazine		Handling and safety problems

$$4CRO_3 + 3Na_2S_2O_5 + 3H_2SO_4 \rightarrow 3Na_2SO_4 + 2Cr_2(SO_4)_3 + 3H_2O)$$

$$Cr_2(SO_4)_3 + 3Ca(OH)_2 \rightarrow 2Cr(OH)_3\downarrow + 3CaSO_4\downarrow$$

$$2CrO_3 + 6FeSO_4 \cdot 7H_2) + 6H_2SO_4 \rightarrow 3Fe_2(SOf4)_3 + Cr_2(SO_4)_3 + 48H_2O$$

$$Cr_2(SO_4)_3 + 3Fe_2(SO_4)_3 + 12Ca(OH)_2 \rightarrow 2Cr(OH)_3\downarrow + 6Fe(OH)_3\downarrow + 12CaSO_4\downarrow$$

In wastewater this is an efficient process, but in the treatment of concentrated residues containing high Cr^{6+} levels, it often is not. Various side reactions can occur that waste reagent and/or cause other problems,

Table 6 Reagent Use and Residue Generation with Various Reducing Agents (per 100 Pounds of Chromic Acid in Waste)

	Reducing agent (lb)		
	$Na_2S_2O_5$	$Na_2S_2O_4$	$FeSO_4 \cdot 7H_2O$
Reducing agent	147	261	834
H_2SO_4 (66 Be^0)	80	—	316
$Ca(OH)_2$	111	—	444
Na_2CO_3	159	—	631
NaOH	120	120	480
Sludge (lb) from:			
$Ca(OH)_2$	307	—	1346
NaOH	103	103	423
Na_2CO_3	103	—	423

such as evolution of sulfur oxides in the gaseous form. The later reaction seems to be catalyzed by the soil surface to the extent that it often becomes impractical and dangerous to use the bisulfite reduction method with soils.

$$4Fe^{2+} + O_2 + H^+ \rightleftharpoons 4Fe^{3+} + 2H_2O$$

$$4Fe^{3+} + 3OH^- \rightleftharpoons Fe(OH)_3 \downarrow$$

$$SO_3{}^{2-} + H^+ \rightleftharpoons HSO_3{}^- + H^+ \rightleftharpoons H_2SO_3 \rightleftharpoons H_2O + SO_2 \uparrow$$

Lead

Lead is a member of group IVA of the periodic table. It has two valence states, +2 and +4, with the +2 state being the most common. Pb^{4+} compounds are regarded as being covalent; Pb^{2+} compounds are primarily ionic. Lead is amphoteric and forms anionic plumbites and plumbates as well as both cations. Lead and its compounds are cumulative poisons. Its use in vessels for making wines and grape syrups in Roman times is said to have caused gout, mental retardation, and personality changes in the Roman aristocracy and its offspring [70]. Lead poisoning may be acute or chronic, but the effects are varied and severe. It may occur by ingestion or inhalation. Pb^{2+} forms a soluble nitrate, chlorate, and acetate; a slightly soluble chloride; and a low-solubility sulfate, carbonate, chromate, phosphate, molybdate, and sulfide. It also forms anhydrous and hydrous basic lead salts such as $4PbO \cdot PbSO_4$, and complexed mixed salts such as $2PbCO_3 \cdot Pb(OH)_2$ (white lead).

Lead is a common constituent of plating wastes [50] and often the biggest problem with these materials. Another major source of lead that poses a real challenge to S/S technology is electric arc furnace dust (K061). This waste contains high lead levels, typically 0.5 to 5%, and often in very leachable forms. Lead is also a problem in other air pollution control residues and in bottom ashes from some sources. In certain cases, the lead is known to be speciated as the soluble chloride, in others as oxides and sulfates.

Lead from process wastes other than electroplating is especially prevalent in the petroleum industry, where it originates from refineries, tank bottoms, and other sources. These wastes may contain both inorganic and organic lead compounds. Actually, most waste residuals from industry contain at least small amounts of lead, which in many cases causes no leaching problems. In general, however, it can be said that lead is the most common and widespread problem encountered in metal immobilization in soils. Its pervasive presence and the many species in which it appears make each waste type, even each new source, an individual challenge. The newest regulatory leaching levels have exacerbated the problem.

The literature contains a great many data on the precipitation or stabilization of lead. As with the other metals, most of this comes from the water treatment area, but an increasing body of information is specific to stabilization of lead in S/S systems. The standard methods of lead removal from wastewater are to precipitate it as the hydroxide, carbonate, or basic carbonate, all of which are relatively insoluble at alkaline pH. Lead in wastewater has been precipitated out with lime and ferrous sulfate [71] and with dolomite ($CaCO_3 \cdot MgCO_3$) to yield lead carbonate [72]. Organic lead compounds in wastewaters can be oxidized to yield inorganic lead, which is precipitated at pH 8 to 9.5 in the presence of carbonates. Sulfides have also been used for this purpose, as has ferrous sulfate at pH 10.4 to 10.8. However, as we have seen, wastewater data are often not very useful in S/S work. The complex matrices and high ionic strengths create a very different environment for precipitation, and equilibrium concepts have limited applicability. We do know that pH control is important. Minimum lead leaching in nearly all S/S systems occurs when the pH is maintained between about 8 and 10 in the leachate [3]. A summary of the various processes that have been reported for lead in residues is given in the following list.

Portland cement
Portland cement + aluminum sulfate
Cement/soluble silicate
Potassium silicate
Cement/soluble silicate + sodium sulfide
Cement/soluble silicate + ammonium phosphate
Lime
Sulfide
Lime/fly ash
Kiln dust
Proprietary

IMMOBILIZATION OF ORGANICS IN SOIL MATRICES

There are many different types of organic-containing wastes that might be encountered in solidification/stabilization (S/S). If they are hazardous, organic-based or high organic water–based streams will normally be incinerated, chemically destroyed, or subjected to separation processes. However, aqueous inorganic waste streams containing up to about 1000 mg/L levels of hazardous organics are quite common and will become even more so as a result of EPA's most recent land disposal restrictions (LDRs), which require destruction of most listed organic-based wastes but set maximum leachable levels (Table 1) for characteristic wastes under the toxicity characteristic (TC) rule.

In the last few years, a number of S/S processes, mostly proprietary, have been offered that claim to fix, destroy, or immobilize organic priority pollutants and other organics of environmental concern. Unfortunately, few credible data have been available. Low leaching levels are claimed without any indication of the initial levels in the waste or the effects of dilution or volatilization of the constituents. Claims of actual destruction in a S/S matrix, especially of chlorinated aromatics, have not been supported by the rigorous testing and control programs that are essential to such controversial assertions. If such reactions do occur, there is the possibility of creating other toxic compounds that are not analyzed for in most studies. To correct this situation, two studies using carefully controlled conditions and analyses have recently been completed at Chemical Waste Management's Geneva Research Center, and additional work is in progress. The first study dealt with residues from the treatment of leachate from hazardous waste landfills and was presented in a recent paper [73]. Such leachates contain a variety of contaminants and must be treated before the water can be discharged or reused. The study was initiated both to evaluate leachate treatment processes and to test the effectiveness of several S/S formulations in immobilizing residual organics and metals. The most recent study on soils will be discussed later in this section.

Organic Immobilization Techniques

Before going into more detail about immobilization of organics, there are five distinct types of organic-containing wastes that might be encountered in stabilization:

1. Oil- and solvent-based wastes such as used solvent, distillation bottoms, and refinery wastes, which are hazardous according to RCRA, Appendix VIII [74], the waste listings, the CCW and CCWE tables, the "California list," and the landbans
2. Aqueous wastes containing large amounts (1 to 20% or more) of water-soluble or water-insoluble emulsified organics that are hazardous according to the regulations cited above
3. Aqueous wastes containing large amounts (1% to 20% or more) of water-soluble or water-insoluble emulsified organics that are not

hazardous or are hazardous only by the characteristic of ignitability, or are marginal, like oil

4. Aqueous wastes containing small amounts of nonhazardous organics (less than 1% and usually in the range 10 to 1000 mg/L), which are of interest in stabilization only when they affect cementitious and other reactions of the stabilization system
5. Aqueous wastes containing small amounts of hazardous organics (less than 1% and usually in the range 10 to 1000 mg/L)

The first type is of interest only in very specialized applications in which solidification is required temporarily for safety in transportation or storage, or in spill control work. These wastes will normally be incinerated if they are hazardous. Many of these wastes, and those in the second group, have been effectively removed from possible stabilization treatment by the landban regulations. The third group of organic-containing wastes comprises oily refinery wastes and other industrial residues, where the only question is the containment of the organic in the solid matrix over time. A considerable amount of stabilization work has been done commercially on this type of waste [75] and will probably continue in the future. The fourth waste type listed above affects only the solidification reactions, not immobilization per se.

The last waste group listed above is the one of real interest in the exploration of stabilization technology. Aqueous, inorganic waste streams containing 1 to 1000 mg/L levels of hazardous organics are quite common and will become even more so as a result of landban regulations which require destruction of organic-based wastes but leave some organic residue that may leach above the allowable TC levels shown in Table 1.

Organic Reaction Principles

Before discussing our most recent findings in this area, it is important to understand something about the reactions of organics, particularly those that can take place in the soil environment. The immobilization of organic constituents can be broken down into two primary classifications: (1) reactions that destroy or alter organic compounds, and (2) physical processes such as adsorption and encapsulation. This distinction is useful for discussion, but there is not a sharp line between the two. In some instances, investigators are finding that immobilization which was assumed to be due to adsorption now appears to involve some sort of chemical bonding or even conversion of the constituent into another compound. Analytical techniques other than those commonly used in the stabilization field are now being used to investigate mechanisms.

Reactions

The number of organic reactions that might occur in contaminated soil treatment is almost infinite. In practice, however, inorganic stabilization systems operating at ambient temperatures and pressures in nonexotic aqueous environments can produce only a relative few reaction schemes. Aside from adsorption, volatilization, and biodegradation, the most likely reactions fall into five categories: (1) hydrolysis, (2) oxidation, (3) reduction, (4) compound formation, and (5) fixation on an insoluble substrate. Some general and specific reactions within these categories are shown in Table 7. R indicates any organic grouping and X, a halide. The reaction products are not all shown in cases where they are not pertinent to the discussion, and the stoichiometry is not necessarily as stated.

Table 7 Possible Organic Reactions in Stabilization Systems

Reactants	Products[a]
Hydrolysis	
$RX + H_2O$	$ROH + HX$
Organoaminos	Organics + NH_3
Oxidation	
Phenol + $14H_2O_2 + Fe^{2+}$	$6CO_2 + 17H_2O$
$R{-}CH_3$	$R{-}COOH$
$R{-}CH_2OH$	$R{-}COOH$
$RCHOH{-}CHOHR'$	$R{-}COOH + R'{-}COOH$
$R{-}CHO$	$R{-}COOH$
R_2CH_2	R_2CO
$R_2CH(OH)$	R_2CO
R_3CH	$R_3C(OH)$
$R_3CH + HCR'_3$	$R_3{-}C{-}C{-}R'_3$
$R_2N{-}H + H{-}NR'_2$	$R_2N{-}NR'_2$
$RCH{=}CHR'$	$RCHOH{-}CHOHR'$
$2R{-}SH$	$R{-}S{-}S{-}R$
$R{-}S{-}S{-}R'$	$R'SO_3H + RSO_3H$
Reduction	
$Fe + 2H_2O + 2RCl$	$2ROH + Fe^{2+} + 2Cl^- + H_2$
Salt formation	
Oxalic acid	Calcium oxalate

[a]All reaction products are not necessarily listed; stoichiometry is not complete.

Physical Processes

Work on "physical" immobilization of low-level organics in aqueous systems has centered primarily on four materials and mechanisms: (1) activated carbon sorption, (2) fly ash sorption, (3) clay (modified and unmodified) sorption, and (4) sorption on, or encapsulation in, stabilization reagents and other additives.

Some recent private work by Cote [76] has shown that a variety of organics can be sorbed fairly effectively in cement-based stabilization processes incorporating activated carbon and bentonite additives; fly ash and soluble silicates were less effective. Much of the current work with clays has centered on the use of modified clays. These clays are natural products that have been altered by various organic chemicals, usually quaternary organoammonium compounds. This alteration changes clay surfaces from hydrophilic to hydrophobic, increasing the sorptive capacity for organic constituents in wastes. Many of the modified clays are available commercially. Although most of this work was done with wastewater treatment concepts in mind, the materials may be useful in stabilization if their cost is reasonable. These products have the possibility of immobilizing organics so that they will not desorb under the conditions of use and disposal practices; some appear to catalyze reactions that might detoxify the organics (see earlier discussions in this chapter). One possible problem with commercial organoclays is cost.

Recent Developments and Advances

Chemical Waste Management, Inc. has been engaged in several major programs dealing with organic stabilization. Our most recent organic stabilization research program [77], and the one most pertinent to the subject of this book, deals with contaminated soil and debris (CS&D), a waste type of special interest to EPA at this time. Using clean soil spiked with a mixture of hazardous organic chemicals, including all TC compounds and representatives of all major chemical classes, 12 different stabilization formulations were tested in the first phase of the program. All samples, untreated and treated, were analyzed for both total and leachable organics, comprising most of the 231 RCRA BDAT constituents: volatile organics, semivolatile organics, and pesticides. Leachable constituents were determined as those that are extracted from the solid in the toxicity characteristic leaching procedure (TCLP). Total analyses were used to help determine the fate of organic compounds retained in the matrix.

Uncontaminated soil was obtained from the surface horizon of a fine-silty, mixed, mesic Mollic Hapludalf (Batavia silt loam, 2 to 5% slope). Portions of the clean soil were spiked with approximately 50

Table 8 TCLP Leachate Analysis (μg/kg): Admixtures A to D

Compound	Spiked sample	Admixture[a]				
		None	A	B	C	D
Acetone	120,500	66,000	55,000	49,500	68,500	38,600
Benzene	26,500	4,500	<1,250	6,000	10,400	2,500
2-Butanone	68,000	40,000	6,000	25,000	40,500	20,500
Carbon disulfide	5,000	<2,500	<1,250	<2,500	<2,500	<2,500
Carbon tetrachloride	18,000	<2,500	<1,250	2,500	5,000	<2,500
Chlorobenzene	35,500	19,000	<1,250	10,000	19,000	22,900
Chloroform	48,500	65,00	63,000	4,500	13,000	<2,500
1,2-Dichloroethane	68,500	15,500	13,800	17,000	25,000	<2,500
1,1-Dichloroethylene	<2,500	<2,500	<1,250	<2,500	<2,500	5,500
Ethyl acetate	29,500	<2,500	<1,250	<2,500	<2,500	<2,500
Ethyl benzene	11,000	6,500	<2,500	<2,500	7,500	10,700
Ethyl ether	<5,000	<5,000	<2,500	<5,000	<5,000	<5,000
Methylene chloride	15,000	<2,500	2,500	<2,500	<2,500	<2,500
4-Methyl-2-pentanone	104,500	58,500	<2,500	<5,000	66,500	36,500
Tetrachloroethylene	22,000	8,500	<1,250	15,000	13,500	13,900
Toluene	33,500	11,000	<1,250	10,000	17,500	<2,500
1,1,1-Trichloroethane	34,500	4,000	1,500	4,000	9,000	<2,500
Trichloroethylene	44,500	8,000	<1,250	6,000	15,500	5,500
1,1,2-Trichloro-1,2,2-trifluoroethane	<2,500	<2,500	<1,250	<2,500	<2,500	<2,500
Total xylenes	11,000	7,000	<1,250	4,500	7,000	11,800
n-Butanol	55,000	<100	60,500	<100	70,000	<100
Cyclohexanone	<100	<100	<100	<100	<100	<100
Isobutyl alcohol	<100	<100	<100	<100	<100	<100
Methanol	232,960	115,360	50,400	96,000	1,218,000	81,000
Anthracene	32	21	<20	<20	<20	<20
Bis(2-ethylhexyl)phthalate	516	155	316	588	<20	<20
1,2-Dichlorobenzene	2,100	2,350	<20	393	1,040	3,030
1,4-Dichlorobenzene	1,860	2,000	<20	1,390	1,050	2,730
2,4-Dinitrotoluene	6,260	10,600	<20	426	5,220	37,600
Hexachlorobenzene	<20	<20	<20	<20	<20	<20
Hexachlorobutadiene	92	101	<20	58	27	126
Hexachloroethane	364	467	<20	302	119	679
2-Methylphenol	<100	<100	<100	<100	<100	<100
4-Methylphenol	12,200	10,500	220	4,500	10,000	11,600
Naphthalene	2,050	2,370	<20	252	1,410	3,340
Nitrobenzene	11,600	6,070	54	3,580	2,500	38,500
Pentachlorophenol	2,200	5,700	<200	<200	8,400	12,100
Pyridine	26,700	33,900	28,200	34,500	9,140	6,860
2,4,5-Trichlorophenol	6,300	11,700	<500	530	22,600	13,200
2,4,6-Trichlorophenol	10,800	16,200	280	1,120	20,000	16,700

Table 8 Continued

Compound	Spiked sample	Admixture[a] None	A	B	C	D
Phthalic anhydride (as acid)	<1000	<1000	<1000	<1000	<1000	<1000
1-Naphthylamine	<20	4,230	<20	<20	3,700	10,600
γ-BHC (Lindane)	1,500	2,300	<400	<400	407	3,000
Heptachlor	400	<8	<8	<8	<8	60
Endrin	300	30	<20	<20	<20	30
Methoxychlor	<10,000	<10,000	<10,000	<10,000	<10,000	<10,000
Chlordane	<30	<30	<30	<30	<30	<30
Toxaphene	<500	<500	<500	<500	<500	<500
2,4-D	<10	<10	<10	<10	<10	<10
2,4-5-TP (Silvex)	<1	<1	<1	<1	<1	<1

[a]All admixtures bound with cement.

organic compounds and mixed in a sealed mixer. Sufficient spiking material was applied so that the spiked soil would contain approximately 100 mg/kg of each constituent after mixing. Actual levels varied between about 5 and 2400 mg/kg as measured in the TCA analysis. Total organic level in the soil was between 0.5 and 1.0%. Combinations of cement alone, and cement plus one of 11 different admixtures, were mixed with portions of the spiked soil. The admixtures selected were those employed previously for the stabilization of organics or believed to be useful for that purpose. All formulations contained cement at 20% by weight (expressed as weight of reagent to weight of raw soil). Admixtures were made at 10% by weight. The spiked soil samples and the stabilization treatment products were completely characterized by both TCA and TCLP.

We found that stabilization decreased leachable organic constituent levels in most cases, but the degree of success depended heavily on the formulation used. The TCLP results for the study are given in Tables 8 and 9. Because it is very confusing to compare these column of numbers, we found that the best way to evaluate results with respect to the ability of stabilization to immobilize low-level organics is to look at *reduction factors*[†] rather than the absolute quantities. The reduction factors for TCLP levels are given in Tables 10 and 11 for organic constituents

[†]The reduction factor is the ratio of constituent before stabilization to constituent after stabilization, taking dilution into account.

Table 9 TCLP Leachate Analysis (μg/kg): Admixtures E to K

Compound	Spiked sample	Admixture[a]						
		E	F	G	H	I	J	K
Acetone	66,000	<500	18,250	17,250	26,750	14,750	20,250	8,850
Benzene	4,900	<250	6,500	9,950	7,750	5,300	<250	2,250
2-Butanone	64,000	<500	17,000	17,250	19,000	13,250	13,250	9,800
Carbon disulfide	8,200	<250	<250	<250	<250	<250	<250	<250
Carbon tetrachloride	32,000	<250	7,800	8,200	4,350	3,250	300	600
Chlorobenzene	<250	5,500	30,250	20,500	12,750	<250	<250	8,700
Chloroform	<250	<250	6,000	11,000	9,200	5,300	1,100	1,450
1,2-Dichloroethane	6,550	<250	13,500	21,000	26,000	15,500	3,450	5,300
1,1-Dichloroethylene	<250	<250	<250	250	<250	<250	<250	<250
Ethyl acetate	11,500	<250	<250	250	<250	<250	<250	<250
Ethyl benzene	16,000	3,050	<17,000	9,250	5,350	7,850	1,250	2,800
Ethyl ether	27,250	<500	1,450	1,900	900	<500	<500	<500
Methylene chloride	12,000	<250	500	<250	800	<250	450	<250
4-Methyl-2-pentanone	77,000	<500	30,000	30,500	40,000	28,000	3,650	28,250
Tetrachloroethylene	24,000	1,200	19,750	9,750	5,950	9,750	<250	2,250
Toluene	32,250	1,450	15,250	13,750	9,000	9,250	<250	4,450
1,1,1-Trichloroethane	13,250	<250	<250	<250	2,050	<250	<250	<250
Trichloroethylene	40,000	<250	10,750	12,750	8,450	7,550	<250	2,400
1,1,2-Trichloro-1,2,2-trifluoroethane	3,050	<250	<250	<250	<250	<250	<250	<250
Total xylenes	11,500	2,900	14,250	7,450	4,200	6,150	<250	2,300
n-Butanol	43,750	<500	31,400	31,450	61,250	30,150	11,950	25,700

Cyclohexanone	11,100	<500	7,100	5,700	9,300	<5,000	<100	5,250
Isobutyl alcohol	47,700	<500	32,050	28,450	50,300	26,900	38,500	26,050
Methanol	268,800	14,493	119,100	105,042	128,686	77,583	74,872	64,868
Anthracene	<20	<20	226	<20	<20	<20	<20	<20
Bis(2-ethylhexyl)phthalate	<20	<20	220	<20	<20	22	<20	<20
1,2-Dichlorobenzene	1,960	630	3,160	714	450	486	<20	526
1,4-Dichlorobenzene	1,460	291	2,410	486	333	577	<20	303
2,4-Dinitrotoluene	4,480	1,680	4,370	7,990	2,910	474	<20	8,020
Hexachlorobenzene	<20	<20	196	<20	<20	<20	<20	<20
Hexachlorobutadiene	137	158	235	<20	20	44	<20	<20
Hexachloroethane	378	172	800	94	102	128	30	60
2-Methylphenol	<100	<100	<100	<100	<100	<100	<100	<100
4-Methylphenol	11,600	<100	14,200	11,000	14,800	4,500	<100	11,500
Naphthalene	2,020	1,450	2,610	745	518	406	140	754
Nitrobenzene	6,050	2,960	6,730	4,890	3,510	4,520	<20	4,600
Pentachlorophenol	2,330	790	14,200	6,800	12,200	<200	<200	4,500
Pyridine	30,200	5,320	24,900	41,100	33,800	32,200	25,200	34,700
2,4,5-Trichlorophenol	8,150	140	18,200	<500	18,000	<500	220	11,500
2,4,6-Trichlorophenol	13,700	<100	16,600	<100	17,200	<100	1,180	12,100
Phthalic anhydride (as acid)	<1000	1,510	<100	<1,000	<1,000	<1,000	<1,000	<1,000
1-Naphthylamine	<20	<20	17,900	16,900	8,320	928	<20	11,600
γ-BHC (Lindane)	<200	456	340	<200	<200	<200	<200	<200
Methoxychlor	<200	<200	<200	<200	<200	<200	<200	<200
2,4-D	<1,000	<5,000	<5,000	<5,000	33,000	<5,000	<5,000	<1,000
2,4,5-TP (Silvex)	<100	<500	<500	<500	28,000	<500	<500	<100

[a]All admixtures bound with cement.

Table 10 Reductions in TCLP Levels After Stabilization: Admixtures A to D[a]

Compound	Admixture				
	None	A	B	C	D
Acetone	NSR	2	2	NSR	3
Benzene	6	>21	4	3	11
2-Butanone	NSR	11	3	NSR	3
Carbon disulfide	>2	>4	>2	>2	>2
Carbon tetrachloride	>7	>14	7	4	>7
Chlorobenzene	NSR	>28	4	NSR	NSR
Chloroform	7	NR	11	4	>19
1,2-Dichloroethane	4	5	4	3	>27
Ethyl acetate	>12	>24	>12	>12	>12
Ethyl benzene	NSR	>4	>4	NSR	NR
Methylene chloride	>6	6	>6	>6	>6
4-Methyl-2-pentanone	2	>42	>21	NSR	3
Tetrachloroethylene	3	>18	2	NSR	NSR
Toluene	3	>27	3	NSR	>13
1,1,1-Trichloroethane	9	23	9	4	>14
Trichloroethylene	6	>36	7	3	8
Total xylenes	NSR	>9	2	NSR	NR
Methanol	2	5	2	NR	3
Bis(2-ethylhexyl)phthalate	3	NSR	NR	>26	>26
1,2-Dichlorobenzene	NR	>105	5	2	NR
1,4-Dichlorobenzene	NR	>93	NSR	NSR	NR
2,4-Dinitrotoluene	NR	>313	15	NSR	NR
Hexachlorobutadiene	NR	>5	NSR	3	NR
Hexachloroethane	NR	>18	NSR	3	NR
4-Methylphenol	NSR	215	3	NSR	NSR
Naphthalene	NR	>103	8	NSR	NR
Nitrobenzene	NSR	55	3	NSR	NSR
Pentachlorophenol	NR	>11	>11	NR	NR
Pyridine	NR	NR	NR	11	NR
2,4,5-Trichlorophenol	NR	>13	12	NR	NR
2,4,6-Trichlorophenol	NR	39	10	NSR	NR
γ-BHC (Lindane)	NR	>4	>4	4	NR
Heptachlor	>50	>50	>50	>50	7
Endrin	10	>15	>15	15	10

[a]Cement binder. NR, no reduction, constituent TCLP concentration for sample exceeded that in spike; NSR, no significant reduction in constituent TCLP concentration as compared to spike.

Table 11 Reductions in TCLP Levels After Stabilization: Admixtures E to K[a]

	Admixture						
Compound	E	F	G	H	I	J	K
Acetone	>120	4	4	2	4	3	7
Benzene	>20	NR	NR	NR	NR	>20	2
2-Butanone	>128	4	4	3	5	5	7
Carbon disulfide	>33	>33	>33	>33	>3	>33	>33
Carbon tetrachloride	>128	4	4	7	10	107	53
1,2-Dichloroethane	>26	NR	NR	NR	NR	NSR	NSR
Ethyl acetate	>46	>46	>46	>46	>46	>46	>46
Ethyl benzene	5	NR	NSR	3	2	13	6
Ethyl ether	>55	19	14	30	>55	>55	>55
Methylene chloride	>48	24	>48	15	>48	27	>48
4-Methyl-2-pentanone	>154	3	3	NSR	3	21	3
Tetrachloroethylene	20	NSR	2	4	2	>96	11
Toluene	22	2	2	4	3	>129	7
1,1,1-Trichloroethane	>53	>53	>53	6	>53	>53	>53
Trichloroethylene	>160	4	3	5	5	>160	17
1,1,2-Trichloro-1,2,2-trifluoroethane	>12	>12	>12	>12	>12	>12	>12
Total xylenes	4	NR	NSR	3	NSR	>46	5
n-Butanol	>9	NSR	NSR	NR	NSR	4	NSR
Cyclohexanone	>2	NSR	NSR	NSR	>2	>2	2
Isobutyl alcohol	>10	NSR	NSR	NR	NSR	NSR	NSR
Methanol	19	2	3	2	3	4	4
1,2-Dichlorobenzene	3	NR	3	4	4	>98	4
1,4-Dichlorobenzene	5	NR	3	4	3	>73	5
2,4-Dinitrotoluene	3	NSR	NR	NSR	9	>224	NR
Hexachlorobutadiene	NR	NR	>7	7	3	>7	>7
Hexachloroethane	2	NR	4	4	3	13	6
4-Methylphenol	>116	NR	NR	NR	3	116	NR
Naphthalene	NSR	NR	3	4	5	14	3
Nitrobenzene	2	NR	NSR	NSR	NSR	>30	NSR
Pentachlorophenol	3	NR	NR	NR	>12	>12	NR
Pyridine	6	NSR	NR	NR	NR	1	NR
2,4,5-Trichlorophenol	58	NR	>16	NR	>16	37	NR
2,4,6-Trichlorophenol	>137	NR	>137	NR	>137	12	NSR

[a]Cement binder. NR, no reduction, constituent TCLP concentration for sample exceeded that in spike; NSR, no significant reduction in constituent TCLP concentration as compared to spike.

found in the TCLP extract of the spiked samples at levels above the limit of quantitation. The larger the TCLP reduction for a given constituent, the more effective the immobilization, although volatility should be consider for the volatile constituents. Admixtures A and J are effective for organic immobilization for volatiles, semivolatiles, and pesticides. Admixture E is effective for volatiles; admixture G is effective for semivolatiles. Other admixtures seem effective for certain specific organic compounds or compound classes.

Certain constituents appear to be more difficult than others to immobilize. These include: acetone, 1,2-dichloroethane, ethyl benzene, tetrachloroethylene, *n*-butanol, methanol, and pyridine. Other constituents were easily immobilized by all formulations: carbon disulfide, ethyl acetate, methylene chloride, 1,1,2-trichloro-1,1,2-trifluorethane, 1,2-dichlorobenzene, 1,4-dichlorobenzene, hexachlorobutadiene, and hexachloroethane. However, all constituents except pyridine could be treated to meet the TC requirements.

With respect to total concentrations of organics (total constituent analysis or TCA) in stabilized soil, one would not expect any reduction except for volatilization of the VOCs and, perhaps, destruction of reactive constituents. We found, however, that several admixtures—G and K—did appear to be effective at reducing the measurable total levels of semivolatile constituents. These reductions may only reflect changes in the extraction efficiency of the analytical methodology as effected by the formulation, but it is also possible that chemical bonding or other changes in some constituents may have occurred. TCA reductions for the volatile constituents are difficult to interpret. TCA reduction of carbon disulfide appears to be due to volatilization, while ethyl acetate, because of its reactivity, may be chemically altered. No TCA reduction is seen for some of the more volatile constituents, suggesting that these compounds are strongly sorbed on the soil, additive, and/or cement, reducing their partitioning into the vapor phase.

The lack of TCA reduction for many volatile organics contradicts some previous studies [78]. The most apparent difference between the present study and many other studies is that here the organic constituents were mixed into the soil prior to stabilization treatment. This suggests that organic compounds sorbed onto or associated with soil particles, which is the way that organic-contaminated soils occur in practice, may be less susceptible to volatilization during stabilization than thought previously. Further study into the volatilization of organics from contaminated soil and debris during stabilization is necessary.

CONDUCTING STABILIZATION TREATABILITY STUDIES

Although much has been learned about the chemistry of S/S in the last 10 years, it is still largely an empirical science. Therefore, formulation requires a treatability study to select and optimize the S/S system for a particular waste problem. The treatability study, in turn, requires such adjuncts as proper sampling, problem definition, waste handling, waste characterization, and testing of the product.

Sampling

Very often, the most difficult part of S/S testing is to obtain a representative sample of the waste. This is especially true in remediation work, where large sludge ponds and/or poorly defined areas of contaminated are involved. The official methods for sampling are given in the USEPA's manual, better known as SW-846 [79]. Exner describes a sampling strategy that is useful [80]. Other methods and techniques have been described by Brantner [81] and in various ASTM draft documents [82–84]. The sampling plan is generally based on a grid system such as that taken from USEPA SW-846 and shown in Figure 10. Such plans are straightforward; the actual sample taking is not. Waste sites are usually not neatly rectangular, nor are the sides vertical. In many cases it is difficult to determine where the waste ends and the surrounding, uncontaminated native soil or fill begins. In fact, a grid sampling program must often be completed to determine by analyses of the samples the contamination boundaries within which remediation is necessary. Such sites are also normally stratified vertically and are nonuniform horizontally due to flow patterns from the waste outfall, dumping, or spill. For those inexperienced in sampling, there are a number of consulting and engineering organizations that specialize in this area of remediation. They have the necessary equipment and expertise to do a professional job and are listed in various hazardous waste services and equipment guides.

When undertaking a sampling program, even with a professional and experienced consulting firm, it is important to make clear at the outset the purpose of the sample taking. Much of this work has been done for Superfund remedial investigation and feasibility studies (RIFSs), where the primary purpose is waste characterization. Samples taken for this purpose are often not suitable for S/S treatability testing, due to lack of sufficient quantity or the use of preservatives for analytical purposes. For treatability work, at least 5 gallons of waste should be taken from every sampling location, and more if a wide variety of

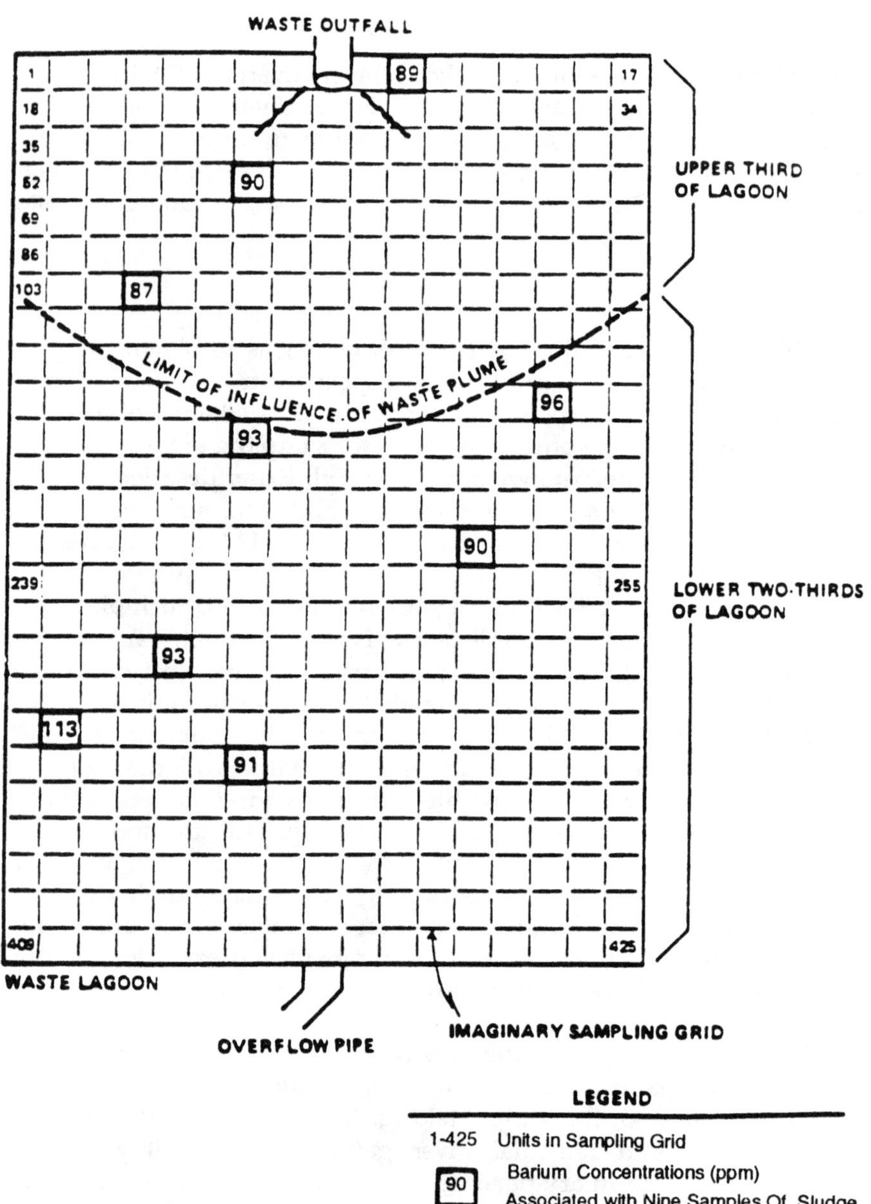

Figure 10 Grid sampling plan.

treatment technologies is to be investigated or if it would be very difficult and expensive to get additional samples later.

It is a sad but common occurrence to find a multimillion-dollar remediation project predicated on one or two grab samples not representative of the problem as a whole. This often happens where a very thorough and expensive characterization had been done, but the samples were not saved or were insufficient for treatability testing. The type of tests conducted and data generated in characterization cannot substitute for actual treatability testing. If the sample subjected to S/S treatability testing is not representative of the waste to be treated, extrapolation of results to a different waste composition is very risky. Finally, the sample must be properly packaged, shipped, and handled, and information about it properly recorded, maintained, and stored. Acceptable procedures for all of this are contained in SW-846 and other USEPA documents.

Waste Characterization

The general information that is required for proper characterization of a waste stream depends not only on the waste itself, but also on the purpose of the S/S (or any other) treatment program. Some of the information is obtainable from observation alone or is specified in the transmittal documents. A complete list of waste characteristics appropriate to S/S treatability testing is given in Table 12. It is often useful to organize all or part of the waste characteristic information shown in Table 12 into a checklist for use by persons who obtain the information.

Stabilization Formulation

S/S formulation procedures are deceptively simple in appearance, involving little in the way of sophisticated measuring devices or complex equipment. There are no "standard" (i.e., USEPA, ASTM, etc.) test methods for S/S work. However, most workers in the field use similar techniques [86] for S/S testing and development, modified or varied to suit the specific process and application. The method used in this phase of the treatability study is to make up a number of formulations on small aliquots of the waste, choosing the mixtures on the basis of waste characteristics, experience from previous S/S treatability testing, and the judgment of the formulator [3]. These experential and judgmental inputs can be aided greatly by the use of a computer database system. The goal at this point of testing is to minimize the number of formulations to be made and tested, especially if the testing is extensive and expensive, as it is when evaluating leachability. Often, the experimental design deliberately contemplates an iterative process, establishing the best

Table 12 Waste Characteristics

Characteristic	Notes[a]
Waste name	(1)
Generating process	
Industry	
USEPA hazardous waste number(s)	(1)
USEPA handling methods number	
Storage	(1)
Treatment	
Disposal	(1)
NFPA hazard identification	
Flammability	(2)
Health	(2)
Reactivity	(2)
Special	(2)
Toxicity rating	
Inhalation	(3)
Dermal	(3)
Oral	(3)
Annual generation rate (gal)	
Quantity stored (gal)	
Type of storage	
Medium (aqueous, oil, solvent, etc.)	
Physical state (solution, sludge, powder, etc.)	
Phases/layering (none, bilayered, multilayered)	
Total solids (%)	(4) Part A
Suspended solids (%)	(4)
Type of solids (organic, inorganic, mixed)	
Bulk density	
Grain size	
Specific gravity	(5)
Bulk density (for solids)	
Grain size distribution	
Viscosity	(6)
Flash point	(7) Method 1010
pH	(7) Method 9040
Total alkalinity/acidity (%)	
Odor	
Color	
Waste analysis	
Metals, total	(7) Method 6010
Metals, leached	(8)

Table 12 Continued

Characteristic	Notes[a]
Inorganics, total	(7) Methods 9030 and 9010
Inorganics, leached	(8)
Organics, total	(7) Methods 8240, 8250, and 8080
Organics, leached	(8)

[a](1) USEPA, various publications in Federal Register.
(2) National Fire Prevention Association.
(3) OSHA.
(4) Standard Methods for the Examination of Water and Wastewater [85].
(5) ASTM D854-83.
(6) See discussion later in this chapter.
(7) USEPA SW-846; also, appropriate methods in Standard Methods of Examination of Water and Wastewater for non-RCRA constituents.
(8) Appendix I to 40 CFR Part 268, 51 Fed. Reg. 40643-40652 (Nov. 7, 1986).

general formulation type in the first round of formulations and tests, and then optimizing it in the second or third rounds. If the formulation system involves two or more components, it is desirable to use a statistical experimental design method such as a factorial or simplex design [87]. In addition to meeting the project specifications listed below, some of the other factors that may be important in establishing boundary conditions for the formulation work are:

Allowable curing time
Maximum volume increase
Allowable cost
Type of mixing: batch or continuous
Heat generation due to chemical reactions
Gas generation by volatilization or chemical reaction
Necessary contact time in the mixer: minimum or maximum
Restrictions due to rapid gel reactions
Handling characteristics and hazards of reagents
Handling characteristics of the waste

After curing, the formulations are then tested by a variety of methods, as prescribed by the specifications for the final product. These specifications may include some or all of the following:

Reactivity
Ignitability
Corrosivity

Table 13 Chemical Tests

SW-846	pH (Method 9045)
40CFR 261.21	Ignitability
40CFR 261.22	Corrosivity
40CFR 261.23	Reactivity
ASTM G-21, 22	Biological activity
ASTM C-186	Heat of hydration
	Total solids and ash (Method 209G)
	Acidity/Alkalinity
Canadian [88]	Sequential chemical extraction
ASTM D-1498-76	Oxidation-reduction potential

Leachability
Physical strength and stability
Permeability
Special requirements

Chemical Tests

The tests listed in Table 13 are, with several exceptions, standard methods familiar to those in the environmental fields. The exceptions are tests for biological activity, which have not been applied in S/S but may be in the future, and heat of hydration, which is applicable to the testing of reagents such as kiln dusts, fly ashes, and quick lime.

Leaching Tests

The primary objective of stabilization is to immobilize (or destroy, in the case of certain inorganic anions and organics) constituents in the waste. Leachability testing is used to predict the degree to which this objective has been accomplished; it was discussed earlier in this chapter. In view of the variety of possible landfill scenarios, it is not surprising that no single leachability testing procedure or protocol can duplicate all possible field conditions. Ideally, the treated waste would be leach tested with the actual surface, ground, or rainwater that is present at the site. In practice this is rarely possible, both because of lack of definitive knowledge about what the conditions really are and because of regulatory philosophy. Therefore, standard leachability tests have been developed and promulgated by USEPA and several states. Also, Canada and other countries have developed their own methodologies. A number of these test methods are listed in Table 14.

Table 14 Leaching Tests

USEPA test procedures
EP toxicity test (EPT)
Toxicity characteristic leaching procedure (TCLP) [89]
Multiple extraction test MEP [90]
Oily waste extraction test (OWEP) [91]
State and province test procedures
California waste extraction test (WET) [92]
Ontario Province test
ASTM test procedures
D-3987
Other
Diffusion modeling tests
ANS 16.1
MCC Tests
Static leaching test (DLT) [93]
Sequential batch extraction tests
Solid waste leaching procedure (SWLP)
Flow-through tests
European and Japanese test procedures
Special and experimental tests
Equilibrium leach test (ELT)

Analysis

The analytical methods used almost universally in the United States are contained and described in SW-846.

Physical Tests

There are a wide selection of ASTM, U.S. Department of Transportation (USDOT), U.S. Army Corps of Engineers (USACOE), and other methods that could be applied to physical testing in the hazardous waste area. These methods were designed to test the properties of a material for a *specific* purpose, usually not related directly to hazardous waste. Therefore, they must be applied with great caution, and modifications are usually required both in the methodology and in the specification values that may be set. Unfortunately, regulatory agencies and users have sometimes specified inappropriate methods or achievement levels. This may have a number of serious ramifications: excessive cost, unnecessary volume increase resulting in excessive use of limited landfill space, and production of landfill with undesirable operating

Table 15 Physical Tests

ASTM D-2166	Unconfined compressive strength of cohesive soil
ASTM D-1633	Unconfined compressive strength of non-cohesive soil
ASTM C-109-86	Method for molding samples for UCS testing [94]
EM 1110-2-1906	USACOE unconfined compressive strength
	California bearing ratio
	Pocket penetrometer
ASTM D-3080	Shear test
ASTM D-2573	Shear test
ASTM D-4318-84	Atterberg limits
ASTM D-698	Compaction
ASTM D-422	Grain size analysis
EM-1110-2-1906	USACOE grain size analysis
ASTM D-698	Water content, moisture-density
ASTM D-2206	Water content, moisture-density
ASTM D-2216	Water content, moisture-density
ASTM D-558-82	Water content, moisture-density
EM-1110-2-1906	USACOE porosity/void ratio

characteristics. The following methods, categorized into types, have been or might be used in S/S testing.

Compressive or Bearing Strength

As measured by procedures such as ASTM D-2166, unconfined compressive strength is meaningful only on cohesive materials. Therefore, its use for specification of suitable bearing strength in a waste landfill is generally improper. If, however, the treated waste were to be used for embankments, retaining walls, or other applications requiring a certain level of shear strength, this kind of test could be applicable, along with other test methods listed in Table 15.

Physical Stability

A controversial area of physical testing of S/S products involves evaluation of how well a material holds up under repeated wet–dry and freeze–thaw cycling. While this property is obviously important in road and building construction, its relevance to waste treatment and disposal is questionable. Today, even after S/S treatment, hazardous wastes are normally disposed of in landfills specifically designed for this purpose. Properly designed and located landfills are subjected to such cycling only for a limited period during the filling of the cell, if at all. The effect of wet–dry of freeze–thaw cycling would be the breakdown of physical integrity, a possibility that is taken into account in the design of regulatory standard leaching tests such as the TCLP. Therefore, such tests are meaningful only if (1) the product is to be reused, not landfilled, or (2) the landfill is designed or located in such a way that such cycling is a valid operating mode. In these cases, specific physical and mechanical tests would be applied according to the end use. Tests that have been used to measure these properties are ASTM D-559-82 (wet–dry durability test) and ASTM D-560-82 (freeze–thaw durability test).

Permeability

This property is frequently specified in S/S technology. Contrary to general belief, low permeability is not necessarily a desirable property in a landfill, RCRA or otherwise. Very low permeability can result in standing water in the landfill cell while it is being filled, making workability very difficult. Also, permeability is not necessarily related to leachability. Several test methods are ASTM D-2434 (Permeability), EM-1110-2-1906 (USACOE—Permeability), and SW-846 Permeability (Method 9100).

Miscellaneous Physical Test Methods

A variety of other physical tests are applicable to the characterization of hazardous waste, solidified or not. Many of these are commonly used in engineering and economic calculations, especially in doing material balances for a system design or comparing the relative total costs of various technologies. The most common methods are listed in Table 16.

DELIVERY SYSTEMS

Delivery systems are the means by which S/S processes are carried out. They include not only equipment but the complete process of developing, designing, planning, permitting, operating, controlling, and financing a S/S project. The project may be remedial or continuing; the equipment may be mobile/portable or fixed; the regulatory structure

Table 16 Miscellaneous Tests

EM-1110-2-1906	USACOE Appendix II: dry density and bulk density
ASTM D-854-83	Bulk density
APHA, ASTM D-854	Specific gravity
	Total solids
APHA	Moisture (Method 209A)
SW-846	Free liquid (paint filter test, Method 9095)
	Water soak tests
ASTM D-2216-80	Water content

under RCRA or CERCLA. The equipment may be designed, built, owned, and operated by a S/S vendor, a generator, a central RCRA TSD facility, or even a public entity, or any combination of these elements. The ultimate disposal of the S/S product may be part of the delivery system, or it may be totally separate. All of these are considerations that affect every phase of S/S technology, and should be kept in mind when designing equipment or conducting a laboratory treatability study. If the process cannot be operated or the equipment used because of limitations in other areas, the laboratory work has gone for naught.

There are two primary types of delivery systems from the mechanical point of view: on-site operations using mobile/transportable equipment, and fixed installations. The latter category is more familiar in the context of RCRA TSDFs and is discussed only briefly here since we are focusing on remedial activities that normally use mobile/ transportable systems.

Fixed Systems

S/S system operation at a fixed installation is usually only one of a number of technologies used in a central treatment facility. A large central treatment facility may receive nearly every kind of hazardous and nonhazardous liquid waste, with the exception of radioactive materials, explosives, or certain military wastes. An example of a typical system was shown in the process flow diagram of Figure 3.

On-Site Systems: Engineered Approach

On-site systems consist not only of mobile or transportable equipment, installation, labor, and support services, but also involve a number of elements that do not come into play in fixed operations. The difference between *mobile* and *transportable* is more than just semantics, but they are very similar in many respects. Mobile is generally taken to mean that the equipment is on wheels and that the whole operation can be rapidly moved, set up, and put into operation at a new site, often within a few days. Transportable, or portable, operations, on the other hand, may be broken down into a number of segments that must be transported separately and are assembled at the operational site. This process may require weeks, even months, to complete.

At a minimum, a S/S on-site service utilizes a treatment unit containing the chemical storage, metering, and mixing equipment necessary to mix the waste with the S/S reagents and discharge it to a holding or disposal area. It may also provide a means of removing the waste from its storage containment, homogenize it, and convey it to the treatment unit. Alternatively, the latter elements may be leased at the site or subcontracted to another party. The technical and operational activities involved in operating a S/S on-site service are, in normal chronological sequence, the following:

1. Obtaining samples of the waste
2. Preliminary laboratory testing
3. Preliminary quote
4. Meeting with customer, field sampling, and preliminary meeting with the regulatory agency
5. Final laboratory solidification, leaching and physical tests
6. Firm quotation to customer
7. Regulatory approval
8. Mobilization
9. Setup at job site
10. Treatment of the waste
11. Close-down and cleanup at job site and return to home base
12. Final laboratory leaching and physical tests on solid produced in job to satisfy contract requirements and protect warranty
13. Possible follow-up sampling and laboratory testing of solidified waste at various times if required by contract or desired by S/S contractor for information or warranty protection

The process of putting everything together into a system that provides a sound, practical scientific and engineering basis for selection of the appropriate S/S solution for a given problem has been called "the engineered approach" [22,95] because it approaches the project from an engineering point of view: definition, solution, and implementation. It has five basic steps:

1. Definition of the problem and collection of samples and data on the important design parameters and project considerations
2. Analysis of the data and comparison with past experience
3. S/S process testing in the laboratory
4. Regulatory interaction and approval
5. Implementation of the solution

Project Definition

When a potential project is proposed, there is often a tendency to leap right in and start doing treatability tests, or even start designing equipment and planning processing. This is invariably a mistake. Even before the site is sampled, it is essential to determine and clearly state the purpose of the remediation and the boundary conditions under which it is to be accomplished. Important considerations are:

Regulatory program: CERCLA, RCRA, state, local?
Disposal of product: on site or off site?
Required properties of stabilized soil
  Test methods
  Criteria
Ultimate use of site after remediation
Geology, hydrology
Operating constraints: noise, odor, hours of operation, access, etc.
Preconceived constraints: in situ versus ex situ
Timing
Allowable cost range

Waste

We have already discussed the importance of proper sampling of the site and characterization of the waste. However, it is reiterated here because so often the treatability study is begun without full knowledge of waste characteristics or on an inappropriate sample. In addition to treatability of the waste, the user must be aware of any health hazards associated with its handling. Common hazards to consider are inhalation toxicity due to vapors, dermal toxicity, corrosiveness (to both human beings and equipment), pyrophoric activity, exothermic reactions (especially occurring during the mixing of the chemicals with the waste), infection from biological sludges (especially sewage sludge), and unpleasantness associated with foul smell. These problems can all be managed properly by using such personal protective equipment as goggles, masks, gloves, boots, disposable protective clothing, and inhalators, and by proper design of operating equipment. Information on health effects, reactive hazards, and compatibility of different wastes with each other and with the S/S reagents can be obtained from sources such as Sax [96] and Bretherick [97]. Obviously, it is necessary to know the composition of the waste to use this information; this again emphasizes the importance of *complete* waste characterization as the vital first step in choosing a S/S system.

Waste Source

Factors associated with the site from which the waste will come, while most often affecting cost, can sometimes dictate the choice of the S/S system, the disposal site, and even determine whether or not S/S technology is the proper approach. Factors to consider include:

Waste pond, pile, etc.: its size, age, condition, physical geometry, accessibility, availability of utilities, and climate

Transport to S/S system: distance, topography, obstacles (railroad tracks, roads, pipelines, etc.), necessary delivery rate, type of transport system (pump, conveyor, pipeline, etc.)
Process mode: continuous, semicontinuous, batch
Waste removal system: dredge, pump, dragline, clamshell

This list is not all-inclusive, but it outlines areas to consider and also helps indicate the overall complexity of a total S/S system.

Treatment Process: Choosing the Right System

The different types of processes were discussed previously. Starting with that general knowledge, the following factors may be used in making an initial choice as to the specific process or processes to be tested in the treatability study.

Chemical and operational cost
In situ or ex situ mixing
Required redundancy for backup
Space requirements
Labor and personnel considerations
Occupational health considerations (dust, noise, toxicity)
Throughput rate
Waste feed system
Mixing system
Reagent availability
Reagent storage requirements
Reagent feed system
Utility and power requirements
Setting and curing times
Handling of solid after solidification
Volume increase

For example, knowing that minimum volume increase was very important because of disposal space limitations on the site, one would not choose a system that utilizes large amounts of bulking agents, such as fly ash, even if the total reagent cost were lower. Conversely, the availability of large amounts of local fly ash at little or not cost would make cement–fly ash a good candidate for a site with sufficient space. If the job specifications called for a short curing time before testing, larger amounts of reagent or a more expensive reagent system than normally used may be required. If a waste with an odor problem is to be processed at a site close to residential housing, a reagent system with a slow exotherm (slow rate of heat generation from chemical reaction) would be preferred to minimize volatilization. The point here is that there are many factors other than the straight laboratory chemistry to consider, and good project definition and information about the waste and the site are essential.

The treatment process itself usually consists of the following operational steps:

Preparation of a detailed operational plan
Mobilization
Setup of equipment and facilities at the job site
Treatment
  Waste homogenization
  Processing
  Technical control
  Chemical acquisition
  General field purchasing
Close-down, demobilization
Clean up and maintenance

In Situ Versus Ex Situ Treatment

In situ literally means "in place" and in the context of S/S means that the waste is not removed from the storage or disposal area to be processed through a mechanical system (ex situ). Usually, treatment is accomplished by mixing the reagent into the waste storage zone by some mechanical means such as a backhoe, auger, or rotary tilling device. In situ solidification was one of earliest techniques used to physically stabilize sludge lagoons and is still being used for this purpose. Most of the commercial S/S processes that we have discussed can be used with in situ delivery systems, and the characteristics of the product are much the same if the reagent is mixed in properly. The latter, however, is easier said than done.

Most of the hazardous waste management firms in the United States have used this method, and a number of construction and engineering organizations have developed both equipment and expertise for in-situ treatment. The USEPA has encouraged evaluation of this technology for use in the Superfund and other remedial action programs, and several evaluations have been done under the Superfund Innovative Technology Evaluation (SITE) program. A number of papers have been published describing the results of these and other studies [98–100]. One of the largest such projects ever undertaken in the United States was done at one of Chemical Waste Management's facilities in 1985. A total of 250,000 cubic yards of lagoon sludge was solidified with

Figure 11 Large in situ stabilization operation on oily/PCB sludges.

a kiln dust–based system using in situ mixing with backhoes. The solid product was then retained in a pile on site awaiting regulatory approvals for on-site landfill. Many similar projects have been done during the last decade at numerous sites in the United States. The most common device used for mixing is the backhoe, because it is widely available and does a satisfactory job within limits. A large-scale backhoe operation is shown in Figure 11. The reagent is introduced by pneumatic or mechanical conveyance to the surface of the waste, where it is then mixed in with the backhoe. This may be repeated until sufficient reagent has been added to produce an acceptable solid.

Obviously, this arrangement is rather crude. It produces a great deal of dust (from the reagent) and the mixing is generally not very intimate. This has caused S/S vendors and their contractors to move toward more sophisticated reagent feed and mixing systems using various injectors and mixers mounted on backhoes. The reaches and working depths are limited only by the size and power of the backhoe. For shallow ponds and contaminated soil areas, a rotary tiller, tractor mounted, can achieve more intimate mixing. These devices can work in waste depths from several inches to about 5 ft. Another approach consists of a drilling auger containing cutting blades and hollow mixing blades attached to a vertical drive shaft. It works much like an post-hole driller but is much larger, and instead of removing the soil, it churns and mixes it in place. In operation, the auger is advanced to the desired maximum depth, the reagents are injected in slurry form through the mixing blades, and mixing is continued during withdrawal of the auger.

As a general rule, in situ S/S systems of any of the types discussed are less costly than removing the waste for treatment and replacement. Also, in certain situations they can reduce immediate local environmental impact such as from odor, and they sometimes have permitting advantages due to peculiarities in the way regulations are written. The primary question always is whether the method chosen accomplishes the requirements of the project in an environmentally acceptable manner.

Disposal Site

Most solidified wastes are disposed of in landfills, although there are exceptions such as when they are stored for possible future recovery or used for road or other structural base. Landfills may range from "clean" structural landfills through municipal and private sanitary landfills designed for domestic refuse, to the highly secured landfills necessary for hazardous material under RCRA. The design of RCRA landfills has become increasingly sophisticated in recent years.

Some solidified waste may be used in direct water contact applications, such as for diking material and for forming new land from lakes, streams, marine waterways, or low-lying swamp areas. In most cases, however, S/S-processed solidified waste is placed in landfills primarily used for waste disposal purposes, and the following factors should be considered:

Distance from the waste source
Location with reference to industrial and residential areas
Geology and hydrology
Available capacity
Disposal cost per ton or cubic yard of waste disposed
Overall environmental impact on the surrounding land, water, and air
Attitude of neighbors
Land use regulations (zoning, etc.)
Political considerations
Liability and insurance consideration

The entire subject of waste disposal sites, from choice of location to operation of the site, is a complex one dealing not only with technical, engineering, and cost factors, but also with the intangible political, public, and legal ones, which are becoming increasingly sensitive. Perhaps the biggest problems in developing disposal sites are public reaction, financial responsibility, and liability. Both the financial responsibility and liability issues are especially troublesome because insurance for waste disposal operations and facilities is difficult and expensive to obtain. Therefore, both the waste generator and the S/S processor must be as certain as possible that the waste treatment system has been properly chosen for the particular disposal site or, alternatively, that the disposal site chosen is compatible with the characteristics of the treated waste.

Costs and Other Commercial Considerations

Cost is usually the final determining factor in choosing a treatment process and should therefore be kept in mind during every aspect of the decision-making process. The number of factors, variables, parameters, and considerations discussed in this chapter make it obvious that choosing a system is not a simple matter, and considerations of cost cannot be deferred until the technical and engineering choices have been made.

As with any other complex system, it is not possible to give flat costs or selling prices for a specific S/S process any more than it is possible for an engineering construction firm to bid on a water treatment plant without knowing the design parameters of that plant. In comparing costs and prices for different systems and vendors, it is important to determine exactly what operations are included (e.g., dredging, chemical processing, transport, disposal). Typical cost ranges for the various operations involved in a complete disposal system based on S/S are given in Table 17. As indicated, transportation and landfill costs have a very wide range. Transportation cost is primarily dependent on distance and can readily be determined on a "typical case" basis. Landfill cost, on the other hand, depends on many factors and can be accurately determined only by specific quote. Dredging costs are site specific and highly volume dependent.

Table 17 Typical Costs of S/S Unit Operations

Operation	Cost ($/ton)
Dredging	1–20
S/S (nonnuclear)	30–150
Transportation	2–200
Landfill (off-site)	
Nonsecure	10–60
Secure	75–200

Regulatory Approval

The regulatory approval process is an integral part of doing a S/S project. However, there are so many variations on the way in which this proceeds that it is not possible to discuss the subject here except to point it out as a necessary, albeit time-consuming and costly process.

CASE STUDIES

The following case studies of actual stabilization projects on contaminated soils and similar materials at remedial sites show the translation of theory and laboratory studies, which we have been discussing to this point, into practice. They are also illustrative of the many and varied types of on-site treatment problems encountered and the different approaches used to solve them. It should be noted that where costs are given, they are in U.S. dollars *at the time of the project*. Most of these case studies are taken from Conner [3].

Oily/PCB Waste Ponds

This site at Vickery, Ohio, now owned by Chemical Waste Management, Inc., had three ponds full of oily sludge containing low levels of PCBs (less than 500 ppm) and dioxins. After consultation with the USEPA and Ohio EPA, it was determined that the most environmentally acceptable remediation plan was to stabilize the wastes on site and construct a RCRA closure cell, also on site, to contain the product permanently. After the completion of treatability studies and testing of the proposed product, a stabilization formulation was chosen consisting of 100 part of sludge to 15 parts of calcium oxide (dolomitic quicklime) and 0 to 20 parts of cement kiln dust (all by weight). Leach testing using the EPT showed that no PCBs, dichlorobenzidine, or dioxins were detected in any of the leachates. This was followed by a pilot test on 2000 yd^3 of sludge, which verified the laboratory results.

Each of the three ponds containing the sludges was 200 ft wide and 800 ft long, with depths ranging from 3 to 14 ft. The reagents were added using two cranes to deliver reagents to the working face of the sludge, and three backhoes mixed them into the sludge (Figure 11). The backhoes started working from one edge of the pond. As the sludge solidified, it was used to form a dike around more sludge and the mixing continued. After stabilization was complete, the material was removed to the rear of the backhoes, the dike broken, and the sludge allowed to flow to the backhoes, where it was stabilized. This process was repeated until the sludge ceased to flow. The backhoes then moved out onto the stabilized material using crane mats. The entire process was then repeated until the pond was stabilized.

The project took 5 months in 1985 to complete, during which approximately 250,000 yd^3 of solid was produced. Approximately 22,000 tons of reagents were used. All of the stabilized sludge was then placed in a stockpile measuring 620 ft × 460 ft × 46 ft high to await final disposal in the closure cell. The job parameters were:

Waste analysis:
  Oil and petroleum products
  Organic priority pollutants
    PCBs
    Dioxins
    3,3′-Dichlorobenzidine
  Metals
    As, Ba, Cd, Hg, Se, Ag: <1.0 mg/kg
    Cr: <11.0 mg/kg
    Pb: <74.0 mg/kg
  Inert inorganic solids
  Water
Volume treated: 210,000 yd^3
Treatment rate: about 1000 tons/day
Time: 1985
Disposal method: secure landfill (on-site)
Properties of treated solid:
  Cohesion: 300–450 psf
  Friction angle (degrees): 35.8–35.5
  Permeability: 1×10^{-6} cm/s
  Leachability of PCBs, dioxins, dichlorobenzidine: below detection limits

Stratified Polymer Pit

One of the tasks in a "superfund-type" project (run much like a Superfund job, but not paid for out of Superfund) in Ohio in 1982 was the cleanup of a "polymer pit." This concrete block impoundment contained three layers of waste: a floating, gummy organic layer containing a wide range of organic priority pollutants; an intermediate contaminated-water layer; and a heavy sludge layer on the bottom. Because of the variety and high concentration of priority pollutants in the waste, secure landfill was chosen as the disposal method. Solidification was required before landfill, and the options were solidification either on site or at the landfill. On-site solidification was chosen for economic and public safety reasons.

Because there were essentially three distinct waste streams, three different formulations were chosen to minimize volume increase and the associated transportation and disposal costs. A cement-based process was used for the sludge and water layers and a lime kiln dust formulation for the organic layer. The volume was relatively small, and because three different formulations were used, a small, portable, batch-type treatment unit was used. The job parameters were:

Waste analysis:
- Organic polymers
- Organic priority pollutants
- Metals
- Inert inorganic solids
- Water

Volume treated: 210,000 gallons
Treatment rate: 20,000 gallons/day
Time: May 1982
Treatment cost: $0.41 per gallon
Disposal method: secure landfill (off site)
Properties of treated solid:
- Unconfined compressive strength: >1.0 ton/ft^2

Contaminated Soil with Lead

This project remediated a large, former battery-processing site in the downtown area of Omaha, Nebraska. The remedial action was required not only to clean up the area that was contaminated with lead, cadmium, arsenic, and antimony, but to allow both the soil and the site to be used as part of a new park. Lead contamination was the primary problem.

Prior to treatment, the soil was excavated, stockpiled, and screened to minus 0.25 in. Oversized material was crushed to minus 0.25 in. before processing. Screened soil was loaded into a feed hopper with a front-end loader, where dry reagents were added. The soil–reagent mix was then conveyed to a pugmill, where water and a liquid additive were added and the combination thoroughly mixed. The treated product, a soil-like solid, was conveyed to a finished stockpile for curing and storage until the site grading was complete. Quality control samples were taken every 1000 yd^3 to ensure effectiveness of the treatment and compliance with regulations.

Waste analysis:
  Metals:
    Lead, total: 12,000 mg/kg
    Lead, TCLP: 620 mg/L
  Soil
  Water
Volume treated: 48,700 yd^3
Treatment rate: 2000 yd^3/day
Time: 1989
Treatment cost: $67 per cubic yard
Disposal method: engineered backfill on site
Properties of treated solid:
  Unconfined compressive strength: >1.0 ton/ft^2
  Lead, TCLP: 0.4 mg/L
    Regulatory compliance limit, TCLP: 5.0 mg/L

Soil Contaminated with Acid Organic Filter Cake

An old plant site in Massachusetts contained some 10,000 yd^3 of filter cake and earth mixture from former operations. The waste was acidic and contained naphthalene-based compounds. Not until its removal for a mass transit system station was the filter cake and earth mixture found to be hazardous. The options were S/S treatment on site with disposal at local sanitary land fills, or transport to a remote, secure landfill. The S/S option was chosen because the total cost was about one-half that of transporting it to a secure landfill (approximately $1,000,000 was saved). Because the waste was not pumpable, a solids type of treatment unit was used. This unit is shown in Figure 12. The job parameters were:

Waste analysis:
  DAXAD (polymer of naphthalenesulfonic acid): 2.5–10%
  Miscellaneous naphthalene-based hydrocarbons: 10–12.5%
  β-Naphthalene sulfonate: 0.5–2%
  Sodium sulfate: 0–1.5%
  Filter aid (perlite): 20–30%
Volume treated: 10,000 yd^3
Treatment rate: 150 yd^3/day average
Time: spring 1981
Treatment cost: $75.00 per cubic yard
Disposal method: sanitary landfill, daily cover material
Properties of treated solid:
  Unconfined compressive strength: 2.5–5.0 tons/ft^2

Figure 12 Treatment of soil contaminated with acid organic filter cake.

Leaching results (EPT):

Organic priority pollutants:	0.6 ppm
Naphthalene:	7.5 ppm
RCRA metals (except barium):	<0.06 ppm
Barium:	1.0 ppm
Phenol:	0.22 ppm
Total cyanides:	0.050

Uranium Sludge/Soil Pond Removal

This very unusual project involved the solidification of waste from a nuclear fuel processing plant that contained minute amounts of uranium. Because of the uranium content, the waste was classified as a nuclear waste, even though its radioactivity was essentially at background level and was therefore subject to the Nuclear Regulatory Commission and its system rather than RCRA. The project plan called for the solidification of the sludge in drums which would later be shipped to the nuclear waste repository at Barnwell, South Carolina. Solidification with no free water had to be confirmed before the drums could be closed, labeled, and shipped. Leachability was not a consideration.

The waste sludge was removed from the pit by clamshell and delivered to a hopper directly above the mixer. After mixing with the cement formulation contained in the silo, the waste was discharged into a batch hopper below the mixer and metered into drums. Filled drums were rolled by conveyor away from the mixing area and the exterior cleaned with a high-pressure steam jet. After testing for free water and solidity, the drums were capped, labeled, palletized, and wrapped for storage until they could be shipped to the disposal site. This project is a good example of a commercial S/S project because it has its own particular characteristics. Every S/S project has some combination of product requirements, wastes to be treated, and physical/mechanical considerations which makes it unique.

Mud/Soil Dredging Pilot Project in Japan

Dredging operations are performed for maintenance of navigational channels, for land reclamation, and sometimes for pollution control in lakes and seabed areas. An experimental project of particular interest to the dredging industry was conducted in February 1974 as part of an undertaking by the Association for Protection of Japanese Aquatic Resources, comprising tests for the commercial application of technologies for control of the Red Tide problem in the Inland Sea (Japan). A cement-soluble silicate process was chosen as one of the test methods for the purpose of developing a practical, commercial system of seabed mud treatment, so that treated mud could be disposed of on land or used in reclamation projects.

The project was conducted using a pneumatic dredge to minimize contamination of the surrounding water while the dredging operation was under way. Mud accumulation on the sea bottom in the area dredged ranged from several centimeters to several meters in thickness. Solids content of the dredged mud ranged from 14.5 to 24.6%, which was ideally suited for treatment by a mobile unit. Solidification tests were conducted under a variety of weather conditions, including snow, rainfall, and temperatures as low as 35°F. Solidification was not affected by rainfall, and the rain did not cause any breakdown or washing away of this particular material. In general, the solidified material became solid enough to support foot traffic within 24 hours. Samples of solidified material were taken 24 hours after it was poured into the solidifying pond.

Leaching tests were done to determine the environmental properties of the solidified seabed mud, in the case by the standard methods specified in the Japan EPA Notification 13. The results are shown below: Extremely small concentrations of heavy metals were detected in the leachate, indicating that the solidified, treated material is safe for use as landfill. PCB and hexane extract values were also extremely small, indicating that harmful organic substances were entrapped in the solid.

Constituent	Concentration in leachate (mg/L)[a]	
	Raw mud	Treated mud
Mercury, total	1.71	0.0002
Cadmium	8.64	<0.001
Lead	516.0	<0.001
Arsenic	3.80	0.003
Cyanide	0.08	<0.001
PCBs	18.8	<0.005

[a]Distilled water leach test; combination of two samples.

REFERENCES

1. Pojasek, R. B. (ed.), *Toxic and Hazardous Waste Disposal*, Ann Arbor Science, Ann Arbor, MI, 1979.
2. Cullinane, M. J., and L. W. Jones, *Stabilization/Solidification of Hazardous Waste*, EPA/600/D-86/028, U.S. Environmental Protection Agency, Hazardous Waste Environmental Research Laboratory, Cincinnati, OH, 1986.
3. Conner, J. R., *Chemical Fixation and Solidification of Hazardous Wastes*, Van Nostrand Reinhold, New York, 1990.
4. Commissariat A L'Energie Atomique, French patent 1,246,848 (October 17, 1960).
5. Bell, N. E., et al., *Solidification of Low-Volume Power Plant Sludges*, Electric Power Research Institute Project 1260–20, CS-2171, 1981.
6. DeLaguna, W., Radioactive waste disposal by hydraulic fracturing, *Ind. Water Eng.* (October 1970).
7. Winsche, W. E., U.S. patent 3,152,984 (October 13, 1964).
8. Smith, C. L., and W. C. Webster, U.S. patent 3,720,609 (March 13, 1973).
9. Conner, J. R., U.S. patent 3,837,872 (September 24, 1974).
10. Chappell, C. L., U.S. patent 4,116,705 (September 26, 1978).
11. *Resource Conservation and Recovery Act*, PL 94-580, 1976.
12. *Comprehensive Environmental Response, Compensation, and Liability Act of 1980*, PL 96-510, 1980.
13. U.S. Environmental Protection Agency, *14th Annual Research Symposium: Land Disposal, Remedial Action, Incineration and Treatment of Hazardous Waste*, Hazardous Waste Environmental Research Laboratory, Cincinnati, OH, 1988.
14. U.S. Environmental Protection Agency, *Fed. Reg.* 50(105):23250 (May 31, 1985).

15. *Resource Conservation and Recovery Act*, PL 94-580, 1976.
16. U.S. Environmental Protection Agency, *Fed. Reg.* 55(61):11798–11877 (March 29, 1990).
17. *Comprehensive Environmental Response, Compensation, and Liability Act of 1980*, PL 96-510, 1980.
18. Conner, J. R., S. Cotton, and P. L. Lear, Stabilization of mixed-code hazardous waste incinerator residues under the RCRA Landbans, *Incineration Conference*, Knoxville, TN, May 13, 1991.
19. U.S. Environmental Protection Agency, *Fed. Reg.* 56(104):24444–24465 (May 30, 1991).
20. U.S. Environmental Protection Agency, *Fed. Reg.* 56(206):55160–55189 (October 24, 1991).
21. *Fed. Reg.* 51(102):19305–19308 (May 28, 1986).
22. Conner, J. R., *The Modern Engineered Approach to Chemical Fixation and Solidification Technology*, 1981.
23. Conner, J. R., Ultimate disposal of liquid wastes by chemical fixation, *Proceedings of the 29th Annual Purdue Industrial Waste Conference*, May 1974.
24. Darcel, F., Recent studies on leach testing: a review, unpublished draft, Ontario MOE, June 1984.
25. Simulated field leaching test for evaluation of surface runoff for land disposal of waste materials, *I.U. Conversions Systems*, 1977.
26. Cote, P., Contaminant leaching from cement-based waste forms under acidic conditions, Ph.D. thesis, McMaster University, Toronto, 1986.
27. *Safety of Public Water Systems* (*Safe Drinking Water Act*), PL 93-523, Washington, DC, 93rd Congress, December 16, 1974.
28. EP toxicity test procedure, 40CFR Part 261.24, Appendix II, *Fed. Reg.* (May 19, 1980).
29. U.S. Environmental Protection Agency, *Solid Waste Leaching Procedure Manual*, SW-924, U.S. EPA, Cincinnati, OH, 1985.
30. *Fed. Reg.* 47(225):52687 (November 22, 1982).
31. *Fed. Reg.* 49(206):42591 (October 23, 1984).
32. Cote, P. L., and D. Isabel, Hazardous and industrial waste management and testing, *ASTM 3rd Symposium*, Philadelphia, pp. 48–60, 1984.
33. *Fed. Reg.* 45CFR:57332 (August 27, 1980).
34. *Fed. Reg.* 51(114):21686 (June 13, 1986).
35. U.S. Environmental Protection Agency, *Guide to the Disposal of Chemically Stabilized and Solidified Waste*, SW 872, U.S. EPA, Washington, DC, 1982.
36. Lowenbach, W. A., *Compilation and Evaluation of Leaching Test Methods*, EPA-600/@-78-095, U.S. Environmental Protection Agency, Cincinnati, OH, 1978.
37. Bishop, P. L., Leaching of inorganic hazardous constituents from stabilized/solidified hazardous wastes, *Hazard. Waste Hazard. Mater.* 5(2):129–43 (1988).
38. Bause, D. E., and K. T. McGregor, *Comparison of four leachate-generation procedures for solid waste characterization in environmental assessment programs*, EPA 600/7-80-118, U.S. Environmental Protection Agency, Cincinnati, OH, 1980.

39. U.S. Environmental Protection Agency, *Test Methods for Evaluating Solid Waste*, SW-846, U.S. EPA, Office of Solid Waste and Emergency Response, Washington, DC, 1986.
40. Malone, P., and R. Larson, Symposium poster presentation.
41. EP toxicity test procedure, 40CFR Part 261.24, Appendix II, *Fed. Reg.* (May 19, 1980).
42. Moore, J. N., W. H. Ficklin, and C. Johns, Partitioning of arsenic and metals in reducing sulfidic sediments, *Environ. Sci. Technol.* 22:432–437 (1988).
43. Dragun, J., The fate of hazardous materials in soil, *Hazard. Mater. Control*, pp. 41–65 (May/June 1988).
44. Vail, J. G., *Soluble Silicates*, Reinhold, New York, 1952.
45. Iler, R. K., *The Chemistry of Silica*, Wiley, New York, 1979.
46. Patterson, J. W., H. E. Allen, and J. J. Scala, Carbonate precipitation from heavy metals pollutants, *J. Water Pollut. Control Fed.* 12:2397–2410 (1977).
47. O'Hara, M. J., and M. R. Surgi, U.S. patent 4,737,356 (April 12, 1988).
48. Sittig, M., *Pollutant Removal Handbook*, Noyes Data Corp., London, 1973.
49. LeGendre, G. R., and D. D. Runnels, Removal of dissolved molybdenum from wastewaters by precipitates of ferric iron, *Environ. Sci. Technol.* 9(8):744–749 (1975).
50. Swallow, K. C., D. N. Hume, and F. M. M. Morel, Sorption of copper and lead by hydrous ferric oxide, *Environ. Sci. Technol.* 14(11):1326–1331 (1980).
51. Pojasek, R. B., *Toxic and Hazardous Waste Disposal*, 1–4, Ann Arbor Science, Ann Arbor, MI, 1980.
52. Manahan, S. E., and M. J. Smith, The importance of chelating agents, *Water Sewage Works*, 102–106 (1973).
53. Nelson, D. A., *Removal of Heavy Metal Ions from Aqueous Solution with Treated Leather*, American Chemical Society, Houston, 1980.
54. *Chem. Eng.*, pp. 83–84 (May 21, 1979).
55. Wing, R. E., Corn starch compound recovers metals from water, *Ind. Wastes*, pp. 26–27 (January/February 1975).
56. Wing, R. E., U.S. patent 3,294,680 (September 27, 1977).
57. Brennan and Mace, Waste treatment of chemical cleaning wastes in the power industry, *Proceedings of the Purdue Industrial Waste Conference*, 1977, pp. 899–907.
58. Bell, W. E., Chelation Chemistry, *Mater. Protect.*, p. 79 (February 1965).
59. *Chem. Eng.*, p. 17 (February 15, 1988).
60. *Fed. Reg.* 47(225):52687 (November 22, 1982).
61. Ottinger, R. S., J. L. Blumenthal, D. F. Dal Porto, G. I. Gruber, M. J. Santy, and C. C. Shih, *Recommended Methods of Reduction, Neutralization, Recovery or Disposal of Hazardous Waste*, EPA-670/2-73-053-f, U.S. Environmental Protection Agency, Washington, DC, 1973.
62. Shively, W. E., and M. A. Crawford, *EP Toxicity ad TCLP Extractions of Industrial and Solidified Hazardous Waste*, CH2M Hill, Boston, 1986.
63. Shively, W., P. Bishop, D. Gress, and T. Brown, Leaching tests of heavy metals stabilized with Portland cement, *J. Water Pollut. Control Fed.* 58(3):234–41 (1986).

64. Chemical Waste Management Inc., *Internal Report on Residue Management.*
65. Chemical Waste Management Inc., *Internal Report on F006 Wastes.*
66. Chemical Waste Management Inc., *Internal Report on Mixing.*
67. Chemical Waste Management Inc., *Internal Report on K061 Waste.*
68. *Fed. Reg.* 53(121):23661–23671 (June 23, 1988).
69. *Fed. Reg.* 51(199):36707–36730 (October 15, 1986).
70. *Environ. Sci. Tecnol.* 17(5):197A (1983).
71. *J. Inst. Munic. Eng.* 84 (November 1957).
72. Volnesenskic, S. A., A. V. Evallanova, and R. V. Suvorova, *Water Pollut. Abstr.* 13:135 (1940).
73. Conner, J. R., A. Li, and S. Cotton, Stabilization of hazardous waste landfill leachate treatment residue, *J. Hazard. Mater.* 24:111 (1990).
74. *Fed. Reg.* 45(98):33119–33133 (May 19, 1980).
75. Delchad, S., Chemical treatment: an inexpensive alternate to handling oily sludge, *Proceedings of the 5th National Conference on Hazardous Wastes and Hazardous Materials.* April 19–21, 1988, pp. 85–88.
76. Cote, P., *Assessment of Solidification Technologies for the Immobilization of Organic Contaminants,* draft by Wastewater Technology Centre, Environment Canada, Burlington, Ontario, April 1987.
77. Lear, P. R., and J. R. Conner, Immobilization of Low Level Organic Compounds in Contaminated Soil, *Proceedings of the 6th National Conference on Hydrocarbon Contaminated Soils,* University of Massachusetts, Amherst, MA, 1991.
78. Weitzman, L., L. E. Hamel, and S. R. Cadmus, *Volatile Emissions from Stabilized Waste in Hazardous Waste Landfills,* USEPA Contract 68-02-3993, Research Triangle Park, NC, August 28, 1987.
79. U.S. Environmental Protection Agency, *Test Methods for Evaluating Solid Waste,* SW-846, U.S. EPA, Office of Solid Waste and Emegency Response, Washington, DC, 1986.
80. Exner, J. H., A sampling strategy for remedial action at hazardous waste sites, *Hazard. Waste Hazard. Mater.* 2(4):503–521 (1985).
81. Brantner, K. A., Priority pollutants sample collection and handling, *Pollut. Eng.*, pp. 34–38 (March 1981).
82. *Standard Guide for General Planning of Waste Sampling,* American Society for Testing and Materials, Philadelphia, 1985.
83. *Standard Practice for Sampling Waste and Soils for Volatile Organics,* American Society for Testing and Materials, Philadelphia, 1987.
84. *Standard Practice for Sampling Non-liquid Waste from Trucks,* American Society for Testing and Materials, Philadelphia, 1987.
85. *Standard Methods for the Examination of Water and Wastewater,* American Public Health Association, Washington, DC.
86. *Immobilization Technology Seminar,* U.S. Environmental Protection Agency, Risk Reduction Engineering Laboratory, Cincinnati, OH, 1989.
87. *American Chemical Society Course on Experimental Design,* ACS, Atlanta, 1985.
88. Tessier, A., P. G. C. Campbell, and M. Bisson, *Anal. Chem.* 51:844–851 (1979).

89. *Fed. Reg.* 51(114) (June 13, 1986).
90. *Fed. Reg.* 47(225):52687 (November 22, 1982).
91. *Fed. Reg.* 49(206):42591 (October 23, 1984).
92. *California Administrative Code,* Tital 22, 66696: 1800.75-84.3, 1985.
93. Cote, P. L., and D. P. Isbel, Application of a dynamic leaching test to solidified hazardous waste, *ASTM Symposium on Industrial and Hazardous Wastes,* March 7–10, 1983.
94. *14th Annual Research Symposium: Land Disposal, Remedial Action, Incineration and Treatment of Hazardous Waste,* U.S. Environmental Protection Agency, Cincinnati, OH, 1988.
95. Conner, J. R., Fixation and solidification of wastes, *Chem. Eng.,* 79–85 (November 10, 1986).
96. Sax, I. R., *Dangerous Properties of Industrial Materials,* Van Nostrand Reinhold, New York, 1979.
97. Bretherick, L., *Handbook of Reactive Chemical Hazards,* CRC Press, Cleveland, OH, 1975.
98. Barich, J. J., J. Greene, and R. Bond, Soil stabilization treatability study at the Western Processing Superfund site, *Superfund '87,* 198–203 (1987).
99. Kuhn, R. C., and K. R. Piontek, A site-specific in-situ treatment process development program for a wood preserving site, *Superfund '87,* pp. 182–186 (1987).
100. Stinson, M. K., and S. Sawyer, *In situ Treatment of PCB-Contaminated Soil,* USEPA Site Papers, pp. 504–507 (1988).

4

Soil Vapor Stripping

David J. Wilson
Vanderbilt University
Nashville, Tennessee

Ann N. Clarke
ECKENFELDER INC.
Nashville, Tennessee

INTRODUCTION

The main objective of this chapter is to provide sufficient information on in situ soil vapor stripping (soil vapor extraction, soil vacuum extraction, soil vacuuming, venting) to give managers and regulators a clear idea of the range of applicability of the technique. A second purpose is to describe the studies that must be done to determine if soil vapor stripping is suitable for any given site. The chapter should also be useful to engineers who are not experts in the field of soil vapor stripping but who need to read proposals, reports, and recommendations involving the technique with some critical understanding. We hope in the near future to write a more detailed (and much more lengthy) discussion of soil vapor stripping directed to the engineer who is involved in the preliminary planning, design, construction, and operation of soil vacuum extraction facilities.

Conceptually, soil vapor stripping is a very simple technology. It is applicable to the removal of volatile organic compounds (VOCs) and some semivolatile compounds (SVOCs) from the vadose zone. The setup is shown schematically in Figure 1. A well is drilled down through the domain of contamination, and a blower is used to suck air down through the contaminated soil, into the well, up through a demister to remove excess water, then through a unit to remove the VOCs (such as an activated carbon bed or catalytic combustion unit), after which the gas is exhausted through the blower to the atmosphere. For

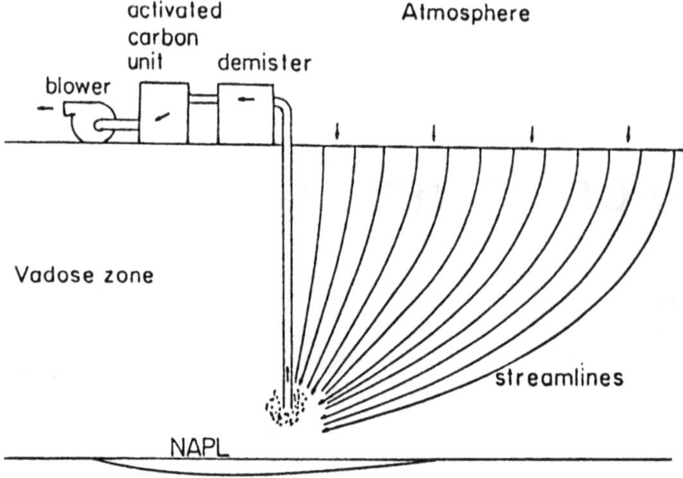

Figure 1 Schematic diagram of a soil vapor stripping well.

larger-scale operations, several wells may be manifolded and served by a single demister, activated carbon unit, and blower. Several other geometries for soil vapor extraction (trenches, horizontal drilling, and soil piles) are shown in Figure 2.

The environmental impact of the technique is low, and site disturbance is generally minimal. Costs are generally a fraction of those of other technologies. Large volumes of soil can readily be treated. Soil vapor extraction (SVE) systems are relatively easy to install and utilize off-the-shelf equipment. Cleanup times are generally fairly short, and the toxic material is removed from the soil and ultimately destroyed, rather than being sequestered and/or relocated. SVE lends itself well to use in combination with other techniques, such as groundwater pump-and-treat operations. These advantages make soil vapor stripping a method of choice in those sites at which it is applicable. Pedersen and Curtis (1991) have recently discussed the advantages of SVE in some detail.

The range of applicability of soil vapor stripping is bounded by the following constraints.

1. The chemicals to be removed must be volatile or at least semivolatile (a vapor pressure of 0.5 torr or greater); it is not feasible to remove metals, most pesticides, and PCBs by vacuum extraction because their vapor pressures are too low.

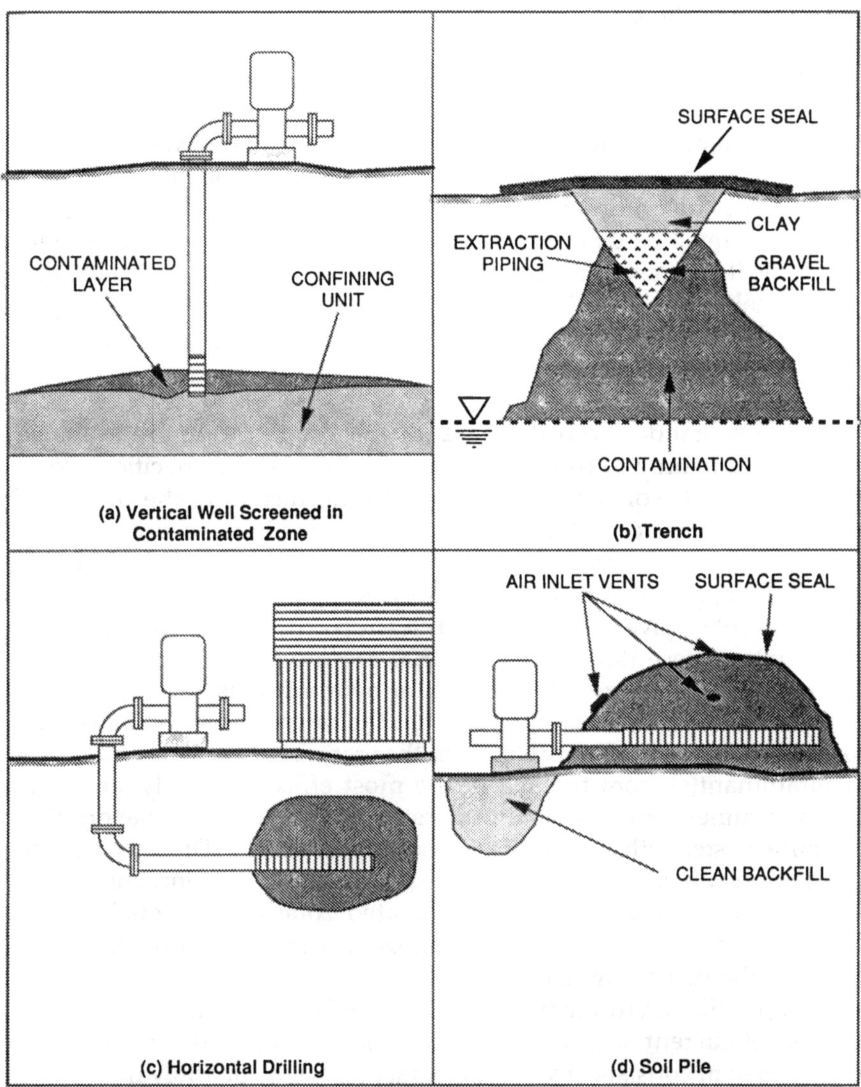

Figure 2 Some soil vapor extraction design system options. (From Pedersen and Curtis, 1991.)

2. The chemicals to be removed must have relatively low water solubility or the soil moisture content must be quite low; generally, it is not feasible to vapor strip chemicals such as acetone or alcohols because their vapor pressures in moist soils are too low.
3. The contaminants to be removed must be above the water table (in the vadose zone) or, in the case of light nonaqueous-phase liquids (LNAPLs), floating on it.
4. The soil must be sufficiently permeable to permit the vapor extraction wells to draw air through all of the contaminated domains at a reasonable rate; a basic principle in soil vapor stripping is that you must be able to move air through the domain if you are going to clean it up.

Very few standardized design and operating criteria are available for SVE, due mainly to the limitations imposed by site-specific factors. These include (1) contaminant characteristics, including the degree of weathering; (2) the extent of the contamination; (3) soil characteristics, particularly permeabilities; (4) depth of the water table; (5) VOC emission control requirements; and (6) criteria set for soil cleanup. Pedersen and Curtis (1991) give a list of the soil cleanup criteria presently adopted by the states; these range from 1 to 500 ppm, with the most common standard being 100 ppm total petroleum hydrocarbons.

Pedersen and Curtis (1991) note that the purpose of the SVE design process is to construct an SVE system that removes the greatest amount of contaminant(s) from the site in the most efficient, timely, and economical manner. This requires understanding of the main factors that determine system effectiveness (Johnson et al., 1989b). These are (1) the composition and characteristics of the contaminants present, and (2) the delivery of flowing air to the contaminated zone at a reasonable rate. SVE system design attempts to maximize the interaction of the vapor flow with the contaminated zone.

Sink (Pacific Environmental Services, 1989) has provided a good description of current soil vapor extraction facilities and their operation. An SVE system consists of vacuum extraction wells, possibly inlet or injection wells, piping headers, vacuum pumps or blowers, vacuum gauges and flow meters, sampling ports, an air/water separator (usually), a VOC control system (usually), and possibly an impermeable cap. The vacuum wells generally fully penetrate the zone of contamination. The screened sections are constructed of slotted plastic pipe in a permeable packing to facilitate gas flow.

VOCs in the zone of contamination move as vapor through the soil into the vacuum extraction well. They are then either discharged to the atmosphere or (more likely) sent to an air pollution control device. These VOCs typically have molecular weights of 200 g/mol or less; larger molecules are insufficiently volatile to move readily as vapor. Inlet (passive) wells or injection wells (in which air is forced into the ground) may be located around the boundaries of the contaminated area to improve airflow in regions in which this may be too slow; injection wells may be used as part of a closed-loop SVE system. Piping is usually plastic (often PVC); headers are either plastic or steel. Large sites may require multiple systems, and it is common practice to have one blower serve several wells. Insulation of pipes and headers may be necessary in cold climates. Air/water separators may be needed in the train before off-gas treatment, particularly if VOCs are being removed with activated carbon.

After the system is turned on, a steady-state flow rate is achieved relatively quickly. The gas flow rate is determined by the wellhead vacuum (controlled by the blower), the pipe size and length, the screening and packing of the well, soil pneumatic permeability, the presence of barriers (building foundations, parking lots, impermeable plastic caps, etc.), and flow control devices. If injection wells are employed, the discharge pressure from the blower may be used to provide positive pressure for air injection. The resulting airflow causes VOC vapor movement to the extraction well(s), where the VOCs are drawn up and captured or destroyed in the VOC treatment device. Often an air/water separator is included in the train before the blower to protect it and to increase the efficiency of VOC removal by activated carbon. Impermeable plastic caps are occasionally used to reduce the volumes of soil in which airflow is excessively sluggish or wastefully fast. These may also limit fugitive VOC emissions from the soil.

Flow rates and off-gas VOC compositions are measured regularly to permit calculation of the rate of VOC removal. Almost always this is initially quite high and then decreases gradually with time. If a significant amount of VOC is present in regions from which it is slowly released by diffusion (clay lenses, for example), the latter portion of the cleanup may be quite prolonged. Unless properly designed, initial measurements will give no warning of this tailing. In the latter stages of cleanup the blower may be cycled on and off or the well run at a reduced flow rate to reduce energy requirements with little loss of removal efficiency. Measurements of soil gas VOC concentration during well operation are not a reliable indication of the extent of cleanup; one should obtain VOC analyses on gas samples taken after the well has been shut down for an extended period to test the extent to which cleanup is complete.

HISTORICAL BACKGROUND AND BRIEF LITERATURE REVIEW

In this section we provide background and historical information on soil vapor stripping. The literature on the topic is sufficiently extensive that a complete review would require excessive space; we therefore present a brief review covering the major topics of interest.

Much of the early work on vacuum extraction was published under the auspices of the American Petroleum Institute (API). In 1980, the API published an analysis and evaluation of the ability of venting (soil vapor stripping) to remove subsurface gasoline vapors from contaminated soil (API, 1980). Another early report (Wootan and Voynick, 1984) dealt with the vapor stripping of gasoline in a large-scale (3 × 3 × 1.2 m) model aquifer. These workers suggested that vapor stripping wells should be deep and slotted only near the bottom to avoid short-circuiting the airflow. They also suggested that impervious covers at the surface of the vented area might improve efficiency. This was followed by an API report on a field demonstration of subsurface venting of hydrocarbon vapors from an underground aquifer. Crow et al. (1985) described data from soil vacuum extraction experiments carried out at a petroleum fuels terminal at which a gasoline spill had taken place. They concluded that the technique removes hydrocarbon vapors from the vadose zone effectively and is also useful in augmenting conventional methods for recovering spilled hydrocarbons from shallow aquifers.

Another early report demonstrated enhanced microbial degradation of gasoline in the presence of an increased oxygen supply (API, 1982). Keech (1989) has summarized the research supported by API in soil vapor stripping (subsurface venting) and noted that API would be publishing a manual on this technique shortly.

Hutzler et al. (1989b, 1990) have recently reviewed soil vacuum extraction; their very comprehensive article includes a listing of representative pilot and field scale soil vapor stripping operations with references to more detailed descriptions of these sites. [Sink (Pacific Environmental Services, 1989) has also published a list of 29 soil vapor stripping operations, including location, pollutants identified, whether VOC emissions are controlled, and references; see Table 1.] Hutzler and co-workers include information on the designs of the facilities, a useful tabulation of Henry's constants for common organic compounds, and a summary of design suggestions. These authors note the importance of diffusion-limited transport and suggest that intermittent well operation is more efficient than continuous operation when vapor stripping from clays and silts. A lab study by Hutzler et al. (1989a) demonstrates the

Table 1 Representative Soil Vapor Extraction Sites

Facility/location	Pollutants identified[a]
Groveland Wells Groveland, Massachusetts	TCE, PCE, MC, DCE, TCA
Service station Wayland, Massachusetts	Gasoline
Waldick Aerospace New Jersey	PCE, petroleum hydrocarbons
Industrial tank farm Puerto Rico	Carbon tetrachloride
Service station San Juan, Puerto Rico	Gasoline
Tyson dumpsite Tyson's Lagoon, Pensylvania	TCE, toluene, ethylbenzene, xylene, trichloropropane, 1,1,1,2-TTCA
Service station Bellview, Florida	Gasoline
AWARE study Nashville, Tennessee	TCE, acetone, chlorobenzene
Petroleum fuels terminal Grainger, Indiana	Gasoline
Seymour facility Seymour, Indiana	1,2-DCA, benzene, vinyl chloride, 1,1,1-TCA, others
Thomas Solvents Co. (Verona Well Field) Battle Creek, Michigan	DCA, TCA, DCE, TCE, PCE, vinyl chloride, chloroform, carbon tetrachloride, benzene, toluene, xylene, ethylbenzene, MEK, MIK
Kimross facility Kimross, Michigan	1,1,1-TCA
Lansing facility Lansing, Michigan	TCE
Bangor facility Bangor, Maine	Toluene, benzene, xylene, ethylbenzene, styrene, ketones, chloroethane, MC
Hillside facility Hillside, Michigan	TCE
Custom Products Stevensville, Michigan	PCE
Twin Cities Army Ammunition Plant New Brighton, Minnesota	TCE, TCA, toluene
Troy facility Troy, Ohio	Acetone, MC, TCE, toluene, xylene
Paint storage warehouse Dayton, Ohio	Acetone, toluene, xylene, ketones

Table 1 Continued

Facility/location	Pollutants identified[a]
Texas Research Inst. Austin, Texas	Gasoline
Waverly facility Waverly, Nebraska	Carbon tetrachloride
Hill Air Force Base Salt Lake City, Utah	Jet fuel
Dowell Schlumberger Casper, Wyoming	Chlorinated hydrocarbons, toluene, xylene, benzene, ethylbenzene
LARCO Casper, Wyoming	Toluene
Southern Pacific spill Benson, Arizona	Dichloropropene
Electronics Co. Santa Clara, California	1,1,1-TCA
Storage tank Cupertino, California	TCA, TCE, DCA, DCE
Well 12A Tacoma, Washington	TCE, PCE, MC, TTCA, DCA, TCA
Ponders Corner Washington	1,2-DCA, TCE, TTCA

Source: Pacific Environmental Services (1989).
[a]DCE, dichloroethylene; TCE, trichloroethylene; PCE, perchloroethylene (tetrachloroethylene); MC, methylene chloride; DCA, dichloroethane; TCE, trichloroethane; TTCA, tetrachloroethane; MEK, methyl ethyl ketone; MIK, methyl isobutyl ketone.

importance of diffusion-limited transport in certain media, in agreement with the field data of Fall et al. (1988) and a number of other workers.

Towers et al. (1989) have presented an analysis of the process of choosing a treatment technique for soil contaminated with volatile halogenated organics at a site in Indiana. Groundwater remediation is necessary at the site. Soil treatment methods considered included excavation and off-site disposal, in situ biodegradation, chemical degradation, soil washing, soil vapor stripping (soil vacuum extraction, SVE), and combined technologies. SVE was substantially cheaper than the other options. A recent paper by Danko (1990) discusses the applicability and limitations of SVE in some detail.

Hoag and his associates have been active in the field for some years; Hoag and Cliff (1985) have described one of their soil venting projects for gasoline removal. In a more recent paper (Baehr et al. 1989) this group described laboratory work, vapor stripping at a field site contaminated with gasoline, and mathematical modeling. They concluded that the technique was extremely effective and that a local equilibrium model was quite adequate for describing vapor stripping at their field site. Other early papers of interest by this group include Marley and Hoag (1984) and Brown et al. (1987). Marley et al. (1989) have discussed the design of soil vapor extraction systems, and Hoag (1989) has summarized his group's work and indicated needs for additional research. Marley et al. (1990) have discussed the use of airflow modeling for both the evaluation of soil properties (the pneumatic permeability tensor) and the design of soil vacuum extraction systems. They note that mass transfer limitations may be important and that the distribution of airflow pathways is always of prime importance; cleanup will not occur if soil gas is not moving through the contaminated soil. The air-permeability tensor is evaluated in situ in the vadose zone by calibrating a steady-state airflow model with pressure measurements made during pumping tests. These air permeabilities and a steady-state airflow model are then used to determine well spacings, depth of wells, length of screened well sections, and type and size of pumps needed to produce the desired gas flow field.

Johnson et al. (1989a,b, 1990) at Shell Development have presented a number of simple mathematical models for use as screening tools in assessing the feasibility of soil vapor stripping on a site-specific basis. One of these is a one-dimensional model which assumes radial symmetry and takes into account the variations in the composition of volatile complex mixtures during the course of a vapor stripping operation. This should be of particular interest in dealing with spills of hydrocarbons, since these normally occur as mixtures of compounds having a relatively wide range of volatilities. The model assumes local equilibrium between the stationary phase and the mobile vapor phase. Another model allows one to estimate the time required for airflow around a vacuum well to approach a steady state.

DiGiulio et al. (1990) have made a number of carefully reasoned recommendations regarding the design of field tests for the assessment of SVE. They note the importance of estimating the effects of diffusion limitation and show that these can be assessed by means of a rather simple field test. A test vapor stripping well is isolated from the bulk of the zone of contamination by surrounding it with passive wells a relatively short distance from it. The vacuum well is then operated for a substantial interval (until the off-gas VOC levels have dropped nearly to zero) and is then shut down. One then monitors the recovery of the VOC concentrations in this well during the shutdown period, which yields information on the extent of diffusion limitation and also on the time constant for this process. These authors provide a formula by means of which effective Henry's constants (or partitioning coefficients) can be related to actual physical and chemical processes. They also point out that one can vapor strip water-soluble organics (such as alcohol, acetone, methyl ethyl ketone, etc.) if one is able to initiate vapor stripping before these have become highly diluted in the soil water in the vadose zone. These authors give a trenchant criticism of the use of overly simplistic models in designing and predicting the behavior of SVE systems.

Stephanatos (1990) has described a rather general model for VOC transport in soils in the vadose zone which includes both fluid flow and vapor transport; the model can be applied to SVE to predict the fate and transport of VOCs. He successfully applied his two-dimensional model to data from the Tyson Site near Philadelphia.

Walton et al. (1990) have presented a rather elaborate set of mathematical models for analyzing a very large, deep, complex site in Idaho at which chlorinated VOCs have contaminated a deep, highly structured vadose zone (basalts interspersed with sedimentary layers) and the underlying zone of saturation. The extremely high cost of carrying out an adequate characterization of a site of such size, depth, and complexity has forced heavy reliance on sophisticated modeling. It will be interesting to see if this effort to substitute increased sophistication for a substantial portion of data is successful.

AWARE (now ECKENFELDER INC.) has published a number of reports and articles on soil vapor stripping. Clarke and Wilson (1988) have described a phased approach to assessing the feasibility of the technique for use at a given site and for subsequent design work. Wilson and co-workers have developed a two-dimensional (axially symmetric) local equilibrium model for field vapor stripping wells and a one-dimensional local equilibrium model for lab column operation; lab column data can then be used to obtain adsorption isotherm parameters needed in the field model (Wilson et al., 1988; Wilson, 1988). The field-scale model was used to explore the effects of well geometrical parameters, the effects of overlying caps and passive wells, and the effects of strata of differing permeabilities and of anisotropic permeabilities, of buried impermeable obstacles, of underlying nonaqueous-phase liquid (NAPL), and of low-permeability lenses (Gannon et al., 1989; Wilson et al., 1989; Mutch and Wilson, 1990; Gomez-Lahoz et al., 1991; Mutch et al., 1989). Major heterogeneities (lenses of high clay content, strata of low permeability, etc.) are of particular interest. Gomez-Lahoz et al. (1991) have examined the effects of low-permeability lenses on the rate of cleanup by means of buried screened horizontal laterals and have developed strategies for reducing the damaging effects of such geological features. Mutch et al. (1989) have used models to interpret piezometer well data in terms of an overlying stratum of low permeability which greatly extended the range of influence of a vacuum extraction well and also to interpret the results of pilot vapor stripping operations at a site in Toms River, New Jersey. Vapor stripping can readily be carried out in soils underneath buildings, parking lots, and other impermeable overlying layers; this is modeled by including a no-flow boundary condition over the appropriate portion of the top boundary of the domain of interest (Gannon et al., 1989). Modeling work also indicates that soil vapor stripping should efficiently remove underlying floating light nonaqueous-phase liquids (Wilson et al., 1989), in agreement with experimental findings reported by Hoag and Cliff (1985).

A major uncertainty at present is the impact of diffusion and/or desorption kinetics on the rate of cleanup of a site by soil vapor stripping. As mentioned above, some sites can apparently be described by means of a local equilibrium model, while others require that mass transport kinetics limitations be included in the model. Wilson (1990) has used a lumped-parameter approach to develop a soil vapor stripping model that includes kinetic effects, and Oma et al. (1990) have examined the impact of these kinetic effects on cleanup times and costs. Analysis of data obtained during the remediation of a 1,3-dichloropropene spill near Benson, Arizona, led Sterrett (1989) to conclude that diffusion kinetics were an important factor in the rate of cleanup of the site. He also noted that high humidity resulted in more rapid release of contaminants (the "wet dog" effect). Megehee and Wilson (1991) and Rodriguez-Maroto and Wilson (1991) have recently developed a steady-state approach to diffusion/desorption-limited transport in soil vapor stripping which greatly increases the speed with which these computations can be carried out. Osejo has carried out lab studies and modeling of the removal of underlying nonaqueous-phase liquid floating on the water table; his results indicate that such material can readily be removed by vapor stripping, provided that the wells are screened near the water table (Osejo and Wilson, 1991). Kuo et al. (1990) have recently published a two-dimensional model for estimating the effective radius of influence of a SVE well.

A significant problem in SVE is the removal of VOCs from the off gas discharged. Pedersen and Curtis (1991) provide a listing of current air discharge standards affecting SVE operations and point out that the states very widely in their air emission regulations, ranging from little or no regulation to contaminant-specific mass discharge rates. Sink (Pacific Environmental Services, 1989) has prepared a report for EPA on the various options available for VOC control from SVE facilities. The report gives a short summary of remediation techniques available for contaminated soil, followed by a list of soil vapor extraction systems in the United States, followed by a detailed discussion of the treatment options available for VOC removal from the off gas from SVE operations. In the past it has often been possible to vent directly to the atmosphere; this is now generally not possible, and VOC control adds substantially to the cost and complexity of SVE operations. At present there are four technologies available for VOC removal: activated carbon absorption, thermal incineration, catalytic incineration, and condensation. Some characteristics of these are summarized in Table 2.

Carbon adsorption is well established for pollution control and solvent recovery. Although it can be used with very dilute VOC streams, its performance is better at concentrations above 700 ppmv (parts per million volume). Although units achieving 99% removal can be designed, actual efficiencies usually range from 60 to 90%, depending on VOC influent concentration, temperature, relative humidity, and maintenance. Dehumidification is usually necessary if the influent relative humidity exceeds 50%, and the stream must be cooled if its temperature exceeds 150°F.

Carbon adsorption handles variable VOC concentrations and gas flow rates better than the other three techniques and performs opti-

Table 2 Key Emission Stream Characteristics for Selecting VOC Treatment Systems at SVE Sites

	Emission stream characteristics			
Vapor treatment system	VOC conc.[a] (ppmv)	Relative humidity (%)	Temp. variation	VOC characteristics
Carbon absorption	>700	<50%	Insensitive <150°F	VOC mol. wt. between 50 and 150 g/mol for best performance
Thermal incineration	>100	—	Sensitive	Controls most VOCs without difficulty
Catalytic incineration	>100	—	Sensitive	P, Bi, Pb, Hg, As, Sn, Zn, Fe oxides, halogenated compounds may poison catalyst
Condensation	>5000	—	Sensitive <200°F	Efficiency limited by vapo pressure–temperature characteristics of VOCs

mally with compounds having molecular weights in the range 50 to 150 (compounds containing 4 to 10 carbon atoms); such compounds are good candidates for SVE. Carbon adsorption is the most commonly used VOC treatment technique in SVE; it is generally chosen unless the gas stream contains a very high concentration of combustible organics, in which case incineration may be cheaper.

Two methods for using activated carbon are employed for SVE work. The fixed-bed regenerative system allows reuse of the carbon; the other system uses carbon canisters which cannot be reused. Fixed-bed systems generally have higher capital and annualized costs. They are often used at sites where the expected cleanup time is fairly long, so that the recurring costs of canister replacement exceeds the capital costs of the regenerative system. The costs of the regenerative system are markedly reduced if steam is readily available at the site.

Thermal incineration is able to control a broader range of compounds than are the other techniques; it gives efficiencies above 99% for concentrations above 200 ppmv, and efficiencies above 95% at VOC concentrations as low as 50 ppmv. (Carbon adsorption is optimal at VOC concentrations above 700 ppmv.) Thermal incinerators are not well suited to streams with variable flow rates (fluctuations exceeding 10%), since these change mixing and residence times from design values and reduce efficiency. The vapor feed stream must be heated to 1500 to 1800°F for 1 to 2 s to oxidize hydrocarbon vapors.

Thermal incineration can be used for streams that are dilute in VOCs, as is the case with most SVE operations during much of their lifetime. Generally, supplemental fuel is required, which increases operating costs relative to the other VOC treatment techniques to the point where thermal incineration is rarely used at SVE installations. It is an economical option if little or no supplemental fuel is needed.

Catalytic incinerators are similar to thermal incinerators except that they use a catalyst to allow the combustion reactions to take place at reduced temperatures, which reduces or eliminates the need for supplemental fuel. Design efficiencies are around 95%, with actual efficiencies running about 90%, depending on operating conditions and maintenance.

Catalytic incinerators are more sensitive to pollutant characteristics and process conditions than are thermal incinerators. Halogen-containing compounds (e.g., chlorinated organics), lead, mercury, bismuth, tin, zinc, iron oxides, and phosphorus may poison the catalyst, ruining its performance. Some newer catalysts are able to tolerate higher concentrations of these substances, and it is unlikely that involatile metal compounds will be extracted in SVE operations. Chlorinated organic solvents may present a problem, although Trowbridge and Malot (1990) have described a system in which a chromium oxide–aluminum oxide catalyst is used which can accept a feed stream whose VOCs are 100% chlorinated hydrocarbons. The vapor feed stream inlet temperature must be 500 to 700°F; often, this preheating can be done by means of a heat exchanger. Catalytic incineration can achieve high destruction efficiencies even at low VOC concentrations, but the technique is sensitive to influent flow-rate fluctuations. At present, it is not as commonly used as activated carbon, but the development of halogen-resistant catalysts will certainly result in its more widespread use for SVE work. Operating costs are generally lower than those for thermal incineration.

Condensers are used as preliminary devices for recovering VOCs present in the feed stream at high concentrations; the stream is then sent to another unit for further treatment. Their requirement of inlet VOC concentrations > 5000 ppmv limits their utility in SVE work, since VOC off-gas concentrations at an SVE site are usually quite high initially but drop off relatively quickly. Removal efficiencies of condensers using chilled water are typically about 50 to 80%; values approaching 90% can be achieved if ethylene glycol or Freon is used as the refrigerant. (This increases costs significantly.) Sink did not find condensers in use at any of the 29 SVE sites he reviewed.

Terra Vac has carried out vapor stripping operations at a number of sites, and descriptions of some of these are available. One of these involved a gasoline leak in Belleview, Florida, which had resulted in contamination of a municipal wellfield. A preliminary report on the vapor stripping operation indicated removal rates averaging about 880 lb/day; removal of this material avoids further long-drawn-out contamination of the underlying aquifer during pump-and-treat operations (Applegate et al., 1987).

The Groveland, Massachusetts, site was used by Terra Vac as a demonstration site; the vadose zone here was contaminated with trichloroethylene (TCE). An 8-week test run resulted in the removal of 1300 lb of VOCs and a marked reduction in soil VOC concentrations in the test area. The author of the evaluating report felt that the results of the test were promising and indicated that the process can remove VOCs from soils of both high and low permeability (Michaels, 1989; Stinson, 1989).

Recently, Trowbridge and Malot (1990) have described three SVE sites at which VOC control was accomplished by catalytic incineration. The catalytic oxidizer removed 98.5% or better of the VOCs, which included a number of chlorinated solvents. These authors felt that catalytic oxidation is economical for the majority of SVE systems, operating at >25 to 50 lb/day VOCs and < 50 mg/L VOC concentration for most of the operation. These authors note the problem presented by diffusion-limited SVE. They report that SVE works very satisfactorily for removing light nonaqueous-phase liquid (LNAPL) floating on top of the water table. The early stages of an SVE operation, when nonaqueous-phase liquid product is present, tend to be controlled by Raoult's law; after this has been removed, an effective Henry's law becomes controlling.

The third site described by Trowbridge and Malot (1990) is the Thomas Solvents (or Verona Wellfield) site in Battle Creek, Michigan; this has been discussed in more detail by Malmanis et al. (1989) and by Danko et al. (1990). Groundwater extraction (pump and treat), started in March 1987, has removed more than 11,000 lb of VOCs, and total groundwater VOC concentrations have dropped from as high as 19,000 μg/L to about 1500 μg/L as of October 1989. The SVE system included 23 wells, a collection manifold, a centrifugal air/water separator, a blower, and an off-gas treatment unit. Initially activated carbon (nonregenerative) was used; this was later replaced by a catalytic oxidation unit. In this latter, the air is preheated to about 430°F in a heat exchanger, after which it enters a burner chamber, where it is heated to about 800°F by a natural gas burner. It then passes through the catalyst bed, where oxidation takes place, including the conversion of chlorinated VOCs to carbon dioxide, water vapor, and hydrogen chloride. The gas leaves the catalyst bed at about 820°F and enters the shell side of the heat exchanger, where it is cooled to about 550°F before being discharged through a stack. Advantages of the catalytic incinerator include no downtime for carbon changing, destruction of the VOCs on site, and substantial anticipated cost reductions. The combination of SVE with another technique (in this case groundwater pump-and-treat) will doubtless become more and more common as more complex and difficult sites are addressed.

Converse Environmental Consultants California has presented a description of a vapor extraction system at a Burbank, California, site in which the vented hydrocarbons were destroyed by combustion or catalytic oxidation (Fall et al., 1988). Gasoline had penetrated to depths of 50 ft below grade. The soil consisted of alluvial sand interspersed with silt and clay. Work at this site demonstrated diffusion (or desorption) kinetics limitation; intermittent operation of the facility was used to maintain combustible concentrations of hydrocarbons in the vented gas.

Fiedler and Shevenell (1990) have used a case study of a cleanup at a gas station by SVE to illustrate estimation of the time required and the cost of an SVE operation. They point out the importance of maintaining high airflow rates through domains of high VOC concentration. One might question their claim that SVE may be effective in remediating contaminated groundwater; diffusion constants in liquids are typically about 10^{-4} times as large as diffusion constants in the gaseous phase, so diffusive mass transport from below the water table into the vadose zone is vanishingly small. The SVE system described in the case study used on-site steam regeneration of activated carbon for VOC control; this was quite satisfactory. These authors use plots of log soil gas VOC concentration versus time to follow the progress of cleanup. This practice, which has been used by others, as well, has been trenchantly criticized by DiGiulio et al. (1990); we strongly concur with their criticism of this overly simplistic procedure.

Pope (1988) has given a summary of SVE technology, together with two brief case histories of SVE remediations at service stations. He notes that the long-term operational costs of VOC control with activated carbon are quite high; those of catalytic incineration can be 25 to 50% less. In one of these cases abatement of explosion hazards in nearby buildings occurred within 3 weeks after startup of the SVE operation.

Patterson (1989) has described a soil venting pilot study to remove benzene, toluene, ethylbenzene, and xylene from contaminated soil at a construction site and the design of a soil venting remediation facility which includes an underslab venting system to remove VOCs from below the building to be constructed at the site. Here SVE is combined with a groundwater pump-and-treat procedure.

Woodward-Clyde Consultants reported on a pilot soil vacuum extraction study near Tacoma, Washington, and provided data which strongly supported their conclusion that vacuum extraction would be effective in remediating the site (Woodward-Clyde, 1985). Roy F. Weston, Inc., reported the results of a pilot study of soil vapor stripping for removal of trichloroethylene (TCE) and other VOCs from contaminated sandy soil at the Twin Cities Army Ammunition Plant, Minnesota. They found the technology effective, but TCE removal from soils containing oily deposits was diminished. They noted that little was known about adsorption isotherms of contaminants at very low concentrations. They suggested high airflow rates and close spacing of venting wells in domains that are heavily contaminated, and noted the importance of identifying these domains in the site assessment (Anastos et al., 1985).

Batelle Columbus has investigated the enhancement of hydrocarbon biodegradation by soil venting; Hinchee, Downey, and DuPont reported on a study at Hill Air Force Base, Utah (Hinchee, 1989; Hinchee et al., 1989). Results indicated that 25% or more of the contaminant hydrocarbon was destroyed by biodegradation. Bench-scale work showed that supplemental moisture and nutrients increased the rate of biodegradation. This and other work (e.g., Dalfonso and Navetta, 1988; Brown and Harper, 1989) indicate a consensus that biodegradation can be made to play a major role in the removal of hydrocarbons.

Bailey and Gervin (1988) have provided a rather detailed description of a pilot study on the use of vapor stripping for the removal of 1,1,1-trichloroethane (TCA) and trichloroethylene (TCE) from the vadose zone at a Michigan site. A very substantial quantity of VOC was recovered during the pilot-scale operation. Mutch et al. (1989) have described a pilot SVE study at Toms River, New Jersey, in which a model for gas flow in the vicinity of a vapor extraction well was used to predict soil gas pressures around the well in the presence of an overlying stratum of low-permeability clay fill. A local equilibrium model for vapor stripping was then used to interpret the results of the pilot vapor stripping runs.

Recently, the U.S. Environmental Protection Agency (USEPA) has published a reference handbook on soil vapor extraction (Pedersen and Curtis, 1991). This includes a discussion of the principles of soil vapor behavior, including a section on air permeability test methods. The procedures to be followed in carrying out site investigations are then presented. The report then discusses system design and equipment selection, system operation and monitoring techniques for the control of vapor emissions, and costs. The handbook closes with 10 research papers on SVE and two appendices on soil cleanup criteria and air discharge criteria.

The remainder of this chapter is mainly concerned with the development of a phased approach to the evaluation and implementation of vacuum extraction on a site-specific basis, and the mathematical models for implementing such an approach. Such a phased approach may permit one to determine that the technique is not feasible at a particular site with a very small investment of time and effort, while the additional work involved if the technique turns out to be feasible is relatively slight. We close with a section on the use of these models for estimating costs of vapor stripping cleanup operations.

PRELIMINARY ASSESSMENT OF VAPOR STRIPPING

Site investigations normally break down into the following steps: (1) a site history review, (2) the preliminary site screening, (3) detailed characterization of the site, (4) contaminant assessment, and (5) pilot testing (Johnson et al., 1989b). The points to be addressed include the following items: (1) the types of quantities of contaminants released (gasoline, oxygenated and chlorinated solvents, jet and diesel fuel, etc.); (2) the extent to which the contaminants have migrated and the pathways for further contaminant movement; (3) product behavior in the soil (is it sorbed, present as NAPL, in the vapor phase, dissolved in groundwater?); and (4) the potential for injury to possible receptors (Pedersen and Curtis, 1991).

Before one invests much effort in assessing the feasibility of soil vapor stripping at a site, there are some simple preliminary exercises that should be carried out. Vacuum extraction requires that the contaminant be removed in the vapor phase. Therefore, the best that the technique can possibly do at a site is limited by the vapor pressures of the contaminants to be removed. The maximum number of kilograms of contaminant VOC that can be removed by 1 m^3 of air is given by

$$m = \frac{\mathrm{MW} \cdot P_0}{760 \cdot 0.08206 \cdot T} = 0.01603\left(\frac{\mathrm{MW} \cdot P_0}{T}\right) \tag{1}$$

where m = maximum mass of VOC removed per m^3 of air, kg/m
MW = VOC molecular weight, g/mol
P_0 = VOC vapor pressure at ambient temperature, torr
T = ambient temperature, K (273.25 + °C)

Thus, for *p*-xylene (molecular weight = 106 g/mol, vapor pressure at 20°C = 6.5 torr), one can remove no more than 0.0377 kg of VOC per cubic meter of air (0.0024 lb/ft^3). A reasonable airflow rate from a vacuum extraction well is on the order of 100 cfm; such a well could therefore remove no more than about 340 lb/day. This would probably be regarded as quite acceptable, so one would then go on to the next phase of the evaluation. A similar calculation for methylnaphthalene gives 3.2×10^{-4} kg/m^3 and a removal rate for a 100-cfm well of 2.9 lb/day. This is rather low, and one might wish to consider other technologies at this point. Vapor pressures for some common organic chemicals at various temperatures are given in Table 3. A useful formula for calculating vapor pressures as functions of temperature is

$$\log_{10} P(T) = A - 0.05223\left(\frac{B}{T}\right) \tag{2}$$

where $P(T)$ = VOC vapor pressure at temperature T, torr
T = ambient temperature, K
A = constant
B = constant (molar heat of vaporization, J/mol)

Values for the parameters A and B for a number of VOCs are given in Table 4. A rough rule of thumb is that vacuum extraction is not likely to be feasible if the vapor pressure of the compound is less than 0.5 torr (Pedersen and Curtis, 1991).

Table 3 Vapor Pressures of Selected VOCs (torr)

Compound	Molecular weight	Torr 1	10	40	100	400	760
		Temperature (°C)					
CCl_2F_2	120.9	−118.5	−97.8	−81.6	−68.6	−43.9	−29.8
CCl_3F	137.4	−84.3	−59.0	−39.0	−23.0	+6.8	23.7
CCl_4	153.8	−50.0	−19.6	+4.3	23.0	57.8	76.7
$CHCl_3$	119.4	−58.0	−29.7	−7.1	+10.4	42.7	61.3
CH_2Br_2	173.8	−35.1	−2.4	+23.3	42.3	79.0	98.6
CH_2Cl_2	84.9	−70.0	−43.3	−22.3	−6.3	+24.1	40.7
CS_2	76.1	−73.8	−44.7	−22.5	−5.1	+28.0	46.5
C_2Cl_4 (PCE)	165.8	−20.6	+13.8	40.1	61.3	100.0	120.8
C_2HCl_3 (TCE)	131.4	−43.8	−12.4	+11.9	31.4	67.0	86.7
cis-1,2-DCE	96.9	−58.4	−29.9	−7.9	+9.5	41.0	59.0
trans-1,2-DCE	96.9	−65.4	−38.0	−17.0	−0.2	+30.8	47.8
1,1-DCE	96.9	−77.2	−51.2	−31.1	−15.0	+14.8	31.7
1,1,1,2-TCA	167.8	−16.3	+19.3	46.7	68.0	108.2	130.5
1,1,2,2-TCA	167.8	−3.8	+33.0	60.8	83.2	124.0	145.9
1,1,1-TCA	133.4	−52.0	−21.9	+1.6	20.0	54.6	74.1
1,1,2-TCA	133.4	−24.0	+8.3	35.2	55.7	93.0	113.9
1,1-DCA	99.0	−60.7	−32.3	−10.2	+7.2	39.8	57.4
1,2-DCA	99.0	−44.5	−13.6	+10.0	29.4	64.0	82.4
Ethyl chloride	64.5	−89.8	−65.8	−47.0	−32.0	−3.9	+12.3
1,1,1-Trichloropropane	147.4	−28.8	+4.2	29.9	50.0	87.5	108.2
1,2,3-Trichloropropane	147.4	+9.0	46.0	74.0	96.1	137.0	158.0
1,2-Dichloropropane	113.0	−38.5	−6.1	+19.4	39.4	76.0	96.8
1-Chloropropane	78.5	−68.3	−41.0	−19.5	−2.5	+29.4	46.4
2-Chloropropane	78.5	−78.8	−52.0	−31.0	−13.7	+18.0	36.5
Pentane	72.1	−76.6	−50.1	−29.2	−12.6	+18.5	36.1
2-Methylbutane	72.1	−82.9	−57.0	−36.5	−20.2	+10.5	27.8
2,2-dimethylpropane	72.1	−102.0	−76.7	−56.1	−39.1	−7.1	+9.5
1,2,3,4-$C_6H_2Cl_4$	215.9	68.5	114.7	149.2	175.7	225.5	254.0
1,2,3,5-$C_6H_2Cl_4$	215.9	58.2	104.1	140.0	168.0	220.0	246.0
1,2,3-$C_6H_3Cl_3$	181.4	40.0	85.6	119.8	146.0	193.5	218.5
1,2,4-$C_6H_3Cl_3$	181.4	38.4	81.7	114.8	140.0	187.7	213.0
1,3,5-$C_6H_3Cl_3$	181.4	—	78.0	110.8	136.0	183.0	208.4
1,2-$C_6H_4Cl_2$	147.0	20.0	59.1	89.4	112.9	155.8	179.0
1,3-$C_6H_4Cl_2$	147.0	12.1	52.0	82.0	105.0	149.0	173.0
1,4-$C_6H_4Cl_2$	147.0	—	54.8	84.8	108.4	150.2	173.9
C_6H_5Cl	112.5	−13.0	+22.2	49.7	70.7	110.0	132.2
Benzene	78.1	−36.7	−11.5	+7.6	26.1	60.6	80.1
Cyclohexane	84.2	−45.3	−15.9	+6.7	25.5	60.8	80.7
Hexane	86.2	−53.9	−25.0	−2.3	+15.8	49.6	68.7
2,2-dimethylbutane	86.2	−69.3	−41.5	−19.5	−2.0	+31.0	49.7
Toluene	92.1	−26.7	+6.4	31.8	51.9	89.5	110.6

Table 3 Continued

Compound	Molecular weight	1	10	40	100	400	760
		Temperature (°C)					
Heptane	100.2	−34.0	−2.1	+22.3	41.8	78.0	98.4
2,2-dimethylpentane	100.2	−49.0	−18.7	+5.0	23.9	59.2	79.2
Styrene	104.1	−7.0	+30.8	59.8	82.0	—	—
Ethylbenzene	106.2	−9.8	+25.9	52.8	74.1	113.8	136.2
2-Xylene	106.2	−3.8	+32.1	59.5	81.3	121.7	144.4
3-Xylene	106.2	−6.9	+28.3	55.3	76.8	116.7	139.1
4-Xylene	106.2	−8.1	+27.3	54.4	75.9	115.9	138.3
Octane	114.2	−14.0	+19.2	45.1	65.7	104.0	125.0
2,2-dimethylhexane	114.2	−29.7	+3.1	28.2	48.2	85.6	106.8
Nonane	128.2	+1.4	38.0	66.0	88.1	128.2	150.8
Decane	142.3	+16.5	55.7	85.5	108.6	150.6	174.1

(Column headers 1–760 are pressures in Torr.)

Source: Weast et al. (1985).

If soil vacuum extraction passes this first test, there is a second that should be investigated. One of the forms by which VOCs are held in soils is in solution in the soil moisture. Such aqueous solutions obey Henry's law, which states that the vapor-phase concentration of VOC in contact with a solution of the VOC is proportional to the VOC concentration in the solution. If one assumes that the VOC is held in the soil only in solution in the soil moisture, one can obtain estimates of the cleanup times required given various limiting assumptions about the vapor stripping process. If this can be regarded as stripping from a column having a very large number of theoretical transfer units (the most efficient possible process), the time required for cleanup is given by

$$t_{100} = \frac{wV_G}{QK_H} \tag{3}$$

where w = volumetric soil moisture content, dimensionless
V_G = volume of soil being treated, m^3
Q = volumetric airflow rate, m^3/s
K_H = Henry's constant of VOC; $c(\text{vapor}) = K_H c(\text{solution})$, dimensionless
t_{100} = time required for complete cleanup, s

A less sanguine assumption for the vapor stripping process is to regard it as analogous to stripping from a well-stirred reactor. This leads to the following expressions for 90%, 99%, and 99.9% cleanup.

Table 4 Vapor Pressure Parameters of Selected VOCs[a]

Compound	Molecular weight	A	B
CCl_2F_2	120.9	7.5356	21,670
CCl_3F	137.4	7.6088	26,853
CCl_4	153.8	7.8584	33,259
$CHCl_3$	119.4	7.9412	32,273
CH_2Br_2	173.8	8.0444	36,525
CH_2Cl_2	84.9	7.9629	30,512
CS_2	76.1	7.5228	28,381
C_2Cl_4 (PCE)	165.8	8.2769	40,006
C_2HCl_3 (TCE)	131.4	7.9635	34,747
cis-1,2-DCE	96.9	8.0250	32,610
trans-1,2-DCE	96.9	7.9335	31,029
1,1-DCE	96.9	7.7616	28,466
1,1,1,2-TCA	167.8	8.1029	39,818
1,1,2,2-TCA	167.8	8.2868	42,724
1,1,1-TCA	133.4	7.8744	33,025
1,1,2-TCA	133.4	8.2994	39,493
1,1-DCA	99.0	7.8442	31,400
1,2-DCA	99.0	7.9571	34,520
Ethyl chloride	64.5	7.6915	26,265
1,1,1-Trichloropropane	147.4	8.2175	38,400
1,2,3-Trichloropropane	147.4	8.5650	46,252
1,2-Dichloropropane	113.0	8.0400	36,075
1-Chloropropane	78.5	7.7875	30,018
2-Chloropropane	78.5	7.4849	27,255
Pentane	72.1	7.6621	28,275
2-Methylbutane	72.1	7.5885	27,095
2,2-dimethylpropane	72.1	7.0864	22,794
1,2,3,4-$C_6H_2Cl_4$	215.9	8.3860	54,846
1,2,3,5-$C_6H_2Cl_4$	215.9	8.1073	51,402
1,2,3-$C_6H_3Cl_3$	181.4	7.8860	47,279
1,2,4-$C_6H_3Cl_3$	181.4	8.1429	48,555
1,3,5-$C_6H_3Cl_3$	181.4	8.0537	47,418
1,2-$C_6H_4Cl_2$	147.0	8.3849	47,034
1,3-$C_6H_4Cl_2$	147.0	8.1411	44,455
1,4-$C_6H_4Cl_2$	147.0	8.1248	44,721
C_6H_5Cl	112.5	8.2666	41,148
Benzene	78.1	7.8948	33,803
Cyclohexane	84.2	7.7703	33,026
Hexane	86.2	7.9399	32,916
2,2-dimethylbutane	86.2	7.5709	28,955
Toluene	92.1	8.2910	39,082

Table 4 Continued

Compound	Molecular weight	A	B
Heptane	100.2	8.1853	37,270
2,2-dimethylpentane	100.2	7.8432	33,285
Styrene	104.1	7.9923	40,713
Ethylbenzene	106.2	8.3462	42,072
2-Xylene	106.2	8.4295	43,453
3-Xylene	106.2	8.4726	43,169
4-Xylene	106.2	8.4075	42,646
Octane	114.2	8.6481	42,878
2,2-dimethylhexane	114.2	8.2728	38,522
Nonane	128.2	8.4220	44,251
Decane	142.3	8.3351	46,208

Source: Calculated from data of Weast et al. (1985).
[a]For use in the equation $\log_{10}P(T)$ (torr) $= A - 0.05223B/T$.

$$t_{90} = 2.303\left(\frac{wV_G}{QK_H}\right) \tag{4}$$

$$t_{99} = 4.605\left(\frac{wV_G}{QK_H}\right) \tag{5}$$

$$t_{99.9} = 6.908\left(\frac{wV_G}{QK_H}\right) \tag{6}$$

For example, if one were proposing to vapor strip *p*-xylene (with a Henry's constant at 15°C of 0.204) from 5000 m^3 of soil having a volumetric moisture content of 0.2 by means of a vacuum well operating at 100 cfm, equation (3) gives $t_{100} = 1.2$ days, while equations (4), (5), and (6) yield $t_{90} = 2.8$ days, $t_{99} = 5.5$ days, and $t_{99.9} = 8.3$ days. From this one would conclude that soil vacuum extraction should still be regarded as a feasible technology for the removal of *p*-xylene at this site.

Table 5 gives the Henry's constants for a number of organic compounds in water at various temperatures. Henry's constants for other compounds can be calculated from the VOC vapor pressure and solubility at the temperature of interest from the following formula:

$$K_H(T) = 0.01603\left[\frac{P_0(T) \cdot MW}{Tc_{\text{sat}}}\right] \tag{7}$$

Table 5 Henry's Constants for Selected VOCs

Compound	10°C	15°C	20°C	25°C
Nonane	17.21	20.98	13.80	16.92
n-Hexane	10.24	17.47	36.71	31.39
2-Methylpentane	30.00	29.35	26.31	33.72
Cyclohexane	4.433	5.329	5.820	7.234
Chlorobenzene	0.1050	0.1188	0.1418	0.1471
1,2-Dichlorobenzene	0.07015	0.06048	0.06984	0.06417
1,3-Dichlorobenzene	0.09511	0.09769	0.1122	0.1696
1,4-Dichlorobenzene	0.09124	0.09177	0.1077	0.1296
2-Xylene	0.1227	0.1527	0.1970	0.2516
4-Xylene	0.1808	0.2043	0.2681	0.3041
3-Xylene	0.1769	0.2098	0.2486	0.3041
Propylbenzene	0.2445	0.3092	0.3662	0.4414
Ethylbenzene	0.1403	0.1907	0.2498	0.3221
Toluene	0.1640	0.2081	0.2307	0.2624
Benzene	0.1420	0.1641	0.1879	0.2158
Methylethylbenzene	0.1511	0.1776	0.2091	0.2281
1,1-Dichloroethane	0.1584	0.1920	0.2340	0.2555
1,2-Dichloroethane	0.05035	0.05498	0.06111	0.05763
1,1,1-TCA	0.4153	0.4864	0.6069	0.7112
1,1,2-TCA	0.01678	0.02664	0.03076	0.03719
cis-1,2-DCE	0.1162	0.1379	0.1497	0.1856
trans-1,2-DCE	0.2539	0.2982	0.3563	0.3863
Tetrachloroethylene	0.3641	0.4694	0.5861	0.6989
Trichloroethylene	0.2315	0.2821	0.3500	0.4169
Tetralin	0.03228	0.04441	0.05654	0.07643
Decalin	3.013	3.540	4.406	4.782
Vinyl chloride	0.6456	0.7105	0.9021	1.083
Chloroethane	0.3267	0.4052	0.4573	0.4946
Hexachloroethane	0.2552	0.2364	0.2457	0.3413
Carbon tetrachloride	0.6370	0.8078	0.9644	1.206
1,3,5-Trimethylbenzene	0.1734	0.1945	0.2374	0.2751
Ethylene dibromide	0.01291	0.02030	0.02536	0.02657
1,1-DCE	0.6628	0.8585	0.9062	1.059
Methylene chloride	0.06025	0.07147	0.1014	0.1210
Chloroform	0.07403	0.09854	0.1380	0.1721
1,1,2,2-TCA	0.01420	0.00846	0.03035	0.01022
1,2-Dichloropropane	0.05251	0.05329	0.07898	0.1459
Dibromochloromethane	0.01635	0.01903	0.04282	0.04823
1,2,4-$C_6H_3Cl_3$	0.05552	0.04441	0.07607	0.07848
2,4-Dimethylphenol	0.3568	0.2850	0.4199	0.2015
1,1,2-$C_2Cl_3F_3$	6.628	9.093	10.18	13.04
Methyl ethyl ketone	0.01205	0.01649	0.00790	0.00531
Methyl isobutyl ketone	0.02841	0.01565	0.01206	0.01594
Methyl Cellosolve	1.898	1.535	4.822	1.263
CCl_3F	2.307	2.876	3.342	4.128

Source: Adapted from Howe et al. (1986).

where $P_0(T)$ = VOC equilibrium vapor pressure at temperature T, torr
MW = VOC molecular weight, g/mol
T = ambient temperature, K
c_{sat} = VOC saturation concentration in water at temperature T, g/L
$K_H(T)$ = Henry's constant of VOC at temperature T, dimensionless

A number of extensive data compilations are available (Verschueren, 1983; Mackay and Shiu, 1981; Weast et al., 1985; Lyman et al., 1982; Brookman et al., 1985; Howe et al., 1986; Danko, 1989; Montgomery and Welkom, 1989). Generally soil vacuum extraction is not feasible for compounds having dimensionless Henry's constants of less than about 0.01 (Danko, 1989; Pedersen and Curtis, 1991).

The two tests of vapor stripping described above may clearly eliminate the technique for use at a particular site. They do not, however, demonstrate that vacuum extraction is in fact feasible. VOCs are held in soils by other mechanisms than solution in the soil moisture—adsorption on clays and humic materials, for example. These mechanisms may reduce the effective Henry's constant (or linear isotherm parameter) of a VOC by a factor of 1/100 or less. The stratigraphy of the site may be such that an adequate airflow cannot be maintained in some portions of the contaminated zone. The USEPA notes (Pedersen and Curtis, 1991) that SVE is most effective if the hydraulic conductivity of the soil is above 0.001 cm/s; contaminant removals have been demonstrated in soils having hydraulic conductivities as low as 10^{-6} cm/s, however (Danko, 1989). Diffusion or desorption kinetics may be rate limiting if the soil contains lumps or lenses of porous but low-permeability clay or porous fractured rock. Or some of the VOC may be contained in buried drums. These matters are addressed in the second and third phases of the feasibility assessment.

Typically, a VOC in the vadose zone will be presented in three or four different forms: (1) as vapor in the interstitial soil gas, (2) dissolved in the soil water, (3) adsorbed on sorption sites in the soil, and (possibly) (4) as nonaqueous-phase product, in the form of droplets or ganglia. Pedersen and Curtis (1991) report that several workers have found residual NAPL occupying on the order of 10% of the available pore space after the wetting front moves through the soil. If the adsorption isotherm for the VOC in the soil is known, one can calculate the equilibrium partitioning of the VOC between these phases as follows. We have

$$c_{tot} = c_v + wc_w + c_{adG} + c_{napl} \tag{8}$$

where c_{tot} = total VOC concentration in the soil, kg/m^3 of bulk soil
c_v = VOC concentration in the soil gas, kg/m^3 of soil gas
c_w = VOC concentration in the soil water, kg/m^3 of soil water
c_{ads} = adsorbed VOC concentration in the soil, kg/m^3 of bulk soil
c_{napl} = concentration of VOC present as nonaqueous-phase liquid, kg/m^3 of bulk soil
= air-filled porosity, dimensionless
w = water-filled porosity, dimensionless

Then c_{ads} and c_w are related to c_v as follows:

$$c_w = \frac{c_v}{K_H} \tag{9}$$

$$c_{adG} = f_v(c_v) \tag{10}$$

Here $f_v(c_v)$ is the adsorption isotherm for the VOC in the soil of interest, which describes the partitioning between the adsorbed and vapor phases. This may be a linear, Langmuir, Freundlich, BET, or other isotherm.

One proceeds by initially assuming that $c_{napl} = 0$, then substituting equations (9) and (10) into (8), which yields an implicit equation for c_v,

$$c_{tot} = c_v + \frac{wc_v}{K_H} + f_v(c_v) \tag{11}$$

This equation is then solved for c_v. The solution is trivial if the adsorption isotherm f_v is a linear function of c_v; if f_v is more complex, one can use a numerical method such as binary search. If c_v is less than the equilibrium vapor concentration of the pure VOC, no NAPL-phase VOC is present, the vapor-phase VOC concentration is c_v, and the aqueous and adsorbed phase concentrations can be calculated by equations (9) and (10). If c_v is equal to or greater than the equilibrium vapor concentration of the pure VOC, one calculates c_{napl} as follows. Substitute equations (9) and (10) into equation (8), set $c_v = c_{v0}$, where c_{vo} is the equilibrium vapor concentration of the VOC, and solve for c_{napl}. This yields

$$c_{napl} = c_{tot} - \left[c_{v0} + \frac{wc_{v0}}{K_H} + f_v(c_{v0})\right] \tag{12}$$

Then c_w and c_{adG} are given by setting $c_v = c_{v0}$ in equations (9) and (10). Usually, the available data are insufficient to justify the use of any adsorption isotherm more sophisticated than a simple linear expression, which is why effective Henry's constants are commonly used.

Note that the analysis above assumes that the menu of VOCs present at the site can be adequately approximated by a single representative compound. If one is dealing with complex mixtures such as gasoline, this approach may not be adequate, and a model based on Raoult's law and the assumption that the bulk of the VOC mixture is present as NAPL may be more suitable, at least during the early stages of the remediation. The USEPA's reference handbook on SVE discusses the SVE of gasoline and similar mixtures in some detail (Pedersen and Curtis, 1991).

The second phase of vapor stripping feasibility assessment involves a laboratory study. Soil samples are taken at the site to represent, roughly, the average and highest levels of contamination and the various media present (sand, silt, till, clay, etc.). Portions of these are analyzed for VOCs, and portions are placed in laboratory soil vapor stripping columns. Stripping is carried out until analysis of the gas vented from the columns indicates that VOC removal is nearly complete. The vapor-stripped soil samples are then analyzed for residual VOCs and the percent removals calculated. A mathematical model for laboratory column operation [such as is described below and in Wilson et al. (1988)] is used to process the resulting data to obtain estimates of the effective Henry's constants for the VOCs in the media present at the site. This is most conveniently done by constructing a plot of percent removal versus effective Henry's constant for a system having the same parameters (column length and diameter, gas flow rate, soil moisture content, etc.) as the laboratory columns, and the effective Henry's constants yielding the observed percent removals are then read off the plot. Such a plot is shown in Figure 3. These Henry's constants can then be used to estimate cleanup times by means of equations (3) to (6). If these are felt to be acceptable, one is ready to carry out the third phase of the feasibility evaluation.

The laboratory column vapor stripping model is constructed as follows. The differential equation governing the pressure of an ideal gas in steady flow through a porous medium is

$$\nabla K_D \, \nabla P^2 = 0 \tag{13}$$

which for a column packed with soil of uniform permeability gives

$$\frac{d^2(P^2)}{dx^2} = 0 \tag{14}$$

where K_D is the pneumatic permeability of the soil (the Darcy's constant, $m^2 atm \cdot s$), P the pressure (atm), and x the distance measured

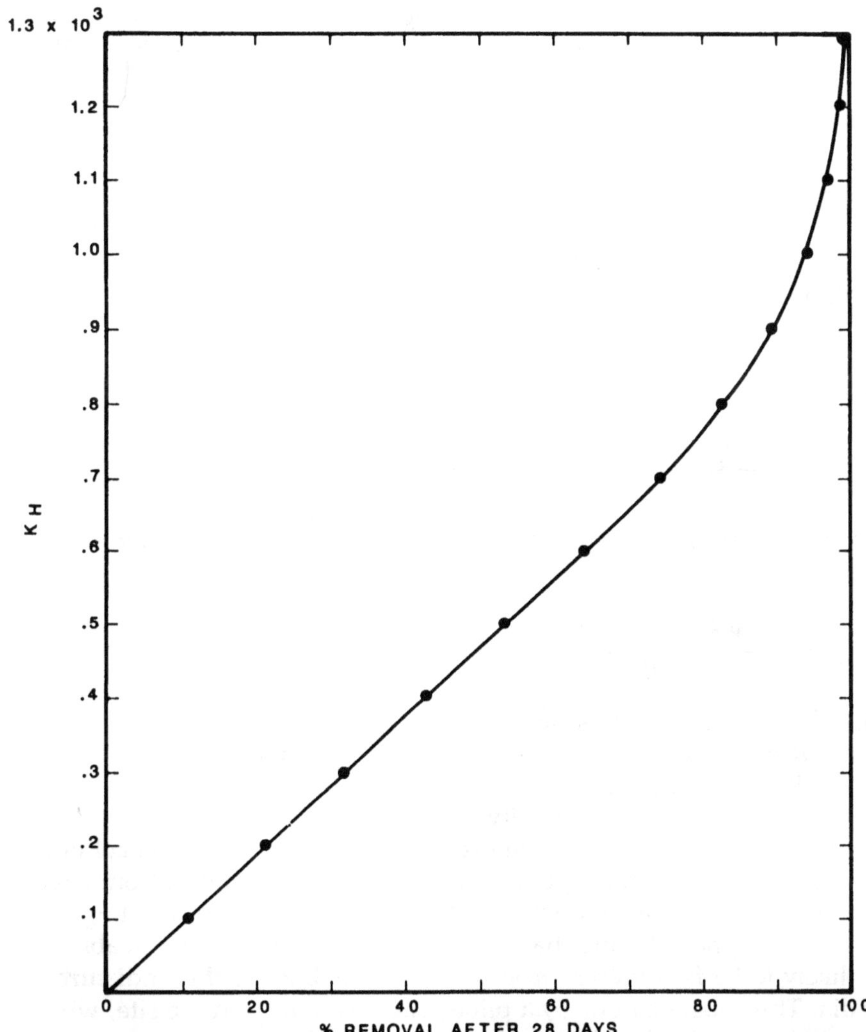

Figure 3 Plot of effective Henry's constant versus percent removal after 28 days of vapor stripping. Lab column simulation: column height, 32.1 cm; column radius, 3.15 cm; airflow rate, 5 mL/min; air-filled porosity, 0.2; volumetric moisture content, 0.2. (From Wilson et al., 1988.)

from the inlet end of the column. Integration of equation (14) and insertion of the boundary conditions at the ends of the column then yields

$$P = \left(P_i^2 - \frac{P_i^2 - P_f^2}{L}x\right)^{1/2} \tag{15}$$

where P_i = column inlet pressure, atm
P_f = column outlet pressure, atm
L = column length, m

Darcy's law,

$$v = -K_D \nabla P \tag{16}$$

then yields

$$v_x = \frac{K_D(P_i^2 - P_f^2)}{2L}\left(P_i^2 - \frac{P_i^2 - P_f^2}{L}x\right)^{-1/2} \tag{17}$$

The molar gas flow rate (mol/s) through the column is readily shown to be

$$Q = \frac{\pi r_c^2 \nu}{RT}\frac{K_D(P_i^2 - P_f^2)}{2L} \tag{18}$$

where r_c = column radius, m
R = gas constant, 8.206×10^{-5} m^3 · atm/mol · deg
T = temperature, K
ν = soil air-filled porosity

It is not advisable to use the pressure drops across the soil columns to estimate the pneumatic permeabilities of the soil samples from equation (18). The samples are sufficiently disturbed during sampling and packing into the columns that the measured pneumatic permeabilities are likely to be far different from the permeabilities of the undisturbed media. This must be done by a pilot-scale field study on the site, which constitutes the third and last phase of the feasibility assessment, or by field tests with a small portable unit. The latter can be done during the course of sampling. The USEPA's reference handbook on SVE (Pedersen and Curtis, 1991) provides discussions of methods for determining permeabilities; Appendix E of the handbook, by P. C. Johnson and his coworkers, gives rather detailed instructions for this.

In modeling the elution of a VOC from the soil column we assume a Henry's law type of isotherm (with an effective Henry's constant K_H) and local equilibrium. The column is partitioned into a set of N volume elements and a mass balance is carried out on the VOC being vapor stripped. Axial dispersion is handled by adjusting the number of volume elements into which the column is partitioned; the larger the number of volume elements, the smaller the axial dispersion. The partial differential equation describing advective transport is

$$\frac{\partial c}{\partial t} = -\frac{\nu K_H}{w + \nu K_H}\frac{\partial v_x c}{\partial x} \tag{19}$$

which we approximate by a set of ordinary differential equations, one for each volume element; this yields

$$\frac{dc_i}{dt} = \frac{\nu K_H}{\Delta x(w + \nu K_H)}[(v_x c)_{i-1} - (v_x c)_i] \qquad i = 1,2, \ldots, N \tag{20}$$

where c = VOC concentration in the column at the point x, kg/m^3
c_i = mean VOC concentration in the ith volume element, kg/m^3
w = soil volumetric moisture content, dimensionless
t = time, s

An initial VOC concentration is selected and equations (20) are then integrated forward in time by a simple predictor–corrector method. The total mass of VOC remaining in the column at time t is given by

$$m_{\text{total}} = \sum \pi r_c^2 \, \Delta x \, c_i(t) \tag{21}$$

Generally, $w >> \nu K_H$, so that removal rates are essentially proportional to the effective Henry's constant. This makes it quite easy to scale results, since one could make a plot using a reduced time $K_H t$.

In the pilot study, one or more vapor stripping wells are placed in or near regions of high contaminant concentration and/or at points where stratigraphic data (well logs, etc.) indicate potential problems may exist (regions of low permeability, buried obstacles, etc.). Preliminary site data can be used in the mathematical models to optimize the design of the pilot study. Piezometer wells can be used around a vacuum well to determine its range of influence and the effects of clay lenses, strata of low and high permeabilities, perched water tables, and so on. Well vacuum and gas flow rate measurements can be used to obtain a mean value for the pneumatic permeability in the vicinity of the well. VOC removal rates can be used to verify the effective Henry's constant values obtained in the lab study. The set of parameters resulting from completion of the third phase of the feasibility assessment is then used in the models to develop an optimized design for the facility and to simulate its operation. From these simulations one can estimate cleanup times and costs, at which point one is in position to make a final assessment of the feasibility of soil vapor stripping at the site. One can also use the results of the simulations to locate those portions of the domain of interest that will be cleaned up most slowly; particular attention may be paid to these during post-cleanup monitoring.

The soil characteristics have a very major effect on the initial transport of VOCs in the soil and on their removal by SVE. Soil porosity and grain size, stratigraphy, and soil moisture content will determine the ability of a vacuum well to deliver airflow to the contaminated soil. The effect of soil moisture content on air and water permeabilities is shown in Figure 4 (Corey, 1957). Soil moisture content and the presence of adsorption sites (on clay or on natural humic materials) will affect the ef-

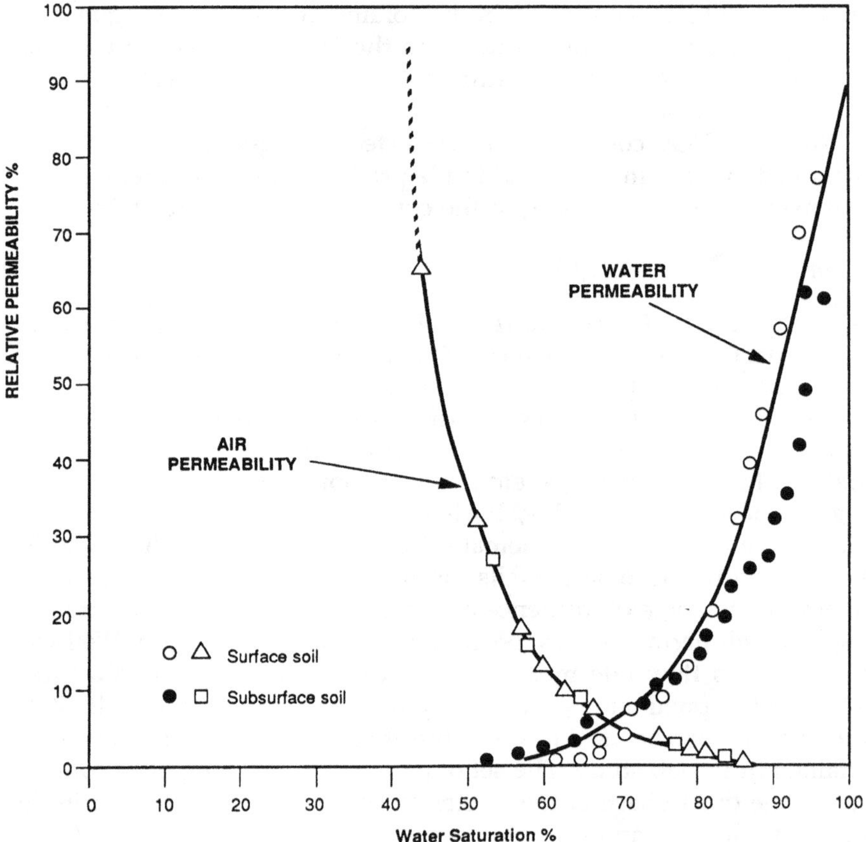

Figure 4 Air and water permeabilities as functions of soil water content. (From Corey, 1957.)

ficiency with which the advecting air will be able to move VOC from the soil. Very low soil moisture contents result in increased binding of VOCs to soil surfaces, thereby reducing SVE efficiency (Davies, 1989).

Frequently, soil vapor stripping will be used in conjunction with other remediation techniques. A pump-and-treat procedure may be necessary to remove contaminants already in the groundwater, with vacuum extraction being used to remove VOCs in the vadose zone, which would slowly recontaminate the groundwater were they not removed. It may be necessary to remove a surface layer of soil contaminated with a very slow-moving nonvolatile compound such as a heavy metal, PCBs, or chlorinated pesticides after one has removed VOCs by vapor stripping. A complex site may require the deployment of a menu of sequential and/or simultaneous technologies for optimally effective cleanup.

As seen above, mathematical modeling plays a major role in the design of the pilot study and in the design of the full-scale remediation facility. In the next section we examine some aspects of these models.

FIELD-SCALE VAPOR STRIPPING MODELS

We describe four models for simulating the operation of field vapor stripping wells. Two model a single vertical well and assume that the domain being vapor stripped is cylindrically symmetrical. The third and fourth model the operation of a horizontal screened lateral pipe. The pneumatic permeability of the soil may be anisotropic, and the domain may contain strata of different permeabilities. The gas being drawn through the soil to the vapor extraction well is assumed to obey the ideal gas law. The adsorption isotherm of the contaminant is assumed to be of the Henry's law type—that is, the equilibrium vapor pressure of the volatile organic compound (VOC) in the soil gas is assumed to be proportional to the total concentration of the VOC in the stationary phase(s), which include the soil moisture and surface adsorption sites. Modification of the models to include a more sophisticated adsorption isotherm could readily be done, but the available experimental data rarely support such a refinement.

In the first and third models the local equilibrium assumption is made—that is, it is assumed that the VOC concentration in the soil gas at any point in the domain of interest is at equilibrium with the VOC concentration in the stationary phase(s). In the second and fourth models, a lumped-parameter method is used to take into account the kinetic effects of mass transport by diffusion from the interiors of blocks of porous media of low permeability, such as clay and till. These models have been described in detail in the literature (Wilson et al., 1988, 1989;

Wilson, 1988; Gannon et al., 1989; Mutch and Wilson, 1990; Mutch et al., 1989; Gomez-Lahoz et al., 1991; Megehee and Wilson, 1991; Rodriguez-Maroto and Wilson, 1991) and have been used extensively by ECKENFELDER INC. in feasibility studies and in the design of soil vapor stripping facilities. Local equilibrium models have also been developed by Baehr et al. (1989), Johnson et al. 1989a,b, 1990; and Stephanatos (1990).

We shall discuss the development of the two models for vertical wells (the most common configuration) in some detail; the modifications to describe horizontal lateral vent pipes are straightforward (Gomez-Lahoz et al., 1991). Horizontal vent pipes are useful when the water table is shallow or if piles of excavated soil are to be treated.

The pressure distribution in the vicinity of the vapor stripping well is calculated by numerical solution of

$$\nabla K_D \nabla P^2 = 0 \tag{22}$$

by means of an overrelaxation method. Boundary conditions are as follows. Here $\mathbf{K}_D$ is the pneumatic permeability tensor ($m^2/atm \cdot s$). At the soil surface (the top of the cylindrical domain)

$$P^2 = 1 \text{ atm}^2 \tag{23}$$

At the water table (the bottom of the cylindrical domain) a no-flow boundary condition is assumed;

$$\frac{\partial P^2}{\partial z} = 0 \tag{24}$$

In these models a no-flow boundary condition is also assumed around the periphery of the domain of interest; this is

$$\frac{\partial P^2}{\partial r} = 0 \tag{25}$$

(If the vapor stripping well is surrounded by passive vent wells, this boundary condition is replaced by $P^2 = P^2_{atm}$.) The pressure inside the volume element containing the screened section of the well is taken as

$$P_s^2 = P_{\text{atm}}^2 - \frac{2r_e}{\Delta x}(1 - P_w^2) \tag{26}$$

which completes the specification of the boundary conditions. Here P_{atm} is the ambient atmospheric pressure, P_w the wellhead pressure, Δx the grid size in the finite-difference representation of the system, and r_e the screened radius of the well.

The soil gas velocity field is then calculated from

$$\mathbf{v} = -\mathbf{K}_D \nabla P \tag{27}$$

A scalar value for the pneumatic permeability K_D can be estimated from the equation

$$K_D = \frac{RTQ}{2\pi\nu(1 - P_w^2)r_e} \tag{28}$$

where R = gas constant, 8.206×10^{-5} atm · m³/mol · deg
T = temperature, K
ν = air-filled porosity of the soil
Q = molar gas flow rate, mol/s

This velocity field is then used in a mass balance equation for the total VOC concentration in the soil, m (kg/m³); for the local equilibrium model this is

$$\frac{\partial m}{\partial t} = -\frac{\nabla(vm)}{1 + w/\nu K_H} \tag{29}$$

where w is the soil moisture volume fraction and K_H is the VOC effective Henry's constant (dimensionless).

For the nonequilibrium model one has the mass balance given by the following equations:

$$m = \nu c^v + wc^s \tag{30}$$

where m = total VOC concentration in the soil, kg/m³
c^v = VOC concentration in the vapor phase, kg/m³
c^s = VOC concentration in the stationary phase, kg/m³

$$\frac{\partial c^s}{\partial t} = -\lambda\left(c^s - \frac{c^v}{K_H}\right) \tag{31}$$

where λ is the rate constant for diffusion transport and/or desorption of VOC into the mobile vapor phase s^{-1}.

$$\frac{\partial c^v}{\partial t} = -\nabla \cdot \mathbf{v}c^v + \frac{\lambda w}{\nu}\left(c^s - \frac{c^v}{K_H}\right) \tag{32}$$

In either model, an initial contaminant concentration distribution is assigned and then the partial differential equations modeling the vapor stripping are approximated by a set of ordinary differential equations defined on a mesh of points spanning the domain of interest. (The spacing of these points can be used to represent dispersion.) These equations are integrated forward in time numerically. The distribution of VOC in the domain of interest can be followed, and the total mass of contaminant remaining is tabulated as a function of elapsed time.

A major problem with using the nonequilibrium model on a microcomputer is the very large amount of computer time that is required to model reasonably realistic cases. This has amounted to as much as 100 hours on a 20-MHz machine operating with a math coprocessor. We have recently developed an approach in which the steady-state assumption is made for the vapor-phase concentration of the VOC; this decreases the computer time required to make a run by a factor of about 1/50 and introduces an error of well below 1%. This method has been implemented for the modeling of soil vapor stripping in lab columns and is being implemented for field-scale wells and horizontal laterals (Rodriguez-Maroto and Wilson, 1991; Megehee and Wilson, 1991)

It is possible to scale the results of calculations with the local equilibrium model to ascertain the effects of changes in several of the parameters in the model. Removal times (90, 99, 99.9%) depend on these parameters as follows:

Parameter	Dependence of removal time
Effective Henry's constant	Inverse
Pneumatic permeability	Inverse
Gas flow rate	Inverse
Wellhead vacuum	Inverse
Contaminant concentration	Independent
Packed radius of well	Inverse

Note that some of these dependences are linked. Gas flow rate is proportional to pneumatic permeability, wellhead vacuum, and packed radius of well. Also, while 99% removal time is independent of contaminant concentration, a domain having a very high contaminant concentration will require a higher percent cleanup.

Useful insight into vacuum extraction design can be obtained by examining the soil gas transit times—these are the times required for gas to move along one of the streamlines from the surface of the soil to the vacuum well. Portions of the domain of interest that are cleaned up by gas having a long transit time will be cleaned up correspondingly slowly. In Figures 5 and 6 transit times are indicated on the streamlines generated on one side of a vapor stripping well. In Figure 5 the well is drilled almost to the water table and is screened at the bottom. In

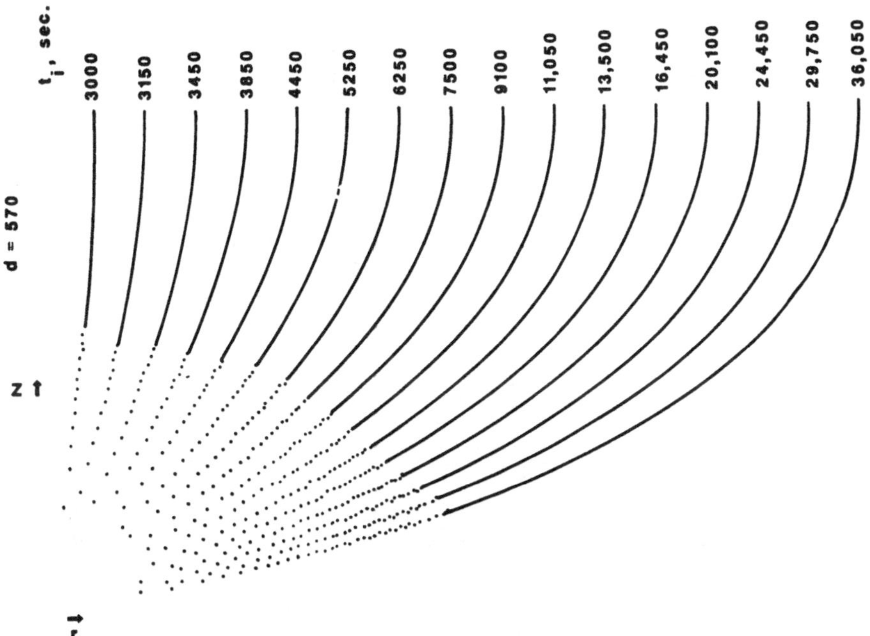

Figure 5 Streamlines and gas transit times t_i (s) in the vicinity of a vapor extraction well at a depth of 5.70 m. Depth of water table, 6.10 m; radius of well screen, 0.127 cm; wellhead pressure, 0.866 atm; permeability, 2.025×10^{-3} m^2/atm · s; air-filled porosity, 0.2; airflow rate, 2.36×10^{-2} m^3/s. (From Wilson et al. 1988.)

Figure 6 the well is drilled only about a third of the way to the water table and is screened at the bottom. The streamlines and the soil gas transit times indicate that contaminant removal by the well shown in Figure 6 will be impeded by long transit times shown out near the periphery of the domain and a virtual absence of gas flow near the bottom of the domain.

Often it is necessary to remediate soils containing buried impermeable obstacles: rocks, pieces of concrete, the remains of drums, and so on. These may have a substantial effect on the shapes of the streamlines and on the gas transit times. Figure 7 illustrates the effect of a buried circular disk 80 cm in diameter on the transit times of gas passing near it. The transit times are increased substantially, and as expected, there are small zones of stagnation above and below the center of the disk. If such obstacles are located in a portion of the domain of interest in which the gas transit times would be long even in the absence of the obstacles,

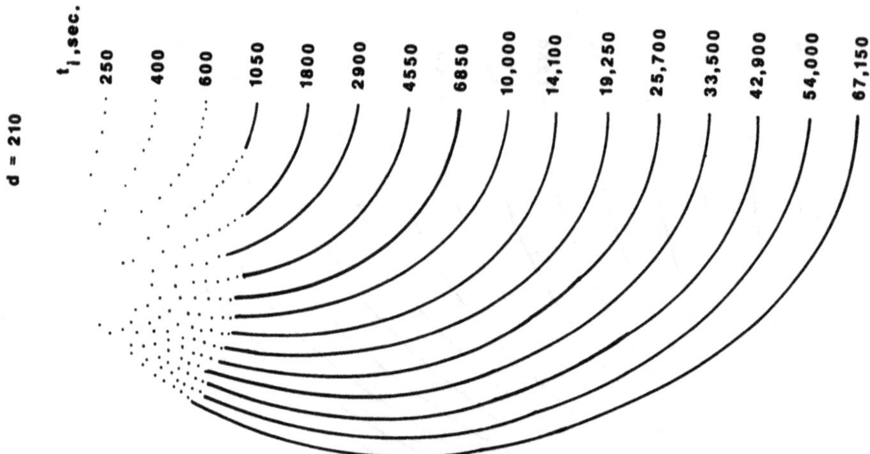

Figure 6 Streamlines and gas transit times in the vicinity of a vapor extraction well at a depth of 2.10 m. Other parameters as in Figure 5. (From Wilson et al., 1988.)

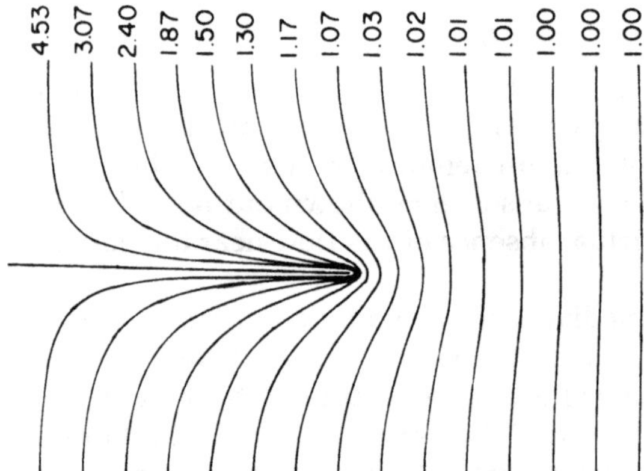

Figure 7 Gas streamlines and relative transit times around a circular disk of 0.80 m diameter in a cylindrical domain of 1.6 m diameter. (From Gannon et al., 1989.)

cleanup times for the site overall might be increased by a factor of 2 to 5. If, however, the well is drilled close to the obstacles and is screened below them, the unimpeded gas velocities along the streamlines originating near the well are sufficiently high that increases in these transit times by a factor of 2 to 5 still leaves them smaller than the transit times for streamlines originating out near the periphery of the domain of interest. Therefore, with this design domains of relatively stagnant gas flow are kept to a minimum, and the time required for overall cleanup of the domain will be affected relatively little by the presence of the obstacles.

The spacing of wells has a profound effect on the rate of cleanup. This therefore has significant impact on the operating and construction costs of vapor stripping operations. If the wells are spaced too far apart, cleanup times will be excessively long. If they are spaced unnecessarily close together, construction costs will be excessively high. If multiple wells are placed on a square grid with a distance a between the wells, the effective radius of a well's domain of influence is about $0.707a$. If the wells are placed on a hexagonal grid (each well with six nearest neighbors a distance a from it), the effective radius of a well's domain of influence is about $0.577a$. Figure 8 shows the effect of well spacing (radius of the domain of influence) on the rate of cleanup. If the distribution of the contaminant is fairly uniform, the wells are drilled through the contaminated zone, and the pneumatic permeability of the soil is constant and isotropic, cleanup rates start to fall off fairly rapidly as the effective radius of the domain of influence is increased above about 1 to 1.5 times the well depth. This is illustrated by Figure 8; in these runs the only parameter changed was the effective radius of the well domain of influence.

It is important that the wells be drilled at least to the bottom of the zone of contamination if this lies in the vadose zone, since this results in the maximum gas transit time along the streamlines having a minimum value, as seen in Figures 5 and 6. The effect of well depth on cleanup rate is seen in Figure 9; here the wells were drilled to within 3, 6, and 9 m of the water table, which was at a depth of 15 m. These results also indicate that well efficiency is probably maximized by screening the wells near the bottom only. Air that is drawn into the upper portion of a long screened section is being drawn through soil that will be cleaned up most rapidly in any case, so this portion of the airflow is wasted and merely increases the amount of air and soil moisture that must be handled by the vapor treatment system. The screened well section must be sufficiently long, however, to allow for expected fluctuations in the height of the water table.

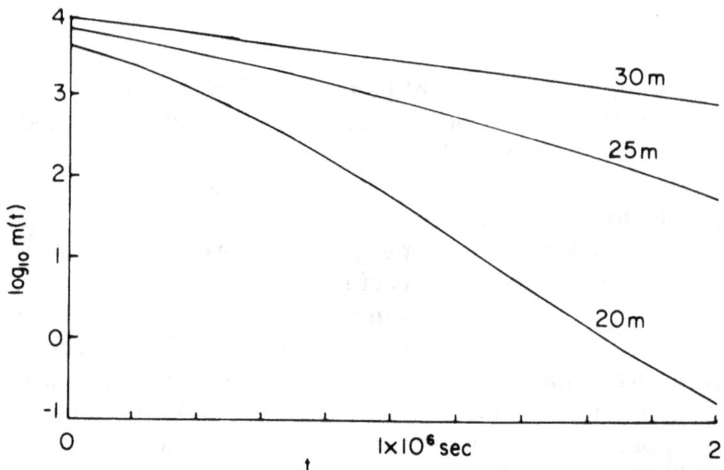

Figure 8 Plots of $\log_{10}$ total contaminant mass versus time, showing the effects of well spacing on the rate of removal. No cap was used in these runs. Depth of water table, 20 m; depth of well, 17 m; radius of zone of influence, 20, 25, and 30 m as indicated; screened radius of well, 0.12 m; molar gas flow rate, 9.1 mol/s; gas-filled porosity, 0.2; specific moisture content, 0.2; effective Henry's constant, 0.01; wellhead pressure, 0.866 atm; permeability, 6.0 m^2/atm · s; soil density, 1.6 g/cm^3; initial VOC concentration, 100 mg/kg. (From Gannon et al., 1989.)

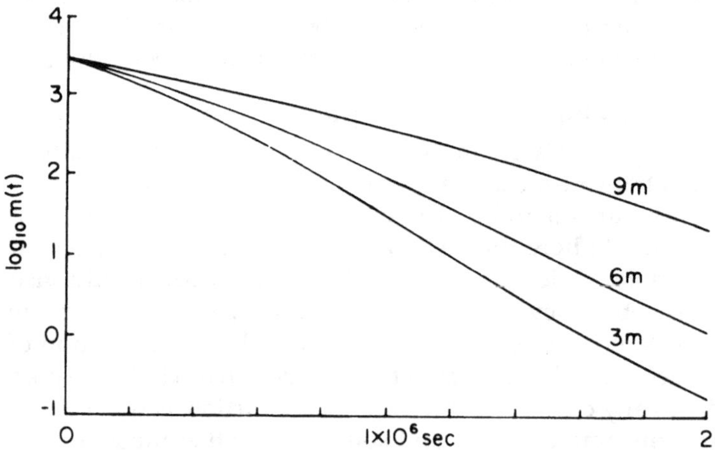

Figure 9 Plots of $\log_{10}$ total contaminant mass versus time; effect of well depth. Depth of water table, 15 m; radius of zone of influence of well, 20 m; height of well above water table, 3, 6, and 9 m. Other parameters as in Figure 8. (From Gannon et al., 1989.)

Table 6 Effects of Well Radius and Permeability on Gas Flow Rate[a]

Well radius (m)	Permeability (m^2/atm · s)	Molar airflow rate (mol/s)
0.12	0.6	0.9076
0.24	0.3	0.8905
0.24	0.6	1.7864
0.12	0.3	0.4543
0.36	0.2	0.8999
0.72	0.1	0.8760

[a]Well depth, 15 m; overlying impermeable cap radius, 15 m; radius of zone of influence, 20 m; well is screened 3 m above the water table; *T*, 25°C; porosity, 0.2; well pressure, 0.866 atm.

Generally, the major portion of the flow resistance to the moving soil gas occurs in the near vicinity of the well, where the pressure gradients are quite large and gas linear velocities are high. If the permeability of the soil is relatively low, one can compensate, within limits, by increasing the radius of the gravel packing surrounding the screened section of the well. Table 6 illustrates the effects of well radius and permeability on the molar gas flow rate of a well. If it is necessary to screen a well in a domain of relatively low permeability, one can increase the airflow rate by this means.

The flow pattern of the soil gas moving to a vapor stripping well can be influenced by the placement of an impermeable cap over the domain of influence of the well and by putting in passive vent wells around the periphery of the domain of influence. If passive vent wells are not present and the well is one in an array, the effects of impermeable caps of various sizes are shown in Figure 10. The presence of an impermeable cap placed coaxially with the well is seen to increase the rate of cleanup by as much as 50%. In many instances the cost and nuisance of the impermeable cap may make this a questionable bargain. However, these results also indicate that the presence of impermeable overlying layers such as parking lots, streets, and building floors can be expected to increase the rate of cleanup in many cases. In such situations vacuum wells should be placed as near the center of the overlying barrier as possible. If multiple wells are used and are operated at the same vacuum, the pressure gradient in the area between the wells will be quite small and cleanup of this area will therefore be extremely slow. If a large gas flow is needed, it should be achieved in this situation either by the use of a single large well with a gravel packing of large radius, or by the

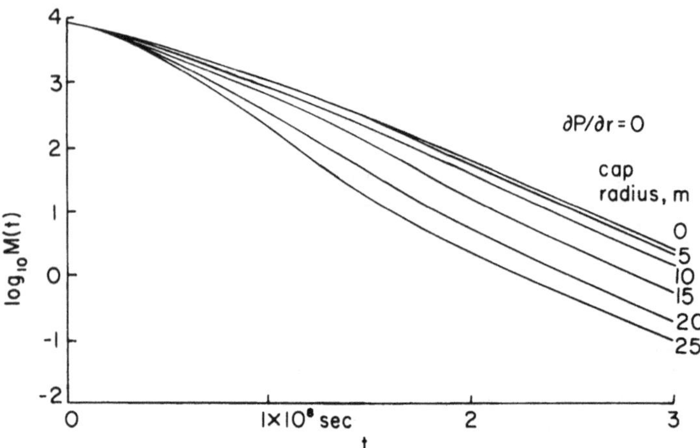

Figure 10 Log plots of total contaminant mass versus time for a field vapor stripping well. Effect of overlying impermeable caps. Permeability, 0.1 m^2/atm · s; radius of influence of the well, 30 m; molar gas flow rate, 0.1744 mol/s (volumetric flow rate = 0.00409 m^3/s). Other parameters as in Figure 8. A no-flow boundary condition is used at the periphery of the domain of influence. Cap radius is 0, 5, 10, 15, 20, and 25 m as indicated. (From Gannon et al., 1989.)

placement of passive wells between the vacuum wells to prevent the occurrence of zones of little or no gas flow.

Passive wells may produce either positive or negative results. Figure 11 shows removals for systems identical to those illustrated in Figure 10 except that the vacuum wells in Figure 11 are surrounded by passive wells screened along their entire lengths. In the absence of an overlying cap the passive wells actually appear to decrease the efficiency of this system, at least along toward the end of the cleanup, as seen in Figure 12. If a 25-m-radius cap is present, however, the presence of passive wells in this system results in a modest improvement in cleanup rate, as seen in Figure 13. In general, if it is apparent that use of passive wells will reduce or eliminate zones in which the soil gas pressure gradients are small without resulting in excessive short-circuiting of gas from other regions, passive wells will be beneficial. Otherwise, they are likely to have relatively little effect.

The presence of strata of different permeabilities can have a large effect on the soil gas pressure distribution around a vacuum extraction well and on its range of influence. Soil gas pressure measurements in the vicinity of a vacuum extraction well in Toms River, New Jersey, demonstrated that the influence of this well extended far beyond what one

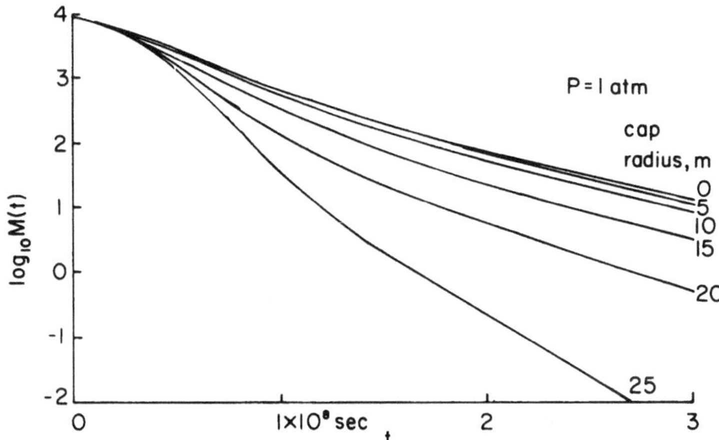

Figure 11 Log plots of total contaminant mass versus time for a field vapor stripping well; effect of impermeable cap radius. The system parameters are as in Figure 10, except that the boundary condition at the periphery of the domain of influence is $P = 1$ atm, corresponding to the presence of passive vent wells around the periphery. (From Wilson, 1990.)

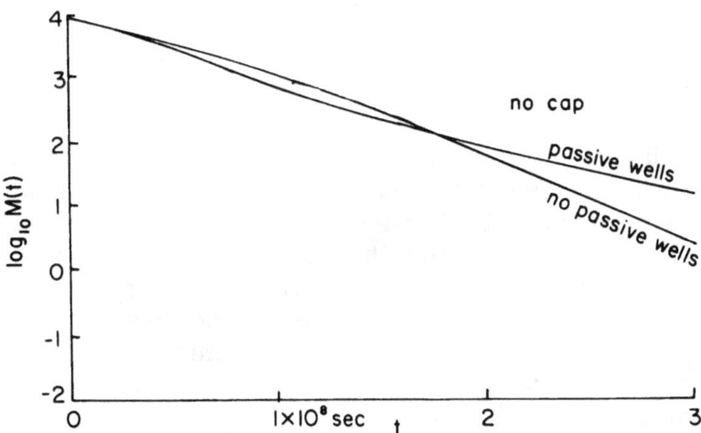

Figure 12 Comparison of systems with and without passive vent wells. The system parameters are as in Figure 10. No impermeable cap is present in these runs. (From Wilson, 1990.)

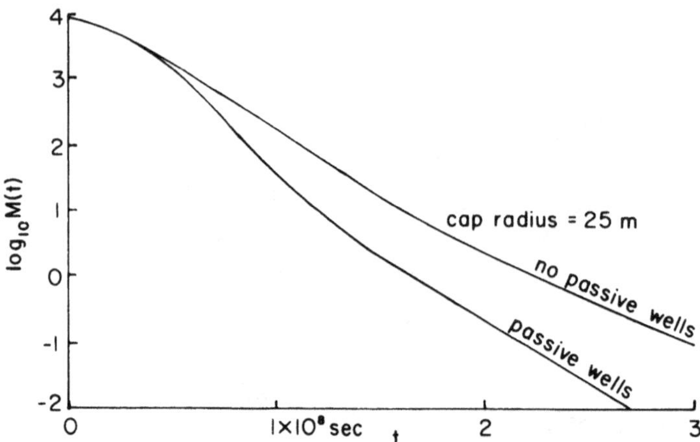

Figure 13 Comparison of systems with and without passive vent wells. The system parameters are as in Figure 10. An impermeable cap of 25 m radius is present in these runs. (From Wilson, 1990.)

would normally expect. Well log data indicated that the top meter of soil was clay and till fill, which was underlain by material that was principally sand and gravel. Assignment of a permeability to the overlying layer which was 1/200 of the permeability of the underlying medium permitted the model to calculate soil gas pressures in the vicinity of the well in good agreement with the experimentally observed values, as shown in Figure 14. The presence of this overlying low-permeability stratum would permit spacing the vacuum extraction wells substantially farther apart than normal, thereby significantly reducing costs.

Soil permeabilities show a good deal of variation from point to point, particularly if structures such as clay or silt lenses are present in the medium. These inhomogeneities can cause substantial changes in the streamlines of the soil gas, which in turn can have a marked effect on cleanup rates. In Figure 15 we see the effects of a low-permeability clay lens in the left side of the domain of interest on the streamlines of soil gas flowing from the surface to a screened horizontal lateral pipe. The gas transit times are quite significantly increased by the presence of the lens, as one might expect. A number of vapor stripping simulations were run with the clay lens located at different places in the domain of interest; plots of $\log_{10}M_{\text{total}}(t)$ for these runs are shown in Figure 16. In run 1 no lens is present. In run 5 the horizontal lateral is screened right in the middle of the lens. In run 4 the lens is located far on the left

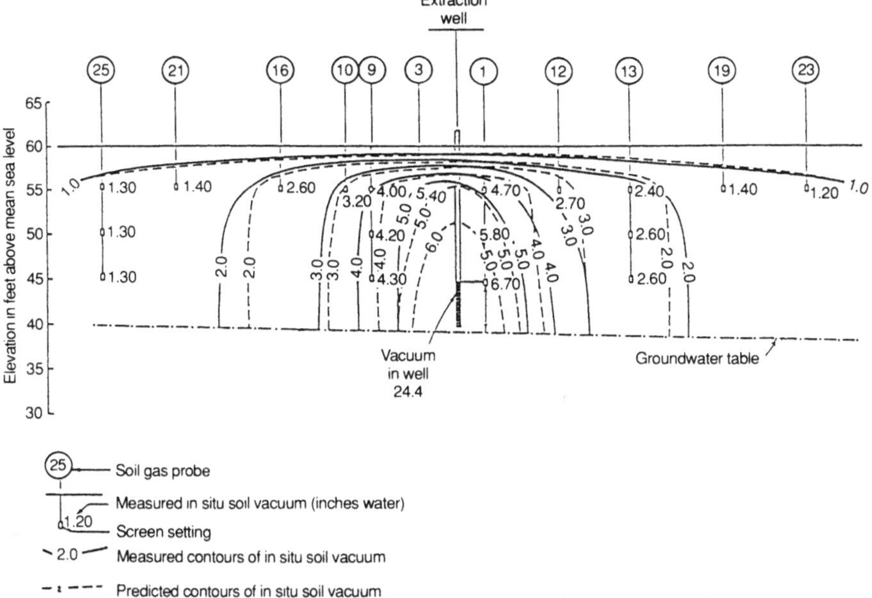

Figure 14 Impact of an overlying 1-m layer of low-permeability clay on soil gas pressures in the vicinity of a vacuum extraction well, Toms River, New Jersey. Cross-sectional in situ soil vacuum contour maps (measured and modeled) (vacuum in inches of water). (From Mutch et al., 1990.)

side of the domain of interest; the streamlines for this run are shown in Figure 15.

These results indicate that uncertainty about the values of the pneumatic permeability function can introduce substantial uncertainties into the calculation of times required for cleanup. Rarely does one have sufficient permeability data for a site to determine the permeability function with any degree of accuracy. One would be fortunate to have enough data to establish a mean value for the permeability, its range and a rough estimate of its standard deviation, the locations of major strata and lenses, and perhaps a correlation length for the permeability. One may then generate a set of permeability functions having these characteristics, model vapor stripping with them as input, and get some idea of the resulting distribution of cleanup times rather than trying to interpret a single value of unknown uncertainty for the cleanup time.

It is also possible to use these results to identify particularly unfavorable vapor stripping configurations. For example, it is evident from Figure 16 that screening a well in a low-permeability zone is disastrous.

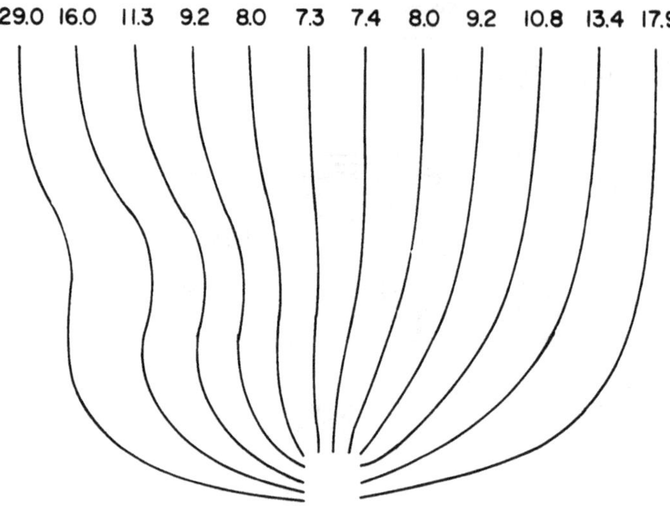

Figure 15 Streamlines in the vicinity of a screened horizontal lateral pipe; effect of a region of low permeability in the far-left portion of the domain of interest. The numbers at the top of the figure (the surface of the soil) are gas transit times in units of 1000 s. The well is located at the center bottom of the figure. (From Gomez-Lahoz et al. 1991.)

This is due to the very low gas flow rates that can be achieved. Having a low-permeability zone out near the edge of the domain of influence of a well is also very damaging (see Figure 16), because this decreases gas flow rates in a critical part of the domain where they are at best rather slow. The permeability functions selected may model the random location of one or several low-permeability lenses, as described above. They may also be Fourier series, chosen to give ranges and correlation lengths corresponding to those estimated from the experimental data; with these the stochastic variation is introduced by random selection of phase factors.

Soil moisture content has two counteracting effects on soil vapor stripping. The first is the "wet dog" effect. Reporting on a vacuum extraction operation in a very dry soil near Benson, Arizona, Sterrett (1989) observed marked increases in VOC concentrations in the soil gas after rains at the site. Davies (1989) has discussed laboratory studies of this effect in the binding of chlorobenzene to soils, and noted that the binding of a VOC to soil may be increased by as much as four orders of magnitude if the soil is very dry. The effect is explained as being due to the competition between VOC molecules and water molecules for ad-

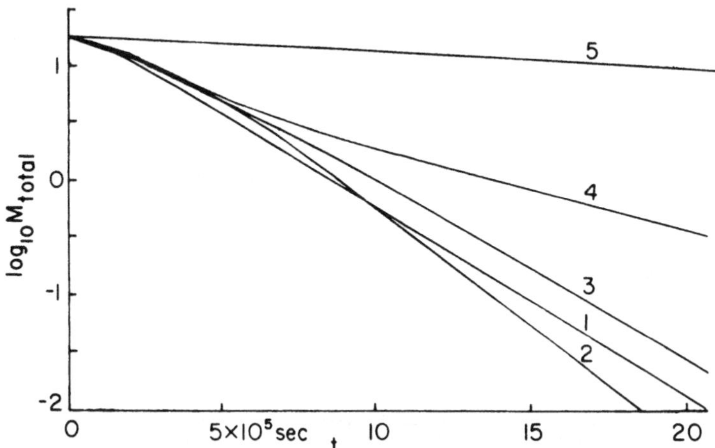

Figure 16 Effect of low-permeability domain location on the removal of VOCs by vapor stripping with a screened horizontal lateral. Log plots of total contaminant mass versus time. Domain length, 13 m; domain depth, 8 m; well located at bottom center of domain; packed radius of lateral, 0.2 m; wellhead pressure, 0.85 atm; soil gas-filled porosity, 0.3; volumetric moisture content, 0.2; permeability of matrix, 0.1 $m^2/atm \cdot s$; soil density, 1.7 g/cm^3; effective Henry's constant, 0.005; initial VOC concentration, 100 mg/kg. The domains of low permeability have cross sections 6 m in width and 3 m in thickness and have permeabilities that are Gaussian in form, with minimum values that are $\frac{1}{20}$ of the permeability of the soil matrix. (From Gomez-Lahoz et al., 1991.)

sorption sites, with highly polar, strongly binding water molecules displacing the VOC. This effect should be kept in mind when planning vapor stripping operations in arid environments, since one may need to increase the soil moisture content artificially in order to improve SVE efficiency.

A second effect of soil moisture content is its impact on the pneumatic permeability of the soil. Millington and Quirk (1961) have proposed a formula relating pneumatic permeability to volumetric moisture content; this can be written as

$$K_D(w) = \left(K_0 \frac{\nu - w}{\nu}\right)^{10/3} = K_0(1 - R_h)^{10/3} \tag{33}$$

where K_0 = pneumatic permeability of dry soil
w = volumetric water content of the soil
ν = total soil voids fraction
R_h = w/ν, the relative moisture content of the soil

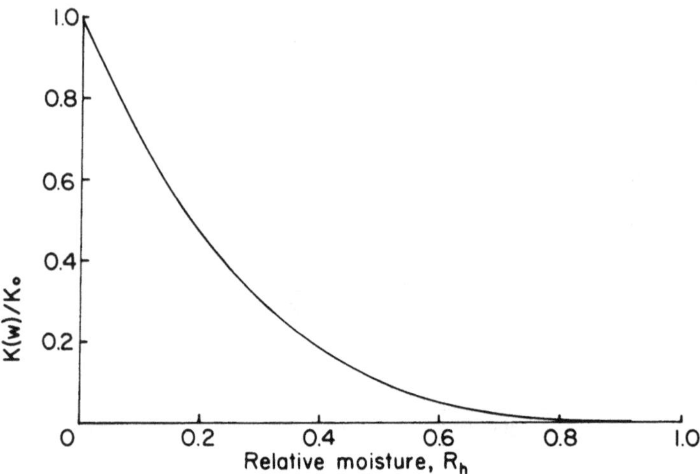

Figure 17 Plot of $K_D\ (w)/K_0$ versus relative moisture content R_h according to equation (33). (From Gomez-Lahoz et al., 1991.)

A plot of $K_D\ (w)/K_0$ versus R_h is shown in Figure 17, in which we see that permeability falls off rapidly with increasing relative moisture content. See also the experimental data plotted in Figure 4. A high soil moisture content also decreases the vapor pressure of VOCs, which are held in aqueous solution according to Henry's law.

Generally, the underlying boundary to the domain of interest in vapor stripping operations is the water table. (It may, however, be bedrock or a stratum of low-permeability clay.) One may therefore expect the relative moisture content to range from 1 (saturation) at the water table to a lower value at the soil surface; this last is determined by recent rainfall events and the atmospheric relative humidity. Note also that the position of the water table may show significant seasonal variation. For purposes of illustration we assume that the relative moisture content is a linear function of depth below the surface of the soil. The impact of this moisture on the streamlines of the soil gas in the vicinity of a screened horizontal lateral is shown in Figures 18 and 19. In Figure 18 it is assumed that the soil is completely dry throughout. In Figure 19 moisture content is taken into account. Notice here the marked increase in gas transit times near the outer portions of the domain of interest and the large increase in the size of the relatively stagnant zones in the lower corners of the domain of interest. These changes in flow pattern result in very marked increases in the cleanup times required. Evidently, in moist climates one would be well advised to exclude as much surface recharge

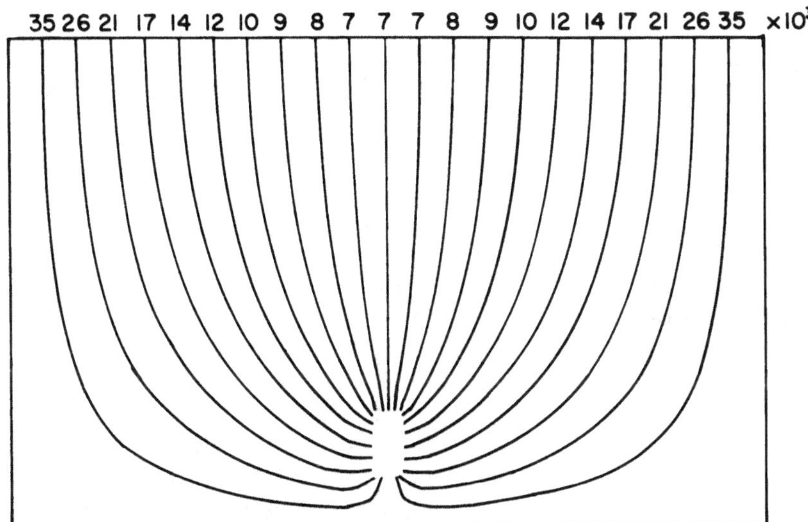

Figure 18 Soil gas streamlines in a homogeneously rather dry soil. (From Gomez-Lahoz et al. 1991.)

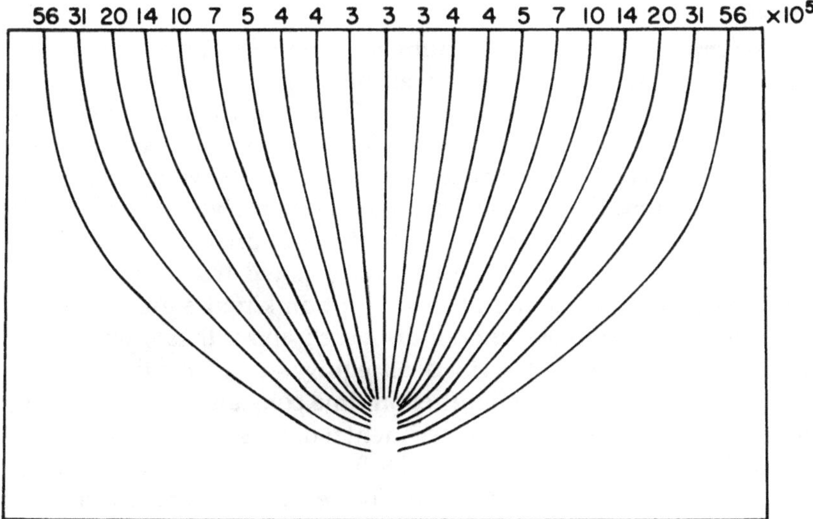

Figure 19 Streamlines in a soil in which the volumetric moisture content increases linearly with depth, approaching saturation at the water table. (From Gomez-Lahoz, et al. 1991.)

water as possible, and might also run pump-and-treat operations (if needed) in such a way as to remove as much water as possible from the contaminated vadose zone. Gomez-Lahoz (1991) has given a more detailed discussion of the effects of soil moisture.

If the soil being vapor stripped is homogeneous and relatively sandy, the local equilibrium assumption is apparently satisfactory (see, e.g., Hoag and Cliff, 1985). On the other hand, Fall (1989), Sterrett (1989) and others have found that soil gas VOC concentrations increase after the wells have been shut down for a period, which is very strong evidence that diffusion and/or desorption kinetics are playing a role and that a nonequilibrium model should be used. Well logs should be carefully examined for indications of the presence of low-permeability domains, and pilot-scale work should include tests to ascertain the extent to which diffusion and/or desorption kinetics may be the rate-limiting factor governing cleanup times. Intermittent operation of the well, with measurement of the rate of recovery of the soil gas VOC concentrations after the well has been turned off, gives information on this point. The response of the system to such intermittent operation is also affected by the distribution of contaminants with respect to the vapor extraction well. Kinetic limitations in some circumstances can be quite severe, leading to much longer cleanup times than would be estimated by local equilibrium models (Wilson, 1990; Oma et al., 1990; DiGiulio et al., 1990).

The assumption of local equilibrium between VOC in the mobile vapor phase and the stationary condensed phase(s) has been questioned by a number of workers, and both lab and field data are available which indicate that there are circumstances under which it is a poor approximation. Bouchard (1989) and Bouchard et al. (1988) have summarized batch and lab column data involving an aqueous mobile phase; the effects of desorption and of diffusion through a stationary liquid layer in soil vapor stripping should be similar. They noted that the effects of slow aqueous phase diffusion and slow sorption kinetics are very similar; in our lumped-parameter model discussed above these two effects have been combined and described as equivalent to diffusion from blocks of a porous medium. The diffusion time constant in such models must be assigned on the basis of experimental data, since it includes desorption as well as diffusion effects. Diffusion through an occluding layer of soil moisture is illustrated in Figure 20; diffusion from low-permeability silt or clay domains into high-permeability sand is illustrated in Figure 21.

Some lab column simulations showing the effects of the diffusion time constant are shown in Figure 22. The parameter set used in these

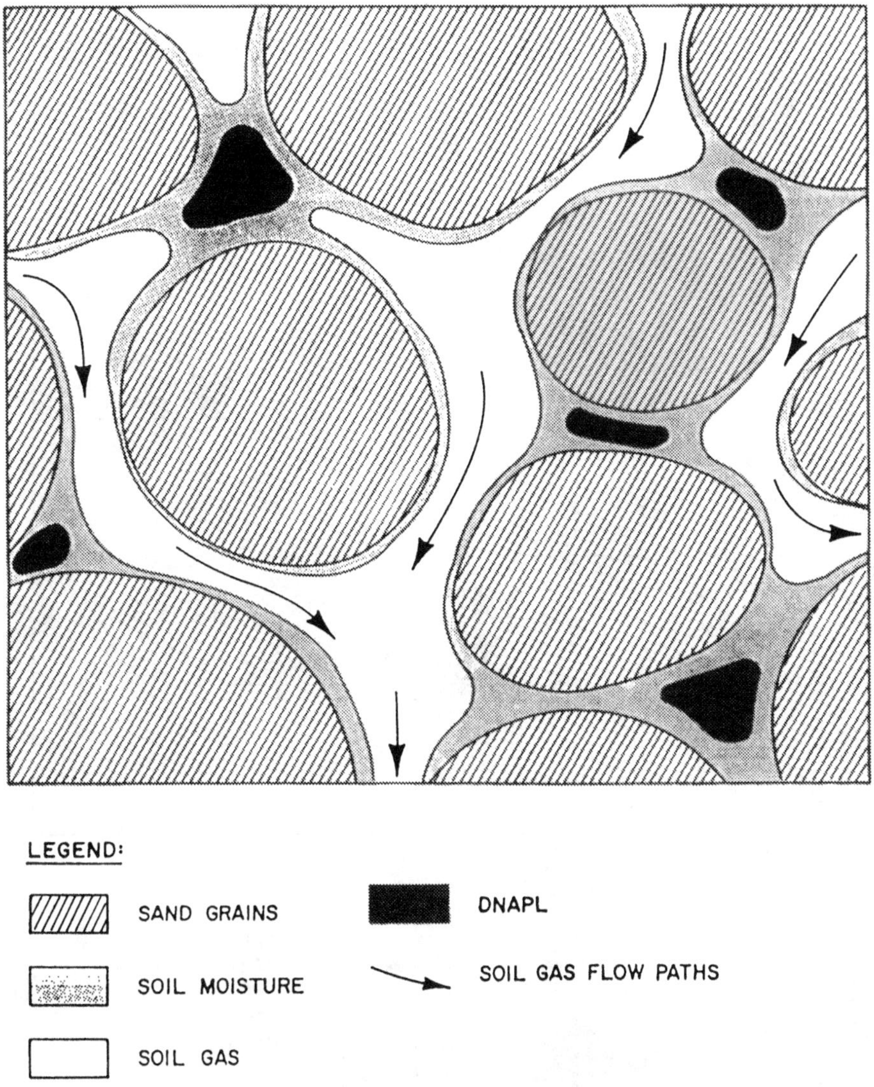

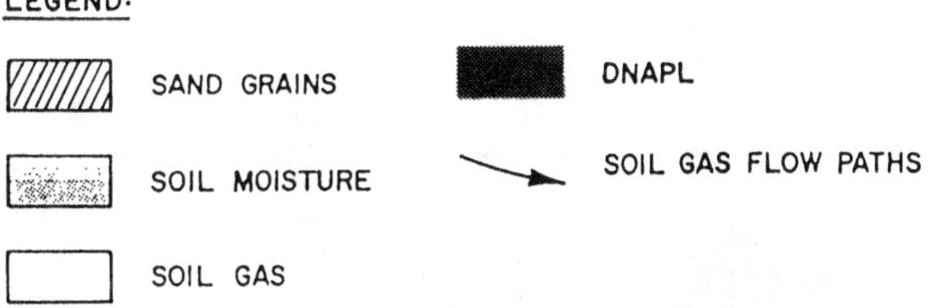

Figure 20 A mechanism for the holding of nonaqueous phase liquid in the vadose zone. VOC must diffuse through a stationary layer of water before reaching the advecting soil gas.

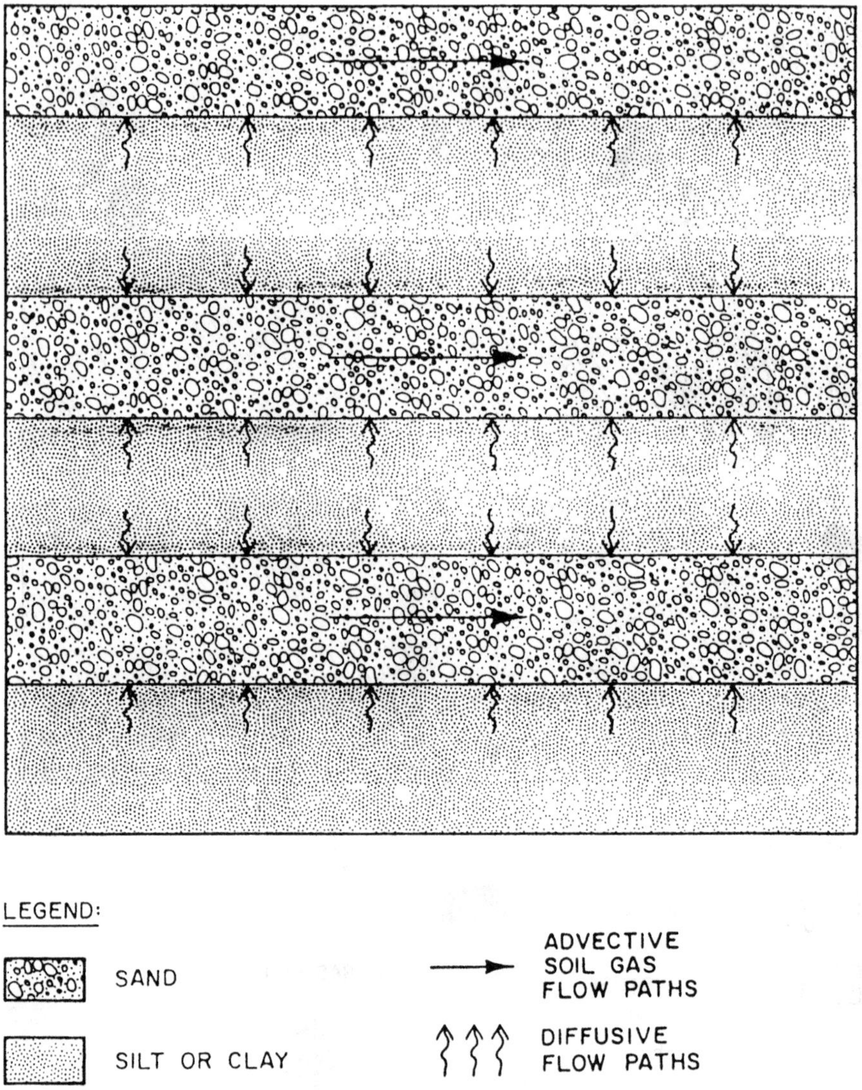

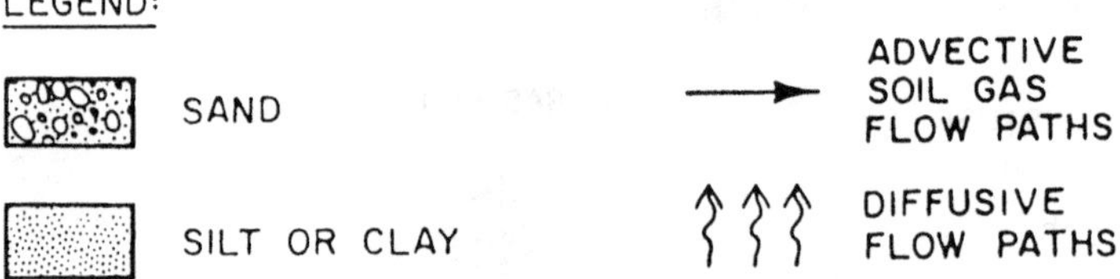

Figure 21 Diffusion and advection of VOC in interbedded formations of low and high permeabilities.

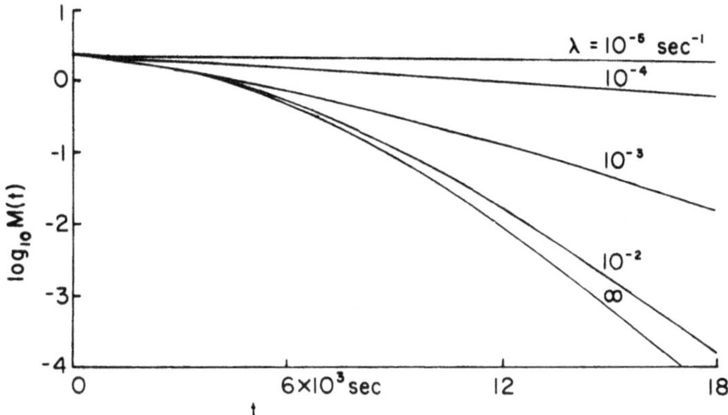

Figure 22 Log plots of contaminant mass versus time, vapor stripping in a laboratory column. Effects of diffusion/desorption rate constant $\lambda = \infty$, 10^{-2}, 10^{-3}, 10^{-4}, and 10^{-5} s^{-1}. Other parameters as in Table 7. (From Wilson, 1990.)

runs is given in Table 7. The curve labeled ∞ corresponds to the local equilibrium approximation. Departures from the local equilibrium approximation can be determined by making runs at different gas flow rates and plotting the residual contaminant mass against the volume of gas passed through the column; the larger the gas flow rate, the greater is the departure of the plot from the local equilibrium curve. An example of this is seen in Figure 23.

Lab column experiments should be quite satisfactory for investigating the effects of desorption kinetics and diffusion from low-permeability porous chunks of medium which are not broken up during the sampling process and the packing of the column. To characterize the

Table 7 Laboratory Column Standard Parameter Set

Parameter	Value
Column length	50 cm
Column radius	10 cm
Number of volume elements into which the column is partitioned	10
Voids fraction associated with mobile gas	0.2
Voids fraction associated with immobile pore liquid	0.2
Gas flow rate	5 mL/s
Effective Henry's constant	0.1
Soil density	1.6 g/cm^3
Initial contaminant concentration	100 mg/kg
dt	1 s

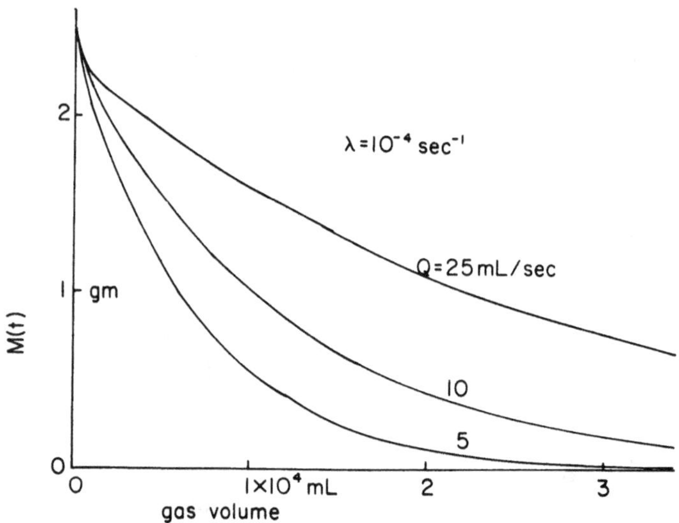

Figure 23 Plots of contaminant mass versus time, vapor stripping in a laboratory column. In these runs $Q = 10^{-4}\ s^{-1}$, and the gas flow rate through the column is 5, 10, and 25 mL/s as indicated. Other parameters as in Table 7. (From Wilson, 1990.)

effects of low-permeability porous domains of larger size (or domains that are disrupted in the preparation of lab column runs), one must carry out pilot-scale field tests. Simulations of such runs are shown in Figure 24; the system parameters for these runs are given in Table 8.

The effects of kinetic limitations on the vapor stripping can be explored in several ways.

1. The well may be operated for a time, shut down for a period, and restarted. If kinetic effects are significant, the VOC soil gas concentration will increase during the period of shutdown; the time dependence of this recovery yields the rate constant for the kinetic processes.

2. The well may be operated at several different flow rates and the VOC soil gas concentrations determined at these flow rates. If kinetic effects are significant, VOC concentrations will be lower when the gas flow rate is higher. The rate constant for the kinetic processes is determined from the relationship between the VOC concentrations and the gas flow rate.

3. Mutch (1990) has suggested that clean air be pumped down through the well and into the surrounding domain. After an interval

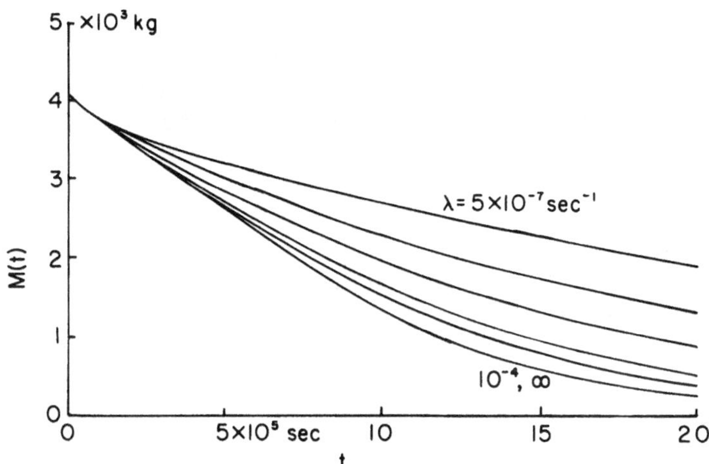

Figure 24 Plots of contaminant mass versus time, vapor stripping by a vacuum extraction well. In these runs, $\lambda =$, 10^{-4} (plots superimposed), 10^{-5}, 5×10^{-6}, 2×10^{-6}, 10^{-6}, and 5×10^{-7} s^{-1}; other parameters are given in Table 8. (From Wilson, 1990.)

Table 8 Field Vapor Stripping Well Standard Parameter Set

Radius of domain of influence	30 m
Depth of water table	20 m
Depth of well	17.5 m
Radius of impermeable cap	25 m
Screened radius of well	0.12 m
Wellhead pressure	0.866 am
Temperature	286 K
Voids fraction associated with mobile gas	0.2
Voids fraction associated with immobile pore liquid	0.2
Pneumatic permeability	1 m^2/atm · s
Effective Henry's constant	0.1
Radius of zone of contamination	20 m
Initial contaminant concentration	100 mg/kg
Soil density	1.6 g/cm^3
Initial contaminant mass	4018.2 kg
Molar gas flow rate	1.6246 mol/s
Volumetric gas flow rate	0.03815 m^3/s
dt	100 s

this gas is sampled and analyzed for VOCs. Several experiments of this type would map out the rate of equilibration of the injected air with the surrounding contaminated soil and yield the rate constant for the kinetic processes. Note, however, that the lumped-parameter method used here replaces what is presumably a rather broad distribution of time constants with a single value. Tests should be designed to estimate a value representative of the long time constants (the small rate constants), since these are what will ultimately control the rate of removal of VOC along toward the end of the remediation. Mathematical analysis of diffusion problems of this sort leads to a spectral distribution of eigenvalues—the inverses of the time constants for the decay of the system toward equilibrium. The smallest eigenvalue yields the longest time constant, which ultimately controls the rate of cleanup (Wilson, 1990).

In some situations one must deal with light nonaqueous-phase liquid which is floating on top of the water table. Osejo (1991) has carried out lab-scale experimental and theoretical work on this, using Scheidegger's (1974) formulas for longitudinal and transverse dispersion and a two-dimensional model for the evaporation of VOC vapor up from the LNAPL surface into the moving soil gas stream. He estimated that a removal rate for hexane from a pool 10 m in diameter and at 25°C of roughly 240 kg/day could be readily achieved. For *p*-xylene (which is much less volatile) at 10°C, the removal rate was estimated to be about 9 kg/day. Osejo also constructed and operated a lab-scale experimental model of a vapor stripping well in which hexane was evaporated from an underlying pool of liquid into sand; removal rates were in good agreement with calculations except at very high gas flow rates, where the rate of evaporation was apparently reduced because of evaporative cooling. Osejo's results were in agreement with earlier modeling work (Wilson et al., 1989) which predicted rapid removal rates of LNAPL pools by suitably designed wells.

CLEANUP TIMES AND COSTS

Oma et al. (1990) carried out modeling calculations for estimating the costs of vapor stripping operations; they paid particular attention to the effects of well spacing and nonequilibrium. The parameters used in the modeling work are given in Table 9, except as indicated in the figures. Simulations were made on an MMG 286 microcomputer (an IBM PC-AT clone) running TurboBASIC at 20 MHz and using an 80287 math coprocessor.

Table 9 Vapor Stripping Well Parameters Used in Modeling for Cost Estimation

Radius of zone of influence of well	6, 8, 10, 12, and 14 m
Depth of water table	8 m
Depth of well	7 m
Screened radius of well	0.3 m
Temperature	14 K
Soil porosity	0.3
Soil volumetric moisture content	0.2
Pneumatic permeability	0.1 $m^2/atm \cdot s$
Effective Henry's constant	0.005
Wellhead pressure	0.9115 atm
Diffusion/desorption time constant, $^{-1}$	5.55 h
Soil density	1.7 g/cm^3

The results of variations in the spacing of wells in a multiple-well array are shown in Figure 25. Recall that for wells in a square array the range of influence of a well is about 0.707 times the well spacing. In a hexagonal grid (each well with six nearest neighbors), the range of influence is essentially 0.577 times the distance between adjacent wells. In these calculations the well depths are all 7 m; generally, one finds rather rapid decrease in cleanup rate as the effective radius of a well is increased much over 1.5 times the depth of the well.

The effects of diffusion/desorption kinetics limitation are shown in Figure 26, in which two runs are plotted. The local equilibrium model is used for one, and a time constant of 5.55 h (rate constant of $5 \times 10^{-5}\ s^{-1}$) is used in the lumped-parameter model for the second. Cleanup times in the second run are about 4.8 times as large as in the first. Evidently, one would be well advised to carry out intermittent flow tests during the pilot-scale study to ascertain whether or not kinetic limitations will be a problem.

Cost analysis was done as follows. Comparison of costs for several vacuum extraction system configurations was done for a 2-acre example site based on the results of the numerical modeling just described. The analysis considered several extraction well spacings, local equilibrium and nonequilibrium kinetics, and two vapor treatment techniques. The vapor treatment techniques examined were granular activated carbon with off-site regeneration of the carbon, and a regenerative thermal oxidizer and acid gas scrubber. These systems are illustrated schematically in Figures 27 and 28. Air and VOCs are drawn from the extraction wells and passed through a dryer or demister to remove some of the water.

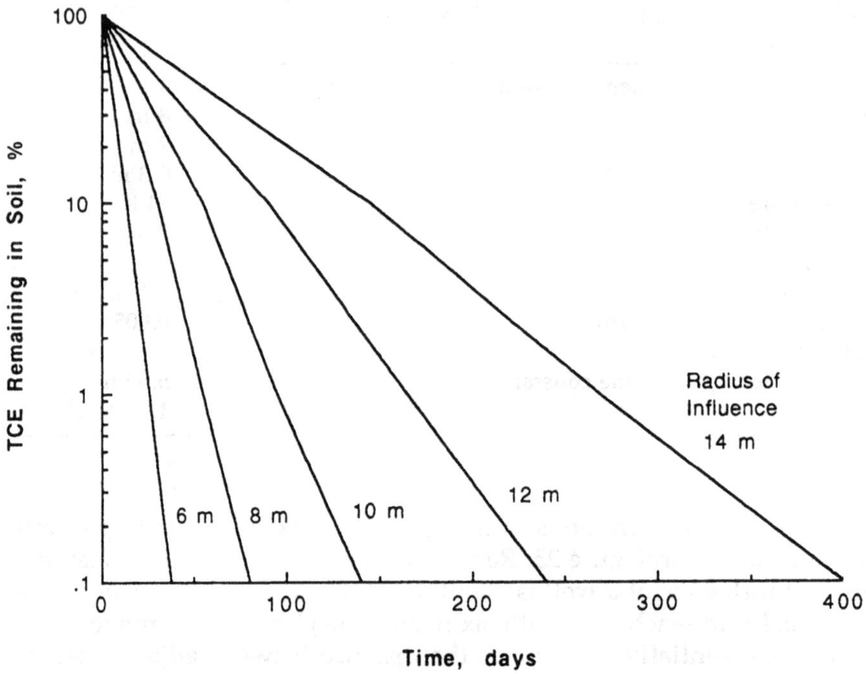

Figure 25 Effect of well radius of influence on TCE removal times for the local equilibrium model. See Table 9 for the parameter values. (From Oma et al., 1990.)

The gas then goes through an induced-draft blower and on to either primary and secondary activated carbon units or the regenerative thermal oxidizer. Cost estimates for in situ vapor stripping (ISVS) were made for both local equilibrium and nonequilibrium conditions.

Costs are divided into (1) capital investment, and (2) operating and maintenance. Capital costs consist of direct costs as follows: purchased equipment, purchased equipment installation, instrumentation and controls, piping, and electrical. Capital costs are amortized over a 5-year period at a conservatively high interest rate of 20%. Indirect costs associated with capital equipment include engineering and supervision, construction expenses, fee, and contingency. The total capital costs were derived by obtaining cost estimates for the purchased equipment, and then applying a ratio factor to estimate the other direct and indirect costs. Table 10 lists the capital investment components and their corresponding ratio factors that were used to develop the capital cost estimates. When purchased equipment cost estimates were not available for

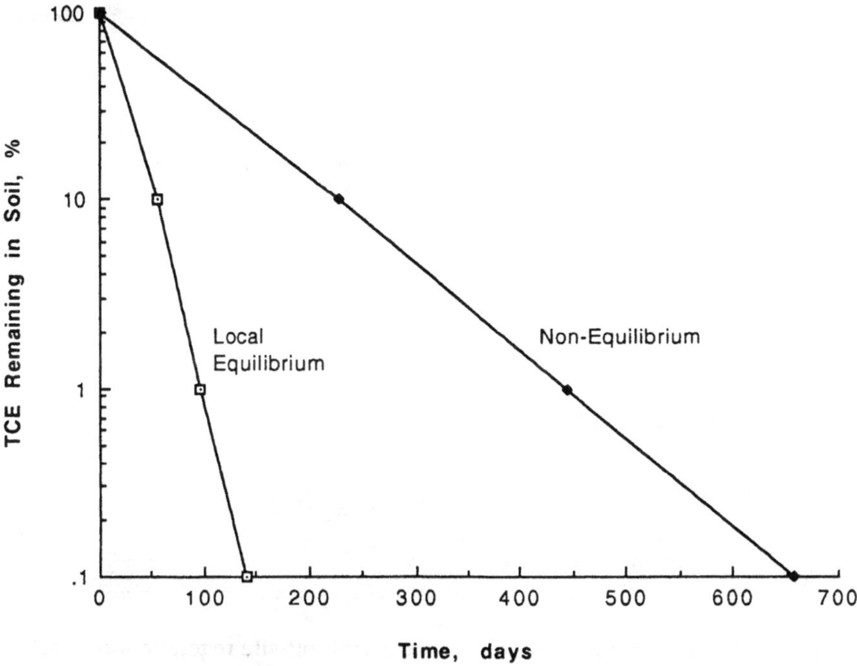

Figure 26 Comparison of TCE removals for the local equilibrium model and the nonequilibrium model. The radius of the zone of influence is 10 m, and $\lambda = 5 \times 10^{-5}\ s^{-1}$ for the nonequilibrium plot. Other parameters are given in Table 9. (From Oma et al., 1990.)

a specific scale of SVE equipment, the six-tenths-factor rule was used to approximate the costs of the new scale (Peters and Timmerhaus, 1968). This rule is as follows:

$$\frac{\text{cost of scale A}}{\text{cost of scale B}} = \left(\frac{\text{flow capacity of scale A}}{\text{flow capacity of scale B}}\right)^{0.6} \tag{34}$$

Assumptions for operating and maintenance costs are given in Table 11. These costs include project labor, analyses, electric power, and activated carbon. Cost estimates for natural gas were included for the system with regenerative thermal oxidation. Costs related to activated carbon in Table 11 are applied only to the SVE system having activated carbon with off-site regeneration. Another cost that is considered separately is that of extraction well installation. This is estimated to be approximately $2800 per extraction well plus a mobilization/demobilization charge of

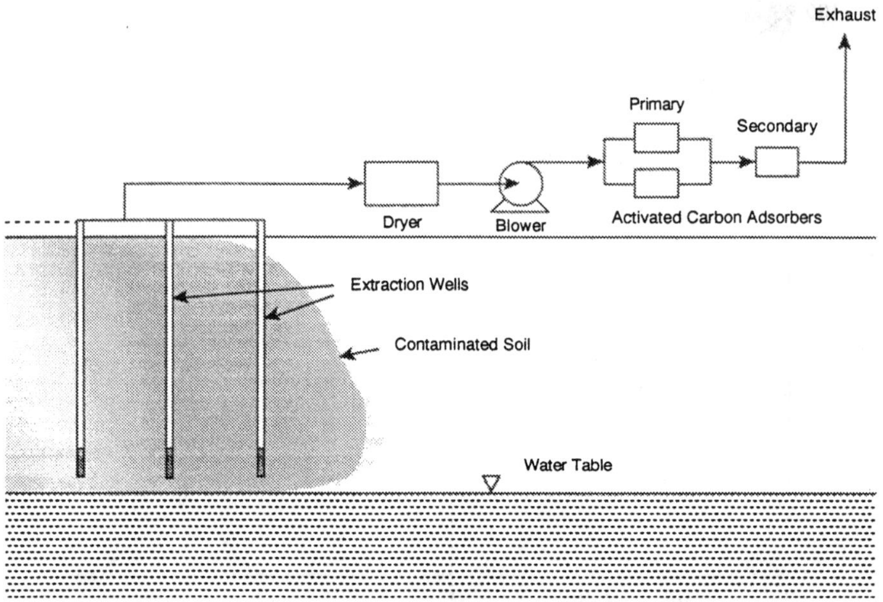

Figure 27 In situ vacuum extraction system with off-site regeneration of activated carbon. (From Oma et al., 1990.)

Table 10 Capital Investment Components

Capital cost component	Ratio factor[a]
Direct	
Purchased equipment	23
Purchased equipment installation	9
Instrumentation and controls (installed)	3
Piping (installed)	9
Electrical (installed)	3
Indirect	
Engineering and supervision	8
Construction expense	8
Fee	3
Contingency	9

[a]Ratio factors derived from Peters and Timmerhaus (1968), p. 104.

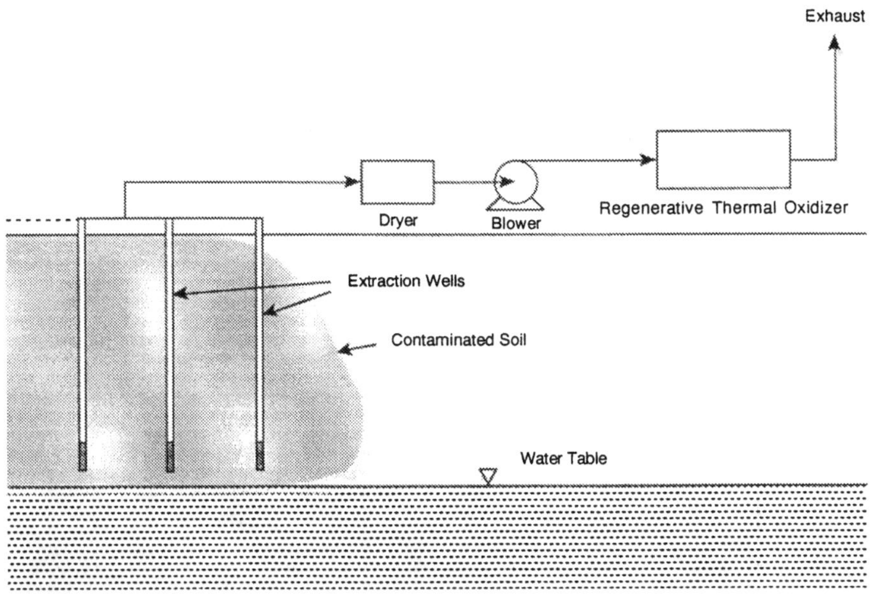

Figure 28 In situ vacuum extraction system with regenerative thermal oxidizer. (From Oma et al. 1990.)

Table 11 Annual Operating and Maintenance Cost Basis

O&M cost component	Cost basis
Labor	
Startup cost (first year)	8% of fixed capital[a]
Maintenance	4% of fixed capital[a]
Operating cost	16 h/week at $50 per hour
Carbon changeout	6 h at $50 per hour per changeout
Analytical	$300 per changeout
Electric power	Assumes $0.065 per kilowatthour
Activated carbon	
Quantity per absorber	9000 lb
Purchase cost	$1.90 per pound
Disposal/regeneration cost	$0.59 per pound
Transportation cost	$0.40 per pound

[a]From Peters and Timmerhaus (1968), pp. 116 and 134.

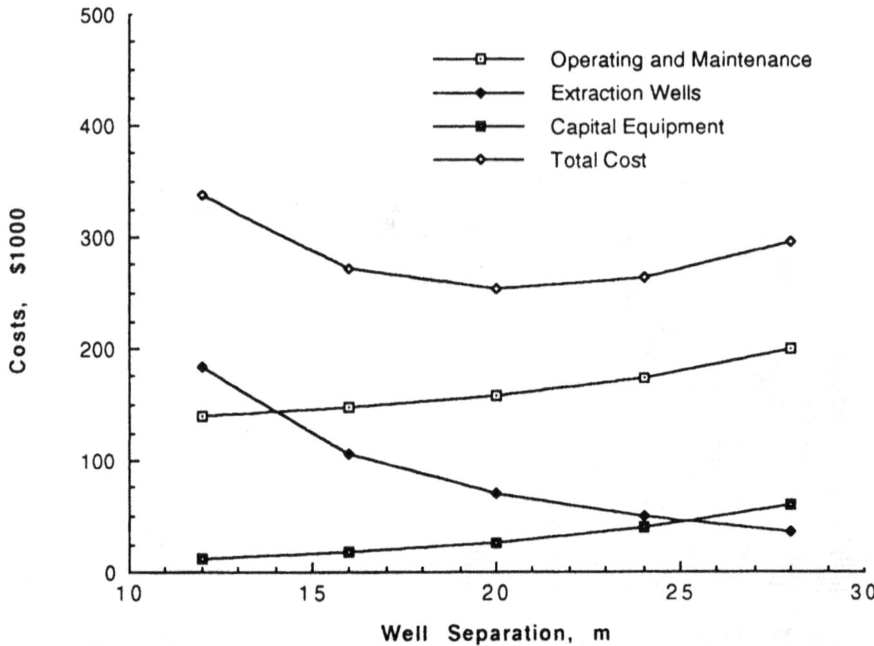

Figure 29 SVE costs for the equilibrium model with off-site regeneration of activated carbon (From Oma et al., 1990.)

$3000. Insurance and permitting costs, which are highly variable, are not included in the calculations.

Figure 29 shows the cost of SVE as a function of well separation for the local equilibrium model with off-site activated carbon regeneration for TCE removal. Figure 30 shows costs when thermal regenerative oxidation is used. For both cases the optimum well separation is about 18 to 22 m. The most cost-effective SVE system employing regenerative oxidation would have a cost of approximately $200,000; optimal cost for a system using off-site regeneration of activated carbon would be about $250,000 for this 2-acre site.

Cost estimates have also been developed for the nonequilibrium diffusion-controlled model; these are shown in Figures 31 and 32. Costs for the SVE system using off-site carbon regeneration are substantially higher for the nonequilibrium model than for the local equilibrium model, due principally to the longer cleanup time required and the resulting decreased carbon efficiency resulting from lower TCE concentrations in the off gas. The least-cost SVE system for the nonequilibrium

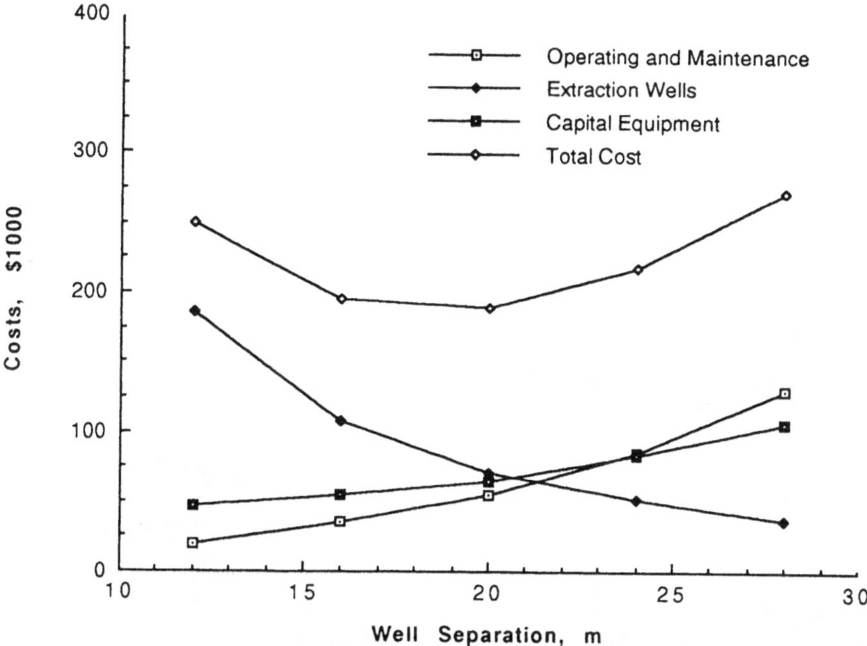

Figure 30 SVE costs for the equilibrium model with thermal oxidation. (From Oma et al., 1990.)

model is about $425,000 using either on-site thermal oxidation or activated carbon with off-site carbon regeneration. The most economical well separation appears to be between 12 and 17 m for the nonequilibrium model. If the optimal well separation suggested by the equilibrium case were employed (18 to 22 m), the cost of the example ISVS system rises to more than $500,000 under the nonequilibrium conditions.

Figure 33 compares the total cost estimates for the different cases that were evaluated. The significant impact that nonequilibrium conditions can have on site remediation is readily seen; costs for cleanup under nonequilibrium conditions for these parameter values are about twice the costs for cleanup under local equilibrium conditions.

The following conclusions can be drawn from this model-based cost analysis.

1. Even modest nonequilibrium (kinetic) effects (a time constant of 5.55 h) can have substantial impact on the rate of cleanup and on the overall cost of an SVE system.

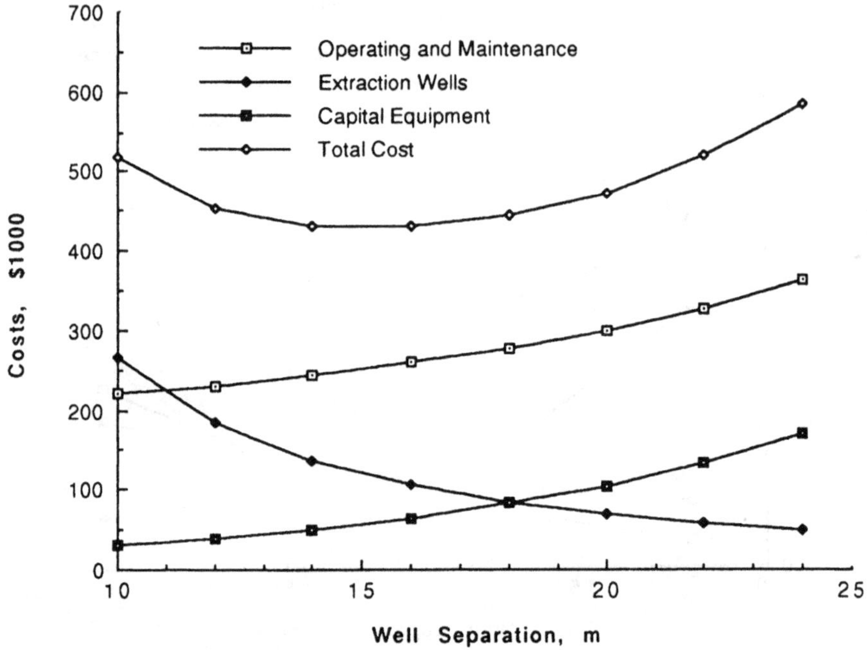

Figure 31 SVE costs for the nonequilibrium model with off-site regeneration of activated carbon. (From Oma et al., 1990.)

2. The extent to which nonequilibrium conditions reduce the rate of cleanup affects the choices of well spacing, pumping rates, and vapor treatment systems. A vapor treatment system that may be most cost-effective where local equilibrium is valid may be less desirable than another system if the cleanup is prolonged due to nonequilibrium effects.
3. Consideration of the possibility of nonequilibrium effects should be taken in the evaluation and design of SVE systems. Pilot-scale studies should be designed to estimate the rate constant for diffusion/desorption-controlled release of VOC. As mentioned above, this can often be accomplished by intermittent pumping or by repeated injection/extraction cycles.
4. The selection of the method of off-gas treatment may have a significant impact on costs. Figure 34 indicates the optimum VOC concentrations for various off-gas treatment technologies.

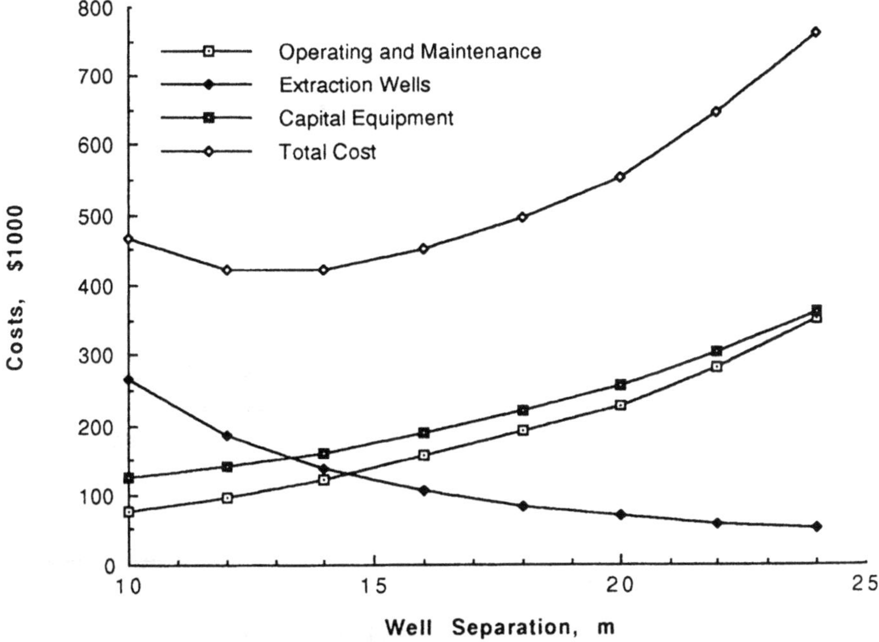

Figure 32 SVE costs for the nonequilibrium model with thermal oxidation. (From Oma et al., 1990.)

For further information on cost estimation for SVE, the reader is referred to the excellent discussion given in EPA's SVE reference handbook (Pedersen and Curtis, 1991).

CONCLUSIONS AND NEEDED RESEARCH

Soil vacuum extraction is not a broad-spectrum technology; for many sites it is not feasible and may not even be possible. On the other hand, when it can be used it has some distinct advantages over many of the competing techniques. When one considers soil vacuum extraction for use at a hazardous waste site, several points must be kept in mind.

1. The technology is applicable only to volatile compounds; it is well adapted for use with many solvents, but cannot be used for metals, PCBs, or most pesticides. It also cannot be used for the removal of water-soluble solvents such as ethanol or acetone, since the vapor

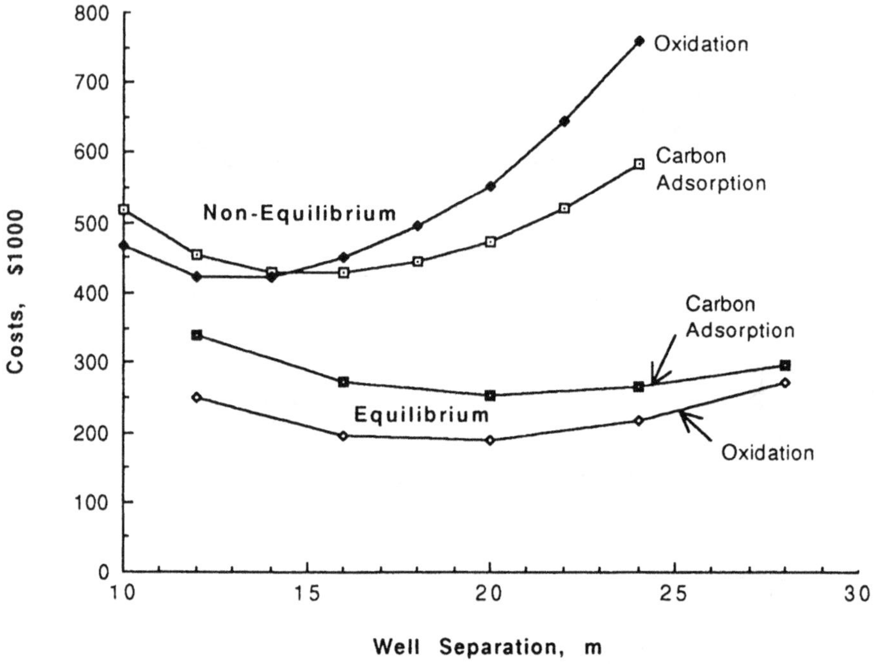

Figure 33 Summary of ISVS cost estimates. (From Oma et al., 1990.)

pressures of these compounds above dilute aqueous solution are too low to permit economical vapor stripping. Breakdown of the latter compounds by biodegradation may well be enhanced, however, by the presence of high oxygen concentrations in the advecting soil gas.

2. Vapor stripping is effective only for contaminants which are above the water table or (in the case of LNAPLs) floating on it. Water tables may be modified by pumping and/or the placement of barriers, which gives one a little flexibility, but one must be able to draw soil gas through the zone of contamination at a significant rate in order to carry out soil vapor stripping. Contaminated domains of low permeability may also be a problem in this regard.

3. The technology is relatively cheap and rapid, has a comparatively low environmental impact, and results in elimination of the contaminated material or its concentration into a small volume of highly concentrated, easily handled waste that may be incinerated or recycled.

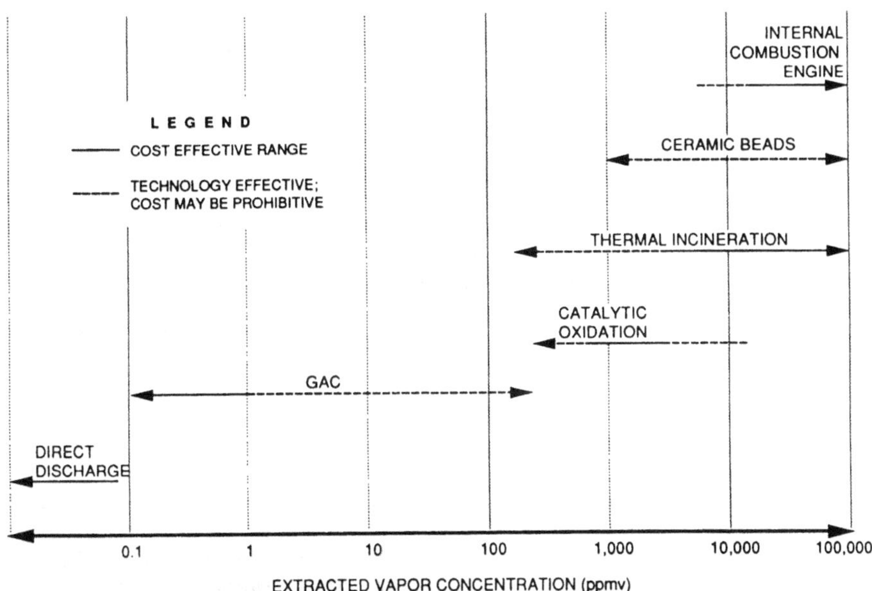

Figure 34 Applicability of SVE off-gas treatment options. (From Pedersen and Curtis, 1991.)

4. A fairly extensive base of models is available for assessing feasibility, designing pilot and full-scale operations, predicting cleanup times, estimating costs, and determining the positions of post-cleanup hot spots. The models can be used to do sensitivity analyses to assess the accuracy with which cleanup times are determined and to determine where expenditure of funds and effort for additional site characterization will produce maximum improvement in the quality of the model estimates.

The decision tree given in Table 12 charts the organization of the process by which SVE is evaluated and planned on a site-specific basis. This is a rather complex process. However, as more and more experience with the technique accumulates, it will be possible to shorten the process by taking advantage of similarities between the site under consideration and other sites at which SVE has been used successfully or unsuccessfully.

There are a number of research needs still unmet. One such need is the determination of better adsorption isotherms for a wider range of VOCs in a broader spectrum of soil matrices. The use of a linear isotherm (common to most work at present) is obviously a rather rough

Table 12 Soil Vapor Stripping Decision Tree

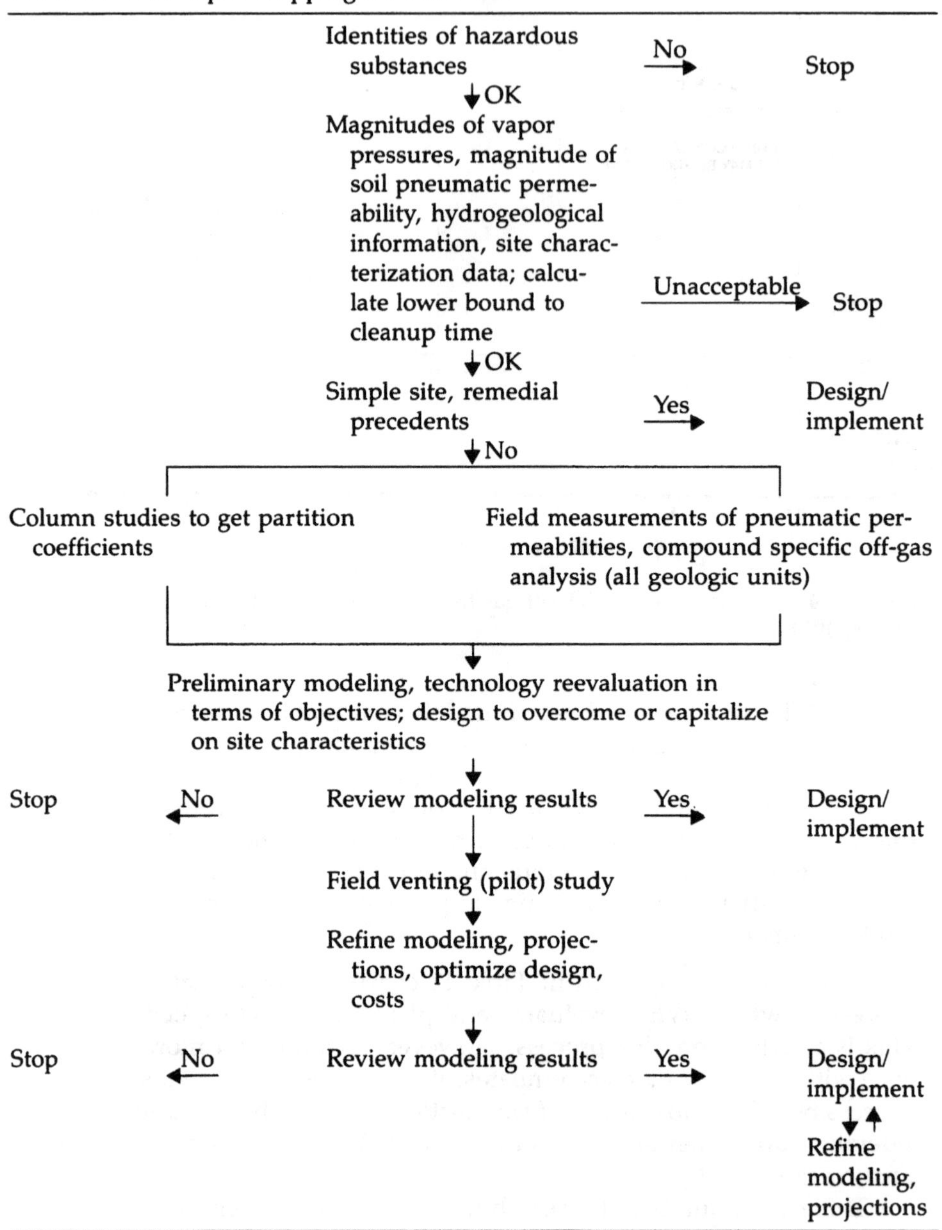
Identities of hazardous substances
No
Stop
OK
Magnitudes of vapor pressures, magnitude of soil pneumatic permeability, hydrogeological information, site characterization data; calculate lower bound to cleanup time
Unacceptable
Stop
OK
Simple site, remedial precedents
Yes
Design/implement
No
Column studies to get partition coefficients
Field measurements of pneumatic permeabilities, compound specific off-gas analysis (all geologic units)
Preliminary modeling, technology reevaluation in terms of objectives; design to overcome or capitalize on site characteristics
Stop
No
Review modeling results
Yes
Design/implement
Field venting (pilot) study
Refine modeling, projections, optimize design, costs
Stop
No
Review modeling results
Yes
Design/implement
Refine modeling, projections

first approximation that is used because generally nothing better is available. Better isotherms could easily be used in the models if the experimental database to support this were available. Intuitively, one would expect that the effective Henry's constant of a VOC would decrease, perhaps very substantially, during the course of the cleanup as the more weakly bound VOC molecules are removed. This raises doubts about the validity of parameters evaluated under conditions in which the soil VOC concentration has not been depleted nearly to completion.

A good many more rate data on diffusion/desorption kinetics of VOCs in a variety of soil matrices are needed to provide a good practical understanding of these kinetic effects. In view of their potential damaging impact on cleanup times, this is necessary if adequate modeling, design, and cost estimates are to be done. Systematic procedures for assigning kinetic parameters should be developed and evaluated.

Many of the sites that are suited to treatment by soil vapor stripping are contaminated with volatile chlorinated organic solvents, particularly trichloroethylene (TCE). In the past, the soil gas exhausted from such a site had to be treated with activated carbon. The high moisture levels and low VOC concentrations in the soil gas made this an inefficient and expensive process. There is great need for improved cheaper methods for removing chlorinated VOCs from the soil gas before it is vented to the atmosphere, although the new catalytic oxidizers look very promising for chlorinated VOCs.

REFERENCES

American Petroleum Institute. 1980. *Examination of Venting for Removal of Gasoline Vapors from Contaminated Soil*. API Publication 4429. API, Washington, DC.

American Petroleum Institute. 1982. *Enhancing the Microbial Degradation of Underground Gasoline by Increasing Available Oxygen*. API Publication 4428. API, Washington, DC.

Anastos, G. J., P. J. Marks, M. H. Corbin, and M. F. Coia. 1985. *In Situ Air Stripping of Soils Pilot Study*. Report AMXTH-TE-TR-85026. Submitted by Roy F. Weston, Inc., to U.S. Army Toxic and Hazardous Materials Agency, Aberdeen Proving Ground, MD, October 1985.

Applegate, J., J. K. Gentry, and J. J. Malot. 1987. Vacuum extraction of hydrocarbons from subsurface soils at a gasoline contamination site. *Proceedings of the 8th National Conference—Superfund*, Washington, DC, p. 273.

Baehr, A. L., G. E. Hoag, and M. C. Marley. 1989. Removing volatile contaminants from the unsaturated zone by inducing advective air-phase transport. *J. Contam. Hydrol.* 4:1.

Bailey, R. E., and D. Gervin. 1988. In situ vapor stripping of contaminated soils: a pilot study. *Proceedings of the First Annual Hazardous Materials Management Conference Central*, Rosemont, IL, March 15–17, p. 207.

Bouchard, D. C. 1989. The role of sorption in contaminant transport, *Workshop on Soil Vacuum Extraction*, April 27–28. U.S. Environmental Protection Agency, Robert S. Kerr Environmental Research Laboratory, Ada, OK.

Bouchard, D. C., A. L. Wood, M. L. Campbell, P. Nkedi-Kizza, and P. C. S. Rao. 1988. Sorption nonequilibrium during solute transport. *J. Contam. Hydrol.* 2:209.

Brookman, G. T., M. Flanagan, and J. O. Kebe. 1985. *Literature Survey: Hydrocarbon Solubilities and Attenuation Mechanisms*. API Publication 4414. American Petroleum Institute, Washington, DC, August.

Brown, R. A., G. E. Hoag, and R. D. Norris. 1987. The remediation game: pump, dig, or treat. *Water Pollution Control Federation Conference*, October 5–8.

Clarke, A. N., and D. J. Wilson. 1988. A phased approach to the development of in situ vapor stripping treatment. *Proceedings of the First Annual Hazardous Materials Management Conference Central*, Rosemont, IL, March 15–17, p. 191.

Crow, W. L., E. P. Anderson, and E. Minugh. 1985. *Subsurface Venting of Hydrocarbon Vapors from an Underground Aquifer*. API Publication 4410. Submitted by Riedel Environmental Services and Radian Corporation. American Petroleum Institute, Washington, DC.

Dalfonso, T. J., and M. S. Navetta. 1988. In situ treatment of contaminated soils using vacuum extraction. *DOE Model Conference Abstracts*, Oak Ridge, TN, October 3–7, p. 59 (abstract only).

Danko, J. 1989. Applicability and limitations of soil vapor extraction for sites contaminated with volatile organic compounds. *Soil Vapor Extraction Technology Workshop*, June 28–29. U.S. Environmental Protection Agency, Risk Reduction Engineering Laboratory, Edison, NJ.

Danko, J. 1990. Soil vapor extraction applicability and limitations. *Proceedings of the 5th Annual Hazardous Materials Management Conference/West*. Tower Conference Management Co., Glen Ellyn, IL.

Danko, J. P., M. J. McCann, and W. D. Byers. 1990. Soil vapor extraction and treatment of VOCs at a Superfund site in Michigan.

Davies, S. H. 1989. The influence of soil characteristics on the sorption of organic vapors. *Workshop on Soil Vacuum Extraction*, April 27–28. U.S. Environmental Protection Agency, Robert S. Kerr Environmental Research Laboratory, Ada, OK.

DiGiulio, D. C., J. S. Cho, R. R. Dupont, and M. W. Kemblowski. 1990, Conducting field tests for evaluation of soil vacuum extraction application. *Proceedings of the 4th National Outdoor Action Conference on Aquifer Restoration, Ground Water Monitoring and Geophysical Methods*, Las Vegas, May 14–17, p. 587.

Fall, E. W., et al. 1988. In situ-hydrocarbon extraction: a case study. *Southwestern Ground Water Focus Conference*, Albuquerque, NM, March 23–25; see also *Hazard. Waste Consultant*, January/February 1989, p. 1–1.

Fiedler, F. R., and T. C. Shevenell. 1990. How to solve the remediation twin dilemnas: how much? and how long? A case study using vapor extraction techniques for gasoline contaminated soils. *Proceedings of the 4th National Outdoor Action Conference on Aquifer Restoration, Ground Water Monitoring and Geophysical Methods*, Las Vegas, May 14–17, p. 587.

Gannon, K., D. J. Wilson, A. N. Clarke, R. D. Mutch, Jr., and J. H. Clarke. 1989. Soil cleanup by in situ aeration. II. Effects of impermeable caps, soil permeability, and evaporative cooling. *Sep. Sci. Technol.* 24:831.

Gomez-Lahoz, C., J. M. Rodriguez-Maroto, and D. J. Wilson. 1991. Soil cleanup by in situ aeration. VI. Effects of variable permeabilities. *Sep. Sci. Technol.* 26, 133.

Hinchee, R. E. 1989. Enhanced biodegradation through soil venting. *Workshop on Soil Vacuum Extraction*, April 27–28. U.S. Environmental Protection Agency, Robert S. Kerr Environmental Research Laboratory, Ada, OK.

Hinchee, R. E., D. Downey, and R. DuPont. 1989. Biodegradation associated with soil venting. *Soil Vapor Extraction Technology Workshop*. June 28–29. U.S. Environmental Protection Agency, Risk Reduction Engineering Laboratory, Edison, NJ.

Hoag, G. E. 1989. Soil vapor extraction research developments. *Soil Vapor Extraction Technology Workshop*, June 28–29. U.S. Environmental Protection Agency, Risk Reduction Engineering Laboratory, Edison, NJ.

Hoag, G. E., and B. Cliff. 1985. The use of the soil venting technique for the remediation of petroleum-contaminated soils. In: E. J. Calabrese and P. T. Kostechi (eds.), *Soils Contaminated by Petroleum: Environmental and Public Health Effects*. Wiley, New York.

Howe, G. B., M. E. Mullins, and T. N. Rogers. 1986. *Evaluation and Prediction of Henry's Law Constants and Aqueous Solubilities for Solvents and Hydrocarbon Fuel Components*, Vol. I: *Technical Discussion*. USAFESE Report ESL-86-66. U.S. Air Force Engineering and Services Center, Tyndall Air Force Base, Panama City, FL, 86 pp.

Hutzler, N. J., D. B. McKenzie, and J. S. Gierke. 1989a. Vapor extraction of volatile organic chemicals from unsaturated soil. *Abstracts, International Symposium on Processes Governing the Movement and Fate of Contaminants in the Subsurface Environment*, Stanford, CA, July 23–26.

Hutzler, N. J., B. E. Murphy, and J. S. Gierke. 1989b. Review of soil vapor extraction system technology. *Soil Vapor Extraction Technology Workshop*, June 28–29. U.S. Environmental Protection Agency, Risk Reduction Engineering Laboratory, Edison, NJ.

Johnson, S. E., G. H. Emrich, and M. A. Apgar. 1986. On-site removal of volatile organic contaminants from soils. *9th Annual Madison Waste Conference*, September 9–10.

Johnson, P. C., M. W. Kemblowski, and J. D. Colthart. 1989a. Practical screening models for soil venting applications. *Workshop on Soil Vacuum Extraction*, April 27–28. U.S. Environmental Protection Agency, Robert S. Kerr Environmental Research Laboratory, Ada, OK.

Johnson, P. C., M. W. Kemblowski, J. D. Colthart, D. L. Byers, and C. C. Stanley. 1989b. A practical approach to the design, operation, and monitoring of in-situ soil venting systems. *Soil Vapor Extraction Technology Workshop*, June 28–29. U.S. Environmental Protection Agency, Risk Reduction Engineering Laboratory, Edison, NJ.

Johnson, P. C., M. W. Kemblowski, and J. D. Colthart. 1990. Quantitative analysis for the cleanup of hydrocarbon-contaminated soils by in situ soil venting. *Ground Water* 28:413.

Keech, D. A. 1989. Subsurface venting research and venting manual by the American Petroleum Research Institute. *Workshop on Soil Vacuum Extraction*, April 27–28. U.S. Environmental Protection Agency, Robert S. Kerr Environmental Research Laboratory, Ada, OK. Most of the work done for API has been carried out by Texas Research Institute, Radian Corp., and Reidel Environmental Services.

Kerfoot, W. B. 1990. Soil venting with pneumatically installed shield screens. *Proceedings of the 4th National Conference on Aquifer Restoration, Ground Water Monitoring and Geophysical Methods*, Las Vegas, May 14–17, p. 571.

Kuo, J. F., E. M. Aieta, and P. H. Yang. 1990. A two-dimensional model for estimating radius on influence of a soil venting process. *Proceedings of the Hazardous Materials Conference '90*, Anaheim, CA, April 17–19, p. 197.

Lyman, W. J., W. F. Reehl, and D. H. Rosenblatt. 1982. Handbook of Chemical Property Estimation Methods: Environmental Behavior of Organic Compounds. McGraw-Hill, New York.

Mackay, D., and W. Y. Shiu. 1981. A critical review of Henry's law constants for chemicals of environmental interest. *J. Phys. Chem. Ref. Data* 10(4):1175.

Malmanis, E., D. W. Fuerst, and R. J. Piniewski. 1989. Superfund site soil remediation using large-scale vacuum extraction. *Proceedings of the 6th National Conference on Hazardous Wastes and Hazardous Materials*, New Orleans, April 12–14, p. 538.

Marley, M. C., and G. E. Hoag. 1984. Induced venting for the recovery/restoration of gasoline hydrocarbons in the vadose zone. *NWWA/API Conference on Petroleum Hydrocarbons and Organic Chemicals in Groundwater*, Houston, November 5–7.

Marley, M. C., S. D. Richter, B. L. Cliff, and P. E. Nangeroni. 1989. Design of soil vapor extraction systems: a scientific approach. *Soil Vapor Extraction Technology Workshop*, June 28–29. U.S. Environmental Protection Agency, Risk Reduction Engineering Laboratory, Edison, NJ.

Marley, M. C., P. E. Nangeroni, B. L. Cliff, and J. D. Polonsky. 1990. Air flow modeling for in situ evaluation of soil properties and engineered vapor extraction system design. *Proceedings of the 4th National Outdoor Action Conference on Aquifer Restoration, Ground Water Monitoring and Geophysical Methods*, Las Vegas, May 14–17, p. 651.

Megehee, M. M., and D. J. Wilson. 1991. Soil cleanup by in situ aeration. VIII. High-speed modeling of kinetic effects. In preparation.

Michaels, P. A. 1989. *Technology Evaluation Report: TerraVac In Situ Vacuum Extraction System, Groveland, Massachusetts*. EPA/540/S5-89/003, May.

Millington, R. J., and J. M. Quirk. 1961. Permeability of porous solids. *Trans. Faraday Soc.* 57:1200.

Montgomery, J. H., and L. M. Welkom. 1989. *Groundwater Chemicals Desk Reference*, Vols. 1 and 2. Lewis Publishers, Chelsea, MI.

Mutch, R. D., Jr., and D. J. Wilson. 1990. Soil cleanup by in situ aeration. IV. Anisotropic permeabilities. *Sep. Sci. Technol.* 25:1.

Mutch, R. D., Jr., A. N. Clarke, and D. J. Wilson. 1989. In situ vapor stripping research project: a progress report. *Proceedings of the 2nd Annual Hazardous Materials Conference/Central*, Rosemont, IL, March 14–16, p. 27.

Oma, K. H., D. J. Wilson, and R. D. Mutch, Jr. 1990. In situ vapor stripping: the importance of nonequilibrium effects in predicting cleanup time and cost. *Proceedings of the Hazardous Materials Management Conferences and Exhibition/International*, Atlantic City, NJ, June 5–7.

Osejo, R. E., and D. J. Wilson. 1991. Soil cleanup by in situ aeration. IX. Removal and underlying LNAPL. *Sep. Sci. Technol., 26*, 1433.

Pacific Environmental Services, Inc. 1989. *Soil Vapor Extraction VOC Control Technology Assessment*. USEPA Report EPA-450/4-89-017, September.

Patterson, J. H. 1989. Case history: soil venting as a construction safety/remediation method for development of contaminated property. *Proceedings of the Seminar on Contamination and the Constructed Project*, Connecticut Society of Civil Engineers/Connecticut Ground-Water Association Hawthorne Inn, Berlin, CT, November 2–3.

Pedersen, T. A., and J. T. Curtis. 1991. *Soil Vapor Extraction Technology Reference Handbook*. USEPA Report EPA/540/2-91/003. U.S. Environmental Protection Agency, Risk Reduction Engineering Laboratory, Edison, NJ.

Peters, M. S., and K. D. Timmerhaus. 1968. *Plant Design and Economics for Chemical Engineers*, 2nd ed. McGraw-Hill, New York.

Pope, J. L. 1988. Abatement/Remediation of volatile organics in the subsurface using soil vapor extraction. *11th Annual Madison Waste Conference*, September 13–14.

Scheidegger, A. E. 1974. *The Physics of Flow through Porous Media*, 3rd ed. University of Toronto Press, Toronto, Ontario, Canada, p. 306.

Stephanatos, B. N. 1990. Modeling the soil venting process for the cleanup of soils containing volatile organics. *Proceedings of the 4th National Outdoor Action Conference on Aquifer Restoration, Ground Water Modeling and Geophysical Methods*, Las Vegas, May 14–17, p. 633.

Sterrett, R. J. 1989. Analysis of in situ soil air stripping data. *Workshop on Soil Vacuum Extraction*, April 27–28. U.S. Environmental Protection Agency, Robert S. Kerr Research Laboratory, Ada, OK.

Stinson, M. K. 1989. EPA SITE demonstration of the Terra Vac in situ vacuum extraction process in Groveland, Massachusetts. *J. Air Pollut. Control Assoc.* 39:1054.

Thornton, J. S., and W. L. Wootan. 1982. Venting for the removal of hydrocarbon vapors from gasoline-contaminated soil. *J. Environ. Sci. Health* A17:31.

Towers, D., M. J. Dent, and D. G. Van Arnam. 1989. Part 1. Choosing a treatment for VHO-contaminated soil. *Hazard. Mater. Consultant*, March/April, p. 8.

Trowbridge, B. E., and J. J. Malot. 1990. Soil remediation and free product removal using in-situ vacuum extraction with catalytic oxidation. *Proceedings of the 4th National Outdoor Action Conference on Aquifer Restoration, Ground Water Monitoring and Geophysical Methods*, Las Vegas, May 14–17, p. 559.

Verschueren, K. 1983. *Handbook of Environmental Data on Organic Chemicals*. Van Nostrand Reinhold, New York.

Walton, J. C., R. G. Baca, J. B. Sisson, and T. R. Wood. 1990. Application of soil venting at a large scale: a data and modeling analysis. *Proceedings of the 4th National Outdoor Action Conference on Aquifer Restoration, Ground Water Monitoring and Geophysical Methods*, Las Vegas, May 14–17, p. 559.

Weast, R. C., M. J. Astle, and W. H. Beyer. 1985. *CRC Handbook of Chemistry and Physics*, 65th ed. CRC Press, Boca Raton, FL.

Wilson, D. J. 1988. Mathematical modeling of in situ vapor stripping of contaminated soils. *Proceedings of the First Annual Hazardous Materials Management Conference/Central*, Rosemont, IL, March 15–17, p. 191.

Wilson, D. J. 1990. Soil cleanup by in situ aeration. V. Vapor stripping from fractured bedrock. *Sep. Sci. Technol.* 25:243.

Wilson, D. J., A. N. Clarke, and J. H. Clarke. 1988. Soil cleanup by in situ aeration. I. Mathematical modeling. *Sep. Sci. Technol.* 23:991.

Wilson, D. J., A. N. Clarke, and R. D. Mutch, Jr. 1989. Soil cleanup by in situ aeration. III. Passive vent wells, recontamination, and removal of underlying nonaqueous phase liquid. *Sep. Sci. Technol.* 24:939.

Woodward-Clyde Consultants. 1985. *Performance Evaluation Pilot Scale Installation and Operation, Soil Gas Vapor Extraction System, Time Oil Company Site, Tacoma, Washington, South Tacoma Channel, Well 12A Project*. Work Assignment 74-ON14.1, Walnut Creek, CA, December 13.

Wootan, W. L., Jr., and T. Voynick. 1984. *Forced Venting to Remove Gasoline Vapor from a Large-Scale Model Aquifer*. API Publication 4431. American Petroleum Institute, Washington, DC.

5

THERMALLY ENHANCED VAPOR STRIPPING

Ann N. Clarke
ECKENFELDER INC.
Nashville, Tennessee

David J. Wilson
Vanderbilt University
Nashville, Tennessee

Paul R. dePercin
U.S. Environmental Protection Agency
Cincinnati, Ohio

INTRODUCTION

Thermal desorption technologies are generally thought of as ex situ treatments of soils, sludges, and other solid waste materials to remove various organic constituents, including higher-boiling compounds (such as PCBs) and volatile metals. There is a variety of technologies that fall into the category of thermal desorption treatment. Application temperatures range from 200 to 1000°F, depending on the specific technique, media, and target constituents (U.S. EPA, 1991a; dePercin, 1991).

There is a growing effort to apply thermally enhanced desorption techniques in situ, coupled with in situ vapor stripping to remove the desorbed constituents. Efforts are on going both in the United States and Europe and range from bench-scale research to field-scale demonstrations. Thus it is felt that these supplemental vapor stripping techniques warrant a separate chapter.

The application of thermal enhancements can be either by direct or indirect contact with the heat source. Although both approaches can be used in ex situ techniques, to date, only direct contact has been employed in the in situ mode. The objective of either mechanism is the same—to increase the rate of vaporization and transport. There are several temperature effects that help improve the removal of volatile and semivolatile constituents. In addition to the increased vapor pressures exhibited by all volatile and semivolatile compounds with increased temperatures, elevated temperatures also increase the rates of diffusion of the vaporized molecules. Another benefit of operating at increased temperatures is the increased solubility of many compounds in water. For those steam injection systems where water transport/movement is the removal mechanism, removal is improved compared to operation under ambient conditions. Additionally, elevated temperatures increase the transfer of constituents across the air/water interface (Thibodeaux, 1979), resulting in the increased removal of contaminants in high-humidity or saturated soil systems. Reduction in residual organic content may also result from hydrolysis and/or oxidation of target compounds. The nonbiological hydrolysis/oxidation of organic compounds is also increased with increasing temperature. To promote oxidation and increase removal in zones of high concentrations, steam stripping has been supplemented by potassium permanganate (U.S. EPA, 1991g).

To date, steam stripping techniques are the most researched for in situ applications. As is the case with many "innovative" remedial technologies, the theory and design are borrowed from other disciplines. In situ steam stripping has been employed for more than a quarter of a century by the petroleum industry to recover hydrocarbons. When steam is used as the heat source to improve volatilization, it also serves as the transport medium for the volatilized materials. Thus, with the addition of an appropriately designed vapor stripping system, the transport (i.e., removal) of the contaminants is also improved.

The petroleum/natural gas industry has other thermally based techniques to improve product recovery, such as hot brine injection and in situ combustion. These technologies are currently considered "conceptual" for site remediation application. The use of steam injection, however, is considered "available" technology for general applications and "developing" for hazardous waste site remediation (U.S. EPA, 1990a).

Radio-frequency heating, where electromagnetic energy causes increased temperature in the molecules of the soil matrix, has also been borrowed from the petroleum industry, which successfully employed the technology in enhanced oil recovery. Radio-frequency heating is considered a "developed" technology for use in site remediation (U.S. EPA, 1990b).

While the following sections concentrate on various hot air and/or steam injection–based technologies in the United States targeted to enhance in situ vapor stripping, radio-frequency heating and European efforts are also discussed. The reader is directed to a recent review article by Houthoofd et al. (1991) which discusses these technologies among other thermal technologies, such as in situ vitrification, ISV, optical fiber solar heating, and electrical resistance heating. While ISV has been demonstrated on the large-field scale, the optical fiber and resistance technologies (as applied to site remediation) are still in the early research stages.

HOT AIR/STEAM-ENHANCED STRIPPING

Simply put, hot air and/or steam is first injected into the soil and then removed, possibly under vacuum, together with the desorbed volatile and semivolatile materials. This gas stream must then undergo treatment. Because of the residual water from the steam, the treatment train contains more processes than are employed for the treatment of the soil gas stripped during basic in situ vapor stripping. Also, the condensed steam must be pumped from the ground and treated. Fortunately, the processes involved (e.g., demisters, scrubbers, condensers, chillers, heaters, etc.) are relatively simple, well-established technologies.

Key parameters during treatability testing and/or operations include temperature and pounds of steam injected, temperature and cfm of hot air injected, duration of operation, and depth of treatment. Care must be exercised to define the zone of influence correctly so that all the stripped materials are captured for treatment. Additionally, the system must be properly operated so that the vadose zone does not become saturated with water and exhibit reduced (or no) permeability for the gases and vapors targeted for removal. Other general site requirements include adequate soil permeability; penetrable soils for insertion of augers, wells, and so on; minimal subsurface obstacles; and ambient temperatures in the range 20 to 100°F (U.S. EPA, 1991g). The subsurface obstacles of concern are those which would interfere with the delivery of the steam flow to the contaminated domains of soil. If mobile treatment units are employed, surficial obstacles also need to be evaluated.

Basic Research

As has occurred with other innovative technologies (see in situ vapor stripping, for example; Wilson et al., 1988; Gannon et al., 1989), implementation at the field scale proceeds before a sound scientific and engineering understanding has been developed. As the understanding develops and matures, the somewhat empirical field designs and operations are improved with resultant increases in efficiency and cost-effectiveness. Hot air/steam-enhanced vapor stripping appears to be one of those technologies where implementation has outpaced the basic research.

One of the major studies of in situ steam stripping is the work carried out for the U.S. Environmental Protection Agency (USEPA) by Lord and his associates at Drexel University (Lord et al., 1987a,b, 1988, 1990, 1991). This group has conducted both laboratory-scale experimental work and theoretical work. In the laboratory studies they have explored the effects of contaminant vapor pressure and polarity, of the proportions of sand, silt, clay, and organic matter (topsoil), and of steam pressure on removal rates. The effect of vapor pressure is that, as expected, decreasing vapor pressures (increasing normal boiling points) result in lower rates of removal of the series octane, decane, dodecane, as well as for butanol and octanol. All these compounds, however, as well as kerosene, were found to steam strip at reasonable rates. The presence of silt, clay, and natural organic material was found to decrease the rate of removal by steam stripping, due both to decreased flow rates of steam and apparently to adsorption processes. Increased steam pressure, resulting in increased steam flow rates, yield more rapid removals, although excessive pressures may result in fracturing of the soil and short-circuiting.

These workers found virtually complete removal of kerosene from contaminated sand after treatment by 126 pore volumes of steam, and also observed removal of dodecane from sand (5% by weight initially) down to about 0.016% by steam stripping at 11.5 psi over a period of 6 h. They found that steam stripping of soil initially dried at 500°C and then spiked was more rapid than vacuum stripping, air stripping, and simple heat stripping; the applicability of these results in the field may be questionable, however, because the very rigorous initial drying of the soil may have removed water from adsorption sites that are occupied by water (and therefore not available to sorb organic compounds) under natural conditions. Greatly increased binding of volatile organic compounds by extremely dry soils and the release of these sorbed compounds in the presence of moisture has been observed in soil vapor stripping operations (Pedersen and Curtis, 1991).

A typical configuration for field-scale vacuum-assisted steam stripping is indicated in Figure 1 (Lord et al., 1990). Under the flexible membrane cap there is a geotextile that provides a path for the steam and off-gases to be collected for treatment. A schematic of the pilot-scale experiment is provided in Figure 2 (Lord et al., 1989). A somewhat more elaborate apparatus, which permits use of a wide range of steam pres-

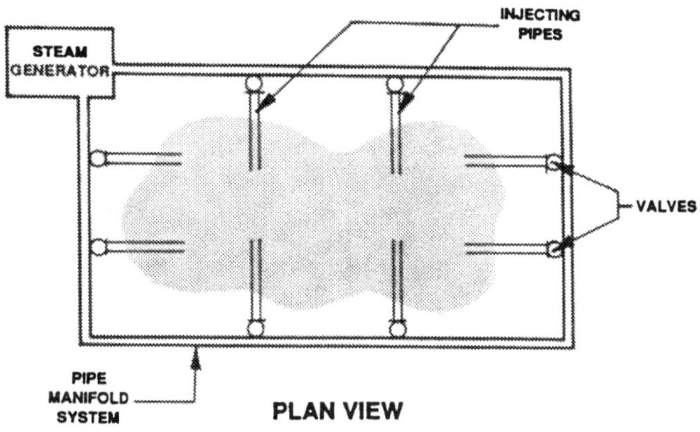

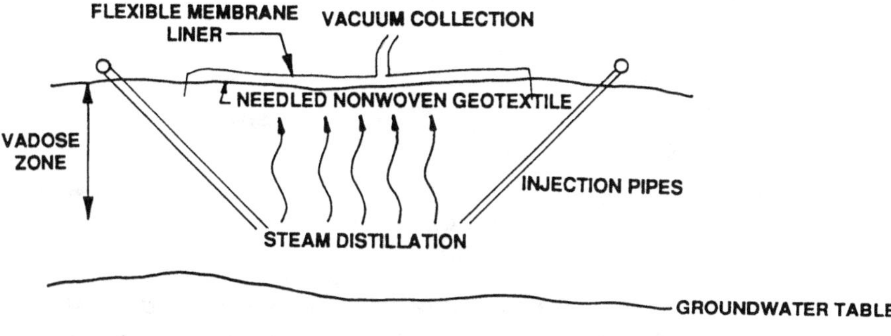

Figure 1 Typical field-scale in situ vacuum-assisted steam stripping configuration. (From Lord et al., 1990.)

sures, is shown in Figure 3 (Lord et al., 1990). Some representative results obtained stripping kerosene from bench sand with the simpler apparatus are shown in Figure 4 (Lord et al., 1989); problems with the analytical method bias these results toward lower reported removals than were probably achieved in fact. A comparison of kerosene removals from 50:50 sand/soil mixtures by steam stripping and air stripping at 100°C is given in Figure 5 (Lord et al., 1990). The authors found that removal of kerosene by simple heating without advection was virtually negligible.

These workers presented a mathematical model for steam stripping in their first paper (Lord et al., 1988), which they used in interpreting

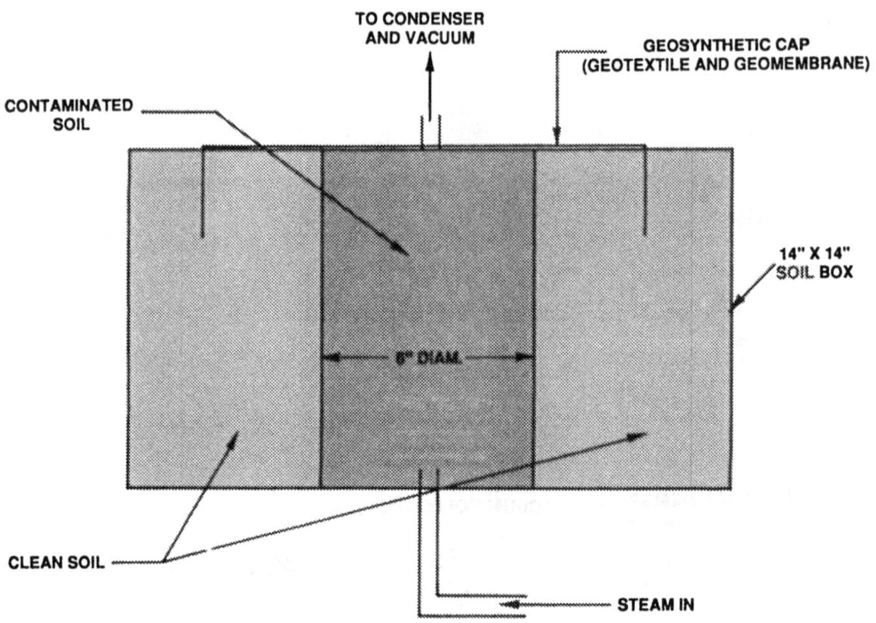

Figure 2 Schematic diagram of pilot-scale experiment for vacuum-assisted, steam stripping using the geosynthetic cap. (From Lord et al., 1989.)

experimental results and in estimating the results of scaled-up steam stripping operations. They noted in a later paper, however (Lord et al., 1990), that the model had been criticized as having some problems, and they were working on a revision. We discuss briefly a few of these problems in the following.

The model assumes that soil gas pressures, P, in the vicinity of a steam injection well can be calculated by a suitably constructed solution to the Laplace equation

$$\nabla^2 P = 0 \tag{1}$$

which is applicable to noncompressible fluids. For gases that may be treated as following the ideal gas law, $PV = nRT$, the correct equation (for a constant, isotropic and an isothermal system) is

$$\nabla^2 P^2 = 0 \tag{2}$$

If the soil gas pressures are all nearly 1 atm, little error results from the use of equation (1); however, in steam stripping, pressures on the order of 10 psi are apparently quite realistic, and steam pressures as high as

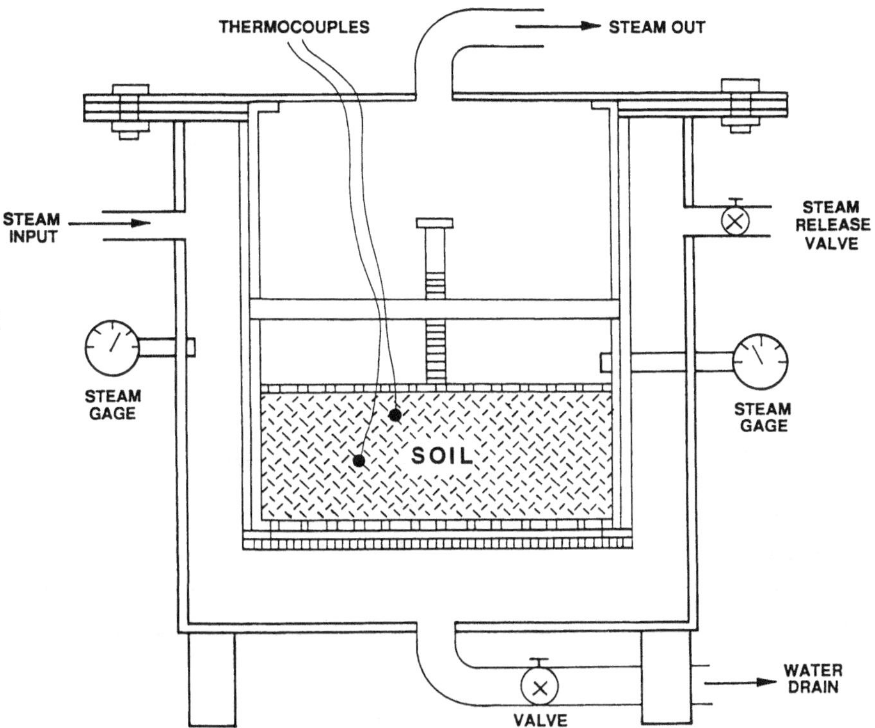

Figure 3 Device designed to steam strip soil samples at various pressures. (From Lord et al., 1990.)

50 psi have been reported, so one can expect significant errors to be introduced by this application.

The method of images is used to obtain a solution to Laplace's equation, but the image potential constructed (which uses a source and a single image) fails to satisfy the no-flow boundary condition that is applicable at the bottom of the vadose zone. If the steam is being injected well down in the vadose zone, reasonably near its lower boundary, this could be a significant source of error. This problem can be eliminated by a somewhat more elaborate array of sources (Wilson et al., 1988), which makes subsequent formulas more complex but still quite manageable on microcomputers.

The bulk of the modeling analysis assumes that transport in the vapor phase is advective; in one section, however, the assumption is made that the steam flows are purely diffusive, which is inconsistent with the earlier discussion of the model. Given the steam pressures under

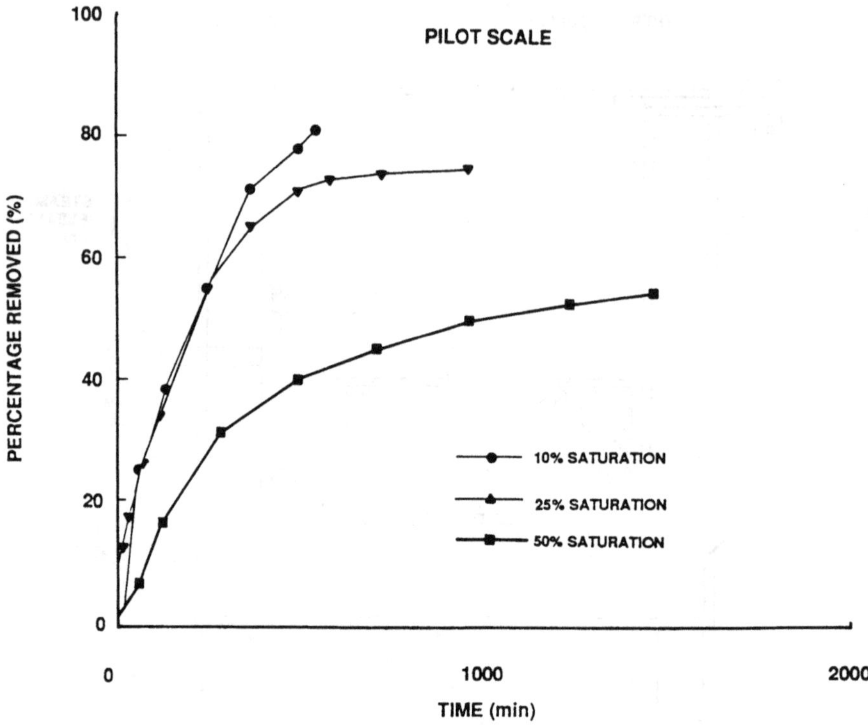

Figure 4 Results for steam stripping kerosene-contaminated beach sand using pilot-scale geosynthetic cap of Figure 3.

consideration and the flow rates involved, one can probably conclude that diffusion is unimportant except within individual soil particles.

Another assumption which may cause some problems is that the permeabilities measured on disturbed samples in laboratory cells are good approximations to permeabilities of undisturbed soil in the field. Of necessity, the soil tested in laboratory cells has been highly disturbed, which should certainly change its permeability compared to that of undisturbed soils. Workers in soil vapor extraction have tended to shy away from the estimation of air permeabilities in the field from laboratory measurements (Pedersen and Curtis, 1991).

We note that a mathematical model for steam stripping would be a useful tool in doing site-specific assessments of technology feasibility, costs, system design, and so on, and we would like to encourage further work in this area.

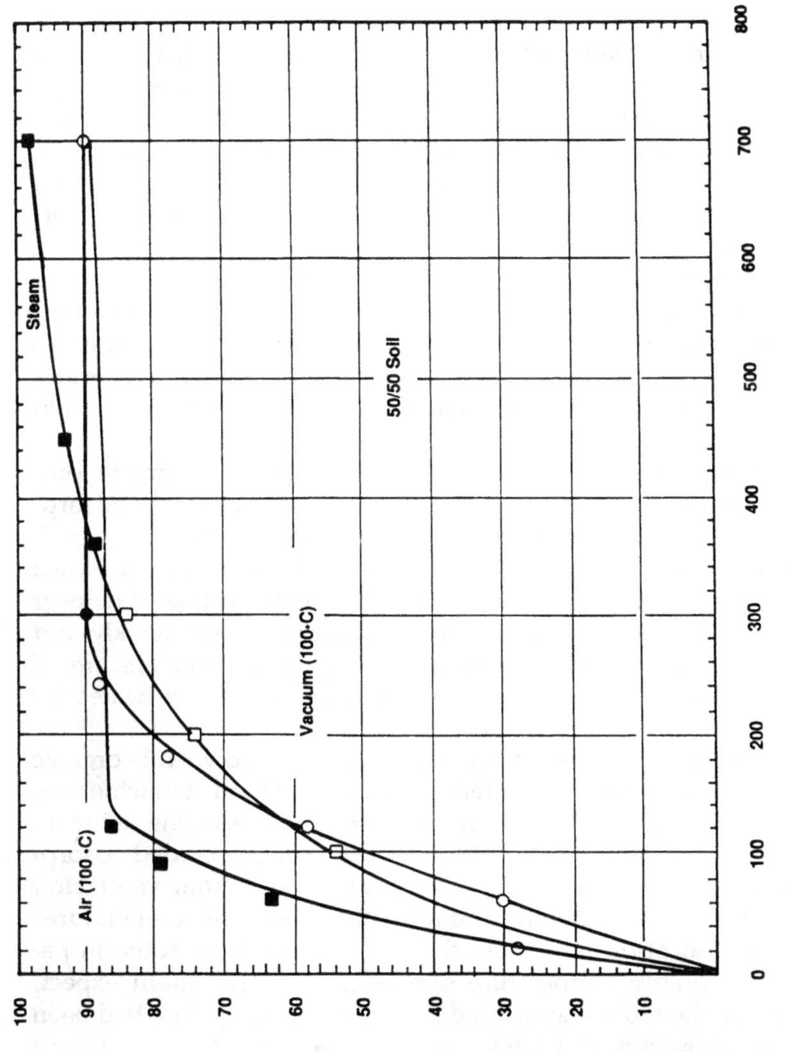

Figure 5 Comparison of decontamination processes (vacuum and air at 100°C). (From Lord et al., 1990.)

Dong and Bozzelli (1990) have presented a gas chromatographic technique for determining adsorption parameters and enthalpies of adsorption for volatile organic compounds on soils over a wide range of temperatures. These authors also developed an unusually complete and detailed theoretical analysis of the processes taking place in the soil columns in the gas chromatograph (advection, axial dispersion within the soil column, inter- and intraparticle diffusion, mass transport through the film around the soil particles, adsorption–desorption equilibrium, and heat of adsorption) which they used in interpreting their data and which should be quite useful as a research tool in laboratory studies. The complexity of the model raises some question as to the feasibility of its extension in all details to problems in two dimensions and its use on the types of microcomputers that are available to most environmental engineers.

Dong and Bozzelli's chromatographic technique appears to provide a very convenient and accurate method for investigating adsorption of volatile organic compounds on soils, a matter crucial to both steam stripping and soil vapor extraction. It provides a tool by which a lot of data on a variety of soils and volatile organic compounds over a wide range of temperatures can be obtained relatively quickly and cheaply. Temperature dependences of adsorption parameters are of particular interest in steam stripping or if the soil is to be heated by other means.

These workers provided data on several chlorinated volatile organic compounds (chloroform, methylene chloride, carbon tetrachloride, and 1,1,1-trichloroethane) and aromatic compounds (benzene, toluene, chlorobenzene, and 1,2,4-trichlorobenzene). The data included adsorption isotherm parameters, enthalpies of adsorption (from van't Hoff plots), mass transfer parameters, and minimum allowable temperatures for complete desorption. They found that the adsorption isotherm parameters were highly temperature dependent, as one might expect. Their experimental work was carried out on soil samples that had been dried by baking overnight at 200°C and then heating to 350°C in the gas chromatograph oven for at least 4 h; this protocol raises some question as to the applicability of the results to soils at field sites, which have not been dried by such rigorous methods. As mentioned previously, the presence of moisture tends to cause soils to bind volatile organic compound much less strongly than they are found to bind in very dry soils. Calculations made using their adsorption parameters can therefore be expected to yield excessively pessimistic assessments for both steam stripping and soil vapor extraction. One may hope that the technique can be used on moist soils as well as dry soils, and that data on the absorption and mass transfer kinetics of volatile organic compounds in moist soils will be forthcoming.

Detoxifier™: NovaTerra Technology

In 1987, a large-scale field test of an innovative technology, Detoxifier™ I, was performed. A prototype unit, Detoxifier™ I, was designed to perform several types of in situ treatment, including (hot) air/steam stripping, neutralization, solidification/stabilization, and/or oxidation (*Hazardous Waste Consultant*, 1987). This field test addressed the removal of hydrocarbons that had been released from a leaking underground storage tank using hot air/steam-assisted stripping.

The Detoxifier™ is a drilling rig designed to deliver (simultaneously) various treatment agents (dry, liquid, slurry, or vapor phase) to the soil and recover any off gases for treatment and/or recycle. The unit can deliver materials to a depth of 25 ft while completely mixing them. Delivery at deeper depths is being studied. The continuous overlapping of the cylinders of treated soil is designed to result ultimately in complete site cleanup.

Steam-treated soils require treatment of the off gas, which is composed of the volatilized constituents, air, and water vapor. Generally, treatment trains include cooling, demisting (scrubbing), condensing, and drying stages to remove water followed by contaminant removal (e.g., activated carbon) as occurs in unenhanced vapor stripping. The various vapors evolving from the ground during this test were trapped in a shroud for treatment. The shroud forms the base of the processing tower. (see Figure 6 for a diagram of the Detoxifier™ I unit.)

The test site in California contained petroleum hydrocarbons in the soil in the range 100 to 1,000 ppm, with hotspots containing more than 15,000 ppm. The treatment site was approximately 10,300 ft^2 and 15 to 22 ft deep, requiring 433 treatment blocks (cylinders with approximately 20% overlap). The hot air/stream reduced the hydrocarbon levels in the soil to less than 100 ppm. This value is based on the analysis soil samples taken at 5, 10, and 21 ft. (Ghassemi, 1988).

Later information on the testing of this technology reports 85 to 99% removal of chlorinated volatile organic compounds from clay soils at another location, the Annex Terminal in San Pedro, California. The major constituents were trichloroethene, tetrachloroethene, and chlorobenzene. The demonstration was part of the USEPA SITE program. It was performed in September 1989 at depths of 6.5 to 7.5 ft (U.S. EPA, 1991d). It was reported that less than 100 ppm of contaminant remained after

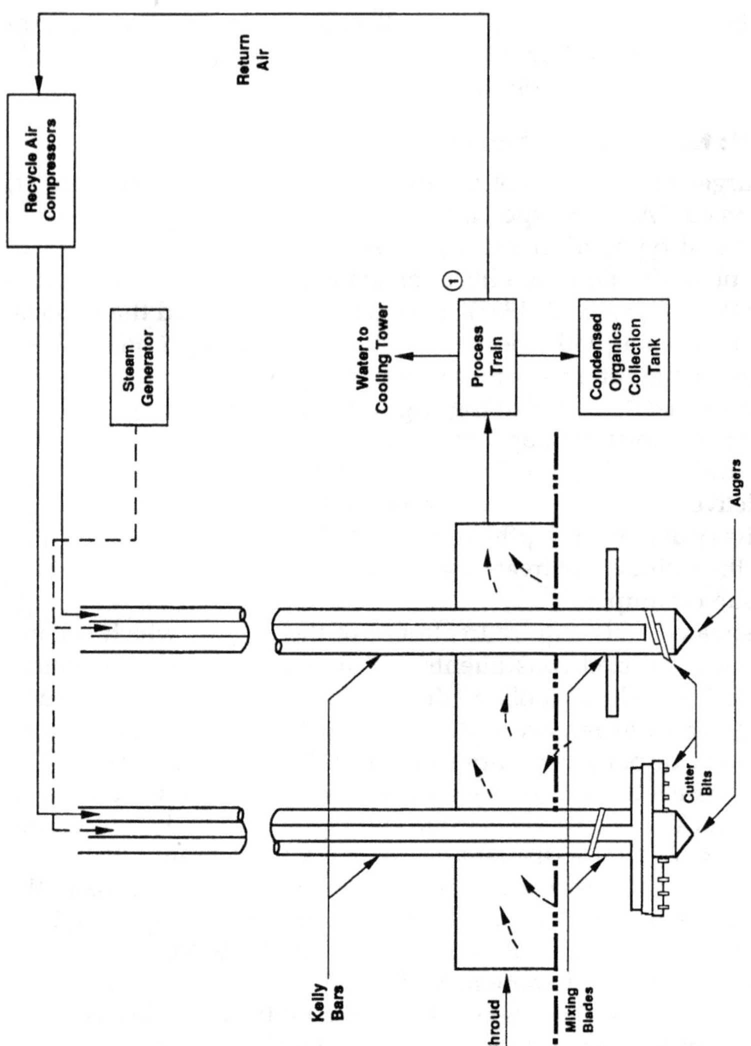

① Process Train Includes: Condensation and separation; activated carbon treatment/compression (vapor phase); distillation and storage (liquid phase)

Figure 6 Diagram of NovaTerra Technology Detoxifier™. (From U.S. EPA, 1991d.)

treatment. This level of removal was achievable 80% of the time. Mass balances of the off-gas condensate versus chemical-specific analysis of the soil before and after treatment accounted for 89% of the material (LaMori, 1990). Approximately 55% removal of the semivolatile organic constituents, as determined by USEPA Method 8270, was also reported, although the mass balance was not decisive.

During treatment, steam was introduced at 450°F and 450 psig and hot air at 300°F and 250 psig (U.S. EPA, 1991e). Fugitive air emissions from the process were characterized as low and no downward migration of constituents was detected during treatment (based on fluorescein dye tests). The SITE report concludes that the technology is not limited by soil particle size or initial porosity or by chemical concentration. High-clay-content soil resulted in increased treatment time and can affect overall economic feasibility (U.S. EPA, 1991d). Vendor-estimated costs range from approximately \$250 to \$350/yd^3.

There has also been growing efforts to enhance stripping of volatile organic compounds from the water stream generated during pump and treat operations by the addition of hot air/steam (Fair and Dryden, 1990). After the NovaTerra SITE demonstration on the soil in the vadose zone was concluded, the vendors began to evaluate hot air/steam in situ stripping treatment in the saturated zone beneath the site using the same technology.

A 6-week "deep study" was performed in the saturated zone at depths of 10 to 12 ft. The operating (downhole) temperatures were higher for the saturated zone testing (205 to 250°F) compared to the vadose zone test (175 to 180°F). Operating temperatures are limited by the maximum permitted shroud gas temperature (170°F). This higher temperature, coupled with the more sandy soil in the saturated zone, resulted in a more rapid contaminant removal rate. The vendor further postulated that the water caused the treatment to act like a fluidized bed which improved the exchange and removal rates (U.S. EPA, 1991d).

In Situ Steam-Enhanced Extraction (ISSEE): Udell Technologies

Udell Technologies of Emeryville, California, is scheduled to demonstrate its steam-assisted technology for the removal of volatile and/or semivolatile organic compounds under the USEPA SITE program at Hill Air Force Base in Utah. In 1988 the technology successfully completed a pilot-scale study in which 764 lb of mixed volatile organic compounds were removed from 900 ft^3 of soil. The SITE program testing will be above the water table (U.S. EPA, 1991b)[†] and is scheduled for 1994. A second demonstration is tentatively scheduled for LeMoore Naval Air Station, California. This is a joint Navy and USEPA pilot study (but not part of the SITE program). A case study that addresses a gasoline spill at Lawrence Livermore National Laboratory in California is planned. Volatile organic compounds are targeted for removal from both above the water table and below (to a depth of 137 ft) at the Lawrence site.

The In Situ Steam-Enhanced Extraction process (ISSEE) forces steam through the soil by injection wells (see Figure 7). The desorbed organic compounds are recovered through extraction wells under vacuum. The recovered material is either treated by vapor-phase activated carbon or condensed and processed with the groundwater. Implementation of this and other similar technologies must provide for the recovery and removal of liquids. With steam stripping, the aqueous waste stream is increased by the presence of the condensed steam. This loading must be addressed in the design of the groundwater treatment train.

Again, the ISSEE design employs demonstrated, readily available technology in both the design and use of injection/extraction wells and in the extracted waste stream treatment train. However, the technology should not be applied to contaminants near the soil surface unless there is an impermeable cap to trap fugitive vapors for treatment. The treatment of DNAPL compounds with this technology needs to include the presence of a geologic barrier at the site to prevent downward migration of the denser-than-water compounds (U.S. EPA, 1991f).

Steam Injection/Vapor Extraction (SIVE): Hughes Environmental Systems Inc.

The SIVE process employs steam injection and vacuum extraction to remove volatile and semivolatile components from the soil in an in situ mode (U.S. EPA, 1991f). As with in situ vapor stripping, the technology can be employed close to and under existing structures and paving. The technology is theoretically applicable to a wide range of soils. The schematic of the process is provided in Figure 8.

Steam is forced into the soil through injection wells, where it enhances the stripping process as discussed previously. The extraction wells serve both to pump and treat groundwater and to transport the steam/air/organic constituent stream to the surface for treatment. The

[†]Earlier reports on the Udell technology refer to it as steam injection/vapor extraction (SIVE). This acronym is now applied to the process demonstrated by Hughes Environmental Systems Inc. (HESI). HESI is located in Manhattan Beach, California. SIVE is discussed later in the chapter.

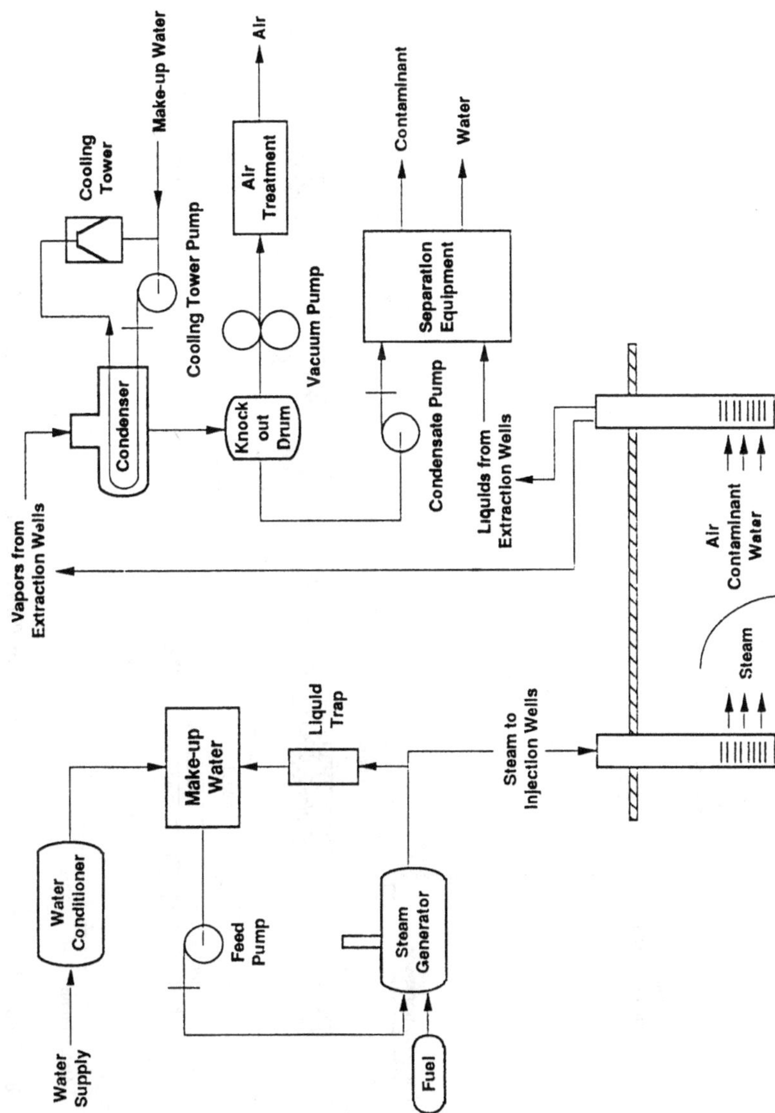

Figure 7 In Situ Steam-Enhanced Extraction (ISSEE) process schematic. (From U.S. EPA, 1991f.)

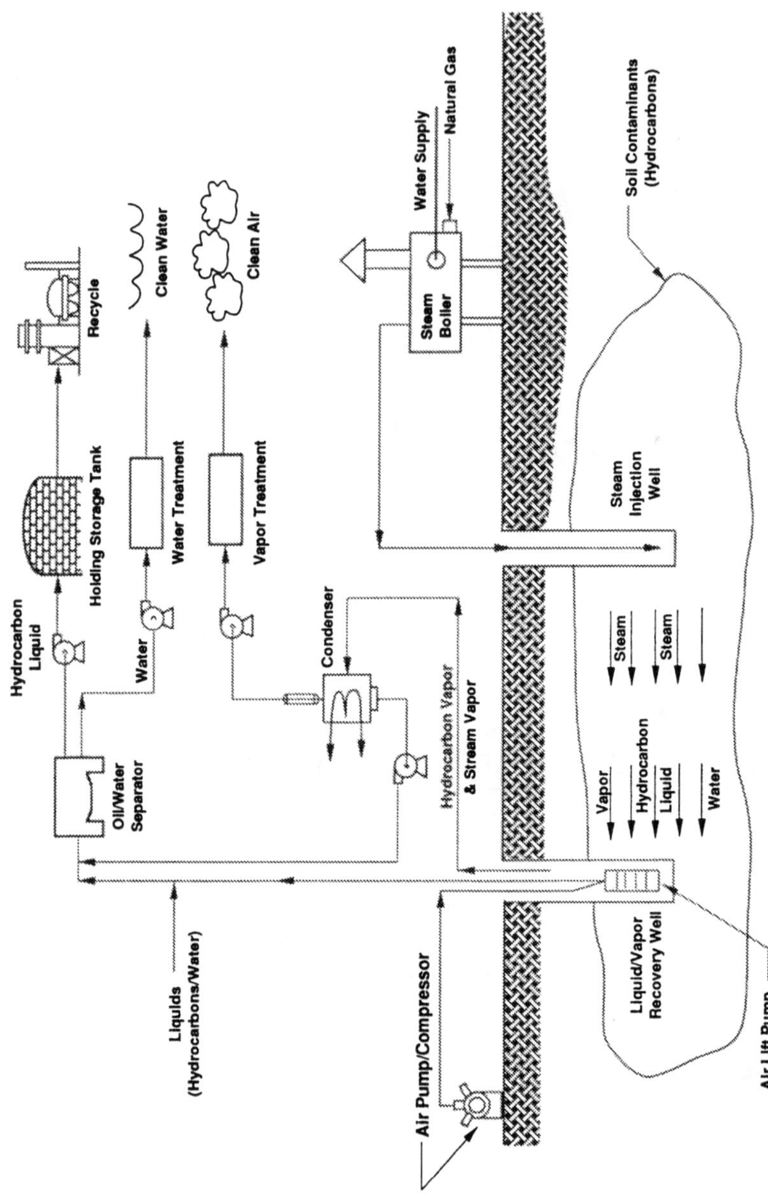

Figure 8 Steam Injection and Vapor Extraction (SIVE) process. (From U.S. EPA, 1991f.)

gas stream is either treated by activated carbon designed for use with the vapor phase or condensed and treated with the groundwater. As with other steam-enhanced stripping technologies, the process train employs demonstrated, available components.

The SIVE process was demonstrated at Huntington Beach, California. The testing began in August 1991 and continued through early 1992. The contamination was diesel fuel from a 135,000-gal spill.

International Technology

Researchers in the Netherlands (U.S. EPA, 1989) have pursued the in situ steam stripping of soils as early as 1983. Tests were made by Heijmans Millieutechniek B.V., the company that now holds the license for the process. Steam at a temperature of 130 to 180°C is injected into the soil at multiple locations. The steam serves both to vaporize the volatile materials and as a medium of transport for removal. The process also places the system under a vacuum, thus lowering the vaporization temperature of the constituents present.

The extracted stream of volatile organic compounds and steam is condensed. Any floating, oily components (if present) are removed. The condensate is then further treated by filtration, air stripping, or activated carbon, as necessary. As with the U.S. systems the condensate treatment is composed of proven established technologies.

The Dutch system is applicable to all types of volatile materials with boiling points between 100 and 250°C. As could be anticipated, the best results are obtained in sandy soils. Although not defined, they report that they have also developed ways to circumvent the difficulties presented by dislocation layers of bog or loam/clay. As with in situ vapor stripping, application of this technology can be alongside and below existing buildings or paved areas.

After abbreviated field testing in 1983, the technique was tested in 1985 at a site containing organic bromide compounds at concentrations as high as 7700 mg/kg in sandy soil. Using steam injected to 4.5 m (groundwater was at 5 m), the concentration was reduced to 220 mg/kg. This is a 97% removal. The organic bromide was converted to an inorganic bromide that was present only in low concentrations in the condensate.

In late 1984, sandy soil (1.8 to 2.6 m deep) at a former gasworks in Mannheim was treated using steam injection to remove benzene (55 mg/kg), toluene (15 mg/kg), xylene and ethyl benzene (2 to 4 mg/kg), and phenol (30 mg/kg). All these constituents were below detection limits after treatment.

However, much of the Dutch experience with this technology encountered problems. Noted problems included overall poor steam stripper performance. Heijmans Millieutechniek, B.V. believes that other limitations will be identified after additional experience is obtained from pilot-scale efforts.

ASSOCIATED TECHNOLOGIES

There are several technologies which do not fit into this chapter or the preceding chapter on in situ vapor stripping. These technologies, however, are similar in part(s) to those already discussed and are of sufficient potential to warrant mention. Therefore, this section which includes brief descriptions of associated technologies is provided.

Integrated AquaDetox®/SVE: AWD Technologies, Inc.

The integrated AquaDetox®/SVE technology was developed by AWD Technologies, Inc. and has been tested at several locations, including a USEPA SITE demonstration. The technology treats groundwater and soil gas simultaneously for the removal of volatile organic compounds. The soil gas treatment is traditional in situ vapor extraction. The groundwater treatment is ex situ with the water stripped in a steam-enhanced moderate vacuum stripping tower. The groundwater stripping tower is maintained at 52°C by the steam. Figure 9 is a diagram of the technology (U.S. EPA, 1991c). The process includes on-site regeneration of the granulated activated carbon units.

The SITE demonstration was performed over a 2-week period in September 1990. Groundwater flow rate and steam flow rate were varied during testing, as was the stripping tower pressure and frequency of carbon bed regeneration. The organic constituents of concern were trichloroethylene (TCE) and perchloroethylene (PCE). Groundwater removals for TCE and PCE combined ranged from 99.92 to 99.99%. PCE exhibited slightly better percent removal rates than TCE. The soil gas removals ranged from 98.0 to 99.9% for the combined chlorinated organics when the activated carbon beds were regenerated every 8 h. (*Note*: The SITE program was targeting the remediation of groundwater and did not specifically track vadose zone cleanup.)

Hrubout® Process

The Hrubout® process is a patented thermal process developed for remediating soil containing petroleum hydrocarbons, although application to other constituents is possible. The focus of the process is the injection of hot air into the contaminated soils (A. Hrubetz, personal

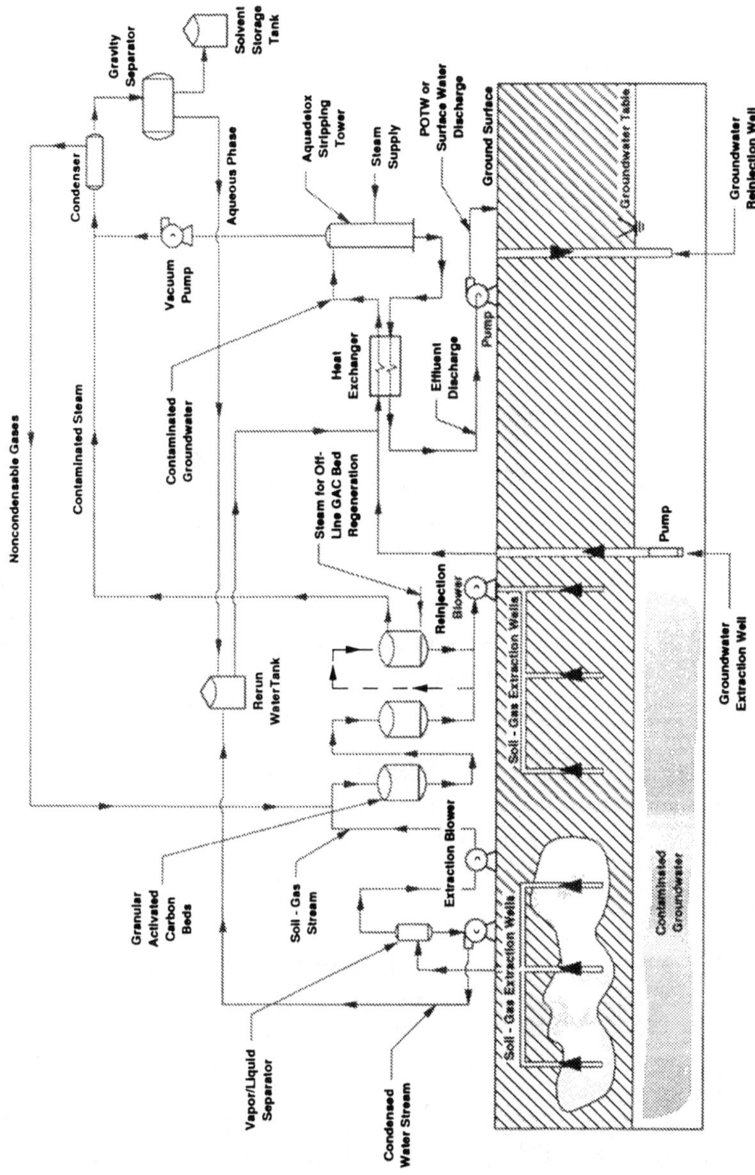

Figure 9 AWD integrated AquaDetox®/SVE system schematic. (From U.S. EPA, 1991c.)

communication, 1991). Air between 250 and 1200°F is introduced under pressure into the vadose in either a vertical or a horizontal mode. The vertical procedure is used for treatment of soil at 10 ft or deeper. The horizontal procedure is used for near-surface remediation as can be necessitated by pipeline leaks. The heat vaporizes the ambient soil moisture and effectively "steam distills" the residual hydrocarbons. Per the Hroubout process literature, gasoline requires temperatures of 250°F; diesel fuel, 500°F, and residual oils, 800°F. This last removal is essentially a slow oxidation process. The water vapor and associated hydrocarbons are extracted with vacuum blowers and incinerated in a dedicated unit at 1500°F.

Radio-Frequency Heating

Radio-frequency (RF) heating is a technique designed for the uniform and rapid heating of large volumes of soil. The soil is heated to a point where volatile and semivolatile organic compounds are vaporized into the interstices of the soil matrix. The heating process does not rely on the thermal conductivity of the soil. Vented electrodes are then used to recover the vapors. The vapor stream is then extracted and the material is recovered, and then incinerated or otherwise treated/disposed (U.S. EPA, 1990b). A negative impact from RF heating, as with any thermally enhanced technology, is the inhibition and/or destruction of soil microbes.

Electromagnetic energy in the frequency of the radio bands is generated and applied to the soil by electrodes. The electrodes are inserted into the soil in previously drilled holes. An RF heating system is composed of an energy disposition array; power generating, transmitting, monitoring, and control systems; a vapor barrier and contaminant system; and a gas/liquid condensate handling and treatment system. The design of the electrode (ex situ) array determines the design and operating parameters of the other systems. The electrodes are placed in three parallel rows that form a central conductor and two outer walls. By appropriate spacing the electromagnetic field can be contained within the boundaries defined by the outer walls.

There has been a field-scale demonstration of RF heating at Volk Air National Guard Base at Camp Douglas, Wisconsin. The site had homogeneous waste and appropriate soil characteristics. The site contained waste hydrocarbons from oils and fuels. The site had served as a firefighting training area for a quarter of a century. Fifty-thousand gallons of hydrocarbons were estimated to be present. The RF process system included a cooler/condenser, a gas/liquid separator, and carbon adsorption for the vapor stream.

After 2 days of applying RF waves to the soil, the temperature reached 100°C; after 8 days, 150°C. The demonstration last 12.5 days. Removal of aliphatic volatile compounds ranged from 98.2 to 99.8% and 88.1 to 98.5% for semivolatile aliphatic compounds. Volatile and semivolatile aromatic compounds were reduced by 98.1 to 99.9%.

REFERENCES

dePercin, P. R. 1991. Thermal desorption attainable remediation levels. *Proceedings of the 17th Annual Hazardous Waste Research Symposium*. Cincinnati, OH, April 1991. EPA/600/9-91/002, pp. 511–520.

Dong, J.-I., and J. W. Bozzelli. 1990. Removal of hazardous organic compounds from soil matrices using thermal desorption with purge. *Proceedings of the Environmental Chemistry Division, American Chemical Society National Meeting*, Boston, April 1990, pp. 1–6.

Fair, G. E., and F. E. Dryden. 1990. Comparison of air stripping versus steam stripping for treatment of volatile organic compounds in contaminated groundwater. *Hazard. Mater. Consultant*, September/October 1990, pp. 18–22.

Gannon, K., D. J. Wilson, A. N. Clarke, R. D. Mutch, Jr., and J. H. Clarke, 1989. Soil clean up by in-situ aeration. II. Effects of impermeable caps, soil permeability, and evaporative cooling. *Sep. Sci. Technol.* 24(11):831–862.

Ghassemi, M. 1988. Innovative in situ treatment technologies for cleanup of contaminated sites. *J. Hazard. Mater.* 17:189–206.

Hazardous Waste Consultant. 1987. In situ treatment technology steam strips contaminated soil. November/December, pp. 1–21 and 1–22.

Houthoofd, J. M., J. H. McCready, and M. H. Roulier. 1991. Soil heating technologies for in situ treatment: a review. *Proceedings of the 17th Annual Hazardous Waste Research Symposium*, Cincinnati, OH, April 1991. EPA/600/9-91/002, pp. 190–203.

LaMori, P. N. 1990. In-situ hot air/steam extraction of volatile organic compounds. *Proceedings of the 2nd Forum on Innovative Hazardous Waste Treatment Technologies: Domestic and International*. Philadelphia, May 1990, pp. 26–43.

Lord, A. E., Jr., R. M. Koerner, V. P. Murphy, and J. E. Brugger. 1987a. Vacuum-assisted in-situ stream stripping to remove pollutants from contaminated soil. *Proceedings of the EPA Conference on Land Disposal, Remedial Action, Incineration and Treatment of Hazardous Waste*, Cincinnati, OH, MAy 6–8, p. 511.

Lord, A. E., Jr., R. M. Koerner, V. P. Murphy, and J. E. Brugger. 1987b. In situ, vacuum-assisted steam stripping of contaminants from soil. *Proceedings of the Superfund '87 Conference (8th National Conference)*, Hazardous Materials Control Research Institute, Washington, DC, November 16–18, pp. 390–395.

Lord, A. E., Jr., R. M. Koerner, V. P. Murphy, and J. E. Brugger. 1988. Laboratory studies of vacuum-assisted steam stripping of organic contaminants from soil. *Land Disposal, Remedial Action, Incineration, and Treatment of Hazardous Waste, Proceedings of the 14th Annual Research Symposium*, Cincinnati, OH, May 9–11. EPA/600/9-88/021, July, pp. 65–92.

Lord, A. E., Jr., R. M. Koerner, D. E. Hullings, and J. E. Brugger. 1989. Laboratory studies of vacuum-assisted steam stripping of organic contaminants from soil. *Proceedings of the 15th Annual Conference on Land Disposal, Remedial Action and Treatment of Hazardous Waste*, Cincinnati, OH, April, pp. 124–136.

Lord, A. E., Jr., D. E. Hullings, R. M. Koerner, and J. E. Brugger. 1990. Vacuum-assisted steam stripping to remove pollutants from contaminated soil: a laboratory study. *Remedial Action, Treatment, and Disposal of Hazardous Waste, Proceedings of the 16th Annual Hazardous Waste Research Symposium*, Cincinnati, OH, April 3–5. EPA/600/9-9/037, August, pp. 377–395.

Lord, A. E., Jr., L. J. Sansone, R. M. Koerner, and J. E. Brugger. 1991. Vacuum-assisted steam stripping to remove pollutants from contaminated soil: a laboratory study. *Proceedings of the 17th Annual Hazardous Waste Research Symposium*, Cincinnati, OH, EPA/600/9-91/002, April, pp. 329–352.

Pedersen, T. A., and J. T. Curtis. 1991. *Soil Vapor Extraction Technology Reference Handbook*. U.S. Environmental Protection Agency, Risk Reduction Engineering Laboratory, Cincinnati, OH. EPA/640/2-91/003, February.

Thibodeaux, L. 1979. *Chemodynamics: Environmental Movement of Chemicals in Air, Water, and Soil*. Wiley, New York, p. 211.

U.S. Environmental Protection Agency. 1989. *Assessment of International Technologies for Superfund Application: Technology Identification and Selection*. EPA/600/2-89/017, May, pp. 246–251.

U.S. Environmental Protection Agency. 1990a. *Technologies of Delivery or Recovery for the Remediation of Hazardous Waste Sites*. RREL, ORD, EPA/600/2-89/066, p. 49–69.

U.S. Environmental Protection Agency. 1990b. *Handbook on In Situ Treatment of Hazardous Waste-Contaminated Soils*. EPA/540/2-90/002, January, pp. 83–85.

U.S. Environmental Protection Agency. 1991a. *Engineering Bulletin: Thermal Desorption Treatment*. OEER, ORD, EPA/540/2-91/008, May, p. 1–8.

U.S. Environmental Protection Agency. 1991b. *Superfund Innovative Technology Evaluation (SITE) Program: Spring Update 1991*. EPA/540/8-91/005, pp. 5, 9.

U.S. Environmental Protection Agency. 1991c. *AWD Technologies Integrated AquaDetox®/SVE Technology: Applications Analyses Report*. EPA/540/A5-91/002, October.

U.S. Environmental Protection Agency. 1991d. *Toxic Treatments In Situ Stream/Hot-Air Stripping Technology: Applications Analyses Report*. EPA/540/A5-90/800, March.

U.S. Environmental Protection Agency. 1991e. *Synopses of Federal Demonstration of Innovative Site Remediation Technologies*. Federal Remediation Technologies Roundtable, Summer 1991, pp. 58–59.

U.S. Environmental Protection Agency. 1991f. *The Superfund Innovative Technology Evaluation Program: Technologies Profiles*, 4th ed. EPA/540/5-91/008, November, pp. 92–93 and 154–155.

U.S. Environmental Protection Agency. 1991g. *Engineering Bulletin: In Situ Steam Extraction*. EPA/540/2-91/005, May.

Wilson, D. J., A. N. Clarke, and J. H. Clarke. 1988. Soil clean up by in situ aeration. I. Mathematical modeling. *Sep. Sci. Technol.* 23(10/11): 991–1037.

6

Thermal Desorption

Richard J. Ayen and Carl R. Palmer
RUST Remedial Services Inc.
Clemson Technical Center, Inc.
Anderson, South Carolina

Carl P. Swanstrom*
CPS Consulting
Naperville, Illinois

CONCEPT OF THERMAL DESORPTION

Soils and sludges contaminated with organic chemicals are a widespread problem, with millions of cubic meters requiring remediation in the United States alone [1]. Cleanup of contaminated soils and sludges is driven by environmental laws, which include the Comprehensive Environmental Response, Compensation, and Liability Act of 1980 (CERCLA or Superfund); the Resource Conservation and Recovery Act (RCRA); and various state laws requiring the cleanup of real estate as a condition of transfer.† The total cost of this cleanup effort is estimated by Cudahy and Troxler [2] to be over $200 billion over the next 30 or 40 years.

A number of processes can be used for treating organics in soils. One of the more popular techniques has been thermal treatment.‡ Thermal treatment can be classified into (1) those that employ incineration of the soil, and (2) those that employ thermal desorption. The latter type of processes generally heats the soil no higher than 550°C in the absence of oxygen. The organics are removed from the soil by volatilization and, in a separate downstream operation, are then either destroyed or condensed. Incineration, on the other hand, heats the soil to

*Consultant to RUST Remedial Services Inc.
†New Jersey's ECRA law is an example.
‡As of early 1990, over 1 million tons of contaminated soil and waste had been thermally remediated by 20 thermal treatment service contractors. [2].

higher temperatures in an oxidizing atmosphere and volatilizes and combusts the organics either simultaneously or in a secondary combustion chamber.

Thermal desorption, which is also sometimes called low-temperature thermal treatment, offers a number of advantages over incineration. This is particularly true for those that employ indirect heating of soils and solids whereby the wastes do not come into contact with combustion flames or products. The gases exiting the volatilization process consist of a purge gas, entrained particulates, evaporated moisture, and organic constituents. The volume of these gases is relatively small in comparison with those produced from an incinerator of equal capacity, enabling downstream air pollution control equipment to be correspondingly reduced. Solids entrainment is minimized, and organics can be recovered by condensation for recycling and further treatment and disposal.

Many recent records of decision[†] issued under Superfund have identified thermal desorption as the treatment of choice, and feasibility studies often list the technology as an applicable treatment technology [3]. Reasons given by the U.S. Environmental Protection Agency (USEPA) for its selection include (1) the ability of the technology to treat effectively a wide range of organic contaminants, (2) availability of mobile commercial systems, and (3) greater public acceptance of the treatment approach, especially in comparison with incineration, thereby facilitating the obtaining of permits. During the past several years, a number of companies have developed thermal desorption processes and have begun treatment on a commercial scale.

FUNDAMENTAL DESORPTION STUDIES

Desorption is the removal of a volatile or semivolatile substance from a solid material. Our interest in this chapter is focused on the desorption of organic contaminants from a solid or sludge. Inevitably, water is desorbed along with the organics. Thermal desorption refers to carrying out this process by the application of heat.

Adsorption is the process whereby a solute, a liquid organic pollutant in this case, accumulates or concentrates on the internal and external surface of a solid; absorption (by a solid), in contrast, refers to the penetration of a substance into the bulk phase of the solid. When it is not clear whether adsorption or absorption is involved, the term sorp-

[†]A *record of decision* is the document by which the USEPA discusses the feasibility of various remedial alternatives for a Superfund cleanup and then decides upon one (40 CFR Part 300).

tion is sometimes used. Desorption is the reversal of the processes of adsorption or absorption. A body of information on the fundamentals of thermal desorption using a variety of experimental techniques has been developed over the past few years. The techniques involved and the key results from basic desorption studies are described in the sections that follow.

Apparatus Employed for Desorption Studies

Work carried out at the University of Utah by Lighty and co-workers [4–6] was aimed at developing an understanding of the fundamental transport phenomena underlying the thermal desorption of organics from contaminated soils. Fundamental experimental studies were carried out using two different types of equipment. Both were designed to provide information relevant for all types of commercial desorption equipment, whether rotary kiln, traveling grate, fluid bed, or other. *Particle characterization reactors* (PCRs) were used for the most fundamental studies, including exploration of intraparticle resistances in the soil. A desorbing nitrogen gas stream was passed through a thin bed of soil under closely controlled conditions, and exhaust gas samples were analyzed by gas chromatography. Two different PCRs were used, one for low temperatures up to 200°C and one for higher temperatures up to 1200°C. The high-temperature version is shown in Figure 1. The heat exchanger assembly heated the desorption gas to the desired temperature, and the gas was then passed upward through the soil bed. The heat exchanger and soil bed were placed inside a cylindrical electrical furnace.

Another experimental device, a packed-bed system termed the *bed characterization reactor* (BCR), was constructed and operated to measure mass and heat transfer effects more accurately in real operating environments. In this apparatus, shown in Figure 2, a desorbing nitrogen gas stream was swept across the top of the bed of organic-contaminated soil while heat was transferred to the bed, and the contaminant was evolved from the soil to the bulk sweep gas. This apparatus was designed to represent a moving grate, multiple hearth, or fluid bed contacting device. Two different sizes of BCRs were employed. One held 1 kg of soil and was used primarily to explore heat transfer and mass transfer considerations through a fixed bed. This was for relatively homogenous soil. A 10-kg BCR was used for the study of more heterogeneous soils, which required a larger sample size.

Desorption studies by Szabo et al. [7] on a surrogate Superfund soil were carried out by spreading the soil on metal trays and placing the soil in a Lindberg furnace, model 51848, for a given period of time. A nitrogen sweep was used, and temperature was measured with two

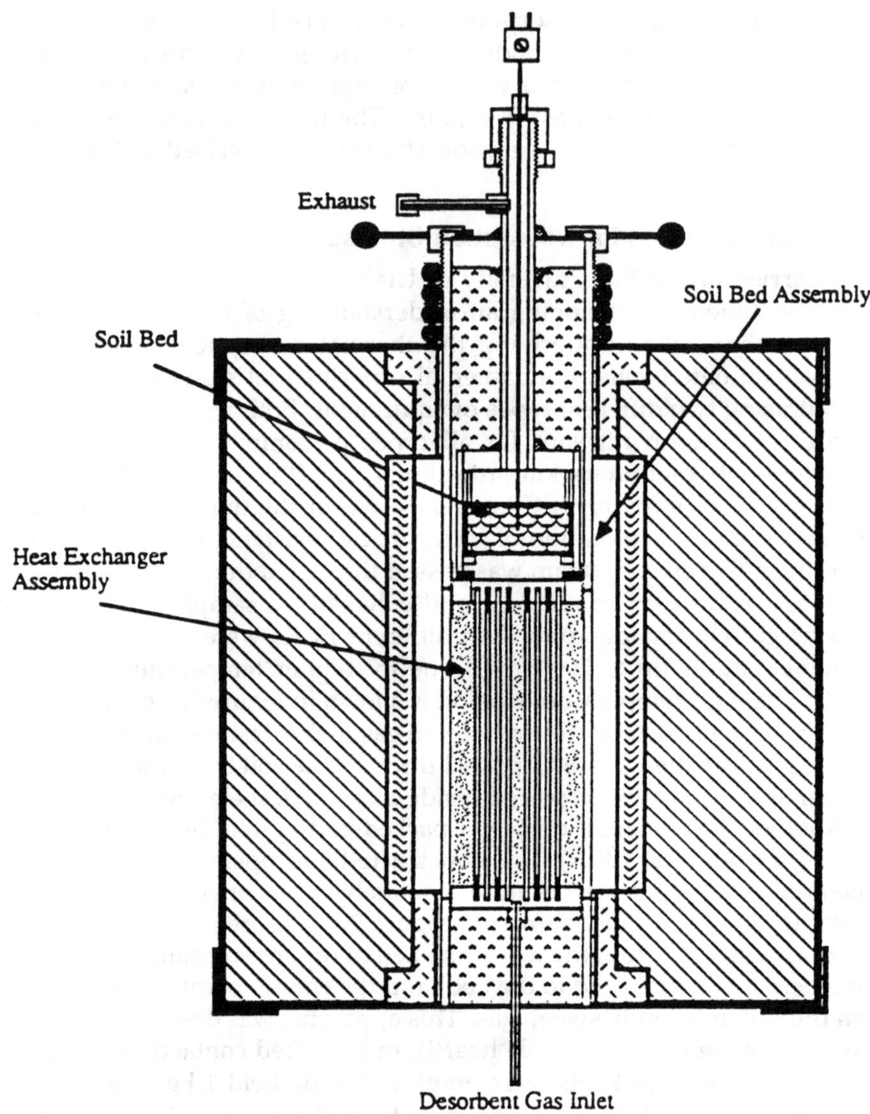

Figure 1 Particle characterization reactor assembly.

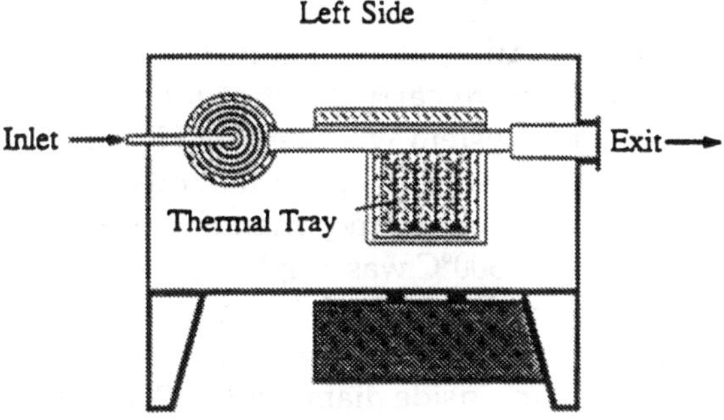

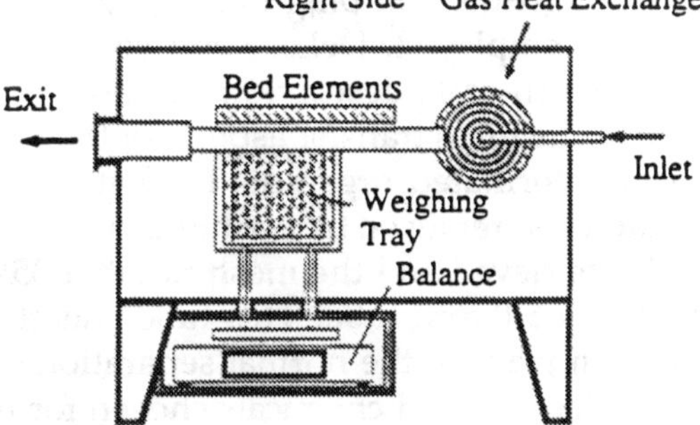

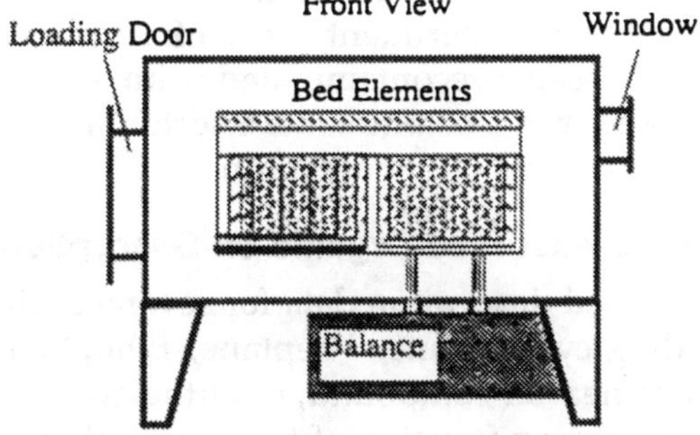

Figure 2 Bed characterization reactor.

thermocouples placed 3 cm above the soil surface at the center of the tray. This system was capable of operating at temperatures of up to 1100°C, although the studies reported were carried out at much lower temperatures of 66 to 288°C. This same system was used by Helsel and Groen [8] and by Helsel et al. [9] in their studies of thermal desorption of organics from soil samples collected from former manufactured gas plant sites. A temperature range of 250 to 500°C was employed. These investigators used, in addition, a Lindberg Mini Mite laboratory tube furnace, model 55035, with a manual temperature controller and an 800-W heater. The furnace chamber size was 1 in. inside diameter by 12 in. long. Soil was packed into a quartz tube and placed inside the furnace chamber. A purge gas was passed through the soil plug during operation. Temperatures of 350 to 500°C were employed. DeLeer and co-workers [10] also used a tube furnace for bench-scale desorption studies.

Dong and Bozzelli [11] studied the mass transfer aspects of the thermal desorption of aromatic and chlorinated organics. They used soil from which the finer particles had been removed by washing. After drying at 200°C overnight, the soil was sieved and the mesh number 35/40 particulate fraction was packed in a 5.0-mm tube. This tube was then installed in a gas chromatograph in place of the normal separation column. Nitrogen was used as a carrier gas, and chemicals chosen for the study were injected onto the head of the soil column.

Studies reported by Varuntanya et al. [12] also used this technique. In addition, they performed desorption experiments on uniformly contaminated soil (i.e., soil that had been precontaminated with known quantities of organics), then packed into a column and inserted into an oven that was connected directly to a gas chromatograph.

Effect of Molecular Weight of the Adsorbed Organic on Desorption

Lighty and co-workers [4–6] obtained desorption data for several of the components of gasoline, including cyclohexane, *i*-heptane, *n*-heptane, and toluene, as well as for a much heavier compound, naphthalene. Desorption rates were found to be a strong function of the chemical composition of the organic. Cyclohexane was volatilized from the clay at a much faster rate than benzene, toluene or *p*-xylene. In one experiment, the desorption rates for naphthalene and *p*-xylene were compared at 315°C. As might be expected, the higher-molecular-weight compound, naphthalene, was desorbed at the slower rate.

As mentioned previously, Szabo et al. [7] conducted desorption studies on synthetically prepared soils intended to simulate contaminated soils found at Superfund sites. These soils were termed synthetic analytical references matrices (SARMs) or synthetic soil matrices

Table 1 Desorption of Volatiles and Semivolatile Organics from SSM-I (High Organics, Low Metals)

Compound	Starting soil concentration (ppm)	Product soil concentration (ppm)		
		66°C (150°F)	177°C (350°F)	288°C (550°F)
Volatiles				
Acetone	4330	20	60.7	71.1
Dichlorobenzene	322	13	0.059	<0.004
1,2-Dichloroethane	384	ND at 0.005[a]	0.008	<0.005
Ethylbenzene	3116	130	0.43	0.009
Styrene	651	ND at 0.005[a]	<0.18	ND at 0.005[a]
Trichloroethylene	423	4.1	0.021	<0.005
Xylene	5277	320	0.95	0.025
Semivolatiles				
Anthracene	7271	5800	1359	<16.1
Bis(2-ethylhexyl)phthalate	2527	2100	1523	<16.1
Pentachlorophenol	381	330	235	58

Source: Ref. 7.
[a]ND, not detected.

(SSMs). Four different SSMs were prepared, corresponding to high and low organics levels and high and low metals levels. Desorption studies were carried out on SSM-I (high organics, low metals) and SSM-II (low organics, low metals) by placing the soil samples on trays in a nitrogen-purged oven for 30 min. In addition to organics, SSM-I contained 20% moisture, and SSM-II contained between 7 and 17% moisture [7].

Key results from the desorption experiments are presented in Tables 1 and 2. Table 1 shows results for SSM-I, the high-organics, low-metals soil. Treatment was carried out at three temperatures: 66, 177, and 288°C. As would be expected, the 288°C condition produced, with one exception, more complete desorption than did the lower temperatures. The exception was acetone, for which the extent of desorption became poorer as temperature was increased. Acetone is an extremely volatile compound, and it should have been desorbed completely. The probable cause of the anomaly was contamination from some other source. The researchers mentioned that the lid used to cover the tray of soil after removal from the oven had been cleaned with acetone.

For all other volatile compounds, concentrations were reduced to below detection limits at 288°C. Even at 66°C, greater than 90% removal was effected for all compounds. Of the three semivolatile compounds tested, anthracene and bis(2-ethyl-hexyl)phthalate were effectively

Table 2 Desorption of Volatiles and Semivolatile Organics from SSM-II (Low Organics, Low Metals)

Compound	Starting soil concentration (ppm)	Product soil concentration (ppm)		
		66°C (150°F)	177°C (350°F)	288°C (550°F)
Volatiles				
Acetone	430	15.0	100	99.2
Chlorobenzene	6.6	0.091	0.017	0.004
1,2-Dichloroethane	1.3	<0.015	0.006	0.008
Ethylbenzene	82.5	1.82	0.18	0.035
Tetrachloroethylene	4.4	<0.028	0.009	0.006
Xylene	155	3.1	0.57	0.1
Semivolatiles				
Anthracene	350	238	23	<9.3
Bis(2-ethylhexyl)phthalate	150	180	54	<9.3
Pentachlorophenol	29	22.3	<9.5	ND at 9.3[a]

Source: Ref. 7.
[a]ND, not detected.

removed at 288°C, while approximately 85% of the pentachlorophenol was desorbed. Note, however, that little of the three semivolatile compounds was removed at the 66°C treatment temperature. Thus, as would be expected, the higher-molecular-weight, less volatile organic compounds are not as easily desorbed as are the more volatile compounds. (See Table 9 for additional data on thermal desorption of organics from SSM-I.)

Similar results were obtained for SSM-II, the low-organic, low-metals soil. As seen in Table 2, with the exception of acetone, excellent removal of all volatile and semivolatile organics was achieved at 288°C, and the effect of molecular weight is demonstrated most dramatically at the 66°C condition. Szabo and co-workers concluded that thermal desorption is a useful technology for the treatment of CERCLA soils contaminated with volatile and semivolatile organic compounds.

These data generated on SSMs are also reported and commented on by Esposito and co-workers [13]. The study by Szabo et al. was part of a larger effort funded by the USEPA and targeted at evaluating a number of treatment technologies for cleanup of hazardous wastes at Superfund sites. In addition to thermal desorption, Esposito presented results on treatment of SSMs by soil washing, chemical treatment with KPEG, incineration, and stabilization/fixation.

Table 3 Minimum Allowable Temperatures for Thermal Desorption

Compound	Boiling point (°C)	MAT (°C)	Fraction remaining[a]
Chloroform	61.7	120	0.026
Methylene chloride	40	100	0.027
Carbon tetrachloride	77	80	0.009
1,1,1-Trichloroethane	74	100	0.034
Benzene	80	120	0.013
Toluene	111	160	0.021
Chlorobenzene	132	160	0.021
1,2,4-Trichlorobenzene	213	240	0.030

Source: Ref. 11.
[a]Fraction of organic remaining in soil after desorption at 1 h at the MAT.

In their desorption experiments, Dong and Bozzelli [11] recorded a minimum allowable temperature (MAT) below which the organic was not desorbed from soil. The results for several organics frequently found in contaminated soil are presented in Table 3. In general, the MATs were higher for higher-boiling-point compounds, such as 1,2,4-trichlorobenzene, chlorobenzene, and toluene than for lower-boiling-point comounds such as chloroform, methylene chloride, carbon tetrachloride, 1,1,1-trichloroethane, and benzene.

Effect of Temperature on Desorption

As shown by the work of Szabo and co-workers [7] discussed in the preceding section, temperature is an important parameter for thermal desorption. Their data showed that treatment at 66°C was ineffective, especially for the semivolatile compounds present. However, much better results were obtained at 288°C.

In the work by Lighty and co-workers [4–6], temperature was shown to be the dominant parameter for desorption. Figure 3 shows data obtained using their bed characterization reactor. Desorption of *p*-xylene is shown as a function of time for desorption temperatures of 175, 240, and 315°C. At 315°C, desorption is essentially complete in 3 h, while only approximately half of the organic contaminant (*p*-xylene) had been removed after 3 h at 175°C. Note that it was necessary to heat the soil to well above the boiling point of *p*-xylene (139°C) to achieve complete desorption in a reasonable time.

Collectively, the data from Lighty and co-workers [4–6] and Szabo and co-workers [7] show that heat source temperatures in the region of 300°C are needed to obtain the degree of semivolatile organic removal required for commercially practical treatment processes. Obviously, the

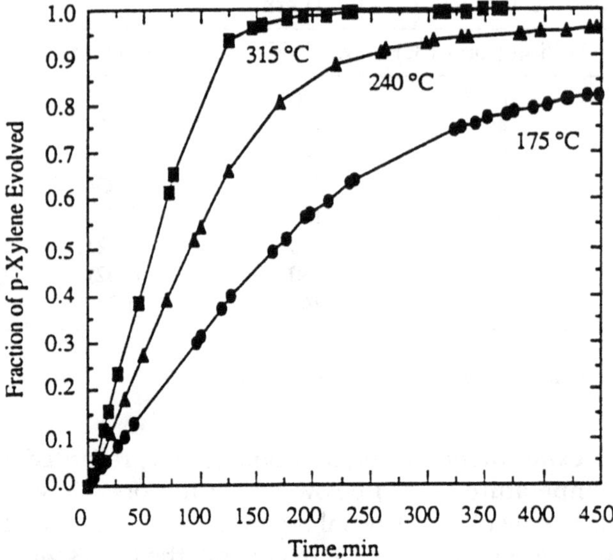

Figure 3 Temperature comparison for bed characterization reactor.

type of equipment involved, soil residence time, and other parameters will also affect the temperature required.

Dong and Bozzelli [11] obtained desorption data for a number of aromatic and chlorinated hydrocarbons over a range of temperatures. They then calculated for each organic an adsorption/desorption equilibrium constant, K_a, defined by

$$C_p = \frac{\theta_p}{1 - \theta_p} K_a C_x$$

where C_p = concentration of organic in soil
θ_p = porosity of particles
C_x = concentration of organic in fluid

The temperature dependence of K_a was then determined, and ln (K_a/T) was found to be a linear function of $1/T$, in accordance with the van't Hoff equation:

$$\frac{K_a}{T} = \left(\frac{K_a}{T}\right)_0 \exp\left(\frac{-\Delta H_0}{RT}\right)$$

where ΔH_0 = heat of adsorption
T = absolute temperature
R = gas constant

These results therefore form the basis for a more systematic approach for presenting or predicting the effect of temperature on thermal desorption.

Effect of Substrate Porosity on Desorption

Lighty et al. [6] explored desorption from various types of substrates. Figure 4 shows the rate of desorption of *p*-xylene from sand, glass beads, clay, and peat. Clay and peat are very porous compared to sand and glass beads. The data are reported as normalized evolution rate versus time, where the normalized evolution rate is defined as the rate of *p*-xylene evolution (g/min) divided by the initial amount (g) of *p*-xylene in the soil. For the nonporous materials, the evolution rate was initially high and then fell off rapidly. For sand and glass beads, the nonporous materials, the *p*-xylene was removed quickly; the rate of desorption fell rapidly to very low levels. For clay and peat, the

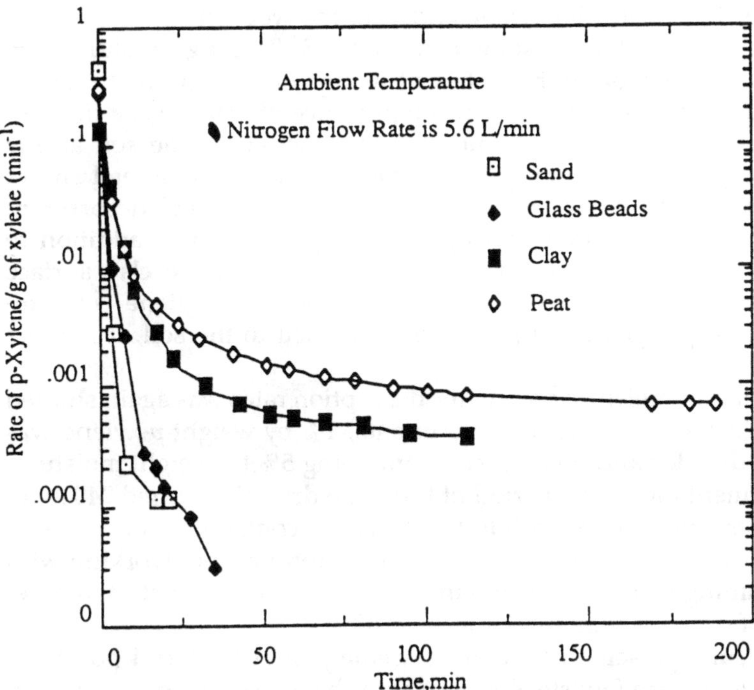

Figure 4 Comparison of the desorption rates of porous clay and peat versus nonporous sand and glass beads.

porous materials, significant amounts of *p*-xylene were still being desorbed after 100 min. The extreme difference in rates of desorption from porous versus nonporous materials can be attributed to the effect of mass transfer resistance through the internal pore structure of the porous materials, the strong adsorption of the organic compound on the higher surface area of the porous materials, or to some combination of these two effects.

Effect of Moisture Content on Desorption

Most studies on desorption of organics have been carried out on substrates from which the moisture had been removed prior to loading the organics onto the substrate. To assess the effect of this practice, experiments were conducted by Lighty et al. [6] on the desorption of *p*-xylene from clay. Three different feedstocks were prepared. One run was made with clay that was dried and then exposed to *p*-xylene to obtain a xylene concentration of 0.5%. A second feedstock was prepared in the same fashion, and approximately 10% by weight of water was then added. A third feedstock was the original moisture-containing soil, having 10% water by weight, with the xylene added to the wet soil.

Desorption results are shown in Figure 5. This figure shows the amount of *p*-xylene left in the soil as a function of time for the three different feedstocks. As can be seen, the presence of water has a dramatically favorable effect on the amount of xylene left in the soil at any given time. The addition of water to the dry clay/*p*-xylene feed shows that steam distillation is an important factor in thermal desorption. The experiment with undried clay/*p*-xylene shows that in addition to steam distillation, water probably occupied sites on the clay surface that would otherwise have been occupied by the *p*-xylene contaminant. With the *p*-xylene not physically adsorbed to the soil, it is more easily desorbed.

The effect of moisture on organic desorption rates was again studied using the BCRs. In one set of experiments, 1% by weight *p*-xylene was loaded on dry clay and then on clay containing 5% by weight moisture. After treatment for a fixed period of time, the dry soil retained 240 μg of *p*-xylene per gram of soil, while the moisture-containing soil retained less than 5 μg per gram of soil. Again, the importance of working with soils containing realistic (natural) moisture contents when studying desorption of organics was shown to be critical.

The results presented here are extremely important and point out the need to prepare feedstock properly for bench-scale studies of thermal desorption. Obviously, the best approach is to work with a soil that has its original moisture content still intact.

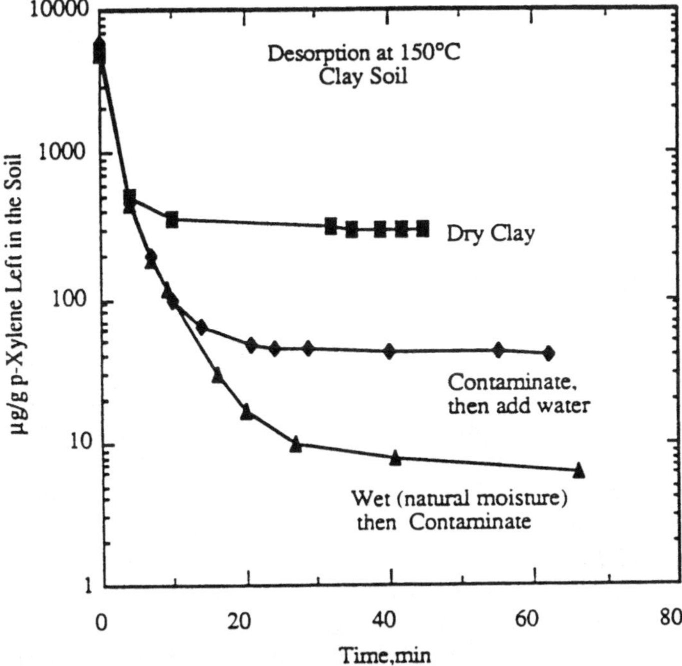

Figure 5 Moisture effects on *p*-xylene desorption at 150°C.

Effect of Age of Waste on Desorption

Using one of their particle characterization reactors, Lighty et al. [6] explored the effect of adsorption time on the desorption rate. Soil was loaded with 0.5% by weight of *p*-xylene, then allowed to age for 1 day, 6 months, and 1 year before desorption studies were carried out. Only very slight differences in desorption rates were observed for the three samples. This puts to rest the often-heard speculation that old contaminated sites will be more difficult to remediate than freshly generated waste, on the premise that the organics have had more time to migrate into the smaller pores or, in some fashion, to become more tightly adsorbed on the soil.

Conclusions from Fundamental Studies

The work by Lighty and co-workers at the University of Utah [4–6] is by far the most comprehensive and most meaningful study carried out to date on thermal desorption of organics from soils. A considerable body of data was generated, and mathematical modeling was applied to these data. For the most part, the modelling was successful and showed that the Freundlich isotherm was the correct adsorption isotherm for these circumstances. Collectively, the fundamental studies indicate that minimum conditions for desorption are a temperature of 300°C and a residence time of 30 min. These conditions will result in the removal of 99% or more of contained polynuclear aromatic hydrocarbons (PAHs) or more volatile compounds.

SMALL-SCALE STUDIES ON ACTUAL WASTES

The Gas Research Institute (GRI) sponsored a series of studies on thermal desorption of organics from soil samples from four former manufactured gas plant (MGP) sites. Under contract to GRI, Helsel and coworkers [8,9] studied removal of phenol, cyanides, and PAHs, including naphthalene, methylnaphthalene, anthracene, and phenanthrene. Static tray tests [8] were performed using temperatures of 250 to 500°C and residence times of 10 to 60 min. At least 99.8% of the phenol was removed at 300°C and a 30-min residence time at more extreme conditions. Cyanide, however, was found to be difficult to remove. Another series of tests [9] studied removal of PAHs from coal tar–contaminated soil from three sites using bench- and pilot-scale equipment. Greater than 99.9% removal of PAHs was obtained at 400°C and 9 min residence time using either a static oven method or a 16.5-cm-inside-diameter indirect-fired rotary kiln. Good agreement was found between the results from the two test methods. Note that the residence time employed, 9 min, was relatively short. In contrast, most investigations of thermal desorption have employed residence times of 30 min or longer, which are more practical for the operation of commercial thermal treatment equipment.

Lighty and co-workers at the University of Utah [4,6] then provided computer modeling and laboratory test results in support of the work of Helsel and co-workers. This information was generated to provide guidelines for the scale-up of indirect-fired rotary kilns to be used as thermal desorbers. In general, it was concluded that soil temperatures of 250°C were marginal for a desorption process, while much better results were obtained at 300 and 350°C.

Measured and calculated temperature profiles for indirect-fired rotary kilns were also presented. The results were then extrapolated to a full-scale indirect-fired rotary kiln with a diameter of 2 m and a length of 13 m. The relative importance of convective and radiative heat transfer was investigated. At kiln wall temperatures of 500 and 800°C, convection between the wall and the soil within the kiln is the dominant mode of heat transfer, although at 800°C radiation becomes important.

The rate of heat transfer can be increased by increasing the rotation rate of the kiln or decreasing the fill fraction within the kiln. A general conclusion from the University of Utah studies is that for reactive, porous soils such as clay, the last monolayer of organic contaminant is tightly bound to the soil requiring treatment temperatures significantly above the boiling point of the organic to ensure adequate removal from the soil.

DeLeer and co-workers [10] studied thermal desorption treatment of soils from former MGP sites in the Netherlands, investigating removal of cyanides and PAHs from three actual wastes. Using a small tube oven, it was shown that iron cyanide complexes were effectively decomposed at 300 and 350°C using 30-min residence times. Concentrations of 1.0 to 2.5 g of CN per kilogram of soil were reduced to 3 to 17 mg/kg. Eleven PAHs were found in measurable quantities, with total PAH concentrations in the range of 30 to 180 mg/kg. Treatment at 300°C for 30 min reduced concentrations to below the limit of detection of 0.01 mg/kg for all compounds. Similar results were then obtained on these same soils using a 500-kg/h pilot-scale indirect-fired rotary dryer.

Data from pilot-scale studies have also been reported by RUST Remedial Services Inc. for its X * TRAX™ process, by the International Technology Crop., by Remediation Technologies, Inc., and by Recycling Sciences, Inc. for its DAVE (desorption and vaporization extraction) process. See the section "Available Technologies" for this information.

TYPES OF PROCESSES

Within the framework of the RCRA and TSCA regulations, three basic processes have been developed for the thermal desorption of organic chemicals from contaminated soils and solids. These are rotary dryers, heated screws, and fluidized bed dryers. These devices serve as the primary equipment to transfer heat to the soil, causing the organic contaminants to vaporize and separate from the solid matrix.

These systems have generally been configured with two types of gas treatment systems attached to the primary soil heating and desorption equipment: a condensation gas treatment system or an afterburner system. In a condensation treatment system, the vaporized gases are cooled, with the distinguishing feature being the recovery of the organic contaminants as a concentrated liquid or sludge. Condensation systems may employ a cyclone separator, baghouse, or filter for particulate solids removal, a carbon adsorber, afterburner, or catalytic oxidizer for residual volatile hydrocarbon emission control; however, their key feature is the recovery of the bulk of the organic material. On the other hand, an afterburner gas treatment system employs a combustion chamber to destroy the organic contaminants removed. An afterburner also requires particulate solids controls and sometimes requires acid gas controls, depending on the waste feed characteristics. Afterburners can be made to work effectively; however, because the organic constituents are destroyed using controlled flame combustion, they may require permitting as incinerators.

In this section we discuss the principles of operation of the three types of processes, along with other emerging technologies. The material presented here is restricted to those systems that have been or could be permitted for RCRA- or TSCA-regulated contaminated soils. The focus of the information is the type of dryer that each system uses. The gas handling equipment does not vary substantially among the various approaches and is discussed in more depth in the section "Available Technologies."

Rotary Dryers (Kilns)

A rotary dryer consists of a large, heated rotating cylinder. It is similar in design to rotating kilns and calciners and is sometimes given these names in the generic sense. Contaminated soil is fed into one end of the cylinder and is conveyed to the other end by the rotation and inclination of the cylinder. As the feed material progresses down the cylinder, it is heated, releasing the organic contaminants and water contained in and on the soil. These vapors are transported to the gas treatment system and the decontaminated soil exits from the dryer to the product handling system.

Heat can be supplied to the soil in the rotating cylinder using either direct or indirect firing. With direct firing, a burner is installed at the end of the cylinder and the flame either impinges or radiates directly on the soil. In this way a direct-fired rotary dryer is similar to an incinerator, and consideration must be given to the formation of products of incomplete combustion (PICs). Because of these considerations, direct-fired dryers are not generally used for thermal desorption of soils where RCRA hazardous waste codes apply or TSCA-regulated PCBs are present. Nonetheless, direct-fired rotary dryers are widely used for the remediation of petroleum-contaminated soils when they are operated under state air permits and do not require EPA, RCRA, or TSCA treatment permits.[†] Direct-fired dryers are not discussed further here because of this restriction to non-RCRA and non-TSCA materials.

[†]Petroleum is generally excluded from the definition of "hazardous waste" under RCRA and would not contain PCBs.

In an indirect-fired rotary dryer, heat is either conducted to the soil through the cylinder wall from external burners, or a heated gas is passed through the dryer, thereby heating the soil, or both. Indirect systems that heat through the dryer cylinder wall do not allow the burner flame and products of combustion to touch either the contaminated soil or the vaporized organic chemicals. This minimizes concerns over the formation of PICs. Also, if "clean" fuels such as propane or natural gas are used for firing the burners, no air pollution control devices are required on the burner stack gases. This greatly reduces the size and complexity of the gas treatment system, which in turn reduces the capital cost for the system, resulting in an economic advantage. This is also generally true of the other indirect-heating methods discussed later.

Figure 6 shows an indirect-fired rotary dryer. The cylinder can be constructed from carbon steel, stainless steel, or alloys. It is supported on two trunnion rolls and rotates on steel riding rings. The cylinder can be driven using either a chain and sprocket or gears. If the cylinder is large enough, it can be rotated by driving the trunnion roll. For transportable equipment, the cylinder will range from 1.5 to 2.5 m in diameter and can be 10 to 18 m long. The residence time for soil in the dryer is generally 1 to 3 h. The residence time is adjusted over about a 60-min range by varying the rotation speed of the cylinder, and over a wider range by varying the angle of inclination of the dryer. Using a carbon steel cylinder material, maximum soil temperatures of up to 500°C can be achieved routinely, and with an alloy cylinder, soil temperatures of up to 800°C can be achieved. The cylinder rotates inside a furnace chamber heated by a number of burners. The temperature of the cylinder shell is controlled by modulating the firing rate of the burners. Heat is conducted through the cylinder wall to the soil, and the burners' combustion gases exit the top of the furnace through stacks. Soil enters and exits the cylinder through stationary breeches. The internal atmosphere

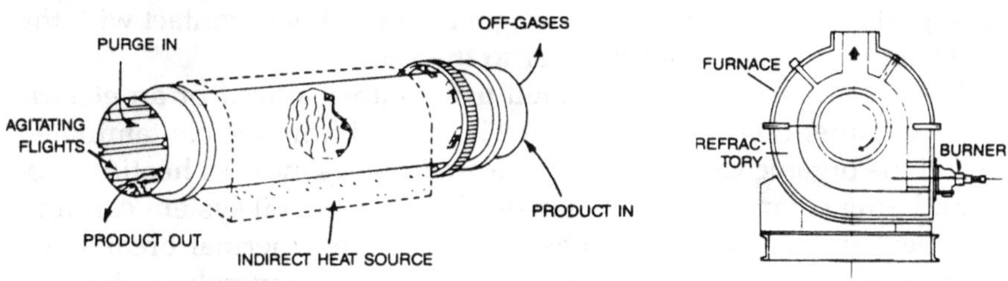

Figure 6 Indirect-heated rotary calciner. (From ABB Raymond, *Bulletin 873*, 1987.)

of the cylinder is maintained using mechanical seals attached to the breeches, preventing both air in-leakage and leakage of the vaporized organic chemicals.

A carrier gas is passed through the rotary dryer to transport the vaporized organic chemicals and water to the gas treatment system. An inert carrier gas such as nitrogen is used to preclude combustion inside the dryer. It is this carrier gas, in combination with the water vapor, that allows the organic chemicals to be desorbed from the solid matrix well below their boiling point, much the same as warm air removes water at well below 100°C in a clothes dryer. Treatment of the desorbed gases varies among the available processes; however, either the condensation or afterburner system discussed previously is used.

Heated Screws (Conveyors)

A variety of heated screw designs is available; these use hot oil, molten salt, or electric resistance elements as the heat source. However, the designs have as a common feature the use of multiple screws, or augers, to heat, mix, and convey the soil inside of enclosed shells or troughs. Figure 7 shows a commonly available heated screw thermal processor. Contaminated soil is fed into one end of the processor, which has a hot oil heat transfer fluid circulating inside the screw shaft, the screw flights, and the outer vessel's shell. Heat is conducted to the soil from the hot oil, and the organic contaminants and water are vaporized. Gas takeoff points are provided at several locations in the outer shell for these vapors, which are ducted to a gas treatment system. The main advantage of heated screws over other primary heating equipment is that there is a relatively high heat transfer surface area for the volume of the equipment, resulting in higher thermal efficiency as well as smaller hardware for the same capacity. A major process restriction is that solids temperature is limited by the maximum allowable temperature of the heat transfer fluid. The heat transfer fluid is pumped into the rotating screws of the thermal processor and through jackets in the conveyor housing. Both the screw shaft and flight are in intimate contact with the soil, resulting in good heat transfer to the soil.

The heat transfer fluid is heated using either a fired or an electric device. In either case, the heat source is isolated from the contaminated soil and the organic vapors, making heated screws indirect heating processes. Using commercial heat transfer fluids, a hot oil system can routinely heat the soil to 275°C. These oils undergo thermal breakdown above 315°C, limiting the maximum operating temperature. Hot oil systems have been demonstrated as being effective for the removal of light solvents, fuel products, and some semivolatile organic chemicals;

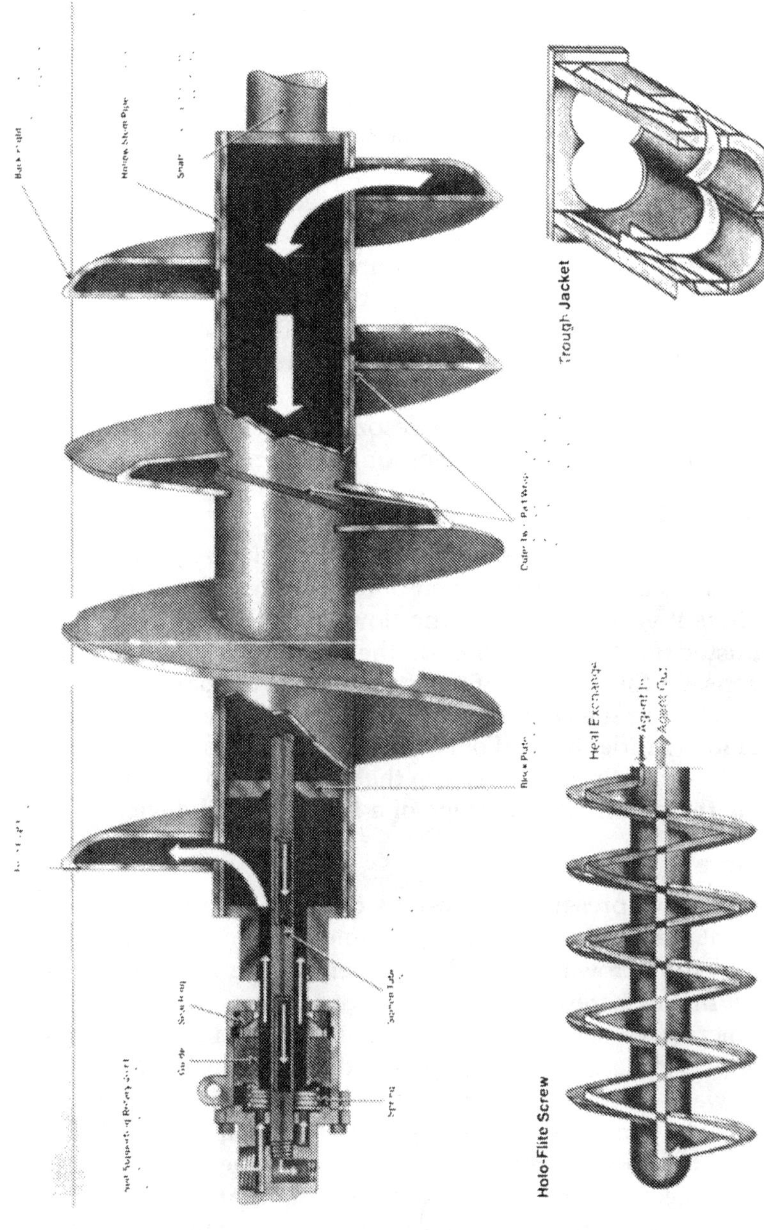

Figure 7 Holo-Flite thermal processor. (From Denver Equipment Co., *Bulletin HF44-B105B*. 1981.)

however, PCBs have not been effectively removed using hot oil systems because of temperature limitations. To reach higher operating temperatures, molten salt heat transfer systems have been used. These systems employ mixtures of inorganic salts that melt at about 140°C and are stable up to 450°C. Using molten salts, soil temperatures of up to 400°C can be achieved.

When electric resistance elements are used for heating, they are attached to the outer wall of screw conveyors. The soil is heated by a combination of conduction and radiation from the heated outer wall. Several such heated screws are manifolded together to make a unit of commercial capacity as shown in Figure 14 (page 303). Because this type of heat transfer is not limited by the properties of a fluid, soil temperatures of up to 1100°C can be achieved if stainless steel or alloy housings are used. As with indirect rotary dryers, the desorbed gases from heated screws can be treated in either condensation or afterburner gas systems.

Fluidized Bed Dryers

A fluidized bed dryer consists of a vertically oriented vessel through which hot gases are circulated from bottom to top. Soil is fed into the vessel, where it is suspended by the flowing gas stream. The gas flow rate is adjusted until the drag force on the soil particles from the flowing gas compensates for the force of gravity, allowing the particles to be suspended in a bed in the center of the dryer vessel. The extremely turbulent conditions inside this bed of solid material results in relatively high heat and mass transfer rates between the fluidizing gas and the bed solids. This is the principal advantage of a fluidized bed device.

Other Dryers

In this section we present descriptions of other dryers that are under development but have not been fully commercialized at the time of writing. These are microwave or radio-frequency type and steam- or air-heated soil mixing screw augers.

In a microwave dryer the soil is heated using microwave or radio-frequency (RF) radiation with equipment that is comparable to a household microwave oven. The dryer consists of a chamber that is connected to a microwave generator by wave guides. The soil is placed into the chamber, and the RF energy is focused on it by the wave guides. The microwave generator is remote from the heating chamber, allowing it to be serviced without exposing workers to a contaminated environment. This is one of the principal advantages of this technology. Also, by using microwaves, the heating energy is focused inside the soil particles. This presents an opportunity to achieve better desorption efficiency (i.e., at a lower average soil temperature) than by heating from the outside in, as with conventional drying technology. However, the microwave generators required to drive commercial capacity systems are relatively expensive, which tends to limit this approach to low-moisture-content materials (i.e., low heat load). This approach has only been proposed; no performance data or constructed systems exist in the hazardous waste industry.

Another application of RF technology is the in-place (in situ) heating of soil. RF source electrodes are placed either in or on the ground in the contaminated area, and energy is transmitted to the soil mass. A fume hood is erected over the entire area, and the vaporized materials are collected and treated in a gas handling system. In this way, large areas of soil can be remediated without excavation, resulting in a potentially substantial reduction in cleanup cost. This technology is early in its application phase to hazardous waste.

Another in situ heating and vaporization technology that is being applied to organic-contaminated soil is the use of stream- or air-heated soil mixing augers. This technology is based on the use of a large, vertical auger common to the construction industry for boring holes for caissons and pilings. Steam or hot air is injected into the soil through the auger as a hole is bored. The vaporized organics are collected in a hood and treated in a gas handling system. Other than this brief description, this technology is outside the scope of this chapter and is not discussed further.

AVAILABLE TECHNOLOGIES

This section presents the thermal desorption treatment systems that are presently available in the soil remediation marketplace. All of these systems have either been constructed in full scale, or detailed full-scale designs have been prepared by the technology developer and supported by pilot or prototype test data.

X * TRAX Process: RUST Remedial Services

RUST Remedial Services has developed a process that utilizes an indirectly fired rotary kiln and condensation of the volatilized organics [14]. The process, termed X * TRAX, was granted U.S. Patent 4,864,942. Development was begun in late 1986 and progressed through extensive operation of a 4.5-metric ton (T)/day pilot plant. A 115-T/day commercial-scale unit has been constructed.

The process flow diagram is shown in Figure 8. If the feed is a soil, it is first screened to a maximum particle size of less than 60 mm. The

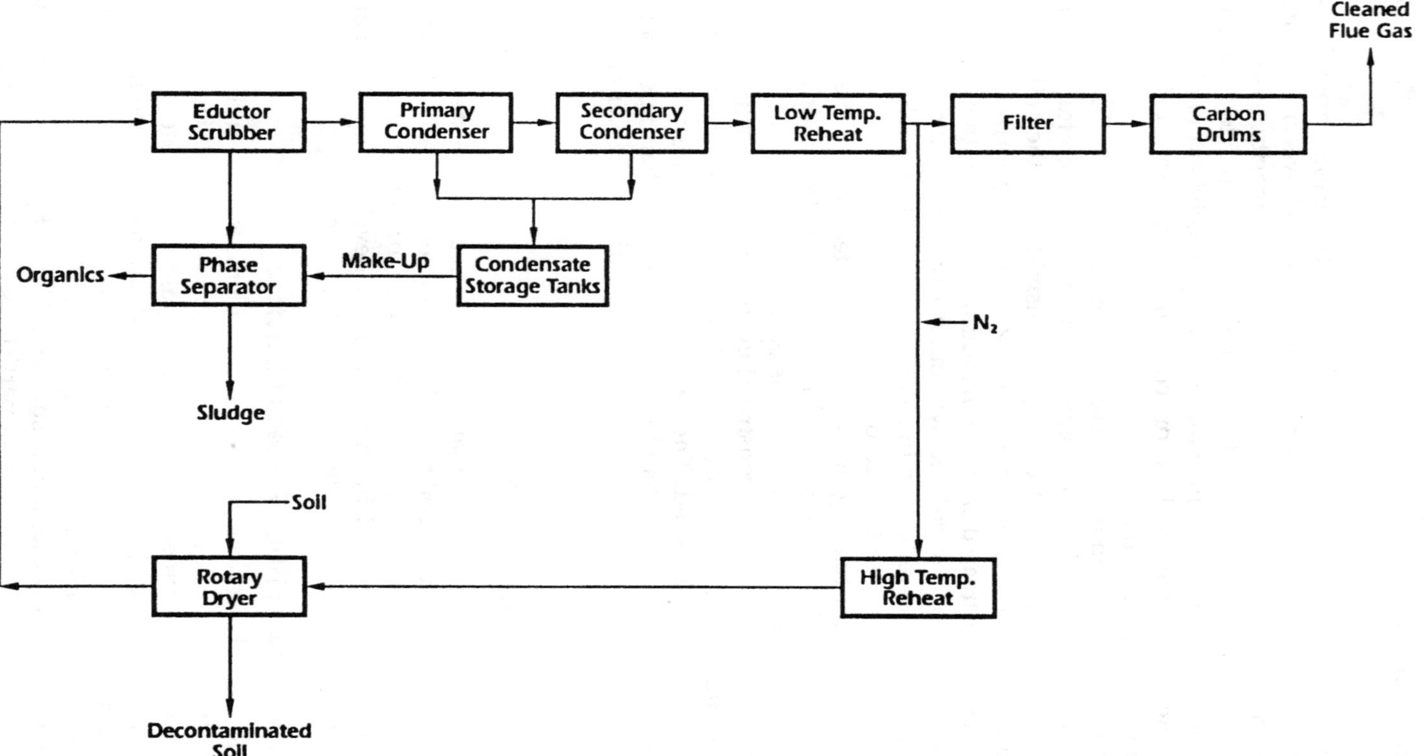

Figure 8 X * TRAX process flow diagram (RUST Remedial Services).

feed material, which otherwise can be either a solid or a pumpable sludge, is conveyed into the indirectly fired rotary dryer. Propane is used as the fuel. Water, volatile organics, and semivolatile organics are vaporized from the solids in the dryer and transported to the gas treatment portion of the system. The hot solids are discharged from the kiln as a powdered or granular dry material. For most applications, water will be mixed with the exiting solids to cool them and to prevent dusting. By adding reagents at this point, metal-containing wastes can be stabilized with the reagent cost being the only additional expense. The water will normally be condensate from the gas treatment system.

Nitrogen is added to the kiln to ensure a nonflammable gas mixture at all times and to aid in desorption of organics. The off gases first pass through a liquid scrubber where entrained solid particles are removed and the gas stream is cooled to its saturation temperature. The scrubber also removes a portion of the volatilized organics. The recirculated scrubber water continuously passes through a phase separator which collects any condensed light organic from the liquid surface and continuously discharges a bottom sludge containing solids, water, and organics. The sludge is dewatered using a filter press. The dewatered solids are either returned to the feed stream or disposed.

The scrubbed gas passes to the first heat exchanger where it is cooled to 5°C above ambient temperature. This heat exchanger will produce the bulk of the liquid condensate. The carrier gas then passes to a second heat exchanger where it is cooled to 4°C. The liquid condensates from both heat exchangers are mixed and allowed to gravity separate. Organics are removed for disposal or further treatment. The condensed water is treated, if necessary, then used to cool and dedust the treated solids exiting the dryer. The 4°C carrier gas now contains some residual moisture and organics that were present in the feed at levels equal to or less than their equilibrium saturation concentration at 4°C. The carrier gas then passes through a reheating step and a recirculation blower. After the blower, 5 to 10% of the carrier gas is vented, and the remainder is heated to 200 to 400°C before returning it to the dryer.

The process vent gas stream passes through a particulate filter (2 μm) and then through a carbon adsorber, where at least 80% of the remaining organics are removed. Actual practice has shown removal efficiencies by the carbon ranging from 89 to 98%. This gas is then vented to the atmosphere. Extensive pilot testing has been carried out on TSCA (PCB-containing) and RCRA hazardous soils and sludges, using either a small pilot unit operated at 1 to 2 kg/h or a larger pilot plant operated at 200 to 400 kg/h [15].

Table 4 Results on PCB-Contaminated Soils: Small Pilot X * TRAX

Feed PCB concentration (mg/kg)	Product PCB concentration (mg/kg)	Percent reduction
4.6	0.94	79.6
5.2	2.0	61.5
12	BDL (1.0)	>91.7
150	3.2	97.9
330	BDL (5.0)	>98.5
770	12	98.4
805	17	97.9
2900	7.0	99.8
2960	170	94.3
3760	164	95.6

Treatability studies on 16 samples of soil from 10 PCB-contaminated sites have been performed using the small pilot X * TRAX unit. The data from several of these treatability studies are presented in Table 4. The data show that a 95% or greater reduction in PCB concentration can be obtained when the contaminated soil contains greater than 100 mg/kg of PCBs. These treatment levels are consistent with the alternate treatability variance levels for PCBs presented in the USEPA's Superfund LDR Guide 6A [16]. For PCB concentrations greater than 100 ppm, a 90 to 99.9% reduction is recommended, and the process meets this requirement.

The larger pilot X * TRAX system has processed 10 different PCB-contaminated soils under a one-time USEPA TSCA R&D permit. The results are summarized in Table 5. Again, at least 95% removal was obtained when the PCB level was greater than 100 mg/kg. Even with 7800 mg/kg PCBs in the feed, the treated soil had less than 25 mg/kg PCBs.

One pesticide-contaminated soil was tested using the small pilot unit. The results are summarized in Table 6. For all identified pesticides at least a 97% reduction was achieved. The TCLP was also performed on the treated soil to assess whether it exhibited toxic organic characteristics (D codes). These results are presented in Table 7. The TCLP concentrations for the three listed pesticides were well below the regulatory level. The samples tested at both the small and large pilot-scale have all had relatively low levels of semivolatile organic compounds. The results for three different compounds are presented in Table 8. When the concentration in the treated soil was measurable, the percentage reduction

Table 5 Results on PCB-Contaminated Soils: Large Pilot X * TRAX

Feed PCB concentration (mg/kg)	Product PCB concentration (mg/kg)	Percent reduction
68	16	76.5
120	3.4	97.2
190	9.6	95.0
630	17	97.3
640	18	97.2
1480	8.7	99.1
1600	4.8	99.7
2800	19	99.3
5000	50	99.0
7800	24	99.7

Table 6 Results on Pesticide-Contaminated Soil: Small Pilot X * TRAX

Contaminant	Feed concentration (ppm)	Product concentration (ppm)	Percent reduction
4,4′-DDE	32	0.57	98.2
4,4′-DDD	320	1.3	99.6
α-Chlordane	100	2.1	97.9
γ-Chlordane	110	3.0	97.3

Table 7 Comparison of Treated Soil to Toxicity Characteristic Regulatory Level: Small Pilot X * TRAX

US EPA HW No.	Contaminant	Treated soil (ppm)	Treated soil TCLP (mg/L)	Regulatory level (mg/L)
D020	Chlordane	5.1	<0.0025	0.03
D031	Heptachlor	<0.03	<0.00025	0.008
D013	Lindane	<0.03	<0.00025	0.4

Table 8 Results on Semivolatile Organics: Large and Small Pilot X * TRAX

Compound	Pilot scale	Feed (ppm)	Product (ppm)	Percent reduction
Bis(2-ethylhexyl)phthalate	Small	3.2	BDL (0.33)	>89.7
	Small	3.9	BDL (0.33)	>91.5
	Large	3.4	0.30	91.2
	Large	9.1	0.18	98.0
Phenanthrene	Small	14	BDL (0.33)	>97.6
	Small	19	0.29	98.5
	Small	30	BDL (13)	>56.7
Naphthalene	Small	34	0.74	97.8
	Small	110	BDL (13)	>88.2
	Small	450	7.9	98.2

was greater than 95%. The data indicated that most semivolatile organic compounds can be reduced to less than 10 ppm and frequently below 1 ppm.

As was the case for semivolatile organics, very few soil samples were received that contained significant quantities of volatile organics. It is not surprising that the most often detected volatile organics have been the BETXs (benzene, ethylbenzene, toluene, and xylene). Several examples of removal rates are presented in Table 9. In most cases the volatile organics were reduced to well below 1 ppm. It is interesting to note that the highest volatile organic contamination levels were in the feed sample of SSM-I, the same material tested by Szabo et al. [7] and for which results were presented earlier. RUST Remedial Services did not participate in the original study since it was initiated before the laboratory X * TRAX system became operational. A sample of the SSM-I soil was obtained from the contractor's archive.

The results for ethylbenzene and xylene were reported in both studies. The X * TRAX results fall between the results reported at 66°C and at 177°C by Szabo et al. Table 10 shows results from tests using the large pilot plant on a refinery waste sludge carrying RCRA waste codes K048, K049, and K050 [17]. The goal was treatment to achieve the best demonstrated available technology (BDAT)† requirements for the organic

†BDAT stands for "best demonstrated available technology" under the RCRA land disposal restriction program. A BDAT level is the performance level that must be met by a treatment technology on a given hazardous waste in order to allow the residuals to be land disposed.

Table 9 Results on Volatile Organics: Small Pilot X * TRAX

Compound	Feed (ppm)	Product (ppm)	Percent reduction
Benzene	0.35	BDL (0.05)	>85.7
	3.0	BDL (0.10)	>96.7
	6.9	1.6	76.8
	30	BDL (0.13)	>99.6
	980	BDL (0.21)	>99.98
Ethylbenzene	0.40	BDL (0.05)	>87.5
	13	BDL (0.10)	>99.2
	50	0.33	99.3
	1600[a]	5.2	99.8
Tetrachloroethylene	150[a]	0.094	99.9
Xylene	0.22	0.030	86.4
	1.6	0.043	97.3
	3.8	BDL (0.050)	>98.7
	7.2	BDL (0.095)	>98.7
	77	BDL (0.10)	>99.9
	130	0.84	99.4
	2400[a]	9.5	99.6

[a]Results on SSM-I.

compounds as given in the Code of Federal Regulations (40 CFR 268.41). As shown in the table, all requirements for organics were met, demonstrating the applicability of thermal desorption for these waste streams. See also the later sections on the processes of Remediation Technologies and Southdown Thermal Dynamics. Note that the BDAT requirements for K048, K049, and K050 also include leaching levels for heavy metals. Thermal desorption will not necessarily meet these requirements, and the residues from thermal desorption will sometimes require stabilization.

A mercury-contaminated soil has been treated using the small pilot plant, with greater than 99.9% reduction measured. The large pilot plant was also successfully operated on a mixed (RCRA hazardous and low-level radioactive) waste at the Oak Ridge National Lab [18,19]. A full-scale production system was constructed for on-site cleanup of contaminated soil. The system is capable of treating 115 T/day of soil with a moisture content of 20%. It is fully transportable, consisting of three semitrailers, one control room trailer, eight equipment skids, and various pieces of movable equipment.

All of the equipment has been designed for over-the-road transport anywhere in the United States or Canada. The kiln is the largest of its

Table 10 Results on K048, K049, and K050 Refinery Waste Sludge: Large Pilot X * TRAX

Component	Total constituent analysis concentration (mg/kg)		
	Feed[a]	Treated product[a]	BDAT for K048–50[b]
Anthracene	9.2	0.37	28
Benzene	BDL (5)	BDL (0.5)	14
Benzo[*a*]pyrene	BDL (100)[c]	BDL (2)	12
Bis(2-ethylhexyl)phthalate[d]	373	BDL (2)	7.3
Chrysene	BDL (100)[c]	1.3	15
Di-*n*-butylphthalate[d]	69	BDL (2)	3.6
Ethylbenzene[d]	40	BDL (0.5)	14
Naphthalene	19	1.4	42
Phenanthrene[d]	44	1.9	34
Phenol	BDL (100)[c]	0.46	3.6
Pyrene[d]	57	1.3	36
Toluene[d]	29	8.7	14
Xylene(s)[d]	203	BDL (0.5)	22
Cyanide	0.5	0.9, 0.2	1.8

[a]BDL(X), below detection limit of X mg/kg.
[b]Non-wastewaters.
[c]BDAT standards are lower than method detection level.
[d]Exceeded BDAT standards in feed.

kind that can be transported over the road. The components are mobilized to the project site and assembled using a relatively small crane. Approximately 3 to 4 weeks are required to install the equipment completely. Site preparation involves grading the site level and providing a firm base such as compacted gravel. Concrete footings are not usually required; however, concrete housekeeping pads may be. All skids or trailers that normally contain liquids have integral liquid containment curbs for spill control.

The system is monitored from the control trailer, which is a heated and air conditioned portable office trailer. All aspects of system operation can be monitored from the control trailer, and all essential process control parameters can be adjusted. The system has been operated on noncontaminated feed materials at rates of over 110 T/day. The commercial unit is scheduled to be transported to North Dartmouth, Massachusetts, during early 1992 to be used to remove PCBs from approximately 32,000 T of soil at the ReSolve Superfund site near New Bedford, Massachusetts.

Deutsche Babcock Anlagen AG Process

Deutsche Babcock Anlagen AG, in Krefeld, Germany, has developed a process that uses an indirectly heated rotary kiln to remove organics from contaminated soils by a combination of pyrolysis and volatilization [20]. The volatile organics are then destroyed in a secondary combustion chamber. In this respect the process uses a combination of thermal desorption and incineration technology. After piloting at the scale of 0.5 to 1.0 T/h, Deutsche Babcock designed and constructed a 7-T/h commercial-scale demonstration plant for Bergbau AG Westfalen near Dortmund, Germany. In 1990 the demonstration plant was in operation at a former coke oven site in Unna-Bonen near Dortmund, Germany.

The process configuration is shown in Figure 9. In the soil pretreatment step, stones, pieces of concrete, and iron are separated from the soil. An impact mill is used for size reduction, with a 50-mm maximum-particle-size specification. Lime is mixed with the soil to control formation of acid gases in the kiln. The soil is then fed to the kiln via a twin-screw conveyor. The conveyor is designed to form an air seal, allowing operation of the kiln at a slight negative pressure. The solids pass into an indirectly fired rotary kiln and are heated to 500 to 650°C, with a maximum temperature of 750°C. Natural gas is used as the fuel. The solids residence time is approximately 1 h. The kiln has a diameter of 2.2 m and is 21 m long.

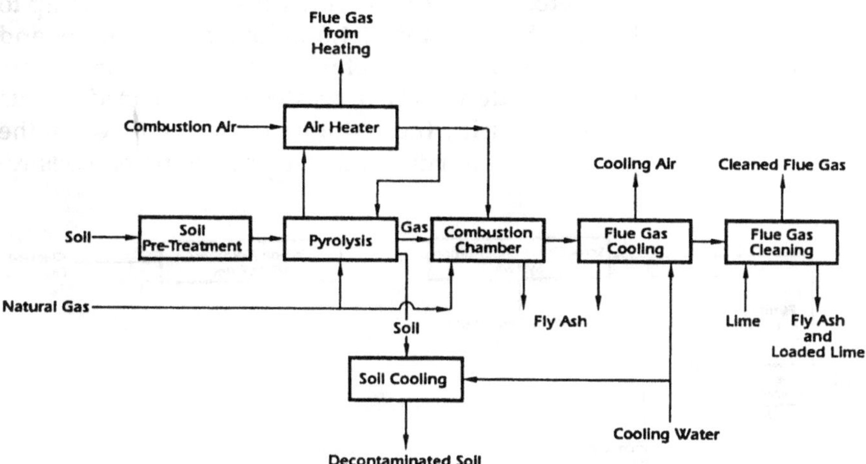

Figure 9 Simplified plant flowsheet for cleanup of contaminated soils by pyrolysis (Deutsche Babcock).

Organics are removed from the soil by a combination of pyrolysis and volatilization. The vaporized organics and moisture pass from the kiln into a secondary combustion chamber. The treated soil falls into a discharge chamber and is then cooled, rewetted, and discharged. The exhaust flue gases from the kiln annulus are passed through a heat exchanger in which combustion air is preheated. This combustion air is then used for firing both the rotary kiln and the secondary combustion chamber (SCC). Natural gas is again used in the SCC to provide combustion temperatures of 1000 to 1300°C. The residence time is approximately 3 s. The flue gas from this operation is cooled to approximately 180°C by water spray/quench. Dust is then removed in a baghouse. Further acid gas removal, if necessary, is effected by injection of lime upstream of the baghouse. The cleaned flue gas, now at a temperature of 120 to 180°C, is then discharged through an induced-draft fan to a stack. For the demonstration at Unna-Bonen, the concentrations of 17 polynuclear aromatic hydrocarbons (PAHs) were measured in the feed and product streams. The total feed PAH concentration was in the range of 2000 to 6000 mg/kg, and greater than 99% removal was obtained.

International Technology Corp. Process

International Technology Corporation (IT) has performed both lab and pilot testing of a proposed thermal desorption system based on the use of an indirect-heated rotary dryer [21,22]. Figure 10 is a simplified schematic of IT's thermal desorption pilot plant [22]. The pilot plant consists of a gas-fired indirect-heated rotary dryer with a stated capacity of up to 91 kg/h. The dryer has a 16.5-cm-diameter 4.3-m-long alloy cylinder and is capable of achieving soil temperatures of up to 600°C. Pilot-scale tests have been conducted on a wide variety of wastes [21]. Selected results are summarized in Table 11. During these studies the feed rate was in the range of 31 to 74 kg/h. Tests were run on dioxin-contaminated soils re-

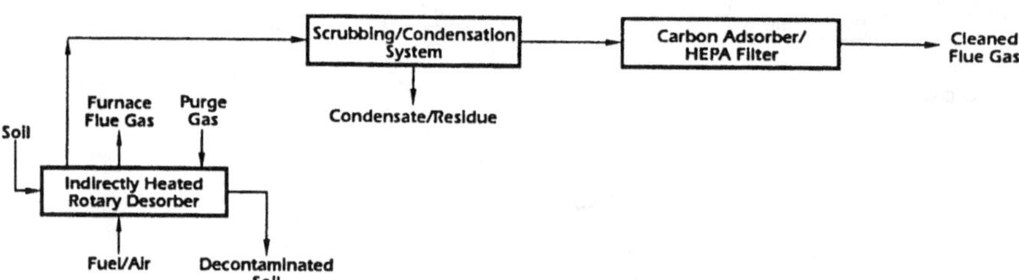

Figure 10 Simplified block flow diagram of thermal desorption pilot plant (IT Corporation).

Table 11 Summary of IT Pilot Tests

Source of waste	Soil treatment conditions		Feed waste		Treated waste[a]	
	Temp. (°C)	Retention time (min)	PCBs (ppm)	2,3,7,8-TCDD (ppb)	PCBs (ppm)	2,3,7,8-TCDD (ppb)
USAF, Mississippi	560	40	—	260	—	ND
	560	19	—	236	—	ND
	560	10.5	—	266	—	ND
	460	24	—	233	—	0.5
USAF, Johnston Island	550	5.6	—	48	—	<0.084
	550	20	—	56	—	0.23
Rosemount Research Center	375	39	216	—	<2.0	—
	371	22	642	—	<2.0	—
	379	39	44,500	—	10.9	—
	377	23	44,500	—	52.3	—
	450	45	44,600	—	3.85	—
	449	22	44,600	—	<2.0	—
	551	21	35,500	—	<2	—

Source: Ref. 21.
[a]ND, not detected.

sulting from leaking drums of Herbicide Orange. These studies were carried out on-site in Mississippi and on Johnston Island in the South Pacific. 2,3,7,8-Tetrachlorodibenzodioxin was reduced from 50–260 ppb to below 1 ppb in the treated soils. Volatilized organics were condensed and destroyed by ultraviolet photolysis.

PCBs were removed from soil from the Rosemount Research Center site in studies at the University of Minnesota using the IT pilot equipment. For soil samples containing 216 to 642 ppm PCBs, treated soil PCB concentrations were below 2 ppm. For feed samples containing 35,500 to 44,600 ppm, treated soils contained from 52 ppm to less than 2 ppm. The IT pilot plant has also been tested on wastes containing 2,4,-D, 2,4,5-T, pentachlorophenol, tetraethyllead, kerosene, mercury sulfide, and creosote. Mixed wastes contaminated with PCBs and with low levels of uranium and technetium were treated, resulting in removal of the PCBs from the soil. The treated soil was then suitable for disposal as a low-level radioactive waste.

IT has achieved low residual contaminant concentrations in their pilot testing program for both PCBs and PAHs. Generally, the residual level was at or near detection levels at the highest soil temperature operating conditions, which ranged from 400 to 560°C. In addition to their pilot testing, IT has performed numerous studies using a tray heating furnace, a fixed quartz tube furnace, and an electric rotary tube furnace. These studies have been conducted to examine key process parameters required to meet cleanup standards. Based on their pilot work, IT has proposed a commercial system with an estimated nominal capacity of 9 T/h that is transportable for on-site remediation. Although the pilot plant has a condensation gas treatment system, IT has proposed either condensation or afterburner systems for full-scale operation.

SoilTech Anaerobic Thermal Process: AOSTRA Taciuk Process

The AOSTRA Taciuk Process (ATP) is a combination thermal desorption and pyrolysis operation that has been under development since 1977 by UMATAC Industrial Processes of Calgary, Alberta, Canada, under funding and ownership agreements with the Alberta Oil Sands Technology and Research Authority (AOSTRA) [24]. AOSTRA holds the patent rights; UMATAC manages the development program and is the primary licensee for use in treating hydrocarbon-bearing wastes. SoilTech, a subsidiary of Canonie Environmental is the licensee operator of the technology in the United States. SoilTech purchased the first commercial-scale plant in 1989.

The central feature of the process is the AOSTRA Taciuk processor shown in Figure 11. Contaminated soil or sludge is fed into a preheat

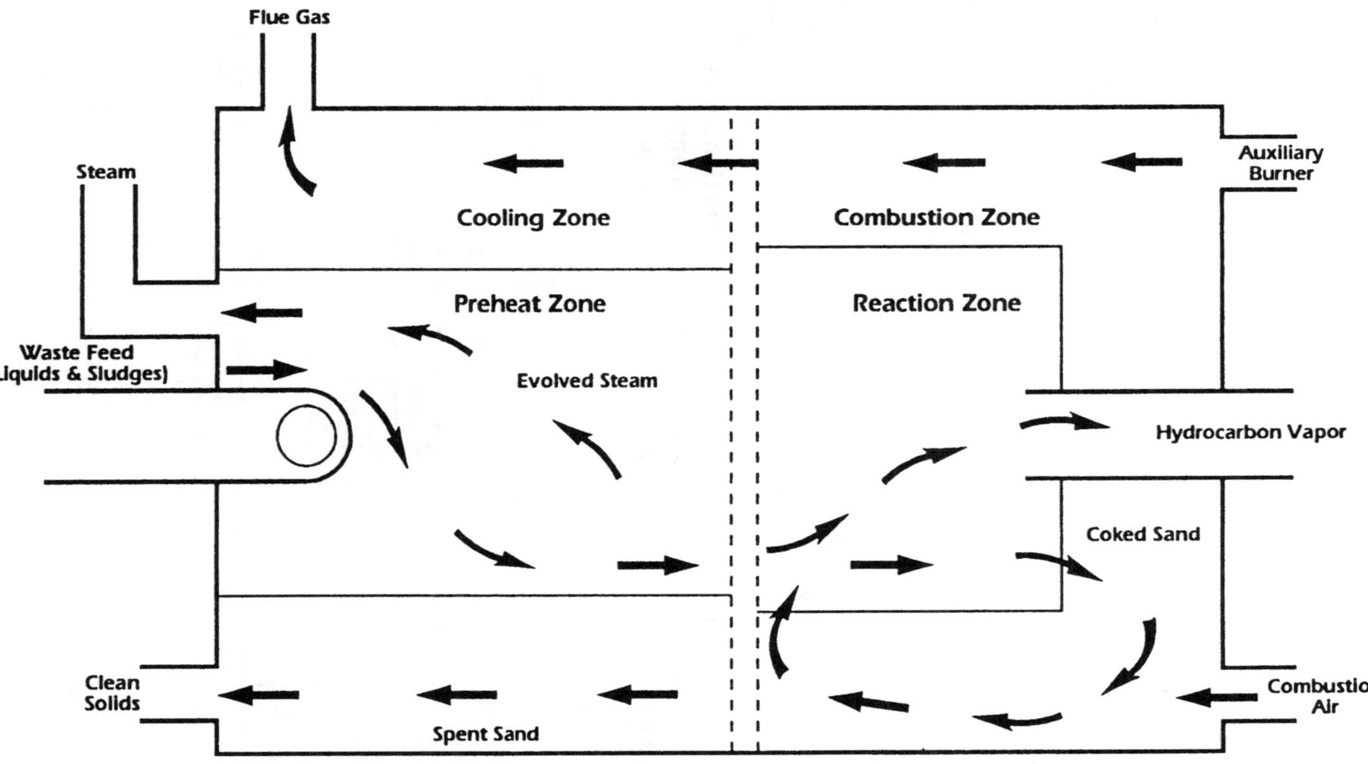

Figure 11 Internal process flows for the AOSTRA Taciuk processor.

zone. The feed is preheated by indirect contact with cleaned solids, and steam and light hydrocarbons are evolved. The feed then enters a reaction zone in which vaporization and/or thermal cracking of heavier hydrocarbons occurs. The hydrocarbons volatilized in this step exit the processor separately. The solids then enter a combustion zone in which carbon residues are burned from the inert solids. Some of the heat requirement for the processor is provided by this operation. Wastes are combusted in the combustion zone, differentiating this technology from most others described in this chapter, which employ indirect heating of the solids. This one feature of the process is shared with incinerators.

SoilTech secured the first contract for remediation of wastes with the ATP at the Wide Beach Superfund site in Brant, New York, about 40 miles southwest of Buffalo. The primary contaminant was PCBs. Figure 12 shows the process configuration used at Wide Beach [25]. Although the thermal processor is capable of removing PCBs from soils by volatilization, the record of decision called for on-site dechlorination. Proprietary dechlorination reagents were therefore added to the feed soil. At the temperatures reached in the processor, it is claimed that these reagents reacted with the PCBs to cause dechlorination. As shown in the figure, flue gases were passed through a cyclone, baghouse, wet scrubber, and a carbon bed prior to exhausting. The steam and light hy-

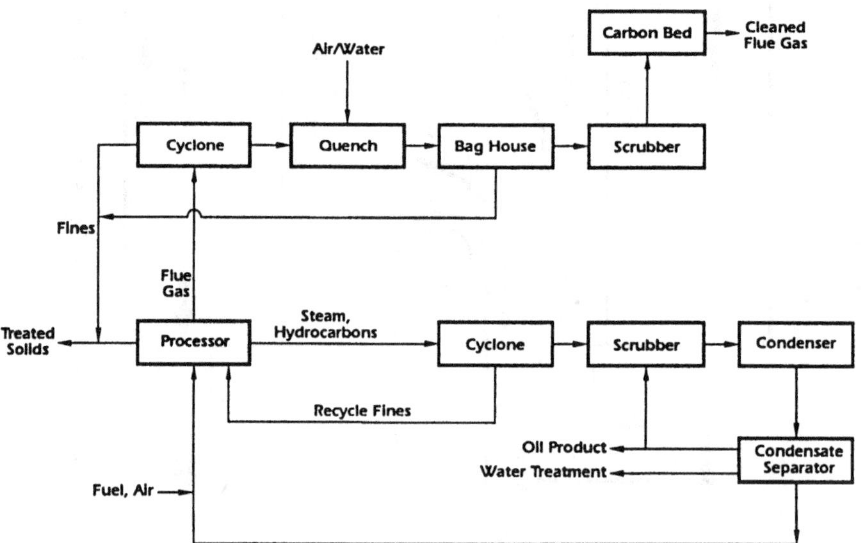

Figure 12 ATP process for oily solid waste management (SoilTech).

drocarbon stream from the preheat zone was combined with the heavy hydrocarbons from the reaction zone, then passed through a scrubber and condenser. Noncondensables were recycled to the burner on the processor. The oil phase from the condenser was mixed with dechlorination reagents and added to the feed soil. PCBs were reduced to below 2 ppm from levels of up to and above 5000 ppm. After completing the Wide Beach project, the unit was scheduled to be relocated to the Waukegan Harbor Superfund Site in Illinois, where it was to be used without the dechlorination feature to remediate PCBs from drained harbor sediments.

Thermal Desorption Process: Remediation Technologies, Inc.

Remediation Technologies (ReTeC) has developed a thermal desorption process based on use of a heated screw for indirect heating of waste and volatilization of contaminants. The specific equipment used is the Holo-Flite screw processor manufactured by the Denver Equipment Company. The processor has a jacketed trough that houses two intermeshing feed screw augers. The augers move the waste through the processor. Hot oil or molten salt is used as the heating medium. This medium is fed through one end of each auger, travels through the hollow flights, and returns through the center of the shaft. The heat transfer medium is also circulated through the jacketed trough to provide additional heat transfer area. Denver Equipment Company has worked with ReTec to incorporate several proprietary improvements in the processor to improve efficiency and economy. These include the use of molten salt as the heat transfer medium, the use of an inert stripping gas, and improved contacting between waste, stripping gas, and heated surfaces. The molten salt is a eutectic mixture consisting of 53% potassium nitrate, 40% sodium nitrite, and 7% sodium nitrate. Its use allows operation at 250 to 450°C, a significantly higher and more useful temperature range than that allowed by use of hot oil.

A process flow diagram is shown in Figure 13 [26]. If necessary the feed is treated to remove free liquids and screened to remove particles larger than about 25 to 50 mm, depending on the size of the processor used. Low-viscosity or watery feeds, < 20% solids, are unsuitable, causing bypassing in the processor and reducing the efficiency of the operation. Waste is fed to the processor through a feed hopper sealed with air-lock valves to minimize leakage of air into the processor. After volatilization of moisture and organics, the residual solids are discharged into a solids cooler through a rotary air lock valve. This cooler is also a Holo-Flite processor, employing a single auger and chilled water to cool the treated solids to 65°C to improve the safety of handling.

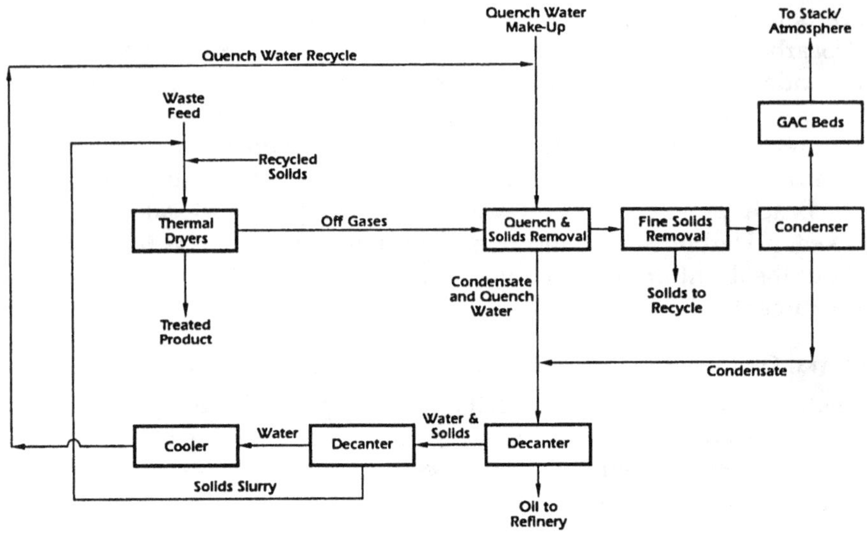

Figure 13 Block flow diagram for a dryer system (ReTeC).

The processor is operated at a slight negative pressure, and an inert gas, such as nitrogen, is added. The off-gas treatment train includes a quench unit to cool the stream and to remove solids. A cyclone is used for additional solids removal. Water and organics are removed in a chilled condenser. The remaining gases are passed through activated carbon and vented. The condensate is phase separated into organic, aqueous, and solids slurry streams. The aqueous stream can be recycled to the quench system or sent to a wastewater treatment system. The solids slurry stream is recycled to the processor. When operating on K048-51, the recovered oil can be returned to the refinery. Treatability tests have been carried out on a number of feeds, including wastes from petroleum refinery, gas utility, wood preserving, and chemical process industries. BDAT requirements can be met for K048-51 refinery wastes.

A 450-kg/h system was started up in 1989 and used for process development and treatability studies. It was designed for use with either hot oil or molten salt. Treatability studies have been carried out on K048-51 refinery filter cakes, manufactured gas plant soils, and other soils and sludges. Results from a treatability study on a refinery K-listed filter cake are shown in Table 12. The treated product met the BDAT requirements. Results for a manufactured gas plant soil are shown in Table 13. Good removal efficiencies were measured for semivolatile organics and cyanides. The 450-kg/h system was used in 1991 for on-site

Table 12 Results on Refinery Filter Cake: ReTeC Process[a]

Compound	Feed (ppm)	Product (ppm)	BDAT standards (ppm)
Benzene	<0.1	<0.1	14
Toluene	3.9	<0.1	14
Ethylbenzene	14	<0.1	14
Xylenes	129	<0.3	22
Naphthalene	250	<0.7	42
Fluorene	192	<0.1	—
Phenanthrene	609	4.6	34
Anthracene	190	<0.6	28
Fluoranthene	2,570	4.1	—
Pyrene	1630	<0.3	36
Benzo[*b*]anthracene	714	0.6	—
Chrysene	291	<0.1	15
Benzo[*b*]fluoranthene	75	<0.6	—
Benzo[*k*]fluoranthene	97	<0.9	—
Oil and grease (%)	23.8	0.3	
Solids (%)	64.2	99	

[a]Treatment temperature 500°F.

feasibility demonstrations. A commercial-scale unit with a 2- to 3-ton/h capacity was constructed in 1991. This system was to be installed at a petroleum refinery for treatment of K wastes, with startup planned for late 1991.

HT5 Thermal Distillation Process: Southdown Thermal Dynamics

The HT5 process employs temperatures of up to 1100°C to thermally desorb volatile and semivolatile organics from solids. Organics are recovered by condensation. The process was invented by Tom F. DesOrmeaux, with development beginning in 1982. Prototype units were tested on oil well drilling cuttings and drilling muds, accumulating over 50,000 h of operating experience. Performance in these applications was cited in U.S. Patent 4,606,283. In 1989, the technology was licensed to Browning Ferris Industries for use in the 48 contiguous states. The license was subsequently acquired by Southdown. Treatability studies have also been carried out on soils contaminated with creosote, hydrocarbons, and mercury.

Figure 14 is a flow diagram for the process as configured for a refinery application [27,28]. The feed, including separator sludges, tank

Table 13 Results on Manufactured Gas Plant Site Soil: ReTeC Process

Compound[a]	Feed (ppm)	Product[b] (ppm)	Percent Reduction
Naphthalene	09	ND	>96.8
Acenaphthylene	92	ND	>95.3
Acenaphthene	130	ND	>96.0
Fluorene	193	ND	>99.7
Phenanthrene	568	ND	>99.7
Anthracene	166	ND	>99.5
Fluoranthene	530	ND	>99.9
Pyrene	892	ND	>99.3
Benzo[*b*]anthracene	366	ND	>99.5
Chrysene	586	ND	>99.3
Benzo[*b*]fluoranthene	154	ND	>99.9
Benzo[*k*]fluoranthene	94	ND	>99.9
Benzo[*a*]pyrene	425	ND	>99.9
Dibenz[*a*,*b*]anthracene	83	ND	>99.6
Benzo[*g*,*h*,*i*]perylene	360	ND	>99.9
Indeno[1,2,3-*cd*]pyrene	90	ND	>85.6
Cyanides	11	ND	>95.0

[a]All semivolatile organics.
[b]ND, not detected.

bottoms, and oily soil, is fed into the process through a dump bin and feed silo. These are not shown on the diagram. As it moves through the silo, air above the waste is replaced with nitrogen, and a nitrogen blanket is then maintained throughout the balance of the process to prevent formation of flammable gas mixtures. The nitrogen also acts as a sweep gas to aid in removing vaporized contaminants from the heated zones.

From the silo the waste is augered to a feed hopper, from which, in turn, it is transferred to one of three parallel banks of heating chambers. Each of the three banks contains three chambers. The waste is moved through the chambers by means of augers. For the prototype units (U.S. Patent 4,606,283), energy was supplied by means of electrically heated rods positioned outside the chamber wall. As the waste travels through the first chamber, it is heated to 200°C, and water and light hydrocarbons are volatilized. The waste then falls by gravity into the second chamber, where it is heated to 480°C and additional organics and the remainder of the water are removed. In the third chamber the waste is heated to the final temperature, which can be as high as 1100°C. The heavier hydrocarbons are removed in this step. The hot, dry solid

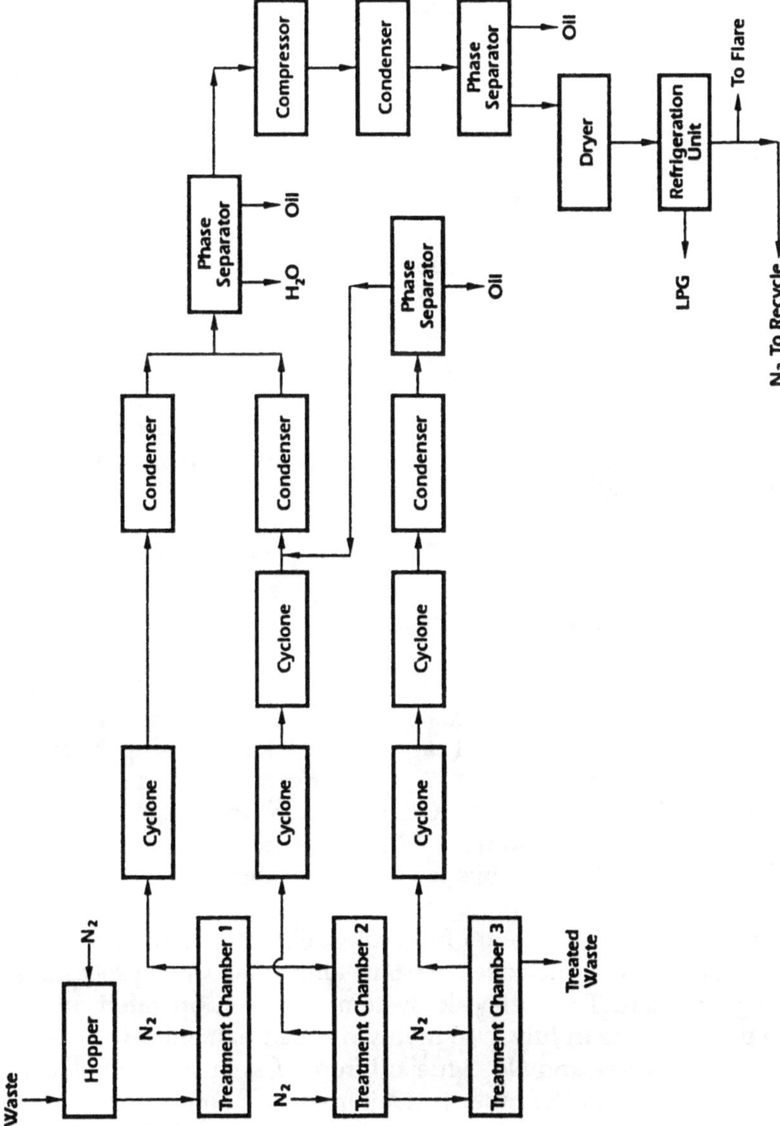

Figure 14 Block flow diagram of the HT-5 thermal distillation process (Southdown Thermal Dynamics).

residues are discharged and transported through a cooling auger to a collection bin.

The volatilized materials and the nitrogen sweep gas pass through cyclones for dust removal. Air-cooled condensers then lower the gas temperature to 10°C above ambient and remove the least volatile materials. The condensate is phase-separated into oil and aqueous fractions. The oil is recycled to the refinery, and the water is sent to the refinery's wastewater treatment system or treated for discharge. The remaining gases are compressed, dried, and cooled to −34°C to recover liquefied petroleum gases. The remaining gas stream, primarily nitrogen, is recycled with a small purge sent to a flare or fuel gas system.

After extensive development work referred to above, a 27-T/day unit was constructed for tests on K048-52 refinery wastes. Volatile and semivolatile organics were below the land disposal restriction requirements for K048-52. On this basis, an 85-T/day unit was designed and installed in 1990 at the Chevron El Segundo refinery.

Low-Temperature Thermal Treatment: Weston Services, Inc.

Weston's LT^3 (low-temperature thermal treatment) system is based around a hot-oil-heated screw. Figure 15 shows a process flow diagram of the LT^3 system. Weston has constructed and tested a pilot system in cooperation with the U.S. Army as part of a USATHAMA demonstration program for the technology. This unit, which uses a 3-m-long twin-screw processor and has a nominal capacity of 68 kg/h, was operated at the Letterkenny Army Depot in 1985 [29]. Performance data from this pilot demonstration showed total VOC reduction from 32,000 ppm to 4.5 ppm, and xylene reduction from 27,200 ppm to 0.55 ppm, at a soil temperature of 160°C. This established the applicability of LT^3 to VOC-contaminated soil [30].

Weston has also constructed a full-scale LT^3 system having a nominal capacity of 6.8 T/h, which uses two 6-m-long quad screw processors that are piggybacked. The full-scale system has been operated on (at least) two projects, one in July 1988 involving the treatment of soil contaminated with gasoline and No. 2 fuel oil from leaking storage tanks in Springfield, Illinois, and the other in 1989 involving soil contaminated with aviation fuel and chlorinated solvents at the Tinker Air Force Base in Oklahoma [30]. During the Springfield project, total BTEX compounds were reduced from 155 ppm to < 0.016 ppm. Data are also reported showing reduction of semivolatile organic compounds; however, the starting level was less than 5 ppm in all cases, which is generally below concern.

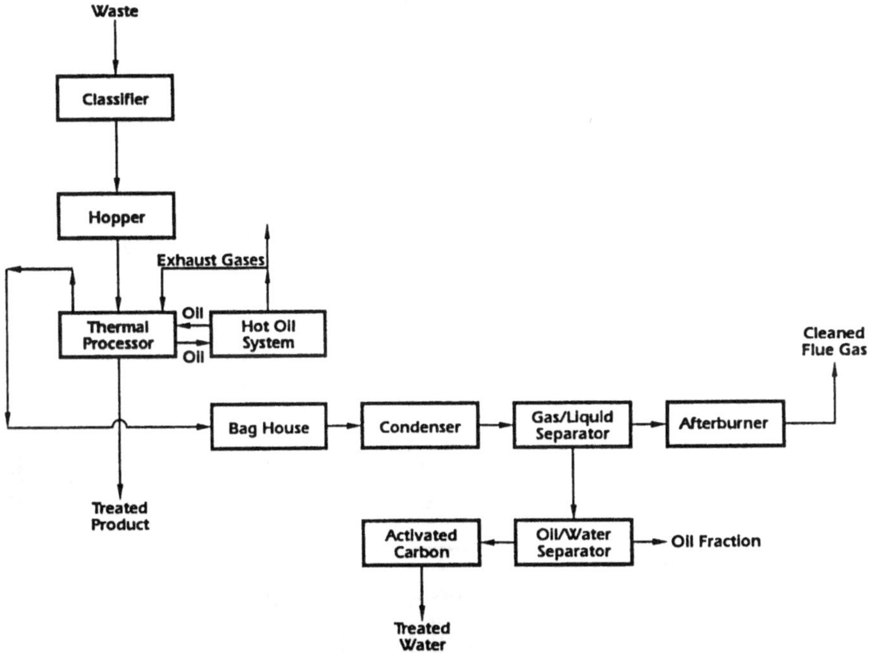

Figure 15 Block flow diagram of the mobile low-temperature thermal treatment (LT^3) system (Weston Services, Inc.).

The thermal processor is operated under a negative pressure from the gas system's induced-draft fan. Oil is heated to a maximum temperature of 340°C in a 1.75-MW (6×10^6 Btu/h) fired heater and then circulated through the processor's shafts, flights, and trough. Maximum soil temperatures of 235°C can be achieved. The vaporized contaminants are swept from the thermal processor using a mixture of air and exhaust gases from the hot oil system's furnace. They are treated with a baghouse to remove particulate carryover and then cooled in a two-stage condenser to remove the bulk of the water vapor and heavy oils. The remaining gas is incinerated in a propane-fired afterburner at 980°C with a 2-s residence time. Because the maximum achievable soil temperature is limited by the use of a hot-oil heat transfer system, the LT^3 system does not appear well suited for removing high-boiling compounds such as PCBs. This is supported by statements made in the public forum. The unit is well suited to VOC contaminated soil, as shown above.

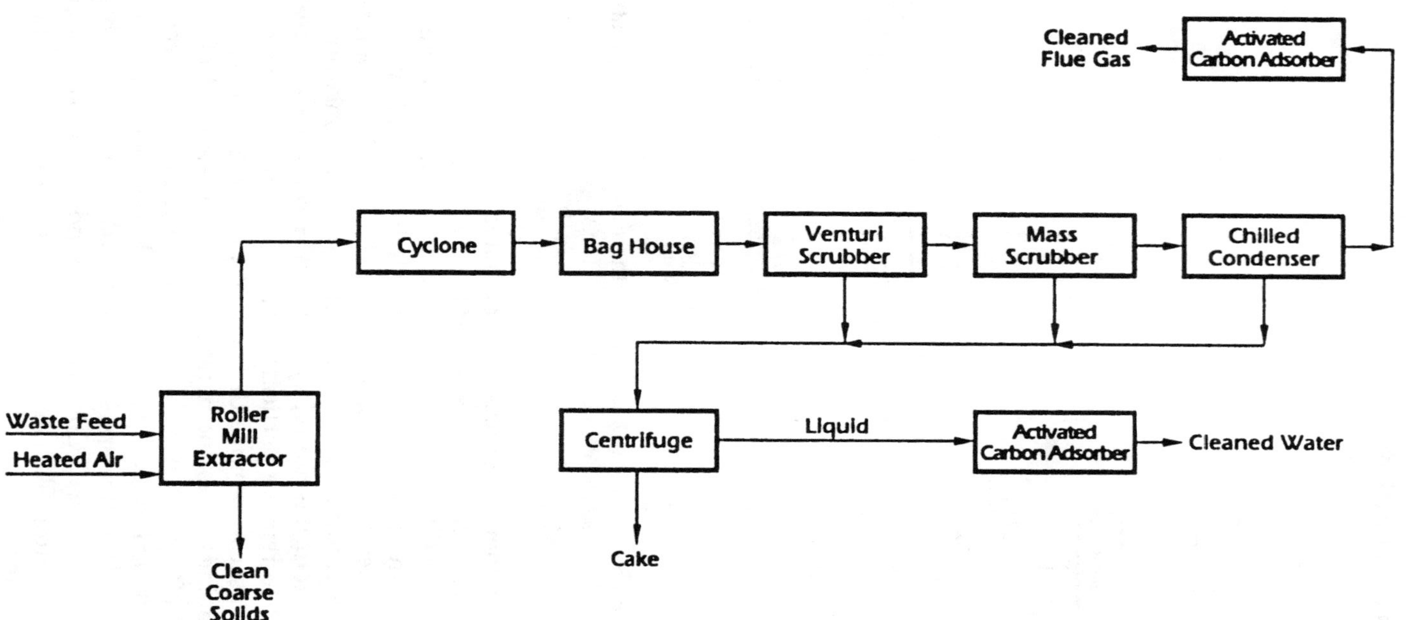

Figure 16 Block flow diagram of the DAVES process (Recycle Sciences International, Inc.).

Desorption and Vapor Extraction System: Recycling Sciences International

Recycling Sciences International (RSI) has developed DAVES (desorption and vapor extraction system), which uses a fluidized bed to volatilize and remove organic compounds from contaminated soils and sediments. RSI was formerly known as American Toxic Disposal (ATD). Figure 16 is a process flow diagram for DAVES [32].

In DAVES the contaminated solids or sediments are fed into a fluidized bed, where they are contacted with hot air from a gas-fired heater at 540 to 760°C. Direct contact of the waste and the hot air results in high drying rates, vaporizing the water and organic contaminants. The unit is operated so that the solids in the fluidized bed reach a temperature of about 160°C, and the solids residence time in the bed is 0.5 to 2 s. The treated solids are discharged from the bottom of the dryer, and the vaporized contaminants, water, and some particulate solids are transferred to the gas treatment system. A cyclone separator and baghouse remove the particulate carryover. The vapors are then treated and cooled in a venturi scrubber, counter current plate column, and chilled heat exchanger, all followed by a gas-phase carbon adsorption system. Residues from the process include a centrifuge sludge containing the separated organic chemicals, and spent carbon from the adsorbers, both requiring off-site disposal. Wastewater and baghouse and

Table 14 RSI Pilot-Scale Test Results on Waukegan Harbor and Hudson River Sediments

Temperature (°C)		PCB concentrations (ppm)	
Gas	Soil	Feed	Product
540	165	44	1.6
760	165	109	1.0
760	150	53	1.5
815	175	37	3.2
870	175	31	0.9
790	205	38	1.5
815	205	28	4.0
870	150	27	1.4
815	165	12.8	0.5
790	135	12.9	0.5
650	130	8.6	1.3
815	135	8.6	1.3
760	135	206	0.8

cyclone dust are also produced and could be recycled to the system according to RSI.

ATD originally developed the fluidized bed dryer as the front end for an innovative incineration system, with the vaporized off gases going to a cyclone separator, baghouse, and a high-temperature combustion chamber (U.S. Patents 4,463,691 and 4,402,274). This work was performed in the early and mid-1980s. The unit was proposed for the treatment of PCB-contaminated harbor and river sediments, and a 9-T/day pilot plant was actually tested under a USEPA R&D permit on PCB sediments from both Waukegan Harbor in Illinois and the Hudson River. The results are presented in Table 14. PCB levels were reduced from a range of 8.6 to 206 ppm in the feed to 0.8 to 4.0 ppm in the treated material. Hot-air temperatures of 540 to 870°C were used, with solids temperatures being 135 to 205°C [3]. Little other public information exists on the DAVES process.

REFERENCES

1. Cudahy, J. J., and A. R. Eicher, Thermal remediation industry: markets, technologies, companies, *Pollut. Eng.*, pp. 76–80 (November 1989).
2. Cudahy, J. J., and W. L. Troxler, Thermal remediation industry update II, *Air and Waste Management Symposium*, Cincinnati, OH, February 1990.
3. dePercin, P., Thermal desorption attainable remediation levels, *EPA Hazardous Waste Symposium*, Cincinnati, OH, March 9–11, 1991.
4. Lighty, J. S., D. W. Pershing, V. A. Cundy, and D. G. Linz, Characterization of thermal desorption phenomena for the cleanup of contaminated soil, *Nucl. Chem. Waste Manage.* 8(22):225–237 (1988).
5. Lighty, J. S., G. D. Silcox, D. W. Pershing, V. A. Cundy, and D. G. Linz, Fundamental experiments on thermal desorption of contaminants from soils, *Environ. Prog.* 8(1):57–61 (1989).
6. Lighty, J. S., G. D. Silcox, and D. W. Pershing, *Investigation of Rate Processes in the Thermal Treatment of Contaminated Soils*. GRI-90/0112, Gas Research Institute, March 1990.
7. Szabo, M. F., R. D. Fox, and R. C. Thurnau, *Application of Low Temperature Thermal Treatment Technology to CERCLA Soils*, USEPA Report 600-9-88-021, 1988, pp. 529–541.
8. Helsel, R. W., and A. Groen, *Laboratory Study of Thermal Desorption Treatment of Contaminated Soils from Former Manufactured Gas Plant Sites*. GRI-88/0161, Gas Research Institute, August 1988.
9. Helsel, R., E. Alperin, and A. Groen, *Engineering-Scale Evaluation of Thermal Desorption Technology for Manufactured Gas Plant Soils*, GRI 89/0271, Gas Research Institute, November 1989.
10. deLeer, E. W. B., M. Baas, C. Erkelens, D. A. Hoogwater, J. W. de Leeuw, P. J. W. Schuyl, L. C. de Leur, and J. W. Graat, Thermal cleaning of soil

contaminated with cyanide wastes from former coal gasification plants, *Proceedings of the International Conference on New Frontiers for Hazardous Waste Management*, Pittsburgh, September 1985, pp. 258–265.

11. Dong, J.-I., and J. W. Bozzelli, Removal of hazardous organic compounds from soil matrices using thermal desorption with purge, *Meeting of the Division of Environmental Chemistry, American Chemical Society*, Boston, April 1990.
12. Varuntanya, C. P., M. Hornsby, A. Chemburkar, and J. W. Bozzelli, Thermal desorption of hazardous and toxic organic compounds from soil matrices, *International Conference on Physiochemical and Biological Detoxification of Hazardous Wastes*, Atlantic City, NJ, May 1988.
13. Esposito, P., J. Hessling, B. B. Locke, M. Taylor, M. Szabo, R. Thurnau, C. Rogers, R. Traver, and E. Barth, Results of treatment evaluations of contaminated soils, *Proceedings of the Air Pollution Control Association Annual Meeting*, Vol. 81, No. 1 (1988).
14. Ayen, R. J., and C. Swanstrom, Development of a transportable thermal separation process, *Environ. Prog.* 10(3) (1991).
15. Swanstrom, C., Determining the applicability of X * TRAX for on-site remediation of soil contaminated with organic compounds, *HazMat Central '91*, Rosemont, IL, April 3, 1991.
16. U.S. Environmental Protection Agency, *Superfund LDR Guide 6A: Obtaining a Soil and Debris Treatability Variance for Remedial Actions*, Office of Solid Waste and Emergency Response Directive 9347.3-06FS, July 1989.
17. Ayen, R. J., and C. P. Swanstrom, Low temperature thermal treatment for petroleum refinery waste sludges, *1991 AIChE Annual Meeting*, Los Angeles, November 1991.
18. Jobson, G. C., L. D. Garner, S. P. duMont III, and R. K. Martin, Low temperature thermal separation of hazardous components from Y-12 plant mixed waste soils, *Proceedings of the 3rd Annual Hazardous Materials Management Conference/Central*, Rosemont, IL, March 13–15, 1990, pp. 650–659.
19. Palmer, C. R., and P. E. Hollenbeck, Sludge detoxification demonstration, *1989 Incineration Conference*, Knoxville, TN, May 1–5, 1989.
20. Schneider, D., and B. D. Beckstrom, Cleanup of contaminated soils by pyrolysis in an indirectly heated rotary kiln. *Environ. Prog.* 9(3):165–168 (1990).
21. Fox, R. D., E. S. Alperin, and H. H. Huls, Thermal treatment for the removal of PCBs and other organics from soil, *Environ. Prog.* 10(1):40–44 (1991).
22. Helsel, R., E. Alperin, and A. Groen, *Engineering-Scale Demonstration of Thermal Desorption Technology for Manufactured Gas Plant Site Soils*, HWRIC RR-038, Hazardous Waste Research and Information Center, November 1989.
23. *The Oak Ridger*, Oak Ridge, TN, 40th Year, No. 157, Thursday, August 25, 1988.
24. Taciuk, W., and R. M. Ritcey, Flexibility in wastes remediation with pyrolysis: the AOSTRA Taciuk process, *84th Annual Air and Waste Management*

Association Meeting and Exhibition, Vancouver, British Columbia, June 16–21, 1991.

25. Vorum, M., "Dechlorination of polychlorinated biphenyls using the SoilTech anaerobic thermal process unit; Wide Beach Superfund Site, New York, *Incineration Conference*, Knoxville, TN, May 16, 1991.
26. Abrishamian, R., Thermal treatment of refinery sludges and contaminated soils, *American Petroleum Institute Committee on Refinery Environmental Control* (CREC), Orlando, FL, May 7–8, 1990.
27. HT-5 thermal distillation process, *Hazard. Waste Consultant* 8.6:4.6–4.8 (November/December 1990).
28. DesOrmeaux, T. F., and B. Horne, Hazardous waste treatment/resource recovery via high temperature thermal distillation, *HazTech International '90 Conference Proceedings*, Institute for International Research, New York, 1990, pp. 5A-477 to 5A-495.
29. Nielson, R. K., and M. G. Cosmos, Low temperature thermal treatment (LT^3) of volatile organic compounds from soil: a technology demonstrated, *Environ. Prog.* (May 1989).
30. Nielson, R. K., and G. Cifolelli, Low temperature thermal treatment (LT^3) of volatile organic compounds from soil, *1989 NPRA Annual Meeting*, San Francisco, March 19–21, 1989.
31. Johnson, N. P., and M. G. Cosmos, Thermal treatment technologies for HazWaste remediation, *Pollut. Eng.* p. 84 (October 1989).
32. *SITE Technology Profile*: Recycling Sciences International, Inc., Desorption and Vapor Extraction System, U.S. Environmental Protection Agency, November 1990, p. 68.

7

ENHANCED BIODEGRADATION FOR ON-SITE REMEDIATION OF CONTAMINATED SOILS AND GROUNDWATER

Ronald E. Hoeppel
Naval Civil Engineering Laboratory
Port Hueneme, California

Robert E. Hinchee
Battelle Memorial Institute
Columbus, Ohio

INTRODUCTION

The Exxon Valdez oil spill during the summer of 1989 did little for the ecology of Prince William Sound, but it brought bioremediation to the forefront of attention. After viewing the beaches treated with oleophilic fertilizer, the director of the U.S. Environmental Protection Agency (USEPA) and other top officials vowed to promote biotechnology and to support additional research and development. Meetings were convened in 1990 to develop "an agenda for action to further advance the environmental applications of biotechnology." One goal of these meetings was to outline research areas that are lacking in using biotechnology to clean up hazardous waste sites.

As of September 1991, 36 Superfund waste sites were being cleaned up using some form of bioremediation. However, as of October 1991, only 13 of the sites were using in situ bioremediation, and only 12 total bioremediation sites were beyond the "project design" stage (U.S. EPA, 1991). Thus there remains a visible absence of in situ and innovative bioremediation methods. One concern of industry and potentially responsible parties is that current federal regulatory barriers are not consonant with the USEPA's enthusiasm for bioremediation. Bioremediation must be viewed in a new light before the engineers are willing to make full use of this emerging technology (*ASM News*, 1991a).

There is no single definition of biodegradation. For this discussion, *biodegradation* is defined as the breakdown of organic compounds by living organisms, especially microorganisms, eventually resulting in the formation of carbon dioxide and water or methane. An alternative definition of biodegradation is the disappearance of environmentally undesirable properties of a substance. The complete breakdown of organic compounds to carbon dioxide is termed *mineralization*.

The term *biotransformation* is often used to describe the conversion of an organic compound into a larger molecular structure, or the loss of some characteristic property of a substance with no decrease in molecular complexity (Rochkind et al., 1986). Biotransformations of organic compounds can alter the toxicity, form, and mobility of the original compound. By definition, inorganic compounds are not biodegraded, although microorganisms can transform inorganic compounds into forms having either greater or lesser mobility and toxicity. *Enhanced biodegradation*, *bioreclamation*, or *bioremediation* refers to the alteration or optimization of environmental factors to stimulate higher biodegradation rates. The addition of nutrients or optimization of soil environmental conditions is often referred to as *biostimulation*, while the addition to a site of microorganisms that have the capability to degrade specific contaminants readily is termed *bioaugmentation* (Mathewson and Grubbs, 1989; Zitrides, 1990).

Some biodegradation and biotransformation processes are promoted by specific microorganisms that attack only a given site on a contaminant molecule. Other microbial processes are quite nonspecific, and most require many different biocatalysts (enzymes) and consortia of microorganisms working together or in a treatment train. Occasionally, the degradative fate of an organic contaminant may be decided by a specific metabolic pathway followed during the degradation process, which results in dead-end metabolic products, as occurs most often with exotic human-made compounds. The metabolic pathways are often dictated by specific environmental conditions, such as the presence or absence of molecular oxygen, pH, or nutrient conditions. Such changing conditions can favor the development and use of specific enzyme systems in a given microorganism or may promote the growth and metabolic activity of other microorganisms that exist in a dormant state in the soil.

Despite the metabolic complexity often shown at the single-organism level, specific groups of compounds tend to require similar treatment conditions. For example, petroleum compounds are most rapidly biodegraded in the presence of molecular oxygen, under aerobic conditions. In contrast, many organic compounds highly saturated with chlorine or other halogens (e.g., perchloroethylene and higher-chlorinated PCBs) seem to require anaerobic conditions for initial rapid removal of halogen groups.

Complete and rapid biodegradation of many contaminants may not only require specific environmental conditions but perhaps changing conditions to satisfy the needs of different microbial consortia. This could be true for some highly chlorinated compounds. Reductive dechlorination may be required initially to remove the first chlorine groups, followed by aerobic conditions to degrade the less chlorinated compounds that tend to accumulate in the absence of molecular oxygen.

Although microbial interactions can be complex, many of the required changes in environmental conditions occur naturally in soil systems. For example, as different food sources and respirable substrates become depleted, shifts in functioning microbial consortia take place. Not all of the natural shifts favor rapid and complete degradation or detoxification of the original contaminants. Methods for modifying the environment to favor accentuated biodegradation will be discussed. The processes, limitations, and usefulness of bioremediation for breaking down soil and groundwater contaminants are discussed in both a basic and an applied context. Because bioremediation of hazardous wastes has just recently gained the attention that it deserves from an engineering viewpoint, some discussion of laboratory studies is required for a current perspective.

MICROORGANISMS AND THEIR REQUIREMENTS

Microbial Diversity

The term *microorganism* encompasses a wide variety of single and multicellular living forms. Included are very simple forms such as bacteria, actinomycetes, and blue-green algae that contain genetic material in a single circular chromosome that is not surrounded by a special nuclear membrane; these microorganisms are termed *prokaryotes*. Each bacterial cell usually reproduces asexually simply by splitting of the parent cell. Exchange of genetic material can occur between related and possibly distantly related bacteria, but this exchange is usually incomplete. Although considered unicellular, bacteria can exist as chains of cells or cellular clusters. Individual and adjacent cells generally function independent of one another, with no complex division of labor. *Actinomycetes* are prokaryotes that usually display branched filaments and a more complex internal structure. *Blue-green algae* are actually photosynthetic bacteria that, like the true algae and higher plants, contain chlorophyll, which serves to convert light energy into cellular tissue and functional energy. Other bacteria also use light for functional energy, but their cells contain light-sensitive pigments not found in higher plants.

Fungi and true algae, termed *eukaryotes,* have single or multiple chromosomes surrounded by a nuclear membrane, like all higher organisms, and their internal cell structure is even more complex and similar to that of multicellular plants and animals. Fungi have an important role in degrading contaminants, as do the prokaryotes. The true algae are less important in bioremediation, because very few of them use organic compounds for energy and growth. However, some photolytic reactions are possible in their presence. Most fungi and microalgae are either filamentous or single celled, with many displaying one form or the other at different stages of their life cycle. Some filamentous forms display cellular specialization, and unicells often show greater intracellular communication than do the prokaryotes. Most eukaryotes also reproduce through a sexual cycle involving chromosome transfer between related single-celled organisms or sex cells of multicellular forms. Microscopic single-celled and multicellular animals, such as protozoans or nematodes, are not considered to be directly responsible for contaminant biodegradation, but they play a secondary role by changing the soil environment and controlling the development of prokaryotes and fungi in their vicinity.

Bacteria

Bacteria comprise a diverse group of microorganisms that obtain their energy from many sources, including: light (*phototrophs* or *photosynthetic* bacteria), the decomposition or oxidation of reduced organic matter (*heterotrophs*), or the oxidation of reduced inorganic compounds (*lithotrophs* or *chemoautotrophic bacteria*). Some bacteria derive energy from more than one source, such as combinations of light and reduced inorganic or organic compounds. Several lithotrophic bacteria and some that use energy source combinations can be important in degrading exotic organic contaminants, such as halogenated compounds. Heterotrophic bacteria are the major group responsible for the biodegradation of organic contaminants. Lithotrophic or photosynthetic bacteria can be important for toxic metal or metalloid transformations and are instrumental in nutrient recycling. All living organisms are ultimately dependent on lithotrophic and photosynthetic (*autotrophic*) bacteria for the cycling of elements and energy.

The mandatory counterpart to energy production and respiration in all living organisms is the creation of reducing power to replenish enzyme systems and maintain the oxidation-reduction power cycle. This component involves the reduction of oxidized compounds by the addition of electrons released from compounds oxidized during energy production. The *electron acceptor* can be either an organic or an inorganic compound. For many bacteria, most fungi, and all higher organisms, the final electron acceptor is oxygen in the process termed *aerobic respiration*. The final reduced substrate in aerobic respiration is water, and the final oxidized compound respired from energy production is carbon dioxide. Many intermediate organic compounds are also oxidized or reduced in various complex metabolic pathways to supply materials required for growth and proliferation, as well as to supply additional energy.

In the absence of oxygen, certain bacterial populations respire other less oxidized inorganic compounds (*anaerobic respiration*) or use only organic compounds (*fermentation*). Anaerobic conditions are often termed *anoxic* if they favor nitrate reduction or denitrification. Denitrification is promoted by bacteria that can thrive under either aerobic or anaerobic conditions (*facultative anaerobes*). Without molecular oxygen, they use nitrate as the terminal electron acceptor for respiration. When oxidation-reduction (redox) potentials within soils are even lower, other oxygenated compounds are used by specific groups of bacteria as terminal electron acceptors. Because oxygen is toxic to these bacteria, they are termed *obligate anaerobes*. Several common alternative electron acceptors and associated bacterial groups include ferric iron (*iron reducers*), sulfate (*sulfate reducers* or *sulfidogens*), and carbon dioxide (*methanogens*).

Contaminated soils typically have a gradation of these different microbial communities, with aerobic respiration and photosynthesis driving the surface soil communities and the other communities forming ordered tiers in the deeper soil profiles as the oxidation-reduction potentials become more negative. Each tier or special group of bacteria is somewhat dependent on another tier or special group for survival. This relationship is termed *commensalism* if one organism or group benefits and the other is unaffected, or *symbiosis* if both organisms or groups benefit from the activities. Collectively, bacteria can feasibly use any nontoxic reducible compound as an electron acceptor for respiration, including perhaps water (Vogel and Grbic-Galic, 1986).

A common misconception is that all contaminants are biodegraded most rapidly and thoroughly under aerobic conditions. Although this is most often true, anaerobic conditions promote some important degradative processes. For example, reductive dehalogenation of highly chlorinated organic compounds is promoted mainly by anaerobic bacteria (Mikesell and Boyd, 1986; Kuhn and Suflita, 1989; DiStefano et al., 1991; Chaudhry and Chapalamadugu, 1991).

Some bacteria produce very resistant internal spores (e.g., *Bacillus* and *Clostridium* spp.), or in the case of the more advanced actinomycetes, profuse aerial spores. These spores tend to form under unfavorable growth conditions and are known to survive for centuries. Some soil actinomycetes are important in degrading recalcitrant and exotic organic compounds.

Fungi

Fungi comprise another diverse group of microorganisms that are actively involved in the biodegradation of organic contaminants. These eukaryotes are all heterotrophic, lack photosynthetic internal organelles (although intimate symbioses with algae are common), and most are aerobic and have a filamentous structure during at least part of their life cycle. *Yeasts* are fungi that exist primarily as independent cells that commonly multiply by asexual budding. Yeasts are usually considered facultative anaerobes and fermenters, although most require low oxygen concentrations (Visser et al., 1990). Some soil yeasts promote organic degradation. The unusual enzymes and degradation pathways of some fungi—rarely found in bacteria—make fungi potentially useful for breaking down exotic compounds. The more advanced fungi, such as the basidiomycetes and deuteromycetes (mushroom formers), seem to have unique enzyme systems that can degrade recalcitrant compounds. Some fungi excrete stable extracellular enzymes that enable them to degrade large organic molecules, which are not readily attacked by most bacteria.

The general belief that fungi function best under acidic conditions where bacteria fail to grow readily is true only with reservations. Certain fungi, especially the common molds, have developed an ability to survive and even grow in very harsh environments (Kirk et al., 1978). Conversely, certain bacteria such as some sulfate-oxidizing thiobacilli, prefer extremely acidic environments that exclude most fungi. Because fungi are so diverse, different types favor different optimal conditions for growth. Under harsh conditions, many common fungi produce large quantities of very resistant asexual spores that can survive for a long time. Composting temperatures of greater than 45°C inhibit the growth of most fungi.

Photosynthetic Microorganisms

Surface soils usually support large populations of photosynthetic prokaryotes (photosynthetic bacteria and blue-green algae) and eukaryotic algae. Light cannot penetrate more than a few centimeters into soils, thus limiting their role in bioremediation. However, many photosynthetic prokaryotes use organic substrates for food, especially in the absence of light, and thus degrade certain organic contaminants. True (eukaryotic) algae, because most do not use organic compounds as food sources, most often play a secondary role in bioremediation. All photosynthetic organisms produce alternative food sources that can increase the quantity of microorganisms that use readily available carbon sources. Their presence favors the rapid degradation of easily degradable contaminants, but may inhibit the breakdown of more recalcitrant contaminants. Some photodecomposition of recalcitrant organic compounds is possible by the presence of algae in surface soils and water. Algae also serve to control and equalize the nutrient supplies reaching the deeper soil profiles, thus preventing excessive nutrient flows to groundwater supplies.

Microfauna

Soil microfauna, such as protozoans, worms, and insects, play primarily an indirect role in degrading organics. They can change the soil environment through the process of bioturbation. Their activities can aerate surface soils, dilute contaminant concentrations through mixing, and modify other environmental variables.

Besides promoting certain populations of microorganisms in their immediate area, many microfauna can control other microbial populations through predation (Wiggins et al., 1987). Ciliate protozoans are especially important microbial predators, but other microfauna are also important for microbial grazing and detritus feeding. Control of microorganisms that rapidly proliferate could aid in many biodegradation schemes, which often involve slower-growing microorganisms. However, extensive grazing, especially of augmented bacteria, could also impede degradation rates (Zaidi et al., 1989). For example, protozoan grazing is often responsible for the acclimation period required to degrade organic compounds in sewage after an upset (Wiggins and Alexander, 1986). Protozoan numbers in sewage can be controlled easily with protozoan-specific antibiotics such as cycloheximide and nystatin

(Wiggins and Alexander, 1986), but such control is more difficult in soils. Also, eukaryotic inhibitors added to contaminated water could suppress biodegradation rates by inhibiting the growth of certain bacteria (Zaidi et al., 1989).

Microorganism Numbers and Activity in Soil

Direct Versus Indirect Counting Methods

The actual numbers of microorganisms present in soils are often based on indirect information, such as the number of bacterial colonies growing on a particular solid growth medium, where each colony [colony-forming unit (CFU)] is meant to represent a single cell. Unfortunately, no single growth medium supports all microorganisms, not even all bacteria. Many bacteria found to be important in biodegrading recalcitrant organics are very fastidious and often fail to grow readily on synthetic growth media. Also, growth media used for total bacterial counts isolate only aerobic heterotrophic populations, and the numbers of anaerobic microorganisms are seldom assessed. This shortcoming is even more important when sampling soils obtained at considerable depth or from anaerobic environments. An illustration of this problem is that microbial counts of hydrocarbon degraders in oil-contaminated soil frequently exceed total heterotroph counts (e.g., Hinchee and Arthur, 1991).

Early studies indicated that both the number and activity of microbes decline drastically with increasing soil depth, with little or no activity found a few meters beneath the earth's surface (Markovetz, 1971). However, recent studies using direct cell count methods instead of plate count methods using solid growth media have found large numbers of active microorganisms in deep saturated and unsaturated subsoils (Wilson et al., 1983; Balkwill and Ghiorse, 1985; Ghiorse and Balkwill, 1985; Bone and Balkwill, 1988). Even aseptic 300-m-deep core samples have shown an abundant and diverse microbial population (Fredrickson et al., 1991a). Most subsurface microorganisms exist in an environment low in organic matter and thus do not grow well, if at all, in conventional, organically rich growth media (Wilson et al., 1986a). In fact, more dilute growth media or soil extracts often prove best for conducting plate counts using subsoils (Ghiorse and Balkwill, 1985).

Microscopic direct counting methods using fluorescent fluorochrome stains have proven to be more meaningful for obtaining microbial numbers in subsoils. One common method involves the addition of acridine orange stain to a soil dilution, to groundwater, or to detached microbes after soil treatment with sodium pyrophosphate (Ghiorse and

Balkwill, 1983). This dye binds with the genetic material of the microorganisms, thus distinguishing between living and nonliving material. Another fluorochrome stain is DAPI (4′,6′-diamidino-2-phenylindole), which does not complex as readily with particulate organic matter and tends to highlight only intact cells (Porter and Feig, 1980). One technique distinguishes between metabolically active and quiescent cells by combining acridine orange staining with INT tetrazolium dye, a stain sensitive to active respiratory enzymes (Zimmerman et al., 1978). Dehydrogenase enzymes convert the water-soluble INT to insoluble formazan granules in the cells, which fluoresce red in contrast to the blue color of acridine orange. A modification of the method of Zimmerman et al. (1978) that is better for groundwater sampling is outlined by King and Parker (1988).

Indirect Measurements of Microbial Biomass

Assays for cellular components of soil microorganisms have also been used to determine microbial biomass and diversity. Phospholipids, which are found only in cell membranes, have been evaluated (White et al., 1979a,b) and found to be comparable to direct count procedures (Balkwill et al., 1988). Another test for microbial biomass involves measuring adenosine 5′-triphosphate (ATP) (Webster et al., 1985). This tedious extraction procedure can be inaccurate for measuring biomass of quiescent microorganisms common in deep soil profiles. However, the ATP test is quite accurate for quantitating actively growing microorganisms.

Microbial Numbers and Activity in Undisturbed Soils

It is important to distinguish between viable and nonviable cells and to separate those that are metabolically active from those that are quiescent. Metabolically active microorganisms can be categorized by whether or not they prefer high concentrations of easily degradable food supplies. *Zymogenous* soil microorganisms are those that respond readily to nutrient supplies. When supplies are plentiful, they readily grow and proliferate. Rapidly growing microorganisms are most numerous in the uppermost soil profile, whereas microbes that grow very slowly and prefer low nutrient concentrations become more important in the deeper soil profiles. The slow-growing microorganisms, termed *oligotrophs,* typically are found in environments deficient in food sources. They are not a distinct taxonomic group, but rather a metabolically diverse group that has adapted to surviving nutrient-starved conditions almost indefinitely. Many quiescent microorganisms are oligotrophic, although most microorganisms can remain viable for many years in a state of suspended animation (Lewis and Gattie, 1991). Oligotrophs have been observed to metabolize very low concentrations of specific contaminants and to persist at very low organic fluxes of less than 1 mg/L·day carbon. Because oligotrophs are the dominant group in deeply buried soils, they are very important in promoting in situ bioremediations and in removing very low contaminant concentrations from groundwater. Many oligotrophs revert to normal metabolism when subjected to high nutrient concentrations (Poindexter, 1981).

Recent studies have shown that microorganisms, especially bacteria, persist in large numbers from surface layers to very deep subsoil and bedrock profiles (Beloin et al., 1988; Bone and Balkwill, 1988; Chapelle and Lovley, 1990; Hirsch and Rades-Rohkohl, 1988; Fredrickson et al., 1991a,b). Shallow aquifer sediments generally contain more than 1 million aerobic bacterial cells per gram, as summarized by Bone and Balkwill (1988). Because both direct and viable counting methods are geared toward detecting aerobic bacteria, it is not surprising that the lowest counts have been found in subsurface clay layers, which can also present cell isolation problems due to sorption on the clays (Beloin et al., 1988). Coarser-grained sediments tended to display higher bacterial counts, especially in the aquifer zone (Beloin et al., 1988), but diversity also seemed to be lowest within the aquifer (Bone and Balkwill, 1988). Subsurface soils from a pristine Oklahoma aquifer contained 10^6 to 10^7 bacterial cells per gram, which was lower than the surface soils, with 10^9 cells per gram (Bone and Balkwill, 1988). A northern German site had comparable total counts but lower viable counts, with 10^3 to 10^4 (versus 10^6) aerobic cells. Higher counts seem to be prevalent in new wells, due to sediment disturbance, which suggests the need to flush new wells extensively prior to microbial sampling (Hirsch and Rades-Rohkohl, 1988).

Microbiological evaluations of samples from 400-m-deep boreholes in 80 million-year-old strata of South Carolina indicate the existence of great microbial diversity in such environments. In these evaluations, 21 physiologically distinct groups of bacteria were isolated (Fredrickson et al., 1991a). Although the overall in situ metabolic rates were found to be low in these deep anaerobic profiles (Chapelle and Lovley, 1990), the populations are capable of degrading a variety of single- and double-ring aromatic compounds as well as a diverse group of organic acids (Fredrickson et al., 1991a,b).

Microbial Numbers and Activity in Contaminated Soils

Contaminants in soil can have either a positive or a negative effect on the numbers and diversity of microorganisms. Application of oily sludges and fertilizer to soil resulted in a significant increase in bacterial colonies (CFUs) but without a comparable increase in the direct count number (Johnston and Robinson, 1982). This discrepancy suggests that many bacteria occur in soils in a dormant state and are stimulated by the addition of suitable food sources. Bacterial counts were also 100 to 1000 times higher in soils contaminated with jet fuel than in adjacent uncontaminated soils (Ehrlich et al., 1985). However, acute toxicity can be a problem, especially when the contaminant is not closely related to natural materials or is moderately to highly soluble in water. Many studies conducted in aqueous laboratory tests have determined the acute toxicity of individual contaminants on microorganisms, as well as on higher organisms (Blum and Speece, 1991; Donnelly et al., 1991; King, 1984; Walker, 1988). Transformation rates have been calculated for some compounds in soils using laboratory respirometric and radiolabeling techniques (Sims and Overcash, 1983), but information on acute or chronic toxicity of contaminants in soils is lacking.

The toxic effects of contaminants can be reduced significantly by their sorption on soil organic matter and clays. Elevated concentrations of many hydrophobic contaminants seem to inhibit biodegradation rates in soils, because the microorganisms either lack the ability to interact with the organic compounds (Miller and Bartha, 1989) or the surface area of contaminant available to them is small (Fogel et al., 1985; Nakahara et al., 1977). This is accentuated by the fact that water does not readily penetrate soils heavily contaminated with hydrophobic compounds, and the microorganisms proliferate mainly at the water interface with the contaminants (Efroymson and Alexander, 1991).

High concentrations of many poorly soluble hydrocarbon contaminants in soils seem to inhibit biodegradation rates but not the numbers or metabolic activity of degradative microorganisms (Watts et al., 1989). However, the same may not hold true for hydrophobic compounds that affect cellular function at very low concentrations or when soluble concentrations in the water phase are high. Latent acute toxicity can occur if a transformation product is more soluble or more toxic than the parent compound (Alvarez-Cohen and McCarty, 1991). Pentachlorophenol, a compound possessing potential acute toxicity to microorganisms, had less effect on microbial numbers when supplementary food sources such as acetate or glucose were added to the soil. However, heavy supplementation with these alternative carbon sources favored rapidly growing bacteria at the expense of the slower growing varieties, which were primarily responsible for degradation of the contaminant (van Beelen et al., 1991). Thus changing environmental conditions can increase microbial numbers while they may simultaneously inhibit biodegradation of the contaminant compounds. Food source supplementation may have the greatest inhibitory effect on degradation rates in subsurface soils and aquifers, where oligotrophic microorganisms are most prevalent.

Methods to Determine Community Structure

Microbial community structure can be determined using specific antibiotics in combination with specific culture conditions and growth media. For example, Fungizone was added to media to prevent fungal growth in bacterial cultures, and streptomycin and tetracycline were added to isolate yeasts and fungi (Walker and Colwell, 1976). The bacterial cell wall components—muramic acid and glucosamine—are extracted with a phospholipid assay in equimolar quantities (Findlay et al., 1983). Because muramic acid is found only in bacterial cells and glucosamine is present in both bacteria and microeukaryotes, the ratio of the two chemicals gives a rough estimate of the prokaryote-to-microeukaryote ratio (Balkwill et al., 1988).

Selective enrichment techniques are commonly used to enumerate specific degrading populations of bacteria in environmental samples (Jain and Sayler, 1987). One technique uses plates of a mineral salts medium suspended in solid agar to which a single organic carbon source is added. The assumption is that microorganisms growing on these plates, excluding those growing on agar plates with no additional organic carbon source, must use that carbon source. Liquid media that exclude agar can also be used, but individual microbial colonies must be isolated.

Transfer of microbial colonies from supplemented agar to unsupplemented agar, usually with a velveteen press, sometimes gives false-positive colonies if considerable numbers of cells are transferred, because these cells can serve as an alternative carbon source. Furthermore, when using complex mixtures of organics such as fuels, microbial growth may be dependent on only a single or a few specific compounds. Also, growth on a single carbon source may result from only partial degradation. Thus this technique cannot be used to verify total degradation of the enriched organics. If substrates are used that contain radiolabeled ^{14}C at specific locations, complete mineralization can be determined by the measurement of $^{14}CO_2$. *Radiolabeling techniques* have been developed to measure not only the mineralization rate but also the uptake of the target organic compound into the cell biomass and metabolites, thus providing a *mass balance* (Dobbins and Pfaender, 1988).

Three other very sensitive and highly specific techniques have been used to detect different microorganisms or genetically engineered bacteria: *immunofluorescence microscopy* (Schieman, 1981), nucleic acid hybridation with *DNA probe techniques* (Holben et al., 1988), and the *polymerase chain reaction* (PCR) DNA amplification method (Saiki et al., 1988). Immunofluorescence techniques are used routinely to detect specific microorganisms deposited on membrane filters. Isolated microorganisms are treated with fluorescein stain attached to specific (monoclonal or polyclonal) antibodies that bind only with specific cell surface antigenic markers (Galfre and Milstein, 1981). The specific tagged microorganisms then selectively fluoresce under ultraviolet light. DNA probes, using radiolabeled DNA, have been used to detect specific degradative populations in soil microcosms (Holben et al., 1988), groundwater (Jain et al., 1987), and PCB-contaminated soil environments (Walia et al., 1990). Although the technique seems to work well with soils and other matrices, combining DNA probes with biodegradation assays is recommended (Walia et al., 1990).

The PCR method and detection of the amplified DNA fragments using gel electrophoresis and radiolabeled gene probes have been used for environmental monitoring of coliform (Bej et al., 1990), luminous bacteria (Wimpee et al., 1991) in water, and genetically engineered bacteria (Steffan and Atlas, 1988). Although the highly sensitive PCR gene amplification technique has had limited use in monitoring soil microorganisms as contaminated sites, it may soon gain wider usage.

Microbial Nutrition and Growth

Microbial Nutrition and Environmental Effects

Microorganisms are composed of combinations of elements that are the components of their genetic material, structural molecules, enzymes, and intracellular plasma. Because of the great diversity among microorganisms, the ratios of these nutrient elements required for proper functioning vary by a rather wide margin. The major nutrient elements that make up microorganisms are hydrogen, carbon, oxygen, nitrogen, phosphorus, and sulfur. Water contains both hydrogen and oxygen but is not readily split into useful components beneath surface soils. Photosynthesis is the major route for breaking water into useful sources of oxygen and hydrogen, but only with light activation. In soils contaminated with most organic compounds, carbon and hydrogen are not limiting, because both elements are major components of hydrocarbons. Sulfur is usually readily available at excessive soluble or bioavailable concentrations. Phosphorus is often plentiful but unavailable because of its poorly soluble natural forms. Nitrogen is plentiful, usually as dinitrogen gas, but is seldom in sufficient quantity in bioavailable forms such as amino acids, ammonium, or nitrate. Thus the major elements that can limit microbial growth in soils are nitrogen and phosphorus, although oxygen is the most deficient nutrient for aerobic microorganisms in contaminated soils. Most bioremediation activities thus frequently include nitrogen and phosphorus supplementation in bioavailable forms and attempt to supply oxygen in a form that supports aerobic metabolism.

Most nutrient supplements attempt to provide a final soil carbon/nitrogen/phosphorus ratio in the range 100:10:1 to 100:10:05, depending on the type of treatment and contaminant form (Torpy et al., 1989). However, lower levels of nitrogen and phosphorus may be ideal at specific sites. For example, a C/N/P ratio of 300:10:1 was adequate at a Florida site contaminated with JP-4 jet fuel, and the addition of nutrients had no significant effect on hydrocarbon degradation rates. Conservative recycling of nutrients in microbial cell biomass, greater conversion of contaminants to energy production, and microbial conversion of nitrogen gas in soils to useful forms through dinitrogen fixation could lower nutrient balance requirements (Miller and Hinchee, 1990).

Supplementing soils with phosphates for in situ bioreclamations should be done with caution. Excessive soluble orthophosphate concentrations added to calcareous soils can cause extensive plugging of the soil around injection wells. Aggarwal et al. (1991) noted that orthophosphate concentrations greater than 10 mg/L can cause soil plugging at certain sites, but noted that other sites require about 20 mg/L for biodegradation enhancement because of soil sorption. Use of polyphosphates was recommended if total available phosphates (including exchangeable) exceeded 20 mg/L. Other nutrients or micronutrients—especially metals such as calcium, magnesium, iron, zinc, and copper—are important for proper cell growth or enzyme functioning. At a given site, any nutrient or micronutrient can limit biodegradation. However, the biological, chemical, and physical implications of adding nutrients must be considered, as illustrated in Figure 1, prior to their field implementation.

Microbial Acclimation and Growth

When all nutrients are present in excess and growth parameters are optimal, most microorganisms undergo *exponential growth*. However, this is usually preceded by a *lag phase* of growth, whereby the cells are adapting to the new environment. Reasons for the length of the lag phase, are

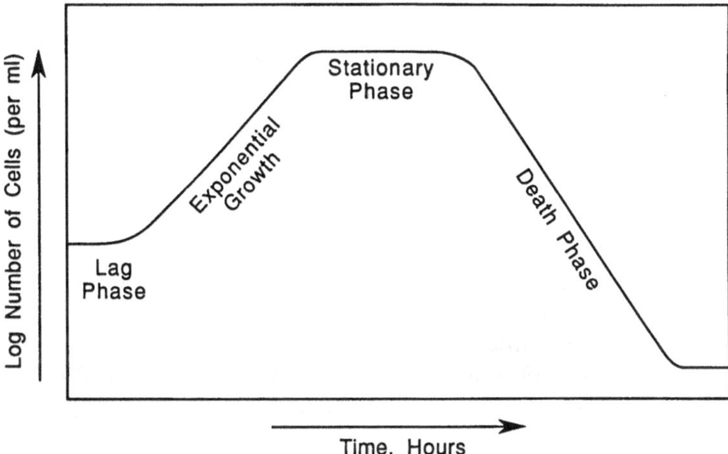

Figure 1 Generalized microbiological growth phases for a batch system into which a new food source is introduced.

not always clear. The term *acclimation* is commonly used when a time interval exists prior to biodegradation (Wiggins and Alexander, 1988). The period of acclimation varies with the material being degraded; anaerobic systems usually require a much longer time interval than do most aerobic systems (Suflita et al., 1982). Enzyme synthesis, which is often not induced until the substrate or similar molecule is present, probably occurs during the acclimation period. Cells that are preadapted to the substrate or growth conditions usually experience a very short lag phase and acclimation period. High chemical concentrations tend to increase the time before biodegradation is initiated (Rossin et al., 1982). Acclimation may also represent the time required to remove other inhibitory compounds or to develop enzyme systems to counter their effects (Atlas and Bartha, 1972; Wiggins et al., 1987). The period required to produce biosurfactants may be important in degrading poorly soluble materials, and the initial removal of competing food sources may be important in soils. A leveling off and ultimate decline in growth usually follows exponential growth as a result of nutrient depletion or the production of toxic metabolic products (Hoeppel, 1989). Certainly, the causes of lag and acclimation periods as well as ultimate growth decline are complex and differ among microbial populations.

With many in situ applications, no acclimation period has been observed (see, e.g., Dupont et al., 1991, and Miller et al., 1991). This is probably the result of the stimulation of indigenous microorganisms that had become adapted to the contaminants prior to treatment.

Cometabolism and Use of Secondary Substrates

Microorganisms, like all living organisms, require the right balance of nutrient elements to grow and metabolize food. Most of these nutrients are released or recycled as a result of microbial metabolism. Organic contaminants are often a primary food source, but in some instances the contaminants are only partially degraded and no significant energy and cell production or direct nutrient recycling results. This process is termed *cometabolism* or *cooxidation*. Two or more organic substrates are required for cometabolism, one being a nongrowth substrate while the other compounds (*primary substrates*) are broken down to useful end products for cell synthesis or energy production. Although the nongrowth cometabolite cannot be used directly by the microbe, other microorganisms can usually further degrade them into useful products (McCarty, 1988).

Many contaminants are used as *secondary substrates* by a given microorganism or microbial consortium. Although these substrates can function as cometabolites, contributing minimally to the growth of the active microorganisms, some serve simply as less desirable food sources for the degradative microorganisms. Biodegradation of secondary substrates is usually impeded or inhibited by the presence of more bioavailable primary substrates, although there are exceptions to this rule (McCarty, 1988). Primary substrates can compete with secondary contaminant substrates for the specific enzymes involved, but toxic intermediate metabolites produced during cometabolic processes can also affect the uptake of primary substrates (Suflita, 1989).

Primary and secondary substrates could feasibly be degraded to lower levels if production of the specific enzymes involved were accentuated, with or without a commensurate increase in microbial biomass. Enzyme production can often be induced by adding a nontoxic substrate that has physical properties or a biochemical pathway in common with the toxic substrate requiring removal. Such enzyme induction techniques are presently being used to aerobically degrade TCE (Chaudhry and Chapalamadugu, 1991), and are proposed for accentuating the degradation of other more natural contaminants, such as monoaromatic hydrocarbons (Alvarez and Vogel, 1991) and polycyclic aromatic hydrocarbons (Ogunseitan et al., 1991).

If a particular organic contaminant is at a trace or very low concentration in the environment, microorganisms usually require alternative food sources to proliferate, even if the minor organic compound is used as an energy source (McCarty, 1975). Microorganisms can use most individual primary substrates down to about part per million levels without growth inhibition. However, if additional primary substrates are present in high concentrations, the primary substrate in low concentration could potentially be degraded to much lower, perhaps parts per billion (µg/L), levels. In this case, the primary substrate is being used like a secondary substrate (McCarty, 1988).

Despite the need for adequate nutrition by microbes, the addition of large quantities of easily degradable organic compounds to a contaminated site may have an inhibitory effect on contaminant degradation rates. For example, the added food source may stimulate rapidly growing, nondegradative microbial populations at the expense of slower-growing active populations. This can alter existing microbial community structure, favoring nondegradative populations. The causes for inhibition can be complex and could include competition for the same enzymes, depletion of electron acceptors required for respiration, toxicity of the competing substrate or metabolic products, and end product enzyme repression (Alvarez and Vogel, 1991).

MICROORGANISM–CONTAMINANT INTERACTIONS IN SOILS

Soil as a Microbial Growth Medium

Several things must occur before a contaminant can be degraded or transformed by microorganisms. The microorganism must be in the immediate vicinity of the contaminant, the contaminant must be available to the microorganism, and the microorganism must have the capacity to partake in some part of the degradation or transformation process. The potential exists in soils for the biodegradation and biotransformation of most organic contaminants and transformation of inorganic compounds. However, microorganisms are not usually ready to engage immediately in biodegradation processes. The genetic potential may exist for degradation of a specific compound, but it is often present within a restricted microbial population and may not be expressed continually to conserve energy. This is especially true for exotic, human-made contaminants, which did not exist prior to recent historical times.

Specific microbial populations prefer or require specific environmental conditions. If these conditions do not exist, these populations tend to become quiescent until more ideal conditions return; specific microorganisms within the populations may even die. Usually, these conditions and changes exist on a microscale, but sometimes large-scale changes occur in population dynamics or community structure. The nature of limiting environmental factors often dictates which bioreclamation procedures to follow. While the subsurface environment can be modified, it is often accomplished with difficulty, incompletely, and at great expense.

Soil Moisture

Biodegradation requires water for microbial growth and diffusion of nutrients and waste by-products. Soil moisture levels can influence the types of microorganisms present. Soil moisture levels can also influence volatilization processes in soils. Although high water saturation can inhibit vapor flux rates through soils (Sims and Bass, 1984), low moisture levels may also decrease the vapor phase of volatile organics by increasing their sorption to soils (Chiou and Shoup, 1985). Torpy et al. (1989) state that the optimal soil moisture for bioremediation is about 1 bar tension; however, Miller and Hinchee (1990) observed that biodegradation associated with soil venting was not greatly inhibited at a soil moisture level of 6.5% by weight.

Soil Temperature

Soil temperature is often one of the most important factors controlling microbial activity and rates of organic matter decomposition in soils. Generally, rates of enzymatic degradation and microbial metabolism double for every 10°C increase in temperature, up until close to inhibitory temperatures, which are usually around 40°C for most soil microorganisms. However, with the exception of subfreezing temperatures, soil microorganisms are capable of degradation at most ambient soil temperatures (Atlas, 1981; Leahy and Colwell, 1990). Temperature can also indirectly influence biodegradation of a compound or mixture by changing its physical properties, bioavailability, or toxicity to microflora (Atlas and Bartha, 1972). Temperature can have a variable effect on volatilization of organics from soils. An increase in temperature usually increases the equilibrium vapor concentration, resulting in an increased volatilization rate, but it can at times also increase sorption to soil particles (Lyman et al., 1982).

Soil pH

Near-neutral soil pH values are usually optimum for biodegrading most organic contaminants. Also, well-plugging problems caused by phosphate precipitates may be reduced in the neutral pH range. The soil hydrogen ion concentration is governed by the types of compounds produced by microbiological activity, and is controlled especially by carbonate–bicarbonate–carbon dioxide equilibria. However, organic compounds and inorganic complexes (e.g., metals and nutrients) are important in removing hydrogen and hydroxide ions from soil solutions (Phung and Fiskell, 1972). Because hydrogen ion transfer is commonly involved in electron transport, pH and redox potential are interdependent.

Soil Redox Potential

The redox potential, often termed E_h, is extremely important in the biotransformation schemes of different organic contaminants and inorganic compounds in soils. Generally, a heavily contaminated site is anaerobic because ongoing microbial respiration has depleted all available oxygen. The resulting anaerobic conditions tend to favor different electron acceptors, with the most oxidized compounds (having higher E_h) being used first. In other words, oxygen gas is favored over nitrate (aerobic respiration), nitrate over sulfate (denitrification), and sulfate over carbon dioxide (sulfate reduction), with methane forming from the ultimate reduction of carbon dioxide or bicarbonate (methanogenesis), as depicted in Figure 2. Thus microbial populations having different degradative potentials can be operative at different times at the same contaminated site as the redox potential varies.

Although E_h measurements can contribute valuable clues about the functioning of soil or sediment geochemical systems, they alone will not give definite information about the chemical species present. They indicate that a potential exists for certain redox reactions, such as those involving microbial respiration. Various organic and inorganic redox reactions cannot be well defined or predicted because of their complexity and different interdependent reaction rates (Blumer, 1967; Stumm, 1984). Redox conditions may be difficult to alter in contaminated subsurface environments (Barcelona and Holm, 1991).

Substrate Utilization and Reaction Rates

Surfactant Effects on Contaminant Bioavailability

Surfactants and emulsifying agents can have major impacts on organic contaminant distributions in different soil phases and on biodegradation rates. These dispersant substances can greatly solubilize poorly soluble contaminants by decreasing surface water tension and increasing the interfacial surface area of the oil or hydrophobic phase through emulsification. However, data show that surfactants can inhibit biodegradation rates for some organic compounds. Efroymson and Alexander (1991) showed the importance of bacterial cell sorption at the oil/water interface on the biodegradation of poorly soluble compounds. When a common surfactant was added, bacterial cells became detached, which

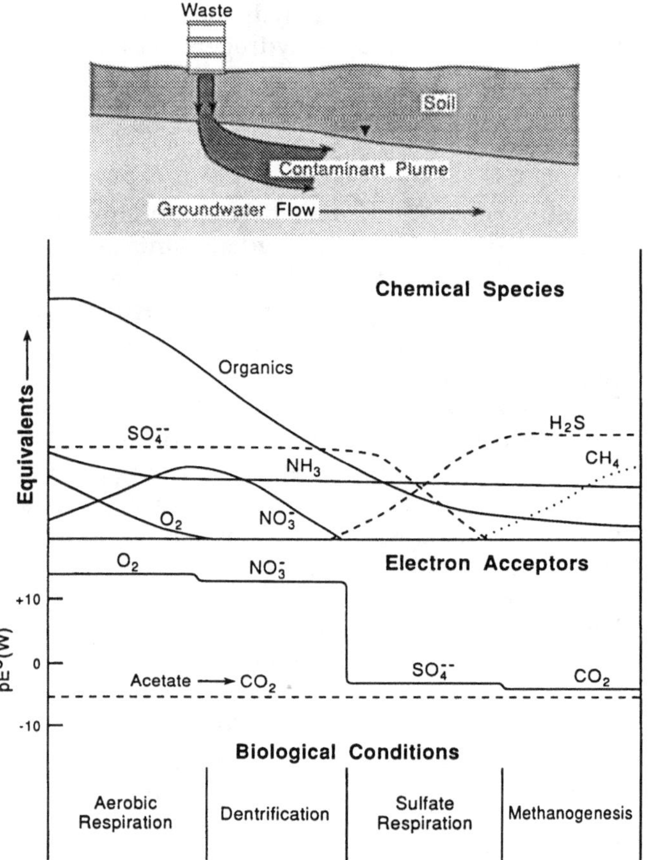

Figure 2 Biological and redox conditions typically underlying a site with contaminated groundwater. (After Bouwer and McCarty, 1984).

inhibited degradation of hexadecane but not the more soluble naphthalene. A study comparing biodegradation rates after treating soils with more than 50 different emulsifiers showed generally inhibitory effects on the biodegradation of heavy fuels (Arthur et al., 1989). Some synthetic dispersants may contain chemicals that are toxic to microorganisms, and potential inhibitory effects must be considered in choosing a particular dispersant and effective concentration.

Bacteria often produce and secrete their own *biosurfactants,* which emulsify hydrophobic compounds in the water phase, thus increasing the hydrocarbon surface area in contact with the biodegrading microbial population. Although biosurfactants are produced and function at very low concentrations, probably only within the immediate vicinity of the secreting cells, the effects may be similar to those of synthetic surfactants if produced or added at elevated concentrations. Inhibition of alkane biodegradation by the biosurfactant emulsan was thought to be caused by interference with hydrocarbon passage through the bacterial membrane, where metabolism occurs (Foght et al., 1989). However, cell desorption from the oil/water interface could have also occurred (Rosenberg et al., 1983). Cellular detachment can inhibit the direct interaction of specialized microbial cell projections with hydrophobic organic compounds (Rosenberg et al., 1982) and, because nutrients accumulate at interfaces, may also result in nutrient-limiting conditions (Atlas, 1991). Yet biosurfactants have also actively promoted biodegradation of long-chain alkane hydrocarbons (Pendrys, 1989; Zajic and Panchel, 1976), thus confirming the complexity of microbial interactions. It appears that considerably more research needs to be conducted with surfactants and biosurfactants before their specific interactions with soil microorganisms and biodegradation processes can be determined and used in practical applications.

Although most microorganisms fare poorly in the presence of high concentrations of water-soluble contaminants, poorly soluble organics present in emulsions may have much less toxicity toward the bioactive microbial populations. Several recently isolated bacteria can grow in low-water-content water-in-oil emulsions (Shabtai, 1991; Inoue et al., 1991). Because hydrophobic contaminants are at low concentrations in the water phase, specialized microbes seem to find a safe haven in the water phase of emulsions. Specific uses for these microbes may lie in the future.

Physical Barriers to Biodegradation

Physical or physicochemical factors can also affect the biodegradation of organic contaminants. Some molecules are recalcitrant to degradation because they are too large to enter the microbial cells, which is usually required for complete breakdown by membrane-bound enzymes. Although some substances are difficult to degrade because the number or locations of functional groups impede enzyme attack, strong sorption on soil particles can also hinder the ability of the cells to attach to, absorb, or enzymatically attack the molecule (Cheng et al., 1983; Voice and Weber, 1983). Although exoenzymes are important in degrading polymers because they are active outside the cells that excrete them, some can actually polymerize or cross-link smaller organic molecules, forming poorly degradable complexes with soil humus (Bollag, 1983).

However, oxidative polymerization reactions have also been observed to be necessary for the final degradation of lignins by some fungi (Reid, 1991).

Sorption in soils and solubility of organic substances are complex, interdependent phenomena that vary with the composition of soils and complex contaminant mixtures. For example, solids concentration was found to greatly affect the partitioning of a substance between air, water, and solid phases of soils (Voice et al., 1983). Additionally, aromatic hydrocarbon concentrations in water extracts of 31 gasoline samples varied over an order of magnitude (Cline et al., 1991). Although gasoline variability could account for part of this, the solubility of each component of mixtures can vary from ideal conditions, with each compound acting as a cosolvent to increase hydrophobic hydrocarbon solubilities (Groves, 1988). Alternatively, organic solvents can affect the sorption of organics on soils (Fu and Luthy, 1986). The effects of cosolvents on biodegradation rates are poorly understood.

Mathematical Estimation of Reaction Rates

Organic contaminants can be used as either primary or secondary substrates by microorganisms. Primary substrates are used directly to produce energy and biomass. Secondary substrates, often through cometabolic processes, are utilized only in conjunction with primary substrates, and energy or biomass production may or may not occur. Biodegradation rates for primary substrates under ideal conditions can be estimated mathematically using Monod kinetics. If microbial cells are allowed to proliferate freely in a homogeneous medium not deficient in nutrients, the rate of cellular growth can be directly correlated with substrate or contaminant disappearance (Rochkind et al., 1986). These mathematical relationships can also be used to define the lower substrate concentration limit, below which no net cell growth occurs (Lawrence and McCarty, 1970). Although such calculations may work well for defining the operation of engineered aboveground bioremediation systems, they do poorly in describing the subsurface environment. Also, utilization rates for secondary substrates and primary substrates at below-threshold concentration limits, which frequently occur in groundwater, are much more difficult to estimate. Mathematical models for predicting biodegradation events in the subsurface are being constantly improved with the input of additional field data.

Various mathematical models have been proposed for describing in situ biodegradation processes. Two conceptual models for aerobic biodegradation have been developed (Borden and Bedient, 1986; Molz et al., 1986). The first model assumed an instantaneous reaction between contaminants and oxygen, and the second model uses a dual-Monod kinetic relationship between the contaminant and oxygen. The instantaneous model theorizes that oxygen transport limits biodegradation, whereas the Monod kinetic model assumes that biodegradation of organic contaminants is limited by the microbial utilization of oxygen and distinguishes between contaminants having different biodegradation rates.

The predictive capabilities of both models were compared by integrating both into a two-dimensional transport model, called BIOPLUME II (Rifai et al., 1987, 1988; Rifai and Bedient, 1990). The more complex Monod kinetic model consists of coupled equations describing substrate transport, utilization of substrate, and dissolved oxygen by microbial consortia. The model requires knowledge of 19 input parameters. The much simpler instantaneous reaction model correlated just as well with actual field data, probably because of the lack of good bioremediation field data needed for model development (Rifai and Bedient, 1990).

BIOPLUME II modeling of aviation fuel plumes is being conducted at sites near Traverse City, Michigan, under aerobic conditions (Rifai et al., 1991) and Charleston, South Carolina, under either aerobic or denitrifying conditions (Widdowson and Aelion, 1991). Initial findings showed some correlation between the model predictions and field data, but they also show the need for accurate input data and clear understanding of the nature of the site. Another model, developed by Kindred and Celia (1989), includes cometabolism, uptake inhibition, and anaerobic conditions. All of these models assume no interaction between soil sorption and biodegradation, nor changing microbial populations.

Additional modeling is being done in laboratory and field bioremediation studies, and many of the problem areas are being addressed. Modeling is especially useful in understanding the mechanisms of interactive transport, sorption, and biodegradation, providing insight into which parameters are most important to study in the field, and rapidly and cost-effectively investigating various potential field remediation remedies. Good models will be developed and verified only after good bioremediation field data have been collected from many different sites (Chiang et al., 1989).

Microbial Movement Through Soils

When important members of the microbial community die, microbes must either directly recolonize the site or their genetic capabilities must be transferred to other microorganisms that can colonize the site. This is also the case when microorganisms having unique metabolic capabilities are applied to surface soils or wells of a contaminated site.

Microbial transport research has been conducted mainly in saturated soil systems, simulating either horizontal movement through an aquifer (Harvey and Garabedian, 1991; Hirsch and Rades-Rohkohl, 1990; Pekdeger and Matthess, 1983) or vertical water movement (Rahe et al., 1978; Schaub and Sorber, 1977; Smith et al., 1985). Such studies indicate that water flow is the major contributor to bacterial movement in soils and is impeded mainly by the size of the soil pores. If the soil pores are smaller than the cells, which for bacteria are usually 0.5 to 2 μm, passive diffusion of the microorganism is not possible. However, movement through cracks and channels in both coarse- and fine-grained soils appears to have an important effect on subsurface colonization (Hagedorn et al., 1981; Rahe et al., 1978; Trevors, et al., 1990).

Many other factors affect the transport of microorganisms through soils. Daniels (1972) reviewed the effects of cell sorption on soils, especially clays and organic films. Mozes et al. (1987) discussed the effect of cell charge on sorption. Some studies indicate that sorption is a minor contributor (Gannon et al., 1991), but others show that it can be a major factor in controlling bacterial mobility (Harvey and Garabedian, 1991). Cell shapes, cell stickiness, and cell motility have also been evaluated (Rosenberg and Kjelleberg, 1986; Sutherland, 1983; Gannon et al., 1991). Despite the complexity of interactions occurring during recolonization, including cell death and grazing or competition for food sources (Hoeppel et al., 1980), the recolonization has been observed to occur fairly rapidly and involves a diverse range of microorganisms (Hirsch and Rades-Rohkohl, 1990).

MICROBIAL METABOLISM

Microbial Metabolic Pathways: General and Specific Nature

Biochemical reactions in microorganisms follow specific metabolic pathways that lead to the degradation of organic compounds for energy production, biosynthesis of small molecules, and biosynthesis of macromolecules and polymers used mainly as structural cell materials. These pathways are significantly more complex and diverse than those of higher organisms. Detailed metabolic pathways are discussed elsewhere (Alexander, 1977; Rochkind et al., 1986). However, it must be stressed that the specific locations and stereospecific orientation of functional groups on organic molecules often determine the degradative pathway followed or whether the molecule can be degraded at all. For example, the orientation of hydroxyl groups when water is added to aromatic molecules determines whether or not the aromatic ring can be broken and degraded.

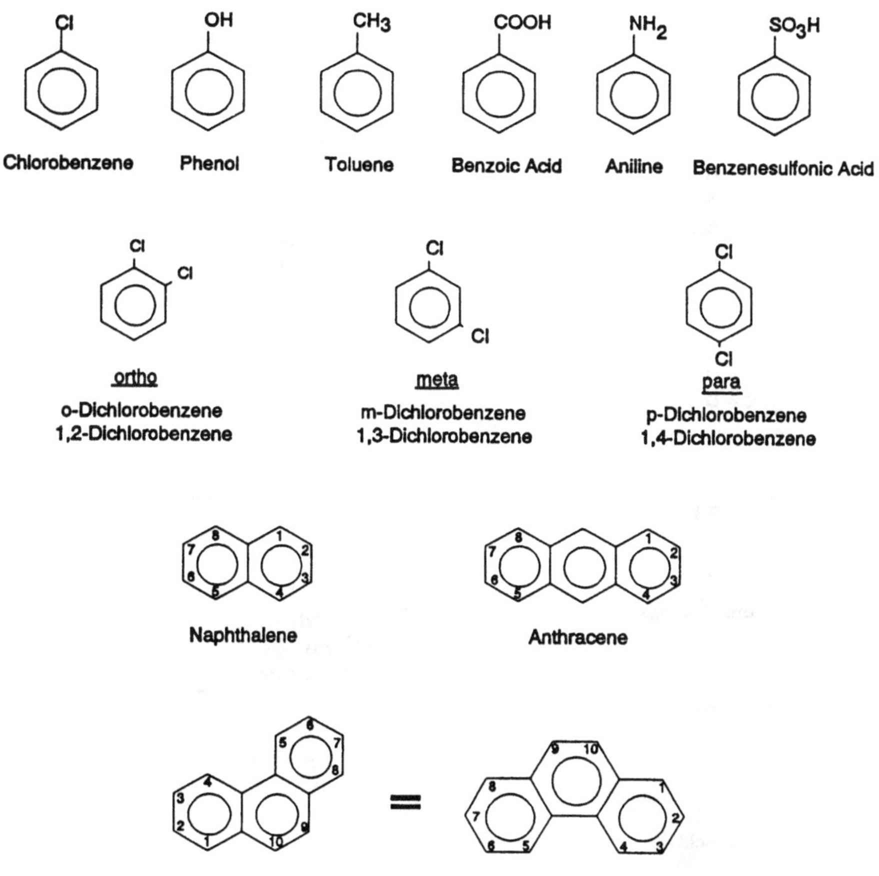

Figure 3 Structures of frequently encountered substituted benzenes.

The three positions on the aromatic ring where hydroxyl groups can be added are termed *ortho* (adjacent to each other), *para* (opposite to each other), and *meta* (between the ortho and para positions), as shown in Figure 3. Bacteria and some fungi can cleave an aromatic ring containing two hydroxyl groups if the groups are located in the ortho or para positions. If the groups are ortho, ring cleavage can occur between the hydroxyl groups (ortho cleavage) or next to one group (meta cleavage), as depicted in Figure 4. The cleavage that occurs depends on both the type of aromatic compound (number of rings and types of side chains) and on the genetic constitution of the organism. The metabolic

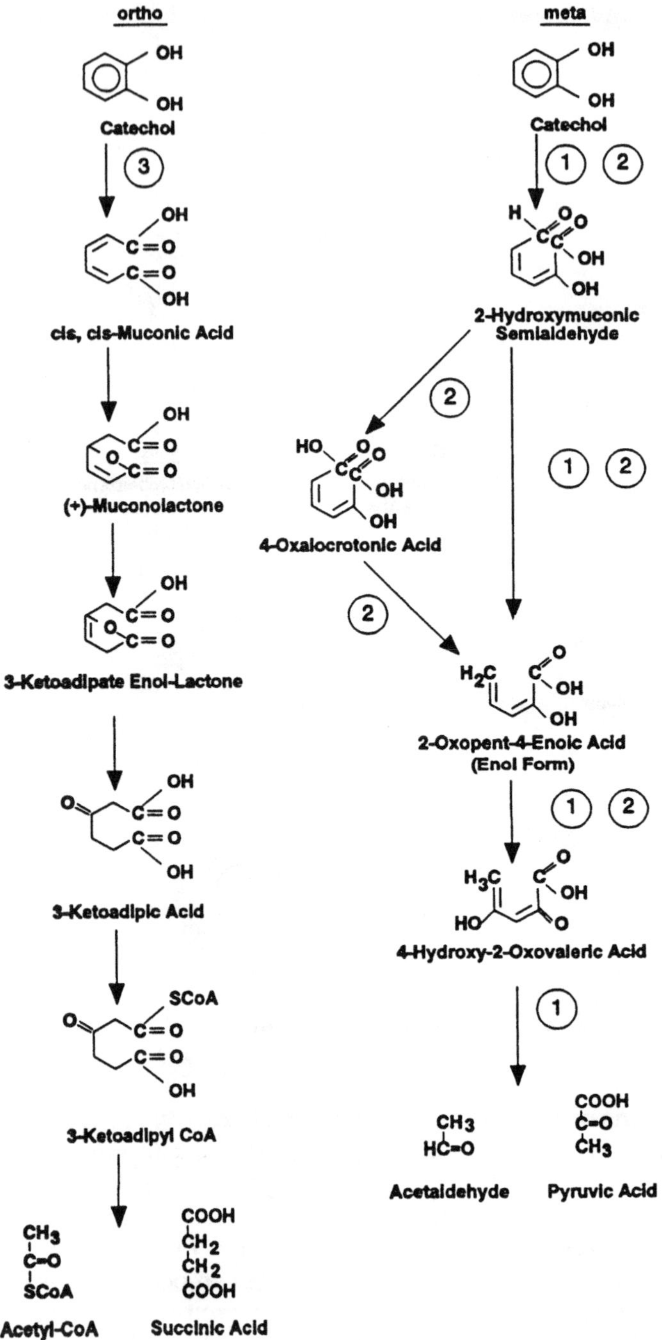

Figure 4 Generally accepted ortho and meta pathways for bacterial metabolism of catechol.

pathway and degradation products that form are dependent on the type of cleavage. Some bacteria and fungi contain the enzymes for both pathways and use each on specific aromatic compounds (Rochkind et al., 1986).

Additionally, the hydroxyl groups must be in the *cis* stereospecific position (hydroxyl groups projecting in the same direction from the molecule) before cleavage can occur. This orientation is achieved only when *dioxygenase enzymes* are involved, which add both atoms of molecular oxygen simultaneously to the aromatic ring. Only bacteria and some fungi produce *cis* stereoisomers while eukaryotes, including most fungi and all higher organisms, produce *trans* stereoisomers. The trans isomer, where the hydroxyl groups project in opposite directions from the molecule, is formed when *monooxygenase enzymes* add single hydroxyl groups sequentially to an aromatic ring and the purpose for the organism is usually to form more soluble conjugated products that can be excreted to the external environment. Figures 5 and 6 outline simple *cis* and *trans* pathways. Although ring cleavage can occur in the absence of dioxygenase enzymes (Shields et al., 1989), which require aerobic conditions to function, the pathways are more selective. Some aromatic compounds cannot be degraded readily, if at all, in the absence of oxygen. Oxygenase enzymes are involved also in the degradation of chlorinated aromatics (Chaudhry and Chapalamadugu, 1991; Rochkind et al., 1986; Vogel et al., 1987) and nitrated aromatics (Spain and Gibson, 1991).

Recent research has shown that monooxygenase and dioxygenase enzymes, having broad substrate specificity and produced by specific groups of aerobic bacteria, can degrade nonaromatic chlorinated aliphatic molecules such as the chlorinated ethylenes (Ensley, 1991; Vannelli et al., 1990; Wackett and Gibson, 1988; Wackett et al., 1989; Wilson and Wilson, 1985). These organisms also readily oxidize nonchlorinated (linear and cyclic) alkanes and nitroalkanes (Britton, 1984; Kido et al., 1975; Trudgill, 1984).

Respiration, which enables the cycling of electrons and provides a source of oxygen used to produce energy, sometimes plays a direct role in contaminant biodegradation. The removal of nitrate from groundwater or reductive dechlorination of chlorinated organic contaminants are important examples of contaminants being used as electron acceptors (Keeney, 1973; Painter, 1970; Vogel et al., 1987). However, the ability of many bacteria to use oxidized organic or inorganic substrates under anoxic and reducing conditions is especially important. Anaerobic respiration in environments devoid of oxygen allows for the decomposition of many contaminants that usually prevail in soils heavily contaminated

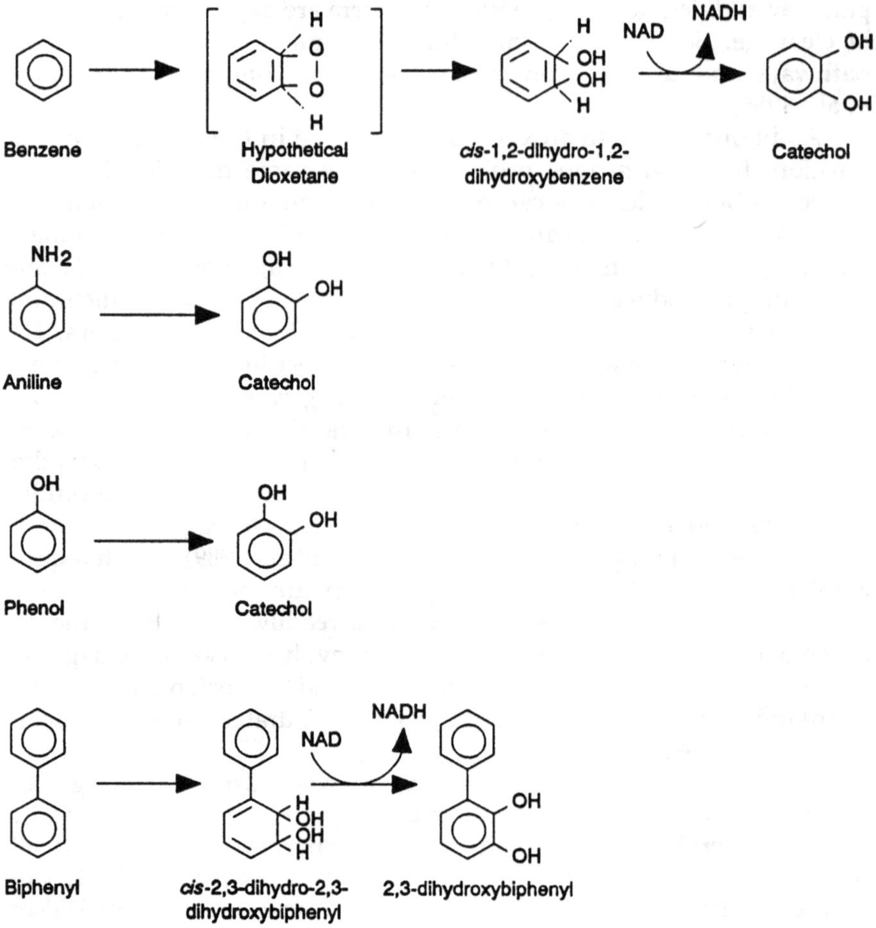

Figure 5 Generally accepted pathways for bacterial oxidation of several substituted benzenes.

with organic compounds. Figures 5 and 6, and pathways 1 and 2 of Figure 7 detail aerobic pathways for microbial degradation of substituted benzenes. Pathway 3 of Figure 7 shows a reductive pathway for benzoic acid.

Bioaugmentation for Bioremediation

Although its effectiveness is frequently in doubt, bioaugmentation has long been used by operators of wastewater treatment plants, who add seed microorganisms to the system at startup following a process upset or to improve treatment. Agronomists are also aware of the value of add-

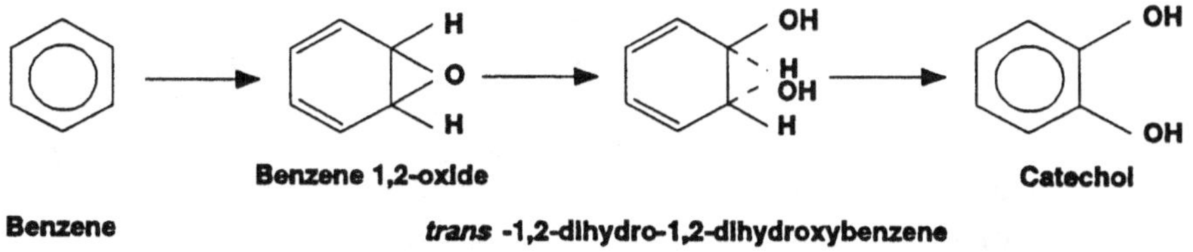

Figure 6 Transformation of benzene to catechol by fungi and yeasts.

ing specific root nodule nitrogen-fixing bacterial strains to their legume seeds. However, when it comes to adding microorganisms such as laboratory-altered bacteria to subsurface soil systems, contrasting viewpoints are often heard. Some say that bioaugmentation is capable of solving most site remediation problems, while others believe that adding microorganisms to a remediation site can confound existing engineering problems (e.g., well plugging) and contribute to higher costs with minimal proven benefit. Unfortunately, the positive viewpoint is propounded widely by the vendors of bioaugmentation products, who seldom have good supporting data from independent parties. At present, there is also the unfortunate involvement of unscrupulous entrepreneurs who hurt bioremediation in general. Good discussions on the pros

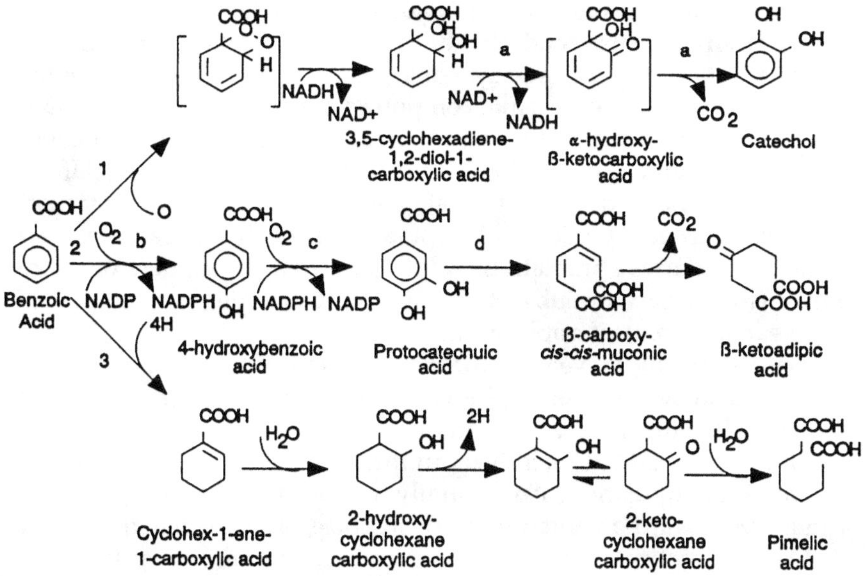

Figure 7 Select pathways for benzoic acid.

and cons of bioaugmentation are presented elsewhere (Goldstein et al., 1985; *Hazmat World,* 1991). Atlas (1991) concluded that although improved biodegradation of petroleum contamination with addition of oxygen and nutrients has been shown, no substantial evidence has been published supporting improvement through bioaugmentation.

Bioaugmentation often involves adding very large numbers of microorganisms (usually bacteria) to a site in an attempt to promote or accentuate biodegradation activities. The numbers of bacteria used to clean up two oil spills on the Texas coast during the summer of 1990 were about 10^{12} per milliliter, which is equivalent to 1 million trillion cells per teaspoonful (*Environment Today,* 1990). This concentration is one to two orders of magnitude greater than levels added to most terrestrial field sites. As of 1989, bioengineered microorganisms had not yet been used in field bioremediations (Nicholas and Giamporcaro, 1989), and all bioaugmentation formulations consisted of natural soil microorganisms, usually sporulating or freeze-dried bacteria. These natural soil isolates are usually made more bioactive toward a specific contaminant or several contaminants through the processes of induced natural mutation and/or gradual culture enrichment with the contaminant. However, when the target contaminant compounds are present in a soil profile for a significant period of time, these same selection processes tend to occur naturally.

The usefulness of bioaugmentation depends on the types of contaminants being degraded and the nature of the site being treated. Terrestrial environments tend to be potentially less responsive to bioaugmentation than aquatic systems, because soils generally contain higher concentrations of organic compounds, greater variety and number of microorganisms, and greater variability in environmental conditions (Bossert and Bartha, 1984; Horowitz and Atlas, 1980). If the contamination has existed at the site for some time and is not exotic (e.g., fuels), an active microbial population probably already exists at the site but requires stimulation. Adding very large numbers of microorganisms to surface or subsurface soils can cause plugging of the soil pores near the point of application.

Survival during movement through the soil of foreign microorganisms must also be considered (Ramadan et al., 1990). The native organisms generally outcompete the introduced strains, but in the short term, large numbers of added microbes can inhibit biodegradation by the native microbial population. Additionally, not all microorganisms of consortia active toward a contaminant or mixture can be isolated from the soil. Many microbes are too fastidious to be cultured on synthetic media, lose their degradative abilities after repeated culturing (probably because plasmids containing the genetic material are lost during storage), or succumb to stresses induced during inoculum preparation (Atlas, 1977; Chaudhry and Chapalamadugu, 1991; Horowitz and Atlas, 1980). Because of such potential problems, controlled laboratory treatability studies should be conducted prior to field application. These should compare the degradation potential of stimulated natural microflora with the microbial inoculum being considered, using soil from different sectors of the treatment site. Any study of bioaugmentation effectiveness must include killed controls. Frequently, the effects observed as the result of the addition of microbial preparations are due only to the nutrients or other material in the preparation, not to the added microorganisms.

Certain site characteristics may make bioaugmentation appear more favorable for site cleanup. The presence of a few exotic or especially recalcitrant contaminants, such as chlorinated hydrocarbons, may be most suitable for bioaugmentation. The decision rests on the results of the treatability studies or a small pilot study. Specially selected and treated microbial isolates could also be effective in degrading contaminants at very high or low levels or at sites containing high levels of heavy metal contamination (*Hazmat World*, 1991). An area now receiving considerable attention—accentuated bioemulsifier production by selected soil isolates—may prove to be a future use of bioaugmentation (Rosenberg et al., 1975).

Bioengineered Microorganisms for Bioremediation

Biotechnology companies engaged in engineering microorganisms have, to date, had little or no impact on bioremediation practice. The main inhibitor is public perception and how it affects regulatory decisions. The regulatory hurdles imposed on biotechnology are discussed in the section "Regulatory Implications for Bioremediation." A very poor public perception of the introduction of genetically engineered microorganisms to the environment, coupled with lack of funding for research, has to date not resulted in development of a novel microbe usable for commercial bioremediation (*Hazmat World*, 1991). Nor as of 1989 had any bioremediation company used for field activities any bioengineered microorganisms that had received foreign genetic material through recombinant techniques (Nicholas and Giamporcaro, 1989). Public perception changes slowly; public and regulatory acceptance may not be achieved for many years. Until then, the applied use of genetically improved microbes may be limited to random chemical and radiation mutagenesis, which has remained largely unregulated (Baskin, 1987). Nevertheless, there is increasing interest in producing recombinant strains with superior potential for bioremediation. Few regulatory hurdles exist that hold back the use of genetic recombinant microorganisms in bioreactors or other closed systems. With regard to field application, the first steps have been taken in the form of well-thought-out guidelines (Tiedje et al., 1989).

Important areas where genetic engineering could be useful to bioremediation include degradation of recalcitrant organic compounds, placement of degrading genes under the control of more useful promoter genes that allow greater process regulation, development of more effective enzyme systems that attack multiple contaminants, improvement of the hardiness of engineered or natural microbes, and production of engineered microbes that eventually self-destruct (*Hazmat World*, 1991). Potentially, molecular biology techniques have these and other uses.

Bioengineering currently is used in high-cell-density bioreactors, with many of the findings relevant to future in situ systems. Because recombinant bacteria compete poorly with native strains, they do not survive for extensive time periods in the presence of high natural microbial populations. This problem could be alleviated by separating the growth from the product formation phases, whereby the recombinant genes would be expressed only in the latter phase. Problems of nutrient starvation, which affects primarily cell growth, would also be reduced. The transfer of "starvation or stress promoters" into bioactive bacteria can confer enhanced resistance to stress on cells as well as a means for controlling longevity of the recombinant microbe (Little et al., 1991).

Longevity of enzyme activity can also be developed in a microorganism. For example, the appropriate genetic material that codes for the enzyme can be transferred from extrachromosomal plasmids to the chromosome. Genetic material on chromosomes is lost less frequently during cell conjugation or division. In some instances, the same bacterium can possess genetic material coding for more than one metabolic pathway, with one pathway controlled by genes in plasmids and the other by genes in the chromosome. Removal of a controlling plasmid may be all that is required to assure both longevity and use of the more desirable metabolic pathway (Harker and Kim, 1990).

Another area of interest is to manufacture microorganisms that produce an important biodegradative enzyme constantly (*constitutively*), in the absence of inducible molecules, especially when the inducible substrate is not present at the site or is a regulated toxic substance. A genetically engineered bacterium that degrades trichloroethylene in the absence of potentially toxic inducible substrates—toluene or phenol—has already been developed (Shields, 1991) and may soon be used in a bioreactor system.

One area of genetic engineering that is significant to the degradation of PCBs and other chlorinated aromatic compounds involves the addition, removal, or modification of genetic control over ortho and meta pathways involved in aromatic ring cleavage. For biphenyls and many aromatic compounds, enzymatic ring cleavage can differ when the rings are chlorinated and unchlorinated or methylated. For example, methyl-substituted aromatic compounds are generally degraded via the meta-cleavage pathway, whereas chlorine-substituted aromatics most frequently follow a modified ortho ring fission. However, intermediate degradation products from the meta pathway inhibit proper functioning of the modified ortho pathway, and thus inhibit the degradation of chlorinated aromatics. In this case, removal of the genetic material responsible for the meta pathway may reroute meta-dependent degradation to the modified ortho pathway (Reineke et al., 1982). Alternatively, bioengineered strains can be developed that can degrade toxic intermediates generated by meta ring cleavage, which should improve the degradation of a wider range of contaminants (Pettigrew et al., 1991).

Success with genetic engineering technology initially requires finding natural microbes that contain enzyme systems or genetic traits of interest. Some "superbugs" already exist in nature as a result of natural mutations, but their survival is often short-lived because their overtaxed genetic systems compete poorly in nature. One of the greatest problems facing biotechnology is developing "superbugs" that will compete effectively with natural soil microorganisms. Alternatively, the main goal could be functionality rather than survivability. Self-destructing microorganisms or engineered genes may be desirable for in situ soil applications, with self-destruction occurring after the site has been remediated (Contreras et al., 1991). In retrospect, genetic robustness and increased cell viability may be desirable traits for high-rate bioreactor development, where nutrient stress and metabolite toxicity could impede efficiency (Little et al., 1991). Ultimately, cost-effectiveness may be the limiting factor in the utility of engineered microorganisms. The cost of maintaining cultures is currently quite high and may not prove cost-effective for bioremediation application.

Mathematical Estimations of Biodegradability

Biodegradation rates have been determined, usually under laboratory conditions, for many organic compounds. However, the number of different organic compounds being developed and used is immense. In an attempt to derive semiquantitative biodegradation rate data for these untested compounds, structure–activity relationships (SARs) have been developed based mainly on their physicochemical properties in relation to structurally related compounds for which biodegradation rates have been measured (Nirmalakhandan and Speece, 1988). Some of these properties include molecular size, molecular configuration, side-chain numbers and locations, and lipophilicity (i.e., oil/water partitioning coefficients). Another SAR technique is to estimate mathematically the first-order biodegradation kinetic constants rather than the rates of biodegradation, which allows for decreasing rate with decreasing contaminant concentration. Desai et al. (1990) used the latter method with the idea of classifying organic molecules according to decompositional groups or fragments. This approach allows the degradative properties of many compounds to be determined from a few hundred degradation fragments. Although SARs appear to be required to handle the endless list of organic compounds, meaningful estimates for field sites may be difficult because of the importance of site conditions in controlling biodegradation rates and the complex interactions of mixed contaminants.

INNOVATIVE BIOREMEDIATION METHODS

Recent literature summarizes innovative technologies in general. Freeman and Sferra (1990) edited a three-volume reference summarizing innovative treatment technologies, with the third volume concerned with biological processes. Chambers et al. (1990) compiled state-of-the-art information on in situ remediation technologies in use in the United States, and Sims (1990) reviewed current approaches to treating soils, stating deficiencies, and recommending improvements. European technologies for treating contaminated soils were outlined and discussed by Pheiffer et al. (1990) and Porta (1991).

In this section we discuss emerging technologies that use biodegradation for the major part of the treatment scheme and emphasize in situ methods. Well-known and frequently used aboveground pump-and-treat technology, which is discussed critically by Mackay and Cherry (1989), will not be evaluated. The new technologies discussed are (1) enhanced bioremediation, a water-driven in situ technique that evolved in the 1980s; (2) bioventing, an air-driven in situ method that has been used in Europe since the mid-1980s; (3) innovative aboveground bioreactors to treat exotic recalcitrant contaminants; and (4) anaerobic–aerobic sequential processes. Many of these processes combine biological with physical or chemical methods for treating wastes.

Enhanced Bioremediation

To date, bioremediation has gained its greatest application in its field application to petroleum hydrocarbon contamination. Conventional in

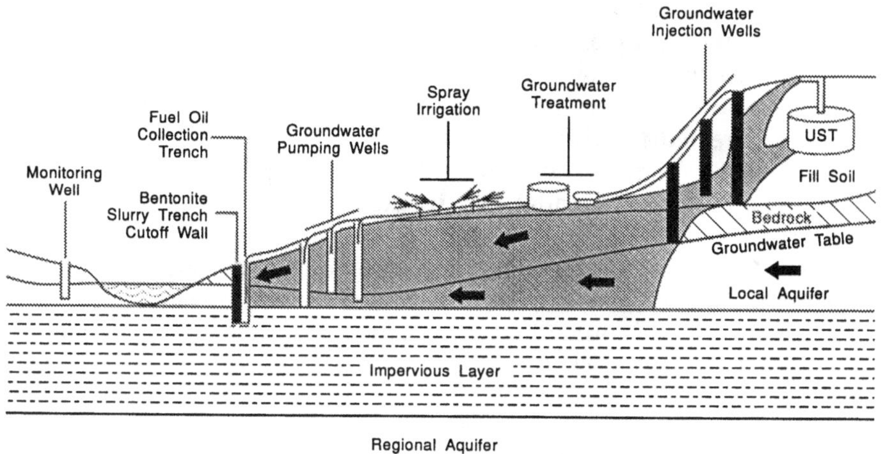

Figure 8 In situ bioreclamation of fuel oil contamination in soil and groundwater.

situ bioremediation of petroleum hydrocarbon–contaminated groundwater and soil involves treating pumped groundwater with nutrients (usually inorganic soluble nitrogen and phosphorus compounds) and an oxygen source (e.g., sparged air or oxygen, stabilized hydrogen peroxide, or nitrate), and reinjecting the water back into the subsurface, as shown in Figure 8.

Usually, the first operation is removing nonaqueous-phase liquid (NAPL) from the subsurface by extracting free product from wells or trenches that intersect the subsurface plume. NAPL lies on the water table if it is less dense than water (e.g., fuels and oils) or concentrates on impervious soil layers beneath the water table if it is more dense than water (e.g., chlorinated solvents). Free product recovery usually involves groundwater pumping, which controls groundwater flow within a sphere of influence, and the depressed groundwater table around the pumping zone eventually accentuates free product accumulation and subsequent removal. At least part of the treated groundwater (or other water source) is treated to remove most contaminants and to add nutrients. Most enhanced bioremediation operations to date have attempted to stimulate aerobic biodegradation, and oxygen was commonly added in the form of hydrogen peroxide or by air or oxygen gas sparging of the groundwater, either above or below ground. These techniques are summarized in greater detail elsewhere (Wilson et al., 1986b; Lee et al., 1988; Riser, 1988; Thomas and Ward, 1989).

Enhanced bioreclamation cost estimates vary widely, from $35 to $270 per cubic yard, because of divergent sources of information and sites being treated (Environmental Solutions, Inc., 1990; IIT Research Institute, 1990; Wetzel, et al., 1987). Another approach to cost is $230 to $300 per gallon of residual fuel in the soil (IIT Research Institute, 1990). However, cost is very site specific and depends on many factors, such as size of the contaminated area, type of contaminant, volume of contaminant spilled, depth of contamination, depth to water table, soil type, and location of contamination. Many past remediations were also experimental and used customized equipment, which would increase overall costs.

Oxygen Limitations in Enhanced Bioremediation

Oxygen-limiting conditions are less frequently experienced in above-ground engineered aqueous treatment systems because the oxygen is transferred rapidly into the water. However, in situ oxygen transport into heavily contaminated soils from air-saturated water is not feasible because water has a very limited capacity to hold and transport oxygen. Air-saturated water holds only 8 to 10 mg/L of oxygen at ambient soil temperatures, and dissolving pure oxygen gas into water only increases the oxygen content of the water fivefold. Because it takes 2 to 3.5 lb of oxygen to aerobically degrade 1 lb of most organic contaminants (Wilson et al., 1986b; Hinchee and Arthur, 1991), about 350,000 lb of water sparged with air would be required to degrade 1 lb of organic contaminant.

For example, in a theoretical contaminated soil matrix of sand, it would take approximately 40 years to degrade a small (1000-gal) subsurface spill using air-saturated water (Hinchee et al., 1987). Upon realization of these limitations (Wilson et al., 1986b; Thomas and Ward, 1989), hydrogen peroxide was added as an oxygen supplement to water.

Hydrogen peroxide contains approximately half its weight as oxygen and, being infinitely soluble in water, can greatly increase the oxygen-holding capacity of water. Unfortunately, even formulations of hydrogen peroxide that have been stabilized with phosphates against breakdown by ferric and ferrous iron appear to be readily attacked by microbial enzymes, such as catalase (Huling et al., 1990, 1991; Spain et al., 1989). Although hydrogen peroxide stability seems dependent on specific site conditions (Lawes, 1991), rapid breakdown was observed during laboratory studies (Aggarwal et al., 1991; Huling et al., 1990, 1991; Morgan and Watkinson, 1992; Soczo et al., 1986; Spain et al., 1989) and a controlled field study (Hinchee et al., 1990). In the field study involving a shallow aquifer, oxygen was released to the atmosphere within a few feet of the injection well, despite the relatively high groundwater velocity (about 1 ft/h).

Hydrogen peroxide was found to be highly toxic to most microorganisms at concentrations above 500 mg/L, but the concentration was increased to 2000 mg/L with only minor bacterial growth inhibition by selecting for H_2O_2-resistant bacteria (Britton, 1985). However, in other studies, concentrations as low as 100 mg/L were found to be toxic to sensitive microorganisms (Huling et al., 1990; Arthur et al., 1989).

Because of the problems noted above, the use of hydrogen peroxide as an efficient carrier of oxygen for aerobic biodegradation needs to be assessed closely before being used at a particular field site, and it may not be usable at many sites. If 500 mg/L of hydrogen peroxide could be adequately stabilized and used efficiently, the maximum oxygen-holding capacity of the water would be only about 250 mg/L. Thus aerobic biodegradation could still require substantial pumping to treat even a small hydrocarbon spill under ideal soil conditions.

Nitrate and Sulfate as Alternative Oxygen Carriers

Many aerobic bacteria use nitrate as a source of oxygen for respiration when oxygen is depleted from soils (Keeney, 1972; Painter, 1970). This process, termed *denitrification*, has been used in the field to promote the degradation of many aromatic compounds found in fuels (Battermann, 1986; Berry-Spark et al., 1986; Hutchins and Wilson, 1991). In the absence of sufficient nitrate to stimulate denitrification, specific anaerobic bacteria can use sulfate as a source of oxygen (reducing power or electron acceptor) for respiration. Although field studies using sulfate to promote the anaerobic degradation of petroleum products are lacking at this writing, field observations (Reinhard et al., 1984) and laboratory evaluations of soils from contaminated field sites (Edwards et al., 1991) substantiate that sulfate-reducing bacteria can effectively biodegrade monoaromatic hydrocarbons. Recent data also indicate that long-chain alkane hydrocarbons can be degraded by sulfate-reducing bacteria (*ASM News*, 1990), although information on reaction rates in soils is lacking.

There are potential advantages to using nitrate or sulfate instead of molecular oxygen in performing in situ bioremediations; these anions are more soluble than oxygen and can therefore be added in sufficient concentration to allow better mass transport to the aquifer. Nitrate may supply nitrogen, if it is limiting. We discuss the degradation of fuels under anoxic conditions in the section "Bioremediation of Fuels and Other Petroleum Products." Technical reviews on anaerobic degradation of aromatic compounds have been published elsewhere (Berry et al., 1987; Evans and Fuchs, 1988).

Many highly chlorinated aromatic and aliphatic hydrocarbons can also be treated under anaerobic conditions. Numerous laboratory and bioreactor studies show that anaerobic degradation of organic compounds highly substituted with chlorine or other halides is feasible, but little published information exists concerning in situ processes. The reductive transformations of halogenated aliphatic compounds in groundwater was demonstrated by Bouwer et al. (1981). Semprini et al. (1991) conducted a small controlled in situ field study at Moffett Field near Mountain View, California, using native bacteria under anaerobic conditions to degrade low levels of carbon tetrachloride, 1,1,1-trichloroethane, Freon-113, and Freon-11. Nitrate and sulfate were present in sufficient concentrations to serve as terminal electron acceptors in the absence of molecular oxygen, and acetate served as a primary food source.

Some degradation was observed under denitrifying conditions, with the more heavily substituted carbon tetrachloride showing greatest removal. However, significantly higher degradation rates for all contaminants occurred after nitrate was depleted from the subsurface, and sulfate was the primary electron acceptor for anaerobic respiration. The poor ability of denitrifying bacteria to degrade carbon tetrachloride, 1,1,1-trichloroethane, and other chlorinated hydrocarbons, as well as the importance of sulfate-reducing bacteria in promoting their degradations, were also noted in laboratory and column studies (Criddle et al., 1990a,b; Klecka et al., 1990; Cobb and Bouwer, 1991). Additional information on anaerobic degradation of halogenated hydrocarbons and potential problems are covered in the section "Halogenated Hydrocarbon Degradation."

Bioventing

Much of the contamination from a petroleum hydrocarbon release remains sorbed to soil particles and occluded in soil pores in the unsaturated zone, above the water table. Usually, only a very small fraction of the released material dissolves in groundwater. The unsaturated soil acts instead as a constant source of groundwater contamination (Mendoza and McAlary, 1990).

Soil venting, also called *soil vacuum extraction*, is gaining acceptance as an in situ technique to treat volatile organic contaminants present in unsaturated zone soils (Bennedsen et al., 1987; Connor, 1988; Hoag and Cliff, 1988; Stransky and Blanchard, 1989). Soil venting works best with compounds having high vapor pressures and Henry's constants, parameters that promote their evaporation from soils and groundwater (Lyman et al., 1982); see Gossett (1987) for a discussion and listing of Henry's constants. The soil gas is often extracted through extraction wells.

The rate of soil gas extraction from a given soil profile depends on how the contaminants are partitioned among the undissolved liquid phase, the dissolved liquid phase, and the soil gas phase. Soil moisture an have an influence on vapor extraction. Dry soils tend to strongly sorb volatile hydrocarbons; thus water vapor in soils tends to increase volatilization rates significantly (Chiou and Shoup, 1985). However, soils having high moisture contents may not vent well because of reduced soil pore volume and restricted vapor flow paths (Sims and Bass, 1984). Because gaseous diffusion rates are many times greater than aqueous ones, soil venting has the potential to remove volatile contaminants from soils from less permeable strata as long as a nearby flow path exists.

Soil venting seems to work well with highly volatile contaminants such as most compounds present in gasoline. It is also economical, with costs as low as $10 per ton ($13 per cubic yard) of soil for large sites not requiring off-gas or wastewater treatment, and averaging $50 per ton ($65 per cubic yard) of soil (IIT Research Institute, 1990). However, surface treatment of the vapors can be expensive. For example, sorption of the extracted vapors on activated carbon can double the cost (Hinchee et al., 1987) and the contaminants have not been destroyed and remain a liability. Possibly a more economical method for aboveground vapor destruction is biodegradation in vapor-phase bioreactors, discussed subsequently.

Low to moderately volatile organic compounds, as often found in diesel or jet fuels, cannot be removed from soils cost-effectively by soil venting. However, many low-volatile compounds found in refined petroleum products can be biodegraded under aerobic conditions. As discussed previously, the main limitation to the success of many water-driven enhanced in situ bioremediations is the inability to aerate the contaminated zone. However, soil venting has been shown to overcome rapidly the oxygen deficits that usually occur in soils heavily contaminated with organics by pulling air from the surface into the soil (Hinchee et al., 1991; Hoeppel et al., 1991). This is due primarily to the fact that air contains approximately 210,000 ppm of molecular oxygen, compared to about 10 mg/L in water. High gaseous diffusion rates also play an important role.

The use of soil venting to promote aerobic biodegradation in soils is termed *bioventing*. Soil aeration and not vapor extraction is the primary purpose of bioventing, potentially making the technique effective in removing, through in situ biodegradation, organic contaminants having both high and low volatilities and water solubilities. Because gases have greater diffusivities than liquids, bioventing can remediate soils with low water permeabilities, such as silty and clayey soils, as long as some airflow paths exist. This was demonstrated during in situ respirometry field testing (Hinchee et al., 1991; Ong et al., 1991). Available data indicate that only 2 to 4% oxygen must be provided in the soil gas to attain maximal biodegradation rates (Miller et al., 1991). Thus, bioventing requires lower soil gas pumping rates than those required for efficient soil venting. Figure 9 illustrates a conceptual bioventing design.

Detailed field bioventing studies have been published elsewhere on low- to high-permeability soils (Dupont et al., 1991; Miller et al., 1991; Urlings et al., 1991; van Eyk and Vreeken, 1989, 1991) and fractured bedrock (Bell and Hoffman, 1991). Combined in situ treatment schemes that integrate air sparging of groundwater with soil venting and bioventing have also been reported in field cleanups (Brown and Sullivan, 1991; Brown et al., 1991). Biodegradation rates measured during bioventing of

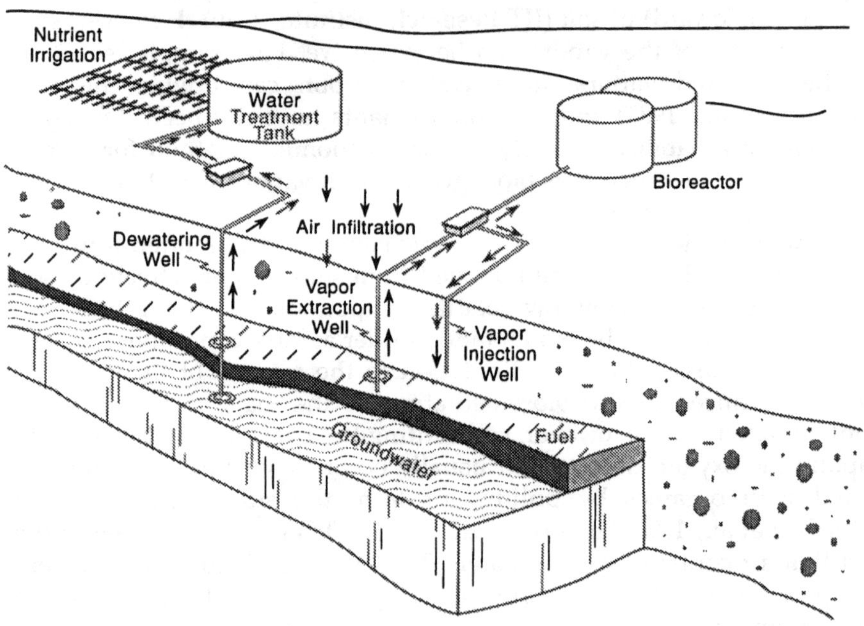

Figure 9 Conceptual design for bioventing and use of vapor-phase bioreactors.

gasoline and middle distillate fuel-contaminated subsoils were between 2 and 20 mg/kg per day, as summarized in Hoeppel et al. (1991).

Costs for bioventing should be comparable to or slightly lower than those for soil venting, excluding aboveground vapor treatment costs. Slightly lower cost is expected because smaller pumps and decreased pumping rates are required to maintain minimal oxygen levels for aerobic respiration than are required by conventional soil venting. Where off-gas vapor treatment is required, bioventing may be substantially less expensive than soil venting.

Innovative Bioreactors

Bioreactors consist of either free suspended microbial cells in a growth medium or cells immobilized on a matrix. Most bioreactor treatments of groundwater contaminant vapors and soil treatments use immobilized cell growth, because most microorganisms in soil environments attach to soil particles. Immobilized cells are genetically more stable, are protected from hydraulic shear stresses and biological predators, and will grow to higher cell densities than will suspended growth cultures.

Matrices commonly used for passive surface immobilization include soil, peat or compost, activated carbon, coke, plastic media, celite, and glass beads. Some of these materials have been used in trickling filter and rotating biological contactor treatment systems, usually to treat wastewaters. Synthetic microporous or cross-linking organic polymers and thermosetting matrices have also been devised to immobilize bacteria within the matrix. Such thermosetting matrices include polyurethane and polyvinyl foams; agarose, potassium carrageenan, calcium alginate, polyacrylamide, and glutaraldehyde/protein, usually formed into beads. Immobilized systems can be run either in batches, as frequently done with soil slurries, or with continuous flow. Continuous-flow systems can have fixed matrices (fixed-film bioreactors) or mobile matrices in a liquid medium (fluidized bed bioreactors) (Cooksey, 1991).

The matrix or bead size determines diffusion distance for oxygen in and end product out of the particles, which may be important factors for controlling degradation rates. Innovative methods recently advanced to get oxygen to the center of immobilizing beads, where it becomes limiting, incorporate within the beads microalgae (and bioreactor illumination) or low concentrations of hydrogen peroxide and its breakdown enzyme, catalase. Such procedures have not yet been scaled up to field use.

Another approach to increase oxygenation uses gas-permeable membranes composed of silicone or polyetherimide macroporous polymer to separate the gas and liquid streams in a flow-through bioreactor. Bacterial colonization of the membrane pits exposes them to oxygen on the inside of the membrane or capillary tubes and to the target contaminants diffusing in from the liquid phase. Matrices with both hydrophilic and hydrophobic components have also been developed to treat hydrophobic chemicals that degrade to water-soluble intermediates; one such matrix is a mixture of polydimethylsiloxane and calcium alginate. Coke offers promise as a nontoxic, inexpensive substitute for activated carbon in fluidized bed bioreactors (Cooksey, 1991).

Vapor-phase bioreactors have been used in Western Europe for almost 10 years, but they are only now being developed in the United States. Bioreactors will be used in soil venting operations and to treat vapors removed from contaminated groundwater by air stripping. The term *biofiltration* refers to a process that uses a fixed bed of porous material or stationary liquid phase, to which active microbial populations attach and through which the vapor stream flows. Solid matrices that have successfully degraded fuel vapors include yard waste compost (Venterea and Findlay, 1991) and peat buffered with calcium carbonate to prevent acidic conditions (Douglass et al., 1991). Such natural matrices are inexpensive, with the peat/$CaCO_3$ costing \$35 to \$70 per cubic meter, depending on quantity. These bioreactors can treat air streams from air stripping with low contaminant loading (Douglass et al., 1991). Effective biofilter operation depends on the interaction of the contaminant with the biofilter matrix and carrier liquid, and on the temperature and structure of the packed bed. A packing material may have excellent removal efficiency, but because of poor packing, may give poor overall contaminant removal rates (Diks and Ottengraf, 1991).

A mobile aqueous phase bioreactor works best in treating chlorinated organic compounds, where hydrochloric acid and toxic intermediate compounds are continuously flushed from the system. A bioreactor can be designed like an air stripping system, but in reverse. The vapor is forced to rise through a solid microbial support matrix while water flows downward through the matrix, thus removing soluble toxic intermediates and end products. The main difference is that the vapors are sorbed to the solid matrix, where they are degraded by the attached microorganisms. A matrix composed of organic compost mixed with polystyrene beads was found to work well (Diks and Ottengraf, 1991). However, if cometabolism is required, as with most aerobically degraded chlorinated organics, extra cosubstrate must be added. A procedure for adding cosubstrates is in the experimental stage.

Anaerobic bioreactors are being developed to treat some contaminants, such as organics that are highly substituted with chlorine (e.g., tetrachloroethylene, carbon tetrachloride, and pentachlorophenol). At present, many of these reactors consist of packed bed or fluidized bed systems using glass or various plastic media for bacterial support. However, a vessel containing suspended sludge to establish reducing conditions was used as part of the treatment scheme for groundwaters containing high levels of organics (Kastner, 1991).

Another way to ensure reducing conditions is to use a precolumn containing a bacterial inoculum before the treatment reactor. The precolumn provides anaerobic conditions through biological use of oxygen (Freedman and Gossett, 1991). The precolumn can also be treated with proper reductants and oxidants (e.g., methane and oxygen) to attain the desired E_h level (Kastner, 1991). These systems have been used only for small-scale treatments. Both batch and continuous-flow reactors have been used, although large-scale units will probably use continuous-flow systems. Because these bioreactors are still being developed, good cost data are lacking.

Reductive dechlorination works best at low E_h levels, such as under sulfate-reducing, fermentation, or methanogenic conditions (Cobb and Bouwer, 1991; Criddle et al., 1990b). It does not work well, and in many cases not at all, under denitrifying conditions (Cobb and Bouwer, 1991; Klecka et al., 1990), with the notable exception of the degradation of carbon tetrachloride, without an accumulation of toxic intermediate compounds (Criddle et al., 1990a). The applicability of this degradation to bioreactor development needs to be assessed. Anaerobic degradation of mono- and dichlorinated alkenes, such as vinyl chloride and dichloroethylene, occurs under denitrifying conditions, because reductive dechlorination is not required for their degradation (Barrio-Lage et al., 1990; DiStefano et al., 1991). However, some mono- and dichlorinated aromatics can be degraded by several different pathways, some involving initial reductive dechlorination (Genthner et al., 1989a). The often-contrasting information from recent laboratory studies indicated that practical use of anaerobic bioreactors for degrading halogenated compounds is still in the future and will probably involve specific microorganisms or microbial consortia.

Anaerobic–Aerobic Processes

Most organic compounds that are highly saturated with halogens (e.g., chlorine and bromine) degrade slowly or not at all under aerobic conditions. However, reductive dehalogenation occurs fairly rapidly under reducing conditions because these compounds act as effective reducing agents or electron acceptors used in respiration (Criddle et al., 1990b). Common contaminants such as tetrachloroethylene (PCE), carbon tetrachloride (CT), and polychlorinated biphenyls (PCBs) do not degrade readily under aerobic conditions. Other common contaminants, such as trichloroethylene (TCE) and pentachlorophenol (PCP), are readily dechlorinated under anaerobic conditions.

A major problem inherent with using anaerobic degradation for PCE and TCE is the accumulation of vinyl chloride, a carcinogen, as a rather stable intermediate compound. However, recent data (Kastner, 1991) show that *cis*-dichloroethylene (DCE) accumulates preferentially if a bioreactor is run under oxygen-limiting conditions, rather than strictly anaerobic conditions. This was accomplished by adding both oxygen and methane to maintain very low oxygen levels. The resulting DCE is less hazardous and more amenable to aerobic biodegradation than vinyl chloride. Aerobic methanotrophic bacteria, which also prosper in the presence of oxygen (for aeration) and methane (as a food source), readily degrade DCE. Thus only minor changes in the feed gases could result in total degradation of recalcitrant chlorinated compounds, which could be performed in the same or tandem bioreactors. This scheme has not yet been scaled up to an operational field unit.

The aerobic degradation of PCBs is generally limited to molecules having five or fewer chlorines and two adjacent unsubstituted carbon atoms (Furukawa et al., 1979; Furukawa, 1982). Most PCBs that are present in common commercial formulations (e.g., Aroclors 1254 and 1260) cannot be degraded aerobically. However, reductive dechlorination of highly chlorinated PCBs has been shown to occur in anaerobic sediments, probably by mixed microbial consortia (Nies and Vogel, 1990; Pettigrew et al., 1990; Quensen et al., 1990; van Dort and Bedard, 1991). Many highly chlorinated PCBs are partially dechlorinated under anaerobic conditions and many of the lower chlorinated PCBs are aerobically degraded. With such PCBs, a logical plan would be to incorporate anaerobic, followed by aerobic bioremediation, into one treatment scheme. Anaerobic dechlorination of PCBs appears to be relatively slow (Anid et al., 1991), so most schemes involve in situ injections of organic nutrients to stimulate the proper anaerobic populations (e.g., methanol), followed by injections of oxidants to promote aerobic conditions in the sediments or soils (Anid et al., 1991). Similar in situ schemes have been proposed for treating PCE, TCE, and other highly chlorinated contaminants by alternately stimulating anaerobes (methanogens) and methanotrophic aerobes (Vira and Fogel, 1991). These sequential anaerobic–aerobic treatments have yet to be conducted in the field.

BIOREMEDIATION OF FUELS AND OTHER PETROLEUM PRODUCTS

There are several good reviews of information on the biodegradation of petroleum hydrocarbons in the environment (Atlas, 1977, 1981, 1984, 1991; Leahy and Colwell, 1990) Riser-Roberts (1992) has provided a detailed review of information prior to 1988 on in-situ and on-site bioremediation of refined oils and fuels. Carlson (1981) reviewed the literature on biodegradation of jet fuels.

Most refined oils and fuels are composed of several hundred individual organic compounds. Gasoline fuels may contain up to 50% alkylbenzenes, with benzene and toluene (the most soluble components) constituting 0.1 to 3.5% and 2.7 to more than 20% of the total composition by weight, respectively (National Research Council, 1981; Sanders and Maynard, 1968). Most nonaromatic compounds are branched-chain alkanes containing 4 to 10 carbon atoms (C_4 to C_{10}) (Myers et al., 1975; Sanders and Maynard, 1968). Middle distillate fuels, such as diesel, kerosene, and jet fuels (JP-4, JP-5, JP-8), generally contain 5 to 12% alkylbenzenes and consist mainly of C_5 to C_{16} straight-chain and some branched-chain alkane hydrocarbons (Smith et al., 1981). The aromatics in fuels tend to be more soluble than the alkanes, but their solubility decreases greatly with the addition of alkyl substituents (Yaws and Yang, 1990; Yaws et al., 1990).

Aerobic Biodegradation

Most organic compounds found in crude oil, refined oils, and fuels are known to biodegrade under aerobic conditions, facilitated by a diverse and large group of aerobic microorganisms, mainly bacteria and fungi (Leahy and Colwell, 1990). Straight-chain *n*-alkanes biodegrade generally more readily than branched-chain alkanes, whereas cycloalkanes usually degrade the slowest. The information is mixed for the aromatic fraction, with both very rapid and relatively slow degradation observed (Atlas, 1981; Leahy and Colwell, 1990). Naphthalenes, with two fused aromatic rings, tend to degrade more slowly than do monoaromatic compounds, but several studies have shown more rapid degradation of naphthalene and methylnaphthalene than for benzene and *n*-hexadecane in sediments contaminated with petroleum and jet fuel (Cooney et al., 1985; Lee and Hoeppel, 1991). In sediments contaminated with crude oil, the monoaromatic fraction degraded faster than the alkane fraction (Jones et al., 1983). The variability in aromatic hydrocarbon degradation may be due to the type of fuel and environmental conditions present, because biodegradability is often linked with availability to the microbial cells (Cheng et al., 1983; Voice and Weber, 1983).

Water-driven in situ bioremediation schemes for refined petroleum products may require large volumes of water and longer times for acceptable cleanup, because these products degrade most rapidly and completely under aerobic soil conditions and are composed primarily of hydrophobic compounds. Air-driven treatment schemes, such as bioventing, have a greater potential for success, but not all field sites are conducive to bioventing because of a shallow groundwater table or large volume of groundwater flow. Additional research with different peroxides, hydrogen peroxide stabilizers, and synthetic and microbial emulsifiers thus seems necessary. For aerobic bioremediation of saturated-zone oiled soils to be more widely applied, satisfactory nontoxic methods must be found to increase the oxygen-holding capacity and penetration capability of water. Hydraulic fracturing is one recent technology that can increase air and water diffusion rates in low permeability soils and bedrock (Davis-Hoover et al., 1991).

Anaerobic Biodegradation

Recent research has shown that nitrate, serving as a terminal electron acceptor in the absence of oxygen, can promote the degradation of many monoaromatic compounds common to most fuels (Evans et al., 1991a,b; Hutchins, 1991; Major et al., 1988). Although most studies have indicated appreciably lower degradation rates under denitrifying versus aerobic conditions (Major et al., 1988), other studies have shown comparable degradation rates for many aromatic compounds (Kuhn et al., 1988). However, with the exception of toluene and oxygen-containing aromatics (e.g., benzoic acids and phenols), degradations under denitrifying (Ball et al., 1991; Evans et al., 1991a; Kuhn et al., 1988), sulfate-reducing (Edwards et al., 1991), iron-reducing (Lovley and Lonergan, 1990), and methanogenic conditions (Grbic-Galic and Vogel, 1987; Wilson et al., 1986a) require the presence of mixed soil microbial populations and the presence of mixed aromatic compounds.

Benzene has been especially recalcitrant to anerobic biodegradation in laboratory studies under denitrifying and sulfate-reducing conditions (Edwards et al., 1991; Evans et al., 1991a; Hutchins, 1991; Kuhn et al., 1988). Yet some microcosm (Major et al., 1988) and in-situ field studies (Battermann, 1986; Berry-Spark et al., 1986; Britton, 1989) have shown the depletion of all common monoaromatic hydrocarbons found in gasoline under denitrifying conditions. Benzene can be biotransformed anaerobically, under methanogenic conditions (Vogel and Grbic-Galic, 1986), but the rate is probably very slow. However, laboratory and field evidence suggests that microbial populations can degrade benzene, as well as other aromatic compounds, more rapidly under denitrifying conditions if very low levels of oxygen are present (Hutchins, 1991). This same phenomenon was also found to be true for alkane hydrocarbons when less than 0.05 mg/L oxygen was present to initiate oxidation (Swain et al., 1978).

Several field in situ bioreclamation studies have been conducted or are under way that attempt to augment the degradation of toxic aromatic compounds in fuel spills under anoxic or anaerobic conditions (Hutchins and Wilson, 1991; Reinhard et al., 1991; Werner, 1991). The results of these studies will determine the applicability of laboratory success stories to field reclamations.

BIOREMEDIATION OF HALOGENATED HYDROCARBONS

Halogenated organic compounds are primarily human-made and thus tend to persist in the environment. Chlorinated compounds are most widely used, comprising solvents, degreasers, dielectric fluids, herbicides, pesticides, plastics, and components to many other commonly used materials. Chlorinated and fluorinated compounds are still being used as refrigerants and fire retardants, and several common pesticides are brominated compounds. Chaudhry and Chapalamadugu (1991) reviewed the biodegradation of halogenated compounds. Other reviews cover more specific groups, such as halogenated aliphatic compounds (Vogel et al., 1987), halogenated aromatics (Reineke and Knackmuss, 1988; Rochkind et al., 1986; Tiedje et al., 1987), and trichloroethylene (Ensley, 1991). These references should be addressed for a comprehensive overview; information pertinent to development of bioremediation systems is presented in this section. Because chlorinated organics are by far the most common hazardous substances, these are emphasized. However, the metabolic pathways for other halogenated compounds are often similar.

Halogenated Aliphatic Hydrocarbons

Chlorinated aliphatic compound (alkanes and alkenes) are the most common and widespread contaminants of groundwater and are important components of contaminated soils and landfill leachates. Many of these compounds have been used as solvents and degreasers in industrial processes. Examples of commonly used one- and two-carbon alkanes, and their common transformation products in soils, include carbon tetrachloride, trichloromethane, methylene chloride, trichloroethanes (TCAs), and dichloroethanes (DCAs). Commonly used alkenes, and their common transformation products in soils, include tetrachloroethylene (PCE), trichloroethylene (TCE), dichloroethylenes (DCEs), and vinyl chloride. Although some abiotic processes can be important in the breakdown of chlorinated aliphatics, the reaction rates are usually much slower than those of biological transformations (Vogel et al., 1987).

Generally, the highly substituted chlorinated aliphatics are rapidly dechlorinated under anaerobic conditions. However, reductive dechlorination reactions proceed progressively more slowly as the number of chlorine or other halogen groups attached to the carbon atoms decreases. In contrast, aerobic biodegradation usually progresses more rapidly as the number of chlorine substituents declines. Several groups of aerobic bacteria that produce rather nonspecific oxygenase enzymes (monooxygenases and dioxygenases) display an exception to this trend. These enzymes can oxidatively dechlorinate highly chlorinated aliphatic compounds such as TCA and TCE, but they appear incapable of degrading chlorinated aliphatics completely saturated with chlorine, such as carbon tetrachloride and PCE. One strain of *Pseudomonas putida* was observed to display both oxidative and reductive pathways for chlorinated ethanes under aerobic conditions (Castro and Belser, 1990). This observation indicates that reductive dechlorination may occur at slow rates under aerobic conditions. Such data further complicate the reactions that may occur in the field.

Oxidative Dechlorination of TCE, TCA, and Metabolites

Aerobic biodegradation of TCE and TCA depends on the cometabolic activity of specific bacterial groups that contain rather nonspecific monooxygenase and dioxygenase enzyme systems. These enzymes oxidatively dehalogenate TCE and TCA only when specific primary food sources are available to the bacteria to promote enzyme production. In these cases the chlorinated aliphatic is degraded fortuitously, in conjunction with degradation and utilization of the primary food sources. These bacterial groups include methane oxidizers or *methanotrophs* that produce *methane monooxygenase* enzyme (Fogel et al., 1986; Little et al., 1988; Wilson and Wilson, 1985); *autotrophic nitrifiers* (*Nitrosomonas* spp.) that produce *ammonia monooxygenase* enzyme (Arciero et al., 1989; Vannelli et al., 1990); specific *aromatic-oxidizing bacteria* that produce *toluene monooxygenase* (e.g., *Pseudomonas cepacia* G4 and *P. mendocina*), *toluene dioxygenase* (e.g., *P. putida* F1), and *phenol-induced monooxygenase* enzymes (e.g., *Alcaligenes eutrophus* JMP134) (Folsom et al., 1990; Harker and Kim, 1990; Shields et al., 1989; Wackett and Gibson, 1988; Winter et al., 1989); *propane-oxidizing bacteria* that produce *propane monooxygenase* enzyme (e.g., *Mycobacterium vaccae*) (Wackett et al., 1989); and *ethylene oxidizers* (Henry and Grbic-Galic, 1989).

One- and two-carbon aliphatic compounds with lower chlorine saturation can be biodegraded by a diverse group of aerobic microorganisms under appropriate environmental conditions, often using the chlorinated hydrocarbon as a primary food source. However, most of the responsible microorganisms in soils have not been isolated and identified (Chaudhry and Chapalamadugu, 1991).

Methanotrophs. Aerobic methanotrophic bacteria use methane, which is formed by anaerobic methanogenic bacteria, as the sole source of carbon energy. They are found naturally in almost all aerated soils and water that receive methane. Although their primary food source is methane, they are capable of partially oxidizing a wide range of organic compounds, including low-molecular-weight chlorinated aliphatics. The responsible enzyme system, methane monooxygenase (MMO), functions intracellularly either attached to cell membranes (particulate MMO) or free in the cellular fluid (soluble MMO).

Recent studies have indicated that TCE and other chlorinated hydrocarbons are degraded primarily by soluble MMO, a form of the enzyme that occurs mainly when copper becomes depleted in the medium. Excess copper, and other nutrients in some methantrophs, promote development of particulate MMO and a decline in the more bioactive soluble form (Dalton et al., 1984; Prior and Dalton, 1985; Stanley et al., 1983; Collins et al., 1991). Soluble MMO in *Methylosinus trichosporium* is also capable of oxidatively dechlorinating dichloromethane, chloroform, dichloroethanes, *cis*-DCE, and *trans*-DCE. Carbon tetrachloride and PCE were not transformed by MMO. MMO synthesis in methanotrophs is induced by methane or methanol, but in high concentrations these can cause competitive inhibition of the oxidative dechlorination process (Oldenhuis et al., 1989). Reducing power must be maintained by the presence of an electron (hydrogen) donor. Formaldehyde and formate can serve as external electron donors for MMO, and their addition to a treatment system has increased the TCE oxidation rate (Alvarez-Cohen and McCarty, 1991; Oldenhuis et al., 1991).

Because methane is required for MMO production by methanotrophs and formate was shown to promote long-term degradation of TCE, Oldenhuis et al. (1989) proposed a two-stage cleanup, with methane added to an initial reactor and formate added to a second bioreactor. However, Henry and Grbic-Galic (1991a) did not observe increased TCE transformation rates after formate addition to all methanotrophic cultures tested; they suggest preassessment testing prior to site treatment with external electron acceptors.

Little et al. (1988) investigated the degradation of TCE by a methanotrophic bacterium and proposed the oxidative pathway shown in Figure

Step 1: Enzymatic Oxidation to Epoxide by Methane Monoxidase

$$\mathrm{Cl_2C{=}CClH} + \tfrac{1}{2}\,\mathrm{O_2} \longrightarrow \mathrm{Cl_2C\!-\!O\!-\!CClH}\ (\text{epoxide ring})$$

TLE Epoxide

Step 2: Spontaneous Degradation of the Epoxide

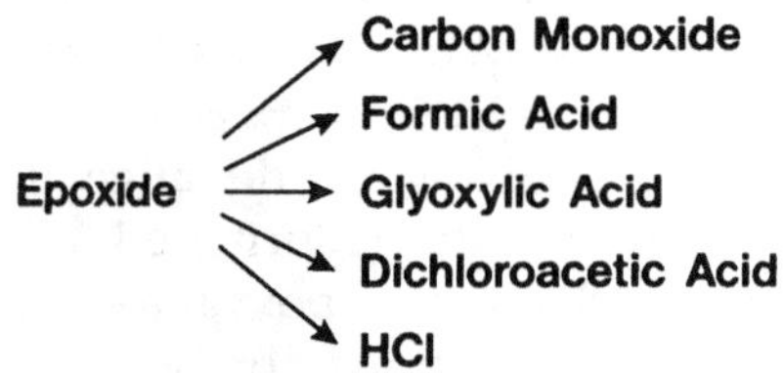

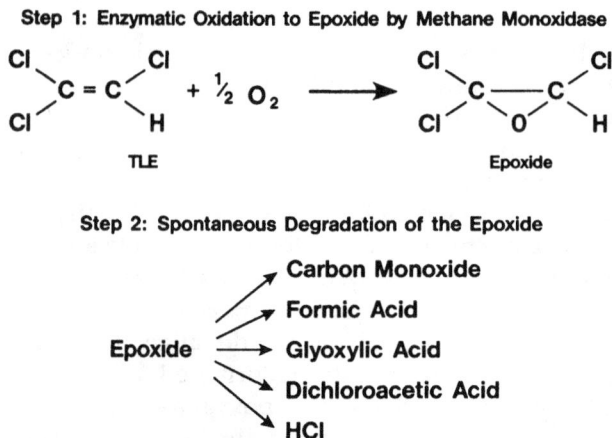

Figure 10 Proposed pathway for oxidative TCE biodegradation in the presence of methanotrophic bacteria.

10. This pathway may be typical for other methanotrophs, because MMO is known to form an epoxy ring at the ethylene double bond. The formate and carbon monoxide could be used by the bacteria, and the other breakdown products could be used by a mixed culture. Little also found that the fastest rate of TCE degradation occurred during active growth on methane, with a declining rate observed shortly after methane depletion. Under the best conditions, her strain degraded only 40% of the TCE added at 0.4 mg/L. However, methane concentrations greater than 2 mg/L have shown competitive inhibition of TCE degradation in dispersed growth studies (Broholm et al., 1990; Brusseau et al., 1991; Oldenhuis et al., 1991).

One unfortunate observation made for methanotrophs is a progressive decline in aliphatic dechlorination activity over time that does not appear to be caused by competitive inhibition by methane of the MMO enzyme system. Similar inhibition has been observed for the other bacterial enzyme systems that oxidatively dechlorinate most chlorinated aliphatic compounds, especially TCE. Most transformations of chlorinated aliphatics seem to produce at least one diffusible inhibitory transformation product.

Suggested inhibitory compounds that form during TCE oxidation include epoxides or acyl chloride (Fox et al., 1990), carbon monoxide (Henry and Grbic-Galic, 1991b), and other epoxide breakdown products (Alvarez-Cohen and McCarty, 1991). The decomposition products of TCE epoxide are known to be pH dependent, with carbon monoxide production favored under basic conditions (Henschler et al., 1979). Carbon monoxide not only is a metabolic toxin but is also a competitive inhibitor and an alternative substrate for the MMO enzyme system (Stirling and Dalton, 1977). Studies with purified soluble MMO have shown that TCE oxidation leads to a progressive loss of MMO enzyme activity. This enzyme inactivation seems to be caused by TCE oxidation products that inactivate general cellular metabolic processes (Fox et al., 1990; Oldenhuis et al., 1991). However, formate added to some methanotrophic cultures restored activity lost by oxidation product toxicity (Henry and Grbic-Galic, 1991b). Additional research is in progress to find better ways of overcoming substrate and intermediate product inhibition.

Autotrophic Ammonia Oxidizers. This group of bacteria, which can perform oxidative dehalogenation with the ammonia monooxygenase (AMO) enzyme (Arciero et al., 1989; Rasche et al., 1990; Vannelli et al., 1990), has received only minor attention to present. These bacteria require both ammonia and molecular oxygen, and appear to be ubiquitous in aerobic soil and aquatic environments. They are also easy to culture and facilitate reaction rate measurements. In nature, AMO oxidizes ammonia to hydroxylamine, which surprisingly is also done by many methanotrophs (Higgins et al., 1984). The only bacterium studied so far is *Nitrosomonas europaea,* an obligate ammonia oxidizer with AMO production occurring continuously. The AMO enzyme seems to be bound to intracellular membranes and exhibits close similarity to membrane-bound MMO in methanotrophs (Hooper and Terry, 1973; Hyman and Wood, 1985). However, the enzyme exhibits greater reactivity with chlorinated compounds than does the particulate MMO system.

The AMO enzyme catalyzes the degradation of TCE, *cis-* and *trans-*DCE, 1,1-DCE, vinyl chloride, 1,1,2,2-tetrachloroethane, 1,1,1-TCA, 1,1,2-TCA, 1,1-DCA, chloroform, and dichloromethane. Similar to the MMO systems, enzyme inhibition was observed. Alkanes with only a single chlorine per carbon atom (chloromethane, chloroethane, and 1,2-dichloroethane) were degraded with no significant enzyme inhibition. Similar to other oxidative dechlorination reactions, carbon tetrachloride and PCE were not dechlorinated (Rasche et al., 1991). Among the alkenes, vinyl chloride is degraded the most rapidly, with rates about five times greater than for TCE. Most one- and two-carbon brominated compounds, except *trans*-dibromoethylene, were also degraded (Vannelli et al., 1990). The reducing power required to dechlorinate aliphatics is supplied by ammonia oxidation or by adding hydroxylamine to the growth medium (Rasche et al., 1991). Additional research is required before the AMO system can be compared with accuracy to other oxygenase systems. However, one advantage to using the AMO system for in situ treatment is that added ammonia could stimulate the growth of other degradative microorganisms, and the resulting nitrification could supply an oxygen source for facultative anaerobic biodegraders.

Aromatic-Oxidizing Bacteria. Several species of *Pseudomonas* possess nonspecific oxygenase enzymes that can oxidatively dechlorinate chlorinated hydrocarbons when the cultures are grown in the presence of toluene or phenol. *P. cepacia* G4 (Shields et al., 1989; Folsom et al., 1990) and *P. mendocina* KR-1 (Winter et al., 1989) are known to possess toluene monooxygenase enzymes as the bioactive component, and *P. putida* F1 (Wackett and Gibson, 1988; Wackett and Householder, 1989) contains a dioxygenase enzyme that elicits the dechlorination. Also, *Alcaligenes eutrophus* JMP134 can degrade TCE by two independent aromatic pathways using monooxygenase enzymes (Harker and Kim, 1990).

Despite some differences, the toluene oxygenases have many properties in common. All display broad substrate specificity, all require an electron-donating factor to function, and (like all oxygenase systems studied) all show decreased activity over time. Dechlorinating activity is stimulated by toluene, and/or phenol, and/or cresol, with the aromatic substrate usually acting both to induce enzyme production and to serve as a primary food source (Ensley, 1991). Toluene and phenol can inhibit dechlorination reactions at elevated concentrations (Nelson et al., 1988).

P. mendocina KR-1 displays a toluene monooxygenase that follows a different pathway than that of *P. cepacia* G4. The time period before observing the usual inhibition was increased by adding the amino acid glutamate to the growth medium. Concentrations of toluene in excess of 1 mmol inhibited TCE oxidation. At noninhibitory toluene concentrations, *P. mendocina* KR-1 oxidized TCE at rates equivalent to those for *P. putida* F1, ranging from 1 to 2 nmol/min/per milligram of bacterial cell protein (Ensley, 1991).

The toluene dioxygenase system of *P. putida* F1 was shown to degrade TCE and *cis*-DCE at equivalent rates, but *trans*-DCE and 1,1-DCE were degraded at much reduced rates. PCE and, surprisingly, vinyl chloride were not measurably oxidized by this dioxygenase system. The decline in activity observed over time was attributed to TCE or TCE metabolite toxicity to the bacterium. However, a mutant strain failed to display significant toxicity effects (Wackett and Householder, 1989). Additional research is under way to investigate the causes and cures for the observed inhibition.

Propane-Oxidizing Bacteria. Three species and five strains of *Mycobacterium*, especially *M. vaccae* JOB5, were shown to oxidize TCE rapidly. Unlike the dioxygenase system of *P. putida* F1, the implicated propane monooxygenase also degraded vinyl chloride, a potent carcinogen, as well as *cis*- and *trans*-DCE. Similar to the other oxygenases studied, PCE was not oxidized (Wackett et al., 1989).

Primary Metabolism of Less-Chlorinated Aliphatics

Generally, only the less-chlorinated alkanes, carboxylic acids, and alcohols can be used as primary food sources by microorganisms. Janssen et al. (1991) determined which chlorinated compounds are used by pure and mixed cultures as sole food sources. They found that all straight-chain alkanes with a single terminal chlorine (1-monohalo-*n*-alkanes) were used as primary substrates by aerobic bacteria. Some dichloroalkanes were also degradable without an alternative food source, including dichloromethane, 1,2-dichloroethane, and 1,3-dichloropropane. Also degraded were 1-bromo-*n*-alkanes. Chloroalkanes that were not used as primary food sources include carbon tetrachloride, chloroform, 1,1,1-TCA and 1,1,2-TCA, 1,1-DCA, and 1,2-dichloropropane. Janssen et al. (1991) also noted that halogenated carboxylic acids and many alcohols, perhaps because of their greater oxidation and polarity, were more easily degraded aerobically than their alkane counterparts. Of the chloroethylenes, only vinyl chloride has been shown to be used as a sole carbon source (Hartmans et al., 1985). Under aerobic conditions, vinyl chloride was readily degraded in groundwater samples taken from a shallow aquifer. More than 99% of the vinyl chloride, at 1 mg/L concentration, was degraded in 108 days with no observed adaptation period (Davis and Carpenter, 1990). An isolated actinomycete was also shown to degrade 67% of radiolabeled vinyl chloride to CO_2 (Phelps et al., 1991). Heavily chlorinated cycloalkanes, such as hexachlorocyclohexane isomers, have also been readily degraded by single bacterial isolates under aerobic conditions (Sahu et al., 1990).

Reductive Dechlorination of Chlorinated Aliphatics

Under anaerobic conditions, certain bacteria can use low-molecular-weight chlorinated aliphatic hydrocarbons as both electron donors (Freedman and Gossett, 1991) and electron acceptors for metabolic processes (Criddle et al., 1990b; Egli et al., 1988). Reductive dechlorination appears to occur because the chlorinated compounds are more oxidized than nonchlorinated hydrocarbons; thus they serve as electron acceptors in respiratory processes. Because chlorinated aliphatic compounds will eventually be dechlorinated to biodegradable food sources, they can also serve as carbon and energy sources for anaerobic bacteria. However, in most cases, additional carbon sources are required for active degradation.

Reductive dechlorination by anaerobic bacteria is important because some highly chlorinated compounds, such as carbon tetrachloride and PCE, are not known to be degraded under aerobic conditions. Reductive dechlorination works most rapidly on compounds having high chlorine (or other halogen) substitution on the aliphatic molecule. The rate seems to decline commensurate with a decrease in chlorine substitution.

One potential problem with the reductive dechlorination of chloroalkenes (e.g., PCE and TCE) is that vinyl chloride, a potent carcinogen, has been observed to accumulate, probably because it is not rapidly dechlorinated further under anaerobic conditions (Barrio-Lage et al., 1986; Vogel and McCarty, 1985). However, recent data have shown that vinyl chloride does not always accumulate in the environment (DiStefano et al., 1991; Freedman and Gossett, 1989; Suflita et al., 1988; Wilson et al., 1986a;). The anaerobic degradation rate and degradation products of vinyl chloride were found to depend on the type and quantity of nutrients added to the growth medium. In the presence of methane, methanol, ammonium phosphate, and phenol, the main biotransformation products were nonchlorinated compounds such as methane and ethylene. If partially oxidized nutrients such as acetate and citrate were added, vinyl chloride was oxidized primarily to CO_2. Four biodegradative pathways were hypothesized, including reductive dechlorination to ethylene, formation of chloromethane, and mineralization to methane and CO_2 (Barrio-Lage et al., 1990).

Reductive dechlorination of PCE has been observed under methanogenic (Fathepure and Boyd, 1988; Freedman and Gossett, 1989; Milde et al., 1988), sulfate-reducing (Bagley and Gossett, 1990; Cobb and Bouwer, 1991), and possibly acetogenic conditions (DiStefano et al., 1991). Denitrifying bacterial populations have not yet been found to be important in PCE or TCE transformations. The incomplete reductive dechlorination of PCE and TCE, resulting in the accumulation of DCE and vinyl chloride, has been observed frequently (Bouwer and Wright, 1988; Parsons et al., 1985; Vogel and McCarty, 1985). However, negligible DCE and vinyl chloride were detected after 4 days when PCE concentrations up to 55 mg/L were added to nonmethanogenic anaerobic enrichment cultures that use methanol as a carbon source. In this case nontoxic ethylene accumulated and dechlorination rates were high (DiStefano et al., 1991). This suggests that reductive dechlorination by certain anaerobic bacteria could be applied to remediate sites contaminated with PCE and

TCE. Sewell and Gibson (1991) also showed that toluene can act as an initial source of the reducing power needed to reductively dechlorinate chloroethenes under anaerobic conditions. Sites containing both chlorinated solvents and gasoline thus may be bioremediated simultaneously.

Carbon tetrachloride (tetrachloromethane) can also be biotransformed under anaerobic conditions. Dechlorination was observed in the presence of specific denitrifying (Criddle et al., 1990a), fermenting (Criddle et al., 1990b), sulfate-reducing, and acetogenic bacteria (Egli et al., 1988). Although carbon tetrachloride has been transformed under denitrifying conditions (Bouwer and McCarty, 1983; Criddle et al., 1990a), other studies indicate that more-reducing conditions favor higher transformation rates (Criddle et al., 1990b; Semprini et al., 1991). While most transformations result in the accumulation of less-chlorinated methanes (chloroform, dichloromethane, and methylene chloride), two acetogens, *Acetobacterium woodi* and *Clostridium thermoaceticum*, were capable of completely degrading carbon tetrachloride to nonchlorinated products and carbon dioxide (Egli et al., 1988). However, a denitrifying bacterium, *Pseudomonas* sp. (strain KC), also produced CO_2 as a major metabolic end product (Criddle et al., 1990a).

Methylene chloride degradation was studied in different soils under aerobic and anaerobic conditions. Surprisingly, degradation was 10 times higher under anaerobic or anoxic conditions following an extensive acclimation period, but the types of anaerobes involved were not determined. Organic carbon and preexposure of the soils to methylene chloride accentuated the degradation rate (Davis and Madsen, 1991).

Applied Developments

Most applications of biodegradation to remove chlorinated aliphatic hydrocarbons (solvents) from soils have involved the soluble MMO system of methantrophs or natural mixtures of methane-, ethane-, and propane-oxidizing bacteria in bioreactors.

One MMO system, using natural methanotrophic bacteria in a sparged continuous-flow glass bead biofilm reactor, degraded 1,1,1-TCA and TCE at maximal rates of 0.3 and 0.4 mg/L·h, respectively. The bioreactor functioned adequately for 6 months, with TCA and TCE removals ranging from 30 to 100% with a 1.5-h liquid retention time. TCE and TCA influent concentrations of 0.5 mg/L (ppm) resulted in effluent concentrations below current drinking water standards of 0.005 and 0.2 mg/L, respectively (Strand et al., 1991). A mixed microbial population isolated from several environments was shown to degrade up to 50 mg/L TCE by more than 99% when methane and other food sources were added (Fliermans et al., 1988). Methanotrophs in mixed cultures are often accompanied by methanol-oxidizing bacteria. Because methane oxidation can result in elevated methanol concentrations, which inhibit oxidative dechlorination, the presence of methanol-oxidizing microorganisms should favor dechlorination reactions (Wilkinson et al., 1974).

Many bioreactors used for treating TCE have a continuous-recycle design. However, a system incorporating both methane- and propane-oxidizing bacteria in an expanded bed batch bioreactor degraded more than 90% of TCE in a 5-day batch operation; the solution was pulsed with 20 mg/L TCE. Pulsed addition of methane and propane with air resulted in the degradation of 1 mol of TCE per 55 mol of substrate used, and competitive inhibition by either substrate did not affect the TCE degradation rate. A small pH shift of 7.2 to 7.5 decreased TCE degradation by 85%, possibly caused by carbon monoxide production and enzyme inhibition, but recovery was rapid when the pH returned to neutral (Phelps et al., 1990). Oxidative dechlorination rates in a mixture of chlorinated aliphatics by *Methylosinus trichosporium* OB3b were determined in both batch and continuous-recycle bioreactor experiments. Formate temporarily accelerated TCE degradation, even when the methane flow was discontinued, but it was speculated that alternate pulsing of methane and formate might result in the fastest degradation rates for most chlorinated aliphatics. Results also indicated that PCE may inhibit the methanotrophic degradation of other chlorinated ethylenes, with total inhibition of TCE dechlorination observed at PCE levels of 0.6 mg/L in a trickling filter bioreactor (Palumbo et al., 1991). This is contrary to the findings for the AMO system of *Nitrosomonas europaea* (Rasche et al., 1991).

Some bioreactors employ a significant gas-phase volume to facilitate mass transfer of poorly soluble methane into the aqueous phase. A rotating biological contactor degraded 95% of the influent TCE and *cis*-DCE with 2% methane in the gas phase. Methane at 15% was inhibitory, probably because of competitive inhibition of the MMO enzyme (Leahy et al., 1989). A methanotrophic-packed bed reactor degraded 1 mg/L TCE and *trans*-DCE with 4% methane in the vapor phase. More than 50% TCE and 90% DCE were degraded in a single pass through the system. However, 2% methane failed to maintain activity after 4 days (Strandberg et al., 1989). In contrast, the soluble MMO of *M. trichosporium* OB3b was not inactivated at methane concentrations below 40% (Oldenhuis et al., 1989). A gas-phase fixed-film bioreactor developed by the USEPA uses propane-oxidizing bacteria to degrade TCE vapors. The maximum TCE loading rate was calculated to be 0.7 μmol TCE/min per kilogram of packing material, above which complete inhibition was observed. Laboratory data indicated that the bioreactor could remove 70 to 80% TCE in a recycle stream having a TCE loading rate of 0.5 μmol/min per kilogram of packing material (Progress report, U.S. EPA, R.S. Kerr Laboratory, Ada, OK).

Aerobic in situ treatment of TCE and metabolites has been performed by stimulating primarily indigenous methanotrophs in the soil. Stanford University personnel conducted a field in situ plot study at Moffett Naval Air Station near Mountain View, California, for 4 years that involved the stimulation of indigenous methanotrophic bacteria. They added alternate pulses of methane and oxygen with reinjected groundwater to a well-characterized isolated aquifer containing low levels of TCE (100 ppb), DCE, and vinyl chloride. With a 1-day detention time, 20% of the TCE, 40% of the *cis*-DCE, 85% of the *trans*-DCE, and 95% of the vinyl chloride were apparently biodegraded (Roberts et al., 1989; Semprini et al., 1990). These results confirm laboratory findings that the less-chlorinated aliphatic compounds are degraded more readily than are the more-chlorinated chemicals.

The results of the Moffett Field study may be used to design a large in situ field demonstration study at St. Joseph, Michigan. Aerobic biodegradation of the less-chlorinated anaerobic degradation products of TCE (DCE and vinyl chloride), which have formed naturally downgradient from a TCE-contaminated aquifer zone, will be one objective of the applied research. Methanotrophs will be stimulated in situ, perhaps by pulsed alternate addition of methane and oxygen. Field stimulation of reductive dechlorination of TCE will be another objective. The proposed treatment zone of 120 by 200 m is 20 m deep. A mathematical model proposes that 480,000 m^3 of aquifer would be treated in 400 days using a groundwater extraction rate of 1514 L/min. A total of 1375 kg of contaminants would be biodegraded, requiring 5200 kg of methane and 19,200 kg of oxygen. These calculations are based on limited data, and one primary objective of this study is to acquire better field data for future modeling (McCarty et al., 1991).

An in situ field trial to bioaugment a site with TMO-producing *P. cepacia* has also been performed. Initial results showed decreases in TCE concentrations from 3 μg/L to 0.5 μg/L within 1 day, with reduction to 0.1 μg/L in a week (*Hazmat World*, 1989).

Halogenated Aromatic Compounds

A diverse group of halogenated, especially chlorinated, monoaromatic compounds can be degraded by microorganisms. Several important groups include chlorobenzenes, chlorobenzoic acids, and chlorophenols. Chlorobenzenes are used as industrial solvents and were added with PCBs to electrical transformers. Chlorobenzoic acids or chlorobenzoates are common degradation products of some herbicides along with

PCBs. Chlorophenols have been used as antifungal agents and preservatives, especially pentachlorophenol (PCP). Less-chlorinated phenols are common transformation products of chlorophenoxy herbicides, such as 2,4,5-T and 2,4-D (Rochkind et al., 1986).

The information given in the following subsections represents a small part of what is known about chloroaromatics and is meant primarily to demonstrate the complexity and specificity of reactions that are possible by microorganisms that transform or degrade them. Although metabolic processes by single bacteria may not be representative of what occurs in soils, there are dominant processes that can lead to the accumulation of toxic intermediates. In other instances, microbial consortia may degrade accumulated intermediates. This research area has potential for human intervention in creating useful microbiological treatment systems, and it is expanding at a rapid pace. Although interesting work with mutant bacterial strains has provided more recent data, much of the information about the degradative pathways for chloroaromatics was acquired during the 1970s and 1980s (Reineke and Knackmuss, 1988; Rochkind et al., 1986).

Oxidative Degradation by Oxygenase Enzyme Systems

Oxidative degradation of halogenated aromatic compounds occurs most favorably under aerobic conditions. The degradation of chlorobenzenes is generally more difficult than for oxygenated and less hydrophobic chlorinated benzoates and phenols.

Halobenzenes. Oxidative degradation of monochlorobenzene by a *Pseudomonas* strain was first demonstrated by Rieneke and Knackmuss (1984). Biodegradation of the three dichlorobenzene isomers was shown by laboratory research (de Bont et al., 1986; Haigler et al., 1988; Oltmanns et al., 1988; Schraa et al., 1986; Spain and Nishino, 1987) and applied studies (Aelion et al., 1987; Bouwer and McCarty, 1985). Trichlorobenzene has also been shown to be oxidatively biodegraded by mixed cultures (Marinucci and Bartha, 1979) and pure strains of bacteria (Sander et al., 1991; van der Meer et al., 1987). Only recent data indicate that benzene with four chlorine groups attached (1,2,4,5-tetrachlorobenzene) can be degraded by pure bacterial cultures (Sander et al., 1991). The responsible bacteria generally require the presence of the chlorinated compound or related compound to induce enzyme activity, but some can use the chlorinated compound as a sole carbon source.

Some strains of *Pseudomonas* can degrade chlorobenzene and the three chlorotoluene isomers if grown with toluene as a cosubstrate (Gibson et al., 1968). These bacteria possess both ortho- and meta-ring-cleavage dioxygenase enzymes. However, the main toluene degradative pathway, involving meta-cleavage, is governed by genes on the chromosome, whereas genes for the ortho-cleavage dioxygenase pathway are located on less stable extrachromosomal plasmids (Grishchenkov et al., 1984). Thus the meta cleavage pathway predominates. If the normal toluene (meta-cleavage) degradative pathway is followed, toxic intermediates (i.e., 3-chlorocatechol and its ring-cleavage product) accumulate in the medium. The ring-cleavage product, a reactive acyl chloride, inactivates the meta-cleavage dioxygenase enzyme that promotes chlorobenzene degradation, thus leading to dead-end metabolism (Bartels et al., 1984). The main toluene degradative pathway must be blocked before these *Pseudomonas* strains can effectively degrade chlorobenzenes. Hybrid *Pseudomonas* strains have been developed that exclude the unwanted meta-cleavage pathway (Sander et al., 1991).

Sander et al. (1991) isolated two strains of *Pseudomonas* (PS12 and PS14) from contaminated soils that degraded 1,2,4-trichlorobenzene and 1,2,4,5-tetrachlorobenzene, as well as all less-chlorinated benzenes. These compounds served as sole sources of carbon and energy to the bacteria, which eventually cleaved the aromatic ring by an ortho-cleavage pathway. Recent gene-cloning experiments have also produced mutant bacterial strains capable of degrading 1,2,4-trichlorobenzene (van der Meer et al., 1991). A pathway proposed for the oxidative degradation of 1,2,4,5-tetrachlorobenzene involves simultaneous removal of a chlorine group with dihydroxylation of the ring, probably by an ortho-cleavage dioxygenase enzyme. The 1,2,4-trichlorobenzene molecule is dihydroxylated prior to any chlorine removal, which subsequently occurs after ring cleavage. Apparently, no inhibitive intermediate metabolites are formed (Sander et al., 1991).

Fluorinated aromatic compounds are difficult to biodegrade, partly because of the toxicity of fluorine to microorganisms. However, *Pseudomonas* sp. strain T-12, which uses the toluene degradative meta-cleavage pathway, simultaneously defluorinates 3-fluoro-substituted benzenes while transforming the aromatic compound to biodegradable 2,3-catechols. Because the dehalogenation occurs prior to ring cleavage, no inhibitory acyl halides are formed (Renganathan, 1989).

Halobenzoates. Chlorobenzoates have been studied extensively for 20 years, with the monochlorobenzoates receiving the greatest attention (Fetzner et al., 1989; Focht and Shelton, 1987; Hartmann et al., 1989; Marks et al., 1984; Miguez et al., 1990; Pertsova et al., 1984; van den Tweel et al., 1986). The metabolism of 3-chlorobenzoate and 4-chlorobenzoate has been shown to proceed through 3-chlorocatechol and 4-hydroxybenzoate, respectively. Less information is available on the oxidative degradation of dichlorobenzoates (Adriaens and Focht, 1991;

Hartmann et al., 1979; Hernandez et al., 1991; Hickey and Focht, 1990) and trichlorobenzoates (Hernandez et al., 1991; Hickey and Focht, 1990). At this time, benzoates that contain four or five chlorine groups are not known to be oxidatively degraded. Some of the more-chlorinated benzoates are thought to be dechlorinated and degraded through a combination of oxidative and reductive pathways (Adriaens and Focht, 1991; van den Tweel et al., 1987).

Pseudomonas aeruginosa JB2 has degraded five different chlorobenzoates (CBa) through primary metabolism: specifically, 2-CBa, 3-CBa, 2,3-CBa, 2,5-CBa, and 2,3,5-CBa. Cometabolism of 2,4-CBa was also observed. Chlorocatechols were intermediates of all CBa metabolic pathways. This strain also used a large group of other mono- and trihalogenated benzoates as growth substrates. The initial attack on the chlorobenzoates was thought to be mediated by dioxygenase enzymes (Hickey and Focht, 1990).

P. putida P111 was capable of growth on 2-CBa, 3-CBa, 4-CBa, 2,3-CBa, 2,4-CBa, 2,5-CBa, and 2,3,5-CBa. However, 3,5-CBa was not degraded and it completely inhibited growth on all ortho-substituted benzoates that were tested. Yet when 3-CBa or 4-CBa were used as cosubstrates with 3,5-CBa, this isomer was degraded (Hernandez et al., 1991).

Chlorophenols. The less-chlorinated chlorophenols are readily biodegraded aerobically by a diverse group of microorganisms, whereas higher chlorination inhibits phenol degradation under aerobic conditions. For this reason pentachlorophenol (PCP) is emphasized in this discussion.

Chlorophenols are often biotransformed to more volatile chloroanisoles, whereby a methyl group is added to the hydroxyl group, forming a methoxy side group (Neilson et al., 1983). This pathway seems to detoxify the chlorophenols for microorganisms, but this product could be more toxic to higher organisms. Both bacteria (Allard et al., 1987) and fungi (Cserjesi and Johnson, 1972) methylate chlorophenols to corresponding methylated derivatives.

Although somewhat recalcitrant to aerobic biodegradation, PCP is decomposed by some aerobic bacterial isolates (Chu and Kirsch, 1972; Edgehill and Finn, 1982; Stanlake and Finn, 1982; Suzuki, 1977; Watanabe, 1973). An important bacterial group consists of unidentified species of *Flavobacterium* (Crawford and Mohn, 1985; Pignatello et al., 1983; Saber and Crawford, 1985). Isolated strains of *Flavobacterium* showed higher PCP degradation rates when glucose and phosphate were added to the medium and nitrogen or sulfate was limiting (Topp and Hanson, 1990). Growth of polyurethane foam-immobilized *Flavobacterium* cells in a semicontinuous batch bioreactor resulted in PCP degradation at initial concentrations as high as 300 mg/L. Initial degradation rates were 3.5 to 4 mg/g foam per day, but these rates dropped by about 0.6% per day. The foam helped to reversibly bind the PCP and protect the bacteria (O'Reilly and Crawford, 1989).

Actinomycetes, *Rhodococcus chlorophenolicus*, and *Mycobacterium* strains are also capable of mineralizing PCP (Apajalahti and Salkinoja-Salonen, 1986; Haggblom et al., 1988). The metabolic mechanism for PCP degradation by *R. chlorophenolicus* PCP-I is initial ring hydroxylation followed by reductive dechlorination to remove chlorine groups (Apajalahti and Salkinoja-Salonen, 1986). Other aerobic degradations may follow similar pathways.

Another unique group of aerobic microorganisms is the white rot fungi. These fungi are so named because they can remove and degrade very complex lignin molecules from wood, leaving behind white-colored cellulose. Several species of *Phanerochaete* (e.g., *P. chrysosporium* and *P. sordida*) are especially effective at transforming and degrading PCP. In a field study, soil containing 250 to 400 mg/kg PCP was supplemented with these fungi and peat as an alternative carbon source. PCP concentrations were subsequently reduced by 88 to 91% in 6.5 weeks. Most of the PCP was transformed to nonextractable soil-bound products, and 8 to 13% was converted to more volatile pentachloroanisole (PCA) (Lamar and Dietrich, 1990). However, the nature of the transformation products, such as their soil extractability, is greatly influenced by soil type (Lamar et al., 1990a). In related laboratory studies, it was shown that *Phanerochaete* species are sensitive to low-PCP concentrations in aqueous media, with *P. sordida* and *P. chrysosporium* being inhibited by 25 mg/kg levels. The greatest mineralization rates were 12% in 30 days with *P. sordida* strain 13 in an aqueous medium. However, less than 2% of PCP was mineralized in soils by either species (Lamar et al., 1990b).

Despite the formation of PCA by *Phanerochaete* spp., methylation does not appear to be an important step in white rot fungal transformations of PCP, and the PCA can eventually be degraded (Lamar and Dietrich, 1990). PCP degradation by *Phanerochaete* spp. seems to be a two-phase process, with the first phase resulting in formation of PCA and the second phase initiating degradation of both the PCA and PCP (Lamar et al., 1990b).

Anaerobic Dechlorination

Anaerobic degradation of chloroaromatic compounds has been studied in sediments (Genthner et al., 1989a; Horowitz et al., 1982), soils (Ide et al., 1972; Murthy et al., 1979), and groundwater (Gibson and Suflita, 1986; Suflita and Miller, 1985). In contrast to most reported aerobic mechanisms, anaerobic pathways generally involve reductive removal of chlorine (or other halogen) groups prior to degradation, which seems to decrease toxicity (Suflita et al., 1982). Hydroxylation reactions also occur in the absence of molecular oxygen, similar to the haloalkanes. Chloride removal by hydroxylation is more common for aromatic compounds having elements other than carbon in their ring structure (heterocyclics) than for homocyclic aromatic compounds (Tiedje et al., 1987).

Reductive dehalogenation reactions are most favored under highly reducing (e.g., methanogenic) conditions (Gibson and Suflita, 1986) but have also been reported under sulfate-reducing conditions (de Weerd et al., 1990; Haggblom and Young, 1990; King, 1988; Kohring et al., 1989). However, high sulfate concentrations (e.g., 30 mmol) have been shown to inhibit reductive dehalogenation in sediments (Gibson and Suflita, 1986, 1990; Kuhn et al., 1990). To date, reductive dehalogenation of aromatic compounds under denitrifying conditions has not been reported, and the addition of nitrate to sediment slurries has also been shown to impede reductive dechlorination (Kohring et al., 1989). Yet a few aerobic bacteria appear to be capable of this process (van den Tweel et al., 1987).

Degradation of monohalogenated benzoates by denitrifying bacteria has been described (Genthner et al., 1989a; Schennen et al., 1983; Taylor et al., 1979), but the mechanism of attack was apparently not reductive dehalogenation. This information suggests that anaerobic degradation or transformation of haloaromatics is facilitated primarily by mixed methanogenic consortia in sediments, but a few active sulfate reducers are also involved. In a practical sense, anaerobic transformations appear not to be useful under nitrate-reducing conditions. Individual bioactive microorganisms remain to be isolated.

Gibson and Suflita (1986) observed that the addition of short-chain organic acids or alcohols to growth media can stimulate the initiation and rate of anaerobic degradation of chloroaromatic compounds. Also, studies with anaerobic fixed-film bioreactors indicate that the addition of a readily degradable alternative carbon source (e.g., glucose) can significantly increase PCP degradation rates (Hendriksen et al., 1991).

Aromatic compounds reported to undergo reductive dechlorination, followed by anaerobic degradation or transformation, are chlorobenzoates (Horowitz et al., 1983; Linkfield et al., 1989) and chlorophenols (Boyd and Shelton, 1984; Bryant et al., 1991; Suflita and Miller, 1985). Reductive dehalogenation has also been reported for halobenzenes (Bosma et al., 1988; Fathepure et al., 1988), chlorinated nitroaromatics

(Tiedje et al., 1987), and chloroanilines (Kuhn et al., 1990). Although reductive dechlorination by anaerobic bacteria is probably an important process in nature, little information exists concerning the microorganisms involved or specific pathways followed (Chaudhry and Chapalamadugu, 1991). The best studied groups at present are the chlorobenzoates and chlorophenols, especially pentachlorophenol.

Chlorobenzoates and Chlorophenols. Genthner et al. (1989b) showed that as a group, chlorophenols were degraded more readily than chlorobenzoates under anaerobic conditions. They also showed that the relative order of degradability for monochlorophenols was ortho > meta > para with regard to the substituent positions on the aromatic ring. For the monochlorobenzoates, the order was meta > ortho > para.

Degradation of monochlorophenols by methanogenic consortia was initiated by ring dechlorination, followed by conversion of phenol to benzoate prior to mineralization (Genthner et al., 1989a). However, distinct phenol- and benzoate-degrading bacteria were suspected, which appeared to be acclimated primarily by the specific location of the chlorine group. Degradation of 3-chlorobenzoate by a denitrifying consortium was also observed, but this was thought to occur through a symbiosis between an anaerobic dechlorinating bacterium and a denitrifier.

Dichlorophenols in freshwater sediments were studied for 1 year. Variables that lowered dechlorination rates were high sulfate and nitrate. Neutral pH was most favorable for reductive dechlorination. Environmental conditions were found to be more important than ring-substitution patterns, but chlorine was preferentially removed from the aromatic ring when the two chlorines were in the ortho position; meta-oriented chlorines were dechlorinated with most difficulty (Hale et al., 1991). The rate-limiting step for the anaerobic degradation of 2,4-dichlorophenol is the conversion of 4-chlorophenol (4-CP) to phenol, with low 4-CP concentrations inhibiting further degradation of accumulated phenol and benzoate. This degradation step needs to be accelerated, because 4-CP is far more toxic than the parent compound (Zhang and Wiegel, 1990).

Pentachlorophenol. Reductive dechlorination rates are generally considered to be dependent on the position and number of chlorines on the compound. The more highly chlorinated compounds are often dechlorinated most rapidly, but they also tend to be most toxic to microorganisms. Mikesell and Boyd (1985) noted that ortho chlorines were removed most rapidly from the more-chlorinated phenols. Anaerobic studies with sewage sludge initially indicated that ortho dechlorination was most prevalent for PCP, yielding 2,3,4,5-tetrachlorophenol and

3,4,5-trichlorophenol (Mikesell and Boyd, 1985; Woods et al., 1989). Further work with different acclimated methanogenic consortia indicated that some dechlorination occurred at all possible chlorine positions, but ortho dechlorination was observed most frequently. Dechlorination of PCP in anaerobic sediments depended on initial preadaptation to chlorinated phenols. Consortia preadapted to 2,4-dichlorophenol initially removed the ortho chlorine, while those preadapted to 3,4-dichlorophenol initially removed the para chlorine from PCP. Nonadapted microbial communities required long lag times (e.g., more than 40 days) before dechlorination commenced. The order of dechlorination by a microbial mixture was para > ortho > meta (Bryant et al., 1991).

Thermophilic (50°C) anaerobic biodegradation of PCP was evaluated using natural media from various sources. Transformation rates were 7.5 μmol/L·B day when 37.5 μmol PCP was added to unacclimated freshwater sediment (Larsen et al., 1991). This is about threefold higher than the rates experienced at ambient temperatures in monochlorophenol-acclimated sludges (Mikesell and Boyd, 1986).

Polychlorinated Biphenyls

Polychlorinated biphenyls (PCBs) are known to have 209 different possible isomeric configurations for the chlorine groups on the biphenyl molecule. Commercial formulations include about 100 isomers. Most highly chlorinated molecules are nearly insoluble in water, and thus have great potential for bioaccumulation into fatty tissues of higher organisms. Although their production is now banned, they are still used in high-voltage electrical equipment, such as transformers and capacitors, and have previously served as components in lubricating oils, hydraulic fluids, plasticizers, and herbicides. Complex mixtures of PCBs have been marketed under various trade names, such as Aroclor in the United States. Aroclors 1242, 1254, and 1260 are most commonly encountered; each contains 42%, 54%, and 60% chlorine by weight, respectively. Aroclor 1242 averages three chlorine groups per biphenyl molecule, whereas Aroclor 1252 averages five chlorine groups but may contain up to 69 different biphenyl isomers; Aroclor 1260 averages six chlorines per molecule (Chaudhry and Chapalamadugu, 1991; Rochkind et al., 1986; Quensen et al., 1990).

Besides their persistence in the environment, PCBs can be partially metabolized to more toxic compounds. They also contain and can be chemically transformed at high temperatures to polychlorinated dibenzofurans (PCDFs), which are considered to be highly toxic chemicals. Pyrolysis of commercial PCBs has yielded about 10% PCDFs (Fortnagel et al., 1990; Rappe, 1984). PCBs with higher chlorination tend to be most persistent, but less-chlorinated molecules tend to be more toxic to bacteria, probably in part because of their greater water solubility (Furukawa et al., 1979). Early observations indicated that PCBs having fewer than five chlorine groups were biodegraded extensively in the environment, since weathered PCBs often contain more than four chlorine groups (Furukawa, 1982). However, anaerobic bacteria tend to promote reductive dechlorination of the more highly chlorinated PCB isomers (Brown et al., 1987).

Aerobic Degradation

The aerobic biodegradation of PCBs depends both on the number and position of chlorines on the biphenyl molecule. Based on information summarized from many studies (Furukawa, 1982; Rochkind et al., 1986), some generalizations can be made regarding aerobic degradation of PCBs:

1. Degradation usually decreases as the number of chlorines per molecule increases.
2. PCBs having only one ring of the biphenyl molecule chlorinated are more easily degraded than those displaying chlorines on both aromatic ring structures.
3. Two chlorines at the ortho positions of a single ring (i.e., 2,6-) or on both rings (i.e., 2,2′-) inhibit degradation.
4. Ring cleavage for PCBs having unequally chlorinated rings will be initiated on the ring having the fewest chlorines.

Aerobic degradation of most components of Aroclor 1242 and some of Aroclor 1254 has been demonstrated but requires a growth substrate that induces enzyme production, such as biphenyl. Little definitive information exists for appreciable aerobic biodegradation of Aroclor 1260 (Bedard et al., 1987; Brunner et al., 1985; Dmochewitz and Ballschmiter, 1988; Kohler et al., 1988). However, some PCB congeners with five and six chlorines have been aerobically biotransformed by pure cultures such as *Alcaligenes eutrophus* H850 and *Pseudomonas putida* LB400 (Bedard et al., 1986, 1987). Despite these findings, no single naturally occurring microorganism has been found that completely mineralizes PCBs containing more than one chlorine group. Yet a culture of two *Acinetobacter* species (strains P6 and 4-CB1) in a continuous aerobic fixed-bed reactor using polyurethane supports degraded 6.5% and 11% of 4,4′- and 3,4′-dichlorobiphenyl, respectively (Adriaens and Focht, 1990).

PCB transformations include intermediate metabolites such as chlorobenzoates, so many of the problems discussed concerning chlorobenzoate and chlorobenzene degradation also apply to the PCBs. Biphenyl is degraded by many bacteria by first initiating meta cleavage of the aromatic rings. However, chlorinated PCBs that undergo aromatic ring cleavage at the meta position produce chlorinated benzoates as intermediates (Furukawa et al., 1979). These chlorobenzoates and their cleavage products are inhibitory to meta-cleavage enzymes, which results in blockage of further PCB degradation. Ortho-cleavage pathways prevent accumulation of these toxic intermediates, but initial hydroxylation of the biphenyl ring, which indicates ring cleavage, is promoted by the meta-cleavage enzymes (Furukawa et al., 1978a). This dilemma was partially solved by Huang (1989), who used the multiple-chemostat method of Krockel and Focht (1987) to create a 3-chlorobiphenyl-degrading strain through enhanced natural genetic recombination of *Acinetobacter* P6 (which degrades PCBs to chlorobenzoates) with a 3-chlorobenzoate using *Pseudomonas*. This was also accomplished using two unidentified pseudomonads (Mokross et al., 1990).

All three isomers of monochlorobiphenyl can be biodegraded by aerobic microorganisms (Bailey et al., 1983; Clark et al., 1979; Furukawa et al., 1978b; Shiaris and Sayler, 1982). Bailey et al (1983) found that each isomer was degraded at comparable rates in river water containing 1 μg/L of a respective isomer. The unchlorinated biphenyl ring was cleaved within 1.5 days' incubation, with the resultant chlorobenzoates showing 50% degradation to CO_2 by day 50. The biodegradation rate of higher concentrations of monochlorobiphenyls was somewhat slower.

Most studies using more-chlorinated PCBs have reported only initial cometabolic transformation by a single microorganism, usually to chlorinated benzoic acids (Adriaens and Focht, 1990; Bedard et al., 1987). However, the white rot fungus, *Phanerochaete chrysosporium*, has been shown to mineralize components of Aroclor 1254 (Eaton, 1985) and individual more-chlorinated PCBs such as 3,4,3′,4′-tetrachlorobiphenyl and 2,4,5,2′,4′,5′-hexachlorobiphenyl (Aust, 1990; Bumpus and Aust, 1987; Bumpus et al., 1985). Because the white rot fungi are a diverse group with apparently unique enzyme systems, they should be studied further for their potential to remediate soils contaminated with PCBs and other recalcitrant aromatic compounds.

Many chlorinated aromatic compounds contain low levels of chlorinated dibenzo-*p*-dioxins (DDs) and dibenzofurans (DFs), and higher levels are formed during their thermal destruction (Rappe, 1984). These compounds are considered to be very toxic to animals and are thus of great environmental concern. However, little information exists concerning the biodegradation of chlorinated and nonchlorinated DDs and DFs. Both the white rot fungus, *P. chrysosporium* (Bumpus, 1989), and aerobic bacteria (Foght and Westlake, 1988; Fortnagel et al., 1990) have been implicated in the slow degradation of DF.

Klecka and Gibson (1980) have described transformation of chlorinated DDs to oxidized products, whereas Ward and Matsumura (1978) showed very slow degradation of 2,3,7,8-tetrachloro-DD (TCDD) in laboratory-model microcosms. *Pseudomonas* strain HH69 metabolized DF to aromatic acids, such as salicylic and gentisic acids, which were slowly degraded. However, several dead-end metabolites also accumulated as a result of dioxygenase enzyme attack. The same bacterial strain oxidatively cleaved DD, forming a 2-phenoxy derivative of muconic acid as the major end product (Harms et al., 1990). These results indicate that like the more-chlorinated PCBs, chlorinated DDs and DFs are only slowly degraded by natural soil microorganisms under aerobic conditions, and natural microbial consortia or genetically engineered microorganisms may be required for their complete degradation.

Anaerobic Dechlorination

PCBs with a higher percentage of chlorine saturation on the biphenyl rings are more susceptible to biotransformation under anaerobic or reducing environmental conditions than under aerobic conditions. This is because reductive dechlorination is favored at a low redox potential. In contrast, PCBs with one and two chlorines seem to remain unaffected for long times under highly reducing conditions, although some dichlorinated isomers (2,3- and 3,4-) may also be dechlorinated (Brown et al., 1987). As discussed previously, anaerobic degradation of most mono- and dichlorobenzenes and benzoates has been observed (Bosma et al., 1988; Genthner et al., 1989a), although an initial oxidative transformation is suspected, as was shown for nonchlorinated aromatics (Vogel and Grbic-Galic, 1986). This may also prove to be true for the less-chlorinated PCBs. Despite the research findings, PCBs with one and two chlorines tend to accumulate in reducing environments (Brown et al., 1987).

The positions of chlorine substituents on the biphenyl rings also seem to be important. For example, many studies have indicated that chlorines in meta and para positions on each ring are preferentially removed, leaving lightly chlorinated ortho-substituted products (Nies and Vogel, 1990; Quensen et al., 1988). However, van Dort and Bedard (1991) showed para-dechlorination in methanogenic sediments contaminated with Aroclor 1260 from Woods Pond in Massachusetts. It appears that many metabolic pathways followed in natural sediments are dependent on site-specific microbial consortia, few of which have yet been studied in detail (Quensen et al., 1990; van Dort and Bedard 1991).

These findings suggest that by combining isolated degradative bacterial consortia from different sites, PCBs could perhaps be degraded to mono- and dichlorinated products under anaerobic conditions. As discussed in the section on anaerobic–aerobic processes, methods are now being developed to accentuate anaerobic dechlorination of PCBs, followed by techniques to aerate these same sediments to promote complete PCB degradation. All lightly dechlorinated PCBs resulting from anaerobic transformations are degradable under aerobic conditions (Bedard et al., 1986).

Most studies indicate that cometabolism is involved in PCB biodegradation under anaerobic conditions, because the PCBs are not used as a food source (Brown et al., 1987; Nies and Vogel, 1990). Aroclor 1242-contaminated sediment containing approximately 300 mg/kg (ppm) of PCBs was incubated anaerobically in batches containing various additional carbon compounds. Batches that were fed methanol, glucose, and acetone showed the highest dechlorination rates, with acetate showing less effect. No significant dechlorination was observed in batches receiving no additional organic substrate (Nies and Vogel, 1990).

BIOREMEDIATION OF POLYNUCLEAR AROMATIC HYDROCARBONS

Polynuclear aromatic hydrocarbons (PAHs) can enter the environment and result in contamination from a variety of sources. The most common sources of PAHs are either incomplete combustion of carbonaceous material or coal gasification processes. PAH contamination may also arise from carbon black, or creosote, and is found in some oils.

PAHs are of primary regulatory concern because some have been shown to cause mutagenesis or carcinogenesis. Regulatory concerns and bioremediation efforts are typically directed at either the 16 PAHs identified by the USEPA as priority pollutants or the subset of suspected carcinogenic PAHs (CPAHs).

Although most PAHs of concern are the result of human activity, many PAHs occur naturally and had been metabolized by microorganisms for many years before the presence of human beings. This has created microbial enzyme systems capable of degrading many PAHs. Based on pure culture studies, it appears that PAHs of three or fewer rings are readily degradable and are used by natural microorganisms as a direct food source. Microbiological transformations of large PAH compounds have been observed (Gibson and Subramanian, 1984). Although microbiological transformations of four- and five-ring compounds have been shown (Keck et al., 1989; Mahaffey et al., 1988), the transformation mechanisms appear to be primarily cometabolic. Heitkamp et al. (1988) have shown that the four-ring PAH pyrene is bacterially mineralized following enzyme induction. Although not widely proven, it appears possible that PAHs larger than four rings may be used as a direct carbon source. The literature on microbial degradation has been summarized by Sims and Overcash (1983) and Atlas (1981).

Biodegradation of PAHs is greatly influenced by environmental redox conditions. PAHs have generally been found to biodegrade under aerobic conditions (Bauer and Capone, 1985; Delfino and Miles, 1985; Heitkamp et al., 1987; Schocken and Gibson, 1984; Mihelcic and Luthy, 1988a,b; Park et al., 1990; Wang et al., 1990). Werner (1991) has shown some PAH degradation under dinitrifying conditions. Studies of PAH degradation in two aerobic sandy loams found PAH half-lives ranging from 2 to 60 days for two-, three-, and four-ring PAHs, and more than 300 days for five-ring PAHs. For the lower-molecular-weight PAHs, as much as 20% of the removal was attributed to volatilization (Matthews, 1989; Park et al., 1990).

In aerobic estuarine sediments, Shiaris (1989) reports biodegradation turnover rates of 13 to 20 days for naphthalene, 8 to 20 days for phenanthrene, and 54 to 82 days for benzo[*a*]-pyrene. In bench-scale experiments Wang et al. (1990), working with a sandy loam contaminated by diesel oil with PAH, found that aerobic biodegradation rates could be increased by adding lime and nutrients. They found that after 12 weeks of treatment, methylnaphthalene concentrations dropped from an initial level of 100 mg/kg to below detection limits; methylphenanthrenes dropped from an initial level of 208 mg/kg to 21 mg/kg without lime added, and to below detection limits with lime added; and C_2-phenanthrenes dropped from an initial level of 191 mg/kg to 22 mg/kg without treatment and below detection limits with treatment.

It is clear that PAH biodegradation may be controlled by a variety of factors. Under aerobic conditions PAH will frequently biodegrade, with the lower-molecular-weight PAH being more readily degradable. It is less clear that under anaerobic conditions PAH biodegradation will occur readily (Bauer and Capone, 1985; Herbes and Schwall, 1978). PAH biodegradation under denitrifying conditions has been documented, and at rates similar to aerobic biodegradation (Mihelcic and Luthy, 1988a,b, 1991). Biodegradation under more reduced anaerobic conditions has not been shown as clearly. Kuhn and Suflita (1989) have shown that some anaerobic biodegradation of nitrogen- and oxygen-substituted PAHs does occur under sulfate-reducing or methanogenic conditions, and in what appears to be an anaerobic aquifer, Godsy et al. (1991) have documented creosote PAH biodegradation.

Bioremediation strategies for PAH contamination typically involve some form of aeration, frequently coupled with nutrient enrichment and/or other environmental modifications. Due to their low aqueous solubility, PAHs typically pose a soil treatment problem. The most common bioremediation practice for PAH treatment is landfarming.

Piotrowski (1991) describes a landfarming effort at the Libby, Montana, Superfund site. This is a former wood treatment site, where soils are contaminated with creosote containing both PAHs and PCP. Over 102 days, Piotrowski (1991) found total PAH removals from 786 mg/kg to 74 mg/kg (91% reduction) and a CPAH reduction from 238 mg/kg to 63 mg/kg (74% reduction). This was the first Superfund site at which the USEPA accepted bioremediation in a record of decision (ROD). In situ PAH bioremediation by the bioventing process has been documented in laboratory experiments by Yong et al. (1991) and in the field by Lund et al. (1991). The white rot fungi (*Phanerochaete* spp.) have also been shown to biodegrade PAHs (Sutherland et al., 1991; Bumpus, 1989; Pothuluri et al., 1990). To the authors' knowledge, no anaerobic processes for PAH bioremediation have yet been demonstrated.

BIOREMEDIATION OF NITROAROMATIC COMPOUNDS

The primary source of nitroaromatic contaminants in the environment is the munitions industry. Contamination may result from disposal of ordnance materials or as a discharge from ordnance manufacture or demilitarization. Compounds of concern include 2,4,6-trinitrotoluene (TNT), 2,4-dinitrotoluene (DNT), 2,3-dinitrobenzene (DNB), 1,3,5-trinitrobenzene (TNB), and hexahydro-1,3,5-trinitro-1,3,5-triazine (RDX).

TNT is probably the most common environmental contaminant of the nitroaromatic group and the most extensively studied. Aerobic TNT biotransformation has been documented for many years. Enzinger (1971) acclimated a consortium of microorganisms from a wastewater treatment plant to TNT. The acclimated consortium reduced initial TNT concentrations of 100 mg/L to 1.25 mg/L in 5 days. Numerous other investigators have found conventional aerobic wastewater treatment processes capable of biotransforming TNT and related compounds (Spanggord et al., 1980).

Studies involving radiolabeled TNT, DNT, and other highly nitrated aromatics have shown that most aerobic and anaerobic transformations do not lead to ring cleavage. Aromatic intermediates are formed, but little or no parent compound is fully mineralized. Carpenter et al. (1978) investigated the fate of ^{14}C-labeled TNT in an activated sludge system and found nearly complete TNT disappearance in 3 to 5 days, but no significant ^{14}C evolved in CO_2. Hoffsomer et al. (1978) found similar results with TNT treatment in an oxidation ditch. Although they observed significant TNT disappearance, they found that less than 0.4% of the ^{14}C evolved as CO_2.

Similar results have been shown with TNT treatment in the solid phase. Osmon and Andrews (1978) investigated the feasibility of TNT disposal using land application and composting. Their laboratory studies indicate TNT biotransformation to a variety of by-products. Kaplan and Kaplan (1982) studied TNT composting and found that most of their radiolabeled TNT was converted to by-products that were "bound to humus fractions."

RDX appears to most readily biotransform anaerobically and has been found to be a cometabolic process (Spanggord et al., 1980). Partial degradation products resulting from RDX biotransformation have also been reported. Both Greene (1985) and McCormick et al. (1984) report mono- and dinitroso derivatives.

The toxicity and environmental effects of the nitroaromatic biotransformation products are not clearly known. Won et al. (1976) found TNT metabolites to be less mutagenic than the parent TNT. Care must be taken in application of bioremediation processes, which may only partially degrade parent compounds.

Recent work with white rot fungis indicates that mineralization is possible. Fernando et al. (1990) found that 15 to 23% of (^{14}C)TNT was mineralized by *Phanerochaete chrysosporium* in a 90-day period. Spanggord et al. (1991) have recently isolated a species of *Pseudomonas* apparently capable of complete mineralization of DNT by a dioxygenase enzyme system under aerobic conditions. Preliminary work with anaerobic biodegradation of TNT by a consortium of bacteria isolated from sheep rumen (forestomach) fluid also suggests mineralization may be possible via anaerobic pathways (Craig, 1991).

The U.S. Army Toxic and Hazardous Materials Agency (USATHAMA) has supported extensive research into the use of composting to treat TNT- and RDX-contaminated soils (Doyle et al., 1986; Isbister et al., 1982). Their results indicate substantial parent compound disappearance, in some cases in excess of 99.9% reduction. Biotransformation rates (first order) with TNT and RDX half-lives in the range of 10 to 25 days have been reported. As with most other field studies, however, significant ring cleavage has not been demonstrated with TNT, and ^{14}C studies indicate that most of the TNT was converted to partially reduced organic intermediates. On the other hand, Doyle et al. (1986) did observe RDX mineralization.

To date, field-scale application of bioremediation to ordnance nitroaromatics has been limited to wastewater treatment (Wiley, 1985) and limited land application and composting (*Chemical Week*, 1988; LeBron 1989; Soviero, 1989). With the exception of the USATHAMA RDX study, it is not clear that any of these field-scale applications has led to mineralization.

BIOTRANSFORMATION OF INORGANIC COMPOUNDS

Metals and other inorganic compounds in soils can be present in various geochemical phases, including those that are water soluble, ion exchangeable, physically sorbed to other inorganic and organic surfaces precipitated compounds (e.g., oxides, hydroxides, carbonates, sulfides), those bound to other precipitates (coprecipitates), and crystalline minerals. Crystalline minerals generally account for about half the total metals in soils and are generally mobilized with difficulty. Inorganic contaminants are not usually in crystalline form unless the contamination is associated with mining. Inorganic precipitates usually occur prior to crystallization, and aging generally promotes more ordered and larger crystalline minerals. Because many precipitates can form rapidly, such as iron hydroxides under aerobic conditions, recent metal contamination is often associated with this phase. The sorbed phase can also play an important role in metal availability (Miller et al., 1986).

Many mineral cycles in undisturbed soils and sediments tend to favor immobilization, usually with only trace quantities remaining in the interstitial water (Lerman and Brunskill, 1971). However, physical disturbance of soils and sediments can promote relatively rapid environmental changes, such as redox potential, which in turn can promote short-term mobilization of inorganic compounds (Lu and Chen, 1977).

Most contaminant heavy metals and other inorganic compounds can be transformed by microorganisms. These transformations depend primarily on biologically induced changes in redox potential, pH, and/or physical sorption. Changes in redox potential are promoted by the redox reactions that occur within or immediately surrounding microbial cells, mainly as a result of aerobic or anaerobic respiratory activities. Microorganisms change the pH of their environment as a result of using or releasing hydrogen or hydroxyl ions during biosynthesis and energy production, and by incorporating hydrogen or hydroxyl ions into inorganic and organic compounds. Microbial modification of nutrient cycles and carbonate equilibria are especially important in controlling pH. Sorption reactions can directly involve cell membranes or can result from adsorption and coprecipitation on microbially induced precipitates. Sorption and precipitation reactions can be passive (caused by changes in the environment) or active (involving enzyme binding or active transport across cell membranes). All of these transformation methods can be interrelated and often result from the activities of many microorganisms present in soils.

Microorganisms can promote the leaching of metals and other elements from soils by converting them from chemical forms having low solubility in soil water to forms having higher solubility. For example, oxidation of ferrous sulfides to ferric sulfate results in temporary mobilization of both iron and sulfur, as well as other elements associated with the solid phase, and is promoted by bacterial species such as *Thiobacillus* (Jeffers, 1991; Lizama and Suzuki, 1988; Lundgren and Silver, 1980). The mining industry has carried bioleaching research beyond the pilot-plant stage (Jeffers, 1991).

Immobilization of inorganic compounds can occur through (1) biosorption (sorption onto microbial cells), (2) bioaccumulation (microbial uptake and concentration from solution), and (3) conversion of inorganic elements to less-soluble forms. Many metals (e.g., mercury, copper, lead, cadmium, zinc, and iron) will tend to be immobilized by anaerobic conditions and activities of sulfate-reducing bacteria (e.g., *Desulfovibrio* and *Clostridium* spp.), which promote the formation of poorly soluble metal sulfides. However, sulfide formation can be slow, and as a result of reduction, metals coprecipitated with ferric hydroxides and oxides (e.g., arsenic, cadmium, cobalt, copper) may tend to become more mobile (Francis and Dodge, 1990).

Sorption of metals and other elements to soils is promoted and controlled primarily by the quantity and types of organic compounds present with the clay fraction. Sorption onto microbial cells in soils can also be important, because microbial biomass is an important component of organics in surface soils and sediments (Flemming et al., 1990; Walker et al., 1989). However, the use of microorganisms for removing inorganic contaminants from leachates and groundwaters has perhaps the greatest remediation potential, because recovery of the contaminants is possible (Beveridge, 1989; Flemming et al., 1990; Jeffers, 1991; Mullen et al., 1989).

Another mechanism for removing inorganic contaminants from soils is volatilization through microbial processes. Volatilization seems to be especially important for removing sulfur and nitrogen compounds from soils and water. Several potentially toxic metals and micronutrients, such as mercury and selenium, can also be microbially converted to volatile forms (Frankenberger and Karlson, 1991; Nakamura et al., 1990).

Bioleaching of Metals

Microorganisms play significant direct and indirect roles in bioleaching reactions by modifying the pH, redox potential, carbonate equilibria, and solubility of inorganic complexes. Complexation of metals with soluble organic compounds excreted by microorganisms (e.g., organic acids and amines) can help solubilize and prevent subsequent precipitation of metals and other inorganic contaminants (McKeague et al., 1986). The activities of several species of primarily chemoautotrophic *Thiobacillus* bacteria are of greatest importance in bioleaching.

T. ferrooxidans is capable of leaching various metals from sulfide minerals, especially those associated with iron sulfides. Both ferrous iron and sulfide are oxidized in the process by the addition of molecular oxygen under aerobic conditions. The rate-limiting step in iron sulfide oxidation is ferrous iron oxidation, which this bacterium accelerates a million-fold over abiotic rates (Singer and Stumm, 1970). The sulfide is eventually oxidized to soluble ferric sulfate, which promotes acidic conditions. Yet this bacterium thrives at pH values below 1.5, with an optimum pH of 2 (Drobner et al., 1990).

T. thiooxidans also aids in the oxidation of sulfide and other reduced sulfur compounds (Lizama and Suzuki, 1988; Lundgren and Silver, 1980). These bacteria are very versatile. For example, *T. ferroxidans* can use ferrous iron, sulfide, thiosulfate, and elemental sulfur as growth substrates for energy, while using carbon dioxide and other inorganic nutrients to produce cellular components (Lizama and Suzuki, 1988). However, transition from one growth substrate to another often requires long adaptation periods, and some strains use only iron and not sulfur compounds (Harrison, 1984). It has also been shown that *T. ferrooxidans* can reduce ferric iron to ferrous iron by using it as an electron acceptor for sulfide oxidation (Suzuki et al., 1990).

The rate of bioleaching by *T. ferrooxidans* is affected by several environmental factors. Lizama and Suzuki (1989) noted inhibition of ferrous iron oxidation by increased concentrations of accumulated ferric iron and bacterial cell biomass. However, this effect was significantly reduced when growth was at lower temperatures, such as 5°C (Kovalenko et al., 1982). Okereke and Stevens (1991) showed inhibition of iron sulfide oxidation at cell biomass concentrations above 0.6 mg/mL. Although low temperatures did not greatly impede reaction rates, higher temperatures accelerated iron oxidation, provided that cell concentrations were below the critical level.

Acidic ferric sulfate solutions resulting from iron sulfide oxidation by *T. ferrooxidans* can assist in commercial recovery of copper and uranium from wastes and mine tailings (Bruynesteyn, 1989). Coal wastes have also been treated by *T. ferrooxidans* to remove leachable sulfur (Monticello and Finnerty, 1985; Olson and Brinckman, 1986). A new moderate acidophilic *Thiobacillus*, named *T. cuprinus*, showed a preference for copper leaching from copper sulfides (Huber and Stetter, 1990), whereas an undescribed species was able to grow by oxidation of lead sulfide (Drobner et al., 1990). Other bacteria recognized for their bioleaching potential include *Thiobacillus prosperous* (Huber and Stetter, 1989) and *Leptospirillum ferrooxidans* (Norris, 1983).

Iron oxides can scavenge soluble transition and heavy metals in contaminated soils and leachates from waste sites. Coprecipitation of metal complexes with iron oxides is an important mechanism for metal immobilization, especially for manganese, arsenic, cadmium, cobalt, chromium, copper, mercury, nickel, lead, zinc, uranium, and selenium. However, various microorganisms, especially anaerobes, are capable of resolubilizing these coprecipitates. This mechanism could pose the problem of promoting metal mobility in contaminated soils, but it may also provide a method for bioleaching anthropogenic metal contamination. Lovley et al. (1991a) showed that enzymatic action by iron-reducing bacteria was responsible for most iron reduction, compared to abiotic effects of redox potential or indirect biological action of organic chelates. A species of anaerobic bacterium, *Clostridium*, solubilized cadmium, chromium, nickel, lead, and zinc that was coprecipitated with iron oxyhydroxide by both direct enzymatic reduction and indirect action due to metabolic products (Francis and Dodge, 1990).

Biopolymers capable of complexing specific metals have been isolated from a diverse group of microorganisms. For example, an exudate from the green alga, *Selenastrum capricornutum* strain Prinz, selectively complexes cadmium, and a biopolymer from the photosynthetic bacterium, *Ectothiorhodospira shaposhnikovii*, very effectively removes zinc from aqueous solution. Enzymes from the yeast *Candida utilis* effectively adsorb cadmium, mercury, and lead, and bioflavanol pigments immobilized in fibers are selective for uranium ion extraction. Polysaccharides excreted by a common sewage bacterium, *Zoogloea ramigera*, have been shown to promote the complexation and flocculation of metals. (*The Bioremediation Report*, 1991).

Bioimmobilizaton of Metals

Microorganisms have a high surface area/volume ratio and, as such, should have a great capacity to sorb metals passively. Bacterial cells seem to be even more efficient than clay minerals in sorbing metals from solution (Walker et al., 1989). Cadmium was sorbed more readily by dead bacterial cells than by montmorillonite clay (Kurek et al., 1982). Several other investigations have shown that large quantities of metal ions can be complexed by algal cells (Laube et al., 1979). This phenomenon has been used commercially, whereby dead algal cells are encased in silica gel polymeric material and the hard granule product used in ion-exchange columns for treating process waters. The granules protect the algal cells from biodegradation, and the system can be regenerated by desorption of the metals from the cell–granule complex by subsequent treatment. Mercury and uranium ions can be removed to very low levels (*ASM News*, 1991b).

Methods proposed for recovering metals from dead algae include changing the pH, adding salt to promote desorption and metal precipitation, and incinerating the metal-containing algal cells. These methods resulted from the observations that sorption of many metals is pH dependent and that increasing salt or particulate concentrations inhibit cell-binding efficiencies for most metals. Maximum binding of chromium by *Chlorella vulgaris* cells occurred at a pH of 3.9, whereas lead was removed most effectively at a pH of 6.3. Although most metals were not sorbed at pH values below 2, silver binding was independent of pH (Sneddon and Pappas, 1991).

Despite the limitations with algae cell sorption, bacteria have been shown to remove mercury and uranium ions to very low levels in water containing different ionic compositions and strengths and high organic residues (*ASM News*, 1991b).

Oak Ridge National Laboratory have developed a system for removing strontium from water that incorporates cells of a bacterium, *Micrococcus luteus*, into gelatin beads. The beads are used in columns, and after their sorption capacity is reached, they are incinerated (*Environment Today*, 1991c).

Microorganisms can adsorb or absorb contaminant metals by a variety of mechanisms, involving either active transport across cell membranes or passive sorption to cell surfaces. For example, several bacterial species were subjected to solutions containing 1 mmol quantities of different heavy metal ions. Although average removal efficiencies for silver, cadmium, copper, and lanthanum were 89, 12, 29, and 27%, respectively, specific bacteria were found to be most efficient in removing different metals, and by diverse mechanisms. Silver formed colloidal precipitates both on and within bacterial cells, whereas lanthanum formed crystalline precipitates on cell surfaces. Cadmium and copper were mainly adsorbed to cell surfaces (Mullen et al., 1989). Zinc can accumulate within cells of a bacterium, *Xanthomonas maltophilia*, probably as an amorphous potassium zinc phosphate precipitate (Sakurai et al., 1990).

Redox reactions facilitated by microorganisms can be used to immobilize some metals. For instance, soluble chromate ions can be reduced to less toxic and poorly soluble chromium hydroxide by the actions of certain bacteria. *Pseudomonas putida* PRS2000 was shown to promote this chromium reduction enzymatically (Ishibashi et al., 1990), but the microbial process that occurs under aerobic conditions involves a membrane-bound enzyme system, whereas a soluble intracellular enzyme appears to be involved in anaerobic environments (Bopp and Ehrlich, 1988; Gvozdyak et al., 1986). The use of microbes to remove uranium ions reductively by converting them to an insoluble form is being evaluated. An iron-reducing bacterium has recently been found capable of this reduction (Lovley et al., 1991b). The iron-oxidizing bacterium, *T. ferrooxidans*, has been used in a fixed-film bioreactor to oxidize ferrous sulfate to ferric sulfate. When the reactor pH was maintained between 1.35 and 1.5, iron precipitation problems were avoided. An activated carbon packed-bed reactor gave the best results, with 78 g of ferrous iron oxidized per liter-hour at a dilution rate of 40 per hour. This represents a hydraulic retention time of 1.5 min (Grishin and Tuovinen, 1988). Such a system could avoid the ferric hydroxide precipitate plugging problems often experienced with conventional iron removal methods.

Selenium is a micronutrient required by animals at very low concentrations, but it can be toxic and mutagenic at concentrations greater than 0.01 mg/L. Methods have been sought to decrease the concentrations of the more soluble and toxic selenate and selenite ions, which are favored in aerobic alkaline environments. One proposed method involves reducing selenate and selenite to poorly soluble elemental selenium and less toxic and soluble selenide compounds by anaerobic bacteria that use the oxidized compounds as terminal electron acceptors in respiration (Doran, 1982; Maiers et al., 1988; Steinberg and Oremland, 1990). Reduction rates in selenate-contaminated waters ranged from 0.07 to 22 μmol/L·h. Although sulfate and selenate are closely related, in many respects, respiratory sulfate reduction appears to involve different microorganisms and enzyme systems. Thus high sulfate concentrations, which are common at selenate-contaminated sites, were found to have only minor inhibitory effect. However, high nitrate concentrations significantly retarded reduction rates, suggesting a selenate and selenite reduction mechanism that is competitive with denitrification (Steinberg and Oremland, 1990). Bacterial selenate reduction in contaminated soils was found to be rapid, with rates estimated at 14 to 155 $\mu mol/m^2 \cdot day$ and varying greatly among different sites (Oremland et al., 1991b). Selenate concentrations of 100 mg/L were not toxic to the indigenous bacteria, which were naturally prevalent in most soils and sediments sampled. However, only 4% of the water samples showed selenate-reducing activity, indicating a potential for bioaugmentation in water requiring treatment. Microbial reduction of 100 mg of selenate per liter was complete after 1 week of incubation. Thus indigenous bacteria seem to play a significant role in the biogeochemical cycling of selenium (Maiers et al., 1988).

A biological process was developed and field tested that very effectively removed selenate and selenite from drainage waters that contained high concentrations of nitrate nitrogen. Because of the inhibitory effect of nitrate on selenate and selenite reduction, a dual process was designed. Initially, planktonic microalgae were grown in high-rate ponds containing selenium-contaminated drainage water. The algae successfully removed part of the inhibitory nitrate, which was reduced and used for algal biomass production. Then the drainage water–algae mixture was released to anoxic pits, where the decomposing algae and remaining nitrate were used to promote large populations of denitrifying bacteria. These bacteria then reduced selenate to selenite and insoluble elemental selenium during denitrification. Addition of ferric chloride to the effluent greatly decreased the soluble selenite concentration, possibly by coprecipitation on ferric hydroxide precipitate, which left final soluble selenium concentrations of about 10 µg/L. High sulfate concentrations did not inhibit the process (Gerhardt et al., 1991).

Volatilization of Metals and Metalloids

Many microorganisms have the capability of increasing the volatility of certain heavy and transition metals, probably as a means to decrease concentrations and toxicity in their environment. Microbial methylation reactions have been recognized for mercury, selenium, arsenic, antimony, tin, lead, tellurium, thorium, platinum, palladium, and gold (Cheng and Focht, 1979; Ehrlich, 1971; Reamer and Zoller, 1980; Wong et al., 1975; Woolson, 1977). However, only mercury and selenium volatilization are understood sufficiently well to have bioremediation potential at this time.

Mercury increases in volatility and bioaccumulation potential in higher organisms when it is methylated. Monomethylmercury is not readily volatilized, as it has strong affinity for soils, but dimethylmercury is moderately volatile (Wood, 1974). Mercury can be methylated under both aerobic and anaerobic conditions, but it appears to take place mainly in anaerobic environments (Regnell and Tunlid, 1991). Sulfate-reducing bacteria play a dominant role in mercury methylation (Compeau and Bartha, 1985, 1987), and thus free sulfides inhibit the reaction rate (Price, 1973). Acidic conditions (pH <5.5) and nitrate also seem to inhibit methylation (Steffan et al., 1988). However, elemental mercury is the most volatile form, and microbial demethylation of methylated mercury is a major source. Demethylation occurs under both aerobic and anaerobic conditions but is most prevalent under anaerobic conditions. The demethylation rate is slow for monomethylmercury, suggesting that anaerobes can only demethylate dimethylmercury (Oremland et al., 1991a). Demethylation reactions seem to be induced by long-term exposure to mercury contamination (Nakamura et al., 1990).

Selenium can be converted by microorganisms to several methylated compounds, especially dimethylselenide and dimethyldiselenide, which seem to have low toxicity. Unlike mercury, elemental selenium forms an insoluble precipitate (Reamer and Zoller, 1980). Dimethylselenide is the dominant methylated form, and although it readily sorbs to soils high in organic matter and clay (Zieve and Peterson, 1985), tilled soils and surface waters seem to show good release to the atmosphere (Thompson-Eagle and Frankenberger, 1990a; Frankenberger and Karlson, 1991). Aerobic conditions, high soluble selenium (selenate and selenite) concentrations, appropriate carbon amendments (glucose, pectin, and casein), and elevated temperatures (35 to 40°C) seem to promote the methylation process (Francis et al., 1974; Frankenberger and Karlson, 1991; Thompson-Eagle and Frankenberger, 1990b; Zieve and Peterson, 1981). Recent information indicates that orange peels plus inorganic nitrogen, zinc, cobalt (as methylcobalamin), and nickel supplementation promote volatilization. The main microorganisms responsible for methylation of selenium appear to be common mold fungi, including species of *Penicillium, Acremonium, Aspergillus, Ulocladium, Fusarium,* and *Alternaria.* These fungi prefer complex plant materials and proteins to simple carbon sources (Francis et al., 1974; Frankenberger and Karlson, 1991).

Cyanide Biodegradation

Microbial degradation of cyanide wastes is economically very attractive, especially when the waste streams contain high organic concentrations. Bioremediation of cyanide-containing wastes has been considered for many years, but most early treatment schemes involved aerobic microorganisms (Harris et al., 1987; Knowles and Bunch, 1986). Aerobic metabolic pathways proposed include hydrolysis to formamide or formate and ammonia (Fry and Millar, 1972; White et al., 1988), and the direct formation of bicarbonate plus ammonia via cyanide oxidation possibly by a dioxygenase enzyme system (Harris and Knowles, 1983). Aerobic processes have already been used in the field to treat natural waters contaminated with cyanide wastes, such as streams contaminated by ore-leaching operations (Jeffers, 1991).

Recent studies have shown that cyanide can be biodegraded effectively under anaerobic conditions. However, very low concentrations of free cyanide inhibit methanogenic bacteria (Fedorak et al., 1986; Yang and Speece, 1986). Fedorak and Hrudey (1989) reported anaerobic degradation of cyanide at 30 mg/L concentrations using semicontinuous batch cultures with a 25-day hydraulic retention time (time for 1 volume replacement). Methanogenesis was inhibited in the system, and the only recognized end product was bicarbonate.

Fallon et al. (1991) recently succeeded in degrading cyanide under methanogenic conditions using upflow anaerobic fixed-bed columns containing activated charcoal. The columns retained metabolic activity at free and weakly complexed cyanide concentrations of up to 300 mg/L and showed consistent long-term treatment at influent concentrations exceeding 100 mg/L. The charcoal was thought to be responsible for the high-capacity treatment by providing microsites for the anaerobic bacteria and sorption (and resultant lower toxicity) of the cyanide in the lower section of the columns. Methanogenesis and high cyanide degradation rates were maintained by adding ethanol, phenol, or methanol as the primary food source. Ammonia and bicarbonate were the major end products. Although anaerobic treatment of cyanide has economic value, the mechanisms of adaptation to and degradation of cyanide in anaerobic systems are not well understood.

REGULATORY IMPLICATIONS FOR BIOREMEDIATION

Federal, state, and local pollution control laws and regulations pose many challenges to pollutant cleanup. Since the mid-1970s, at least a dozen federal laws have been enacted to control the release of pollutants into the air, water, and soil. Certain of these statutes have not only promoted innovative treatment technologies for pollutant cleanup, but have also impeded their implementation. This has been especially true for bioremediation and use of biotechnology to remediate environmental contamination. Several of these laws and the resultant regulations, promulgated by the U.S. Environmental Protection Agency (USEPA) can affect the use of microorganisms to degrade toxic contaminants in the field or in closed systems. Their impacts are summarized below, and in greater detail in Bakst (1991).

The USEPA, under the authority of the Toxic Substances Control Act (TSCA), proposed in 1984 and 1986 to put controls on biotechnology by regulating genetically engineered microorganisms and microorganisms deemed to have a "significant new use." Genetically engineered microorganisms were defined as those resulting from combining genetic material from different taxonomic genera, and their production would have mandated in the proposed ruling extensive "premanufacture notification (PMN) requirements," similar to the requirements for new chemicals. Other microorganisms *could be* regulated through the "significant new use rule" governing the release of microorganisms to "new environments or different exposures." However, these proposals received considerable negative comment from industry, academia, and other government agencies. As a result, the USEPA has yet to publish a formal regulation governing biotechnology.

The recently introduced Omnibus Biotechnology Act, after passage, will mandate the USEPA and the U.S. Department of Agriculture (USDA) to promulgate biotechnology regulations within one year of enactment. It is doubtful, with present impetus, that bioremediation will be affected significantly by TSCA unless genetically modified microorganisms are employed. When biotechnology produces useful genetically engineered microorganisms for bioremediation purposes, it is anticipated that a PMN or permit would be required prior to their release. It is also likely that the PMN reporting requirements would be reduced in the future for certain microbial treatment processes conducted in enclosed systems such as bioreactors. Two trade organizations, the Association of Biotechnology Companies (ABC) and the Applied Biotreatment Association (ABTA), have formed to provide information to the USEPA and the U.S. Congress concerning the benefits of biotechnology for remediating polluted environments.

The Resource Conservation and Recovery Act (RCRA) has promoted on-site treatment of RCRA hazardous wastes through its permitting process, because remediation companies do not require a RCRA permit unless the firm itself receives and treats the hazardous waste. However, the RCRA land disposal restrictions (LDRs) have affected on-site bioremediation by including "placement of wastes on the land" under the definition of land disposal. This definition broadly restricts the placement of contaminated soils in "waste piles and surface impoundments." Because the LDRs remain to be defined clearly with reference to untreated and partially treated wastes, on-site bioremediation methods such as landfarming are often overlooked for treating contaminated soils. Treatment standards must be established for all RCRA-regulated wastes, based on best demonstrated available technologies (BDATs).

Although on-site bioremediation has not been excluded as a treatment option under the LDRs, many remediation companies decline bioremediation in favor of more proven existing technologies because they fear that the mandated RCRA cleanup standards will not be met.

In-situ bioremediation activities have also been affected by the RCRA LDRs. As presently defined, RCRA restricts the reinjection of contaminated groundwater unless (1) it is done as part of a RCRA corrective action (2) the contaminated groundwater is treated to "substantially reduce" hazardous constituents prior to such reinjection, and (3) the corrective action is "sufficient" to protect human health and the environment upon completion (DEVO Enterprises, 1991a). Presently, there are about 3700 RCRA facilities that may need corrective action and more than 1 million potentially leaking underground storage tanks in the United States (DEVO Enterprises, 1991b). Thus clarification of the LDRs is required and appears imminent because of the high cost of presently available technologies.

Although RCRA regulates the management, handling, and disposal practices for hazardous and nonhazardous wastes, the Comprehensive Environmental Response, Compensation and Liability Act (CERCLA or Superfund) deals with the cleanup of toxic and hazardous substances at closed or abandoned waste disposal sites. CERCLA establishes a preference for on-site treatment methods, especially in situ methods that permanently reduce the toxicity and mobility of wastes at cleanup sites. More than 1200 existing sites have been deemed sufficiently hazardous to warrant federal funding for cleanup. These sites have been placed on the National Priorities List (NPL) (DEVO Enterprises, 1991b). However, CERCLA requires that site cleanups comply legally with "applicable or relevant and appropriate requirements" (ARARs) of all federal and more stringent state environmental laws. The use of ARARs, which are criteria involved in choosing a specific cleanup remedy for a CERCLA site, can dissuade the use of bioremediation in a manner similar to the RCRA LDRs. CERCLA corrective actions may also be governed by the RCRA LDRs, especially because most CERCLA wastes are also RCRA wastes. Although the LDRs prevent the placement of wastes on the land, neither leaving restricted wastes in place (e.g., in situ bioremediation) nor moving wastes within an "area of contamination" (e.g., on-site bioremediation) constitutes a placement by CERCLA definition. However, the definition of an "area of contamination" is vague and includes any area at which solid wastes have been "routinely and systematically" placed, and is determined further by the area of continuous contamination. The initial purpose of the LDRs was to promote on-site treatment instead of treatment at a treatment, storage, and disposal facility (TSDF). Certainly a clearer picture is needed of the interrelationship between CERCLA and RCRA so as not to impede the use of bioremediation as a cleanup option.

Several other laws could affect bioremediation activities. The Federal Plant Pest Act, under USDA jurisdiction, regulates the importation and interstate transport of plant pests (i.e., any biological agents that directly or indirectly cause damage to plants or plant parts). Because this definition encompasses a large group of microorganisms, this law could affect bioaugmentation activities and those involving the use of genetically engineered microorganisms. This law may also require permitting from both the USDA and the USEPA when microorganisms not native to a site are involved in a treatment scheme. The Clean Water Act (CWA), Safe Drinking Water Act (SDWA), and Clean Air Act (CAA) are involved in bioremediation activities that discharge contaminants to surface waters, to groundwater, or to the air. Appropriate permits must be obtained to construct and operate potential sources of water and air pollution, which are usually dispensed and monitored by state regulatory agencies.

CERCLA AND RCRA ACTION LEVELS

RCRA solid waste management units and CERCLA areas of contamination require certain cleanup action levels that must be attained before a treatment is considered completed. However, the USEPA does not require the same cleanup levels for all sites. The cleanup action levels are established on a site-specific basis by each USEPA regional administrator. Although the primary goal is that contaminants be treated to the most stringent levels (e.g., water quality standards, if they exist for a contaminant) or a level within a protective risk range (e.g., one human impact in a population of 1 million people), cleanups generally require only that final contaminant levels be "protective of current or future use" of a site. For example, soils and surface waters that are expected to come into contact with human beings would require more stringent cleanup requirements than those planned for industrial sites or uses. Groundwater is generally considered potentially potable, and thus remediation requirements seem to be those established for drinking water irrespective of its actual final use (Bakst 1991).

The USEPA has proposed four criteria that must be considered initially in selecting a treatment remedy. The chosen technology must (1) be protective of human health and the environment, (2) attain a prespecified cleanup standards, (3) control releases so as to reduce or eliminate future releases, and (4) comply with certain standards for management of wastes. If a technology meets these rather vague criteria, the USEPA will then consider five general factors in selecting a remediation technology: (1) long-term reliability and effectiveness of the remedy; (2) reduction of toxicity, mobility, or volume of the wastes; (3) short-term effectiveness; (4) implementability; and (5) cost (Bakst, 1991). Such criteria were meant to be protective of the environment but also strongly promote the use of existing and proven technologies over less proven innovative technologies, even though preliminary data may indicate that a new remedy could be more effective at a considerably lower cost.

INNOVATIVE TECHNOLOGY DEVELOPMENT

The USEPA has financed several programs to develop more effective and economical treatment methods for contaminated soils and groundwater. Two programs are presently funded: the Superfund Innovative Technology Evaluation (SITE) and the Small Business Innovative Research (SBIR) programs. The SITE program has two components: the emerging technologies component, which promotes development of promising bench-scale technologies not yet field tested, and the demonstration technology component, which supports performance and cost evaluations of demonstration field studies not yet ready for commercial use. Emerging technologies can receive up to $150,000 per year for each technology. For the demonstration-scale evaluations, the vendors are usually responsible for the general demonstration costs, while USEPA pays for the scientific evaluation costs. More detailed information is given in Bates et al. (1989).

The SBIR grant program is presently small, but future financial support from the U.S. Department of Energy could make it an important source of funds for testing innovative remediation technologies. California has a similar program, and other states may soon follow. These programs have not as yet given strong support to bioremediation projects, especially in situ methods. However, the number is growing with recent interest in this subject (*Hazmat World*, 1991).

The USEPA, through its Technology Innovation Office, is developing an extensive database with information on more than 500 cleanup projects made accessible through the Alternate Treatment Technology Information Center (ATTIC) computer system. This information will be used to determine the most appropriate methods for Superfund cleanups but should be available to outside sources (*Environment Today*, 1991a). A new USEPA database that should be available in 1992 is the Vendor Information System for Innovative Treatment Technologies

(VISITT). It contains extensive information on vendors and their innovative techniques for treating soil, groundwater, sediments, sludge, and solid wastes, including bioremediation. This database will not be online but will be available on diskette and updated annually. VISITT can be accessed by calling a hotline number, 800/245-4505 (*Environment Today*, 1991b).

SUMMARY AND CONCLUSIONS

Bioremediation is a technology that has been used by human beings for many centuries for removing organic wastes. The recent interest in cleanup of toxic organic contaminants and heavy metals in soils and groundwater has driven the development of innovative bioremediation methods. Biodegradation or biotransformation of these contaminants involves the interaction of many different microorganisms working together or sequentially. Bacteria and fungi are the most important bioactive groups, but their activities are often governed by the presence of other soil or aquatic microorganisms and the environmental conditions they promote. Microorganisms have a profound influence on soil and aquatic environmental conditions, such as pH and redox potential. Changes in these environmental factors can, in turn, have a profound influence on the types of microorganisms present and their metabolic activities.

Surface soils have long been known to contain high numbers of microorganisms, usually greater than 1 million per gram of soil. However, recent information has shown that deep subsurface soils and groundwater aquifers also contain high bacterial populations having diverse metabolic potentials. This knowledge has resulted from new techniques for measuring microbial numbers, biomass, and activity. The importance of community structure in promoting biodegradation potential is also becoming realized as basic research is directed to field cleanup activities.

Frequently, biodegradation of a single organic compound requires the interactions of different groups of microorganisms. Human-made organic compounds, such as chlorinated solvents, are often biodegraded only in the presence of other, often specific, food sources. Such cometabolism is apparently required because human beings have produced new exotic compounds faster than metabolic pathways have evolved for their ultimate destruction. Although cometabolism frequently produces toxic intermediate metabolites that can eventually decrease degradation rates, new engineering methods are being developed to overcome many of these problems.

Enhanced biodegradation involves the input of electron acceptors, nutrients, microorganisms, or growth factors that can either reduce the period required before the desired degree of degradation is attained, or increase degradation rates. Changing environmental conditions in the subsurface can also improve degradation rates. Although environmental manipulations can readily be accomplished in aboveground treatment systems, similar changes can be difficult in situ. Future techniques may allow for environmental manipulations by promoting the growth and activities of specific soil microorganisms. However, today's technology for in situ bioremediation in limited. Numerous problems can arise with mass transport and interactions of chemicals added for in situ bioremediation. For example, the addition of nitrate to treatment sites not only supplies a needed nutrient, nitrogen, but also affects redox potential in the soil and creates conditions that favor a specific group of bacteria, the denitrifiers.

The complete biodegradation of petroleum hydrocarbons found in fuels and crude oil appears to require aerobic soil conditions. Although aeration of aboveground treatment units is manageable, subsurface in situ aeration has become a major challenge. This is because water has usually been the medium for transporting oxygen, despite its extremely poor oxygen-carrying capacity. To overcome this, hydrogen peroxide has been used as the transport vehicle because it is infinitely soluble in water and potentially can degrade to 50% of its weight as molecular oxygen. Unfortunately, hydrogen peroxide lacks sufficient stability at most sites to be a cost-effective oxygen additive. Alternatively, nitrate has been added to aquifers to stimulate denitrifiers, which are capable of degrading the most toxic components in fuels. Nitrate also provides a stable source of oxygen that migrates with the fuel plume.

Recent research has shown that soil venting can rapidly aerate the unsaturated zone, which is where the majority of fuel contamination resides. This technique, coined *bioventing*, appears to promote rapid in situ biodegradation of most fuel and may also be useful for promoting the degradation of other compounds that readily biodegrade under aerobic conditions. Bioventing may also facilitate microbial interaction with contaminants not soluble in water.

Water does not readily mix or penetrate soils containing high concentrations of hydrophobic organic compounds. Surfactants have been used to improve water penetration and surface area/volume ratio of contaminants, which provides for greater interaction of contaminant molecules with microbial cells. Unfortunately, surfactant transport in situ is poorly understood and is not currently in practice. Biosurfactants are produced by many bacteria and at the most effective concentrations.

The biodegradation of halogenated organic compounds involves many complex mechanisms and often very specific groups of microorganisms. Anaerobic conditions and the presence of specific anaerobic bacteria seem to be required to degrade some highly halogenated (e.g., chlorinated) compounds, such as carbon tetrachloride, perchloroethylene, and PCBs containing more than two or three chlorines (e.g., Aroclor 1254 and 1260). Unfortunately, anaerobic bacteria degrade the less-chlorinated compounds with difficulty, resulting in the potential accumulation of toxic intermediate compounds. However, specific aerobic bacteria can degrade these less-chlorinated compounds. For example, methanotrophs, autotrophic nitrifiers, and toluene degraders are capable of effectively degrading trichloroethylene (TCE) and less-chlorinated compounds. Thus sequential anaerobic–aerobic conditions may prove best for degrading many chlorinated solvents. Although such processes may prove straightforward in bioreactor systems, great ingenuity may be required to promote such sequential changes in situ.

Aromatic compounds with fused multiple rings (polycyclic aromatic compounds or PAHs) are generally more difficult to degrade than are single-ringed aromatic compounds. However, most PAHs having fewer than four rings are readily biodegraded under aerobic conditions and without alternative food sources. Some PAHs can also be biodegraded under anaerobic, especially denitrifying, conditions, but the degradation rates are reduced. Creosote wastes often present additional problems in that they are often mixed with heavily chlorinated compounds, such as pentachlorophenol (PCP), that degrade most readily in soils under anaerobic conditions. Such wastes are also often contaminated with metals that are highly toxic to most microorganisms. Potential methods for treating these mixed wastes could include anaerobic–aerobic sequential treatment or use of specific aerobic microorganisms (e.g., white rot fungi and *Flavobacterium* spp.).

Most nitrated organic compounds containing only one or two nitro groups are readily biodegraded. However, a similarity with chlorinated compounds seems to exist in that the highly nitrated compounds are not readily degraded under aerobic conditions. Munitions compounds, such as TNT, are not easily biodegraded, but specific aerobic (e.g., white rot fungi) and anaerobic microorganisms seem to have good biodegradative potential. Both oxidative and reductive mechanisms have been shown to be involved in the degradation of nitrated compounds, but additional research is required with most highly nitrated compounds.

Metal and other inorganic compounds can be either actively or passively transformed by microorganisms. In fact, microbial processes are of major importance in recycling, mobilizing, or immobilizing most metals, metalloids, and inorganic nutrients. Active processes involve active transport across cell membranes, which is usually promoted by enzyme action. Passive processes result from changes in physicochemical conditions in soils that result from microbial growth and metabolism. Bioleaching has already gained respectability in the mining industry, and the techniques are now branching into environmental remediations. Bioimmobilization can result from sorption on or within cells, or by converting inorganic contaminants to geochemical forms having lower potential mobility in soils, such as could occur by changing pH, redox potential, nutrient forms, or carbonate equilibria. Microbially induced volatilization of specific inorganic contaminants, such as mercury and selenium, has also been considered and may be ready for practical use.

Although bioremediation has gained considerable attention lately, several obstacles remain before microorganisms can be used effectively for detoxifying wastes affecting soils and groundwater. Regulatory impediments appear to remain, despite recent changes that should actually promote such innovative, on-site treatment methods. A lack of knowledge, or incorrect knowledge concerning what can and cannot be done with microbial treatment schemes, also has delayed progress. Many unscrupulous salespersons have taken advantage of the misunderstandings and misinformation to push products that appear to be and frequently are too good to be true. This has resulted in unrealistic expectations leading to disappointment and has set back the progress of bioremediation. If the USEPA's goals for cleanup are to be realized, regulatory obstacles must be overcome.

REFERENCES

Adriaens, P., and D. D. Focht. 1990. Continuous coculture degradation of selected polychlorinated biphenyl cogeners by *Acinetobacter* spp. in an aerobic reactor system. *Environ. Sci. Technol.* 24:1042–1049.

Adriaens, P., and D. D. Focht. 1991. Cometabolism of 3,4-dichlorobenzoate by *Acinetobacter* sp. strain 4-CB 1. *Appl. Environ. Microbiol.* 57:173–179.

Aelion, C. M., M. C. Swindoll, and F. K. Pfaender. 1987. Adaptation to the biodegradation of xenobiotic compounds by microbial communities from a pristine aquifer. *Appl. Environ. Microbiol.* 53:2212–2217.

Aggarwal, P. K., J. L. Means, and R. E. Hinchee, 1991. Formulation of nutrient solutions for in situ bioremediation, pp. 51–66. In: R. E. Hinchee and R. F. Olfenbuttel (eds.), *In Situ Bioreclamation: Applications and Investigations for Hydrocarbon and Contaminated Site Remediation*. Butterworth-Heinemann, Boston.

Alexander, M. 1977. *Introduction to Soil Microbiology*, 2nd ed. Wiley, New York.

Allard, A.-S., M. Remberger, and A. H. Nielson. 1987. Bacterial O-methylation of halogen-substituted phenols. *Appl. Environ. Microbiol.* 53:839–845.

Alvarez, P. J. J., and T. M. Vogel. 1991. Substrate interactions of benzene, toluene, and *para*-xylene during microbial degradation by pure cultures and mixed culture aquifer slurries. *Appl. Environ. Microbiol.* 57:2981–2985.

Alvarez-Cohen, L., and P. L. McCarty. 1991. Product toxicity and cometabolic competitive inhibition modeling of chloroform and trichloroethylene transformation by methanotrophic resting cells. *Appl. Environ. Microbiol.* 57:1031–1037.

Anid, P. J., L. Nies, and T. M. Vogel. 1991. Sequential anaerobic-aerobic biodegradation of PCBs in a river model, pp. 428–436. In: R. E. Hinchee and R. F Olfenbuttel (eds.), *On-Site Bioreclamation: Processes for Xenobiotic and Hydrocarbon Treatment.* Butterworth-Heinemann, Boston.

Apajalahti, J. H., and M. S. Salkinoja-Salonen. 1986. Degradation of polychlorinated phenols by *Rhodococcus chlorophenolicus. Appl. Microbiol. Biotechnol.* 25:62–67.

Arciero, D., T. Vannelli, M. Logan, and A. B. Hooper. 1989. Degradation of trichloroethylene by the ammonia-oxidizing bacterium *Nitrosomonas europaea. Biochem. Biophys. Res. Commun.* 159:640–643.

Arthur, M. F., G. K. O'Brien, S. S. Marsh, and T. C. Zwick. 1989. *Evaluation of Innovative Approaches to Stimulate Degradation of Jet Fuels in Subsoils and Groundwater.* Battelle, Columbus, OH. Report to Naval Civil Engineering Laboratory, Port Hueneme, CA, August 1989. U.S. Army Research Office/TCN-88409.

ASM News. 1990. Environmental biogeochemistry meeting report. 56:523–524.

ASM News. 1991a. EPA fortifies bioremediation plans. 57:447–448.

ASM News. 1991b. Microbes help clean heavy metals. 57:296.

Atlas, R. M. 1977. Stimulated petroleum biodegradation. *Crit. Rev. Microbiol.* 5:371–386.

Atlas, R. M. 1981. Microbial degradation of petroleum hydrocarbons: an environmental perspective. *Microbiol. Rev.* 45:180–209.

Atlas, R. M. (ed.). 1984. Petroleum microbiology. Macmillan Publ. Co., New York.

Atlas, R. M. 1991. Bioremediation of fossil fuel contaminated soils, pp. 14–32. In: R. E. Hinchee and R. F. Olfenbuttel (eds.), *In Situ Bioreclamation: Applications and Investigations for Hydrocarbon and Contaminated Site Remediation.* Butterworth-Heinemann, Boston.

Atlas, R. M., and R. Bartha. 1972. Biodegradation of petroleum in seawater at low temperatures. *Can. J. Microbiol.* 18:1851–1855.

Aust, S. D. 1990. Degradation of environmental pollutants by *Phanerochaete chrysosporium. Microb. Ecol.* 20:197–209.

Bagley, D. M., and J. M. Gossett. 1990. Tetrachloroethene transformation to trichloroethene and *cis*-1,2,-dichloroethene by sulfate-reducing enrichment cultures. *Appl. Environ. Microbiol.* 56:2511–2516.

Bailey, R. E., S. J. Gonsior, and W. L Rhinehart. 1983. Biodegradation of the monochlorobiphenyls and biphenyl in river water. *Environ. Sci. Technol.* 17:617–621.

Bakst, J. S. 1991. Impact of present and future regulations on biotechnology and toxic waste degradation. *J. Ind. Microbiol.* 8:13–22.

Balkwill, D. L., and W. C. Ghiorse. 1985. Characterization of subsurface bacteria associated with two shallow aquifers in Oklahoma. *Appl. Environ. Microbiol.* 50:580–588.

Balkwill, D. L., F. R. Leach, J. T. Wilson, J. F. McNabb, and D. C. White. 1988. Equivalence of microbial biomass measures based on membrane lipid and cell wall components, adenosine triphosphate, and direct counts in subsurface aquifer sediments. *Microb. Ecol.* 16:73–84.

Ball, H. A., M. Reinhard, and P. L. McCarty. 1991. Biotransformation of monoaromatic hydrocarbons under anoxic conditions, pp. 458–463. In: R. E. Hinchee and R. F. Olfenbuttel (eds.), *In Situ Bioreclamation: Applications and Investigations for Hydrocarbon and Contaminated Site Remediation*. Butterworth-Heinemann, Boston.

Barcelona, M. J., and T. R. Holm. 1991. Oxidation–reduction capacities of aquifer solids. *Environ. Sci. Technol.* 25:1565–1572.

Barrio-Lage, G., F. Z. Parsons, R. S. Nassar, and P. A. Lorenzo. 1986. Sequential dehalogenation of chlorinated ethenes. *Environ. Sci. Technol.* 20:96–99.

Barrio-Lage, G. A., F. Z. Parsons, R. M. Narbaitz, P. A. Lorenzo, and H. E. Archer. 1990. Enhanced anaerobic biodegradation of vinyl chloride in ground water. *Environ. Toxicol. Chem.* 9:403–415.

Bartels, I., H.-J. Knackmuss, and W. Reineke. 1984. Suicide inactivation of catechol-2,3-dioxygenase from *Pseudomonas* mt-2 by 3-halocatechols. *Appl. Environ. Microbiol.* 47:500–505.

Baskin, Y. 1987. Testing engineered microbes in the field. *ASM News* 53:611–614.

Bates, E. R., J. G. Herrmann, and D. E. Sanning. 1989. The U.S. Environmental Agencies SITE Emergency Technology Program. *J. Air Pollut. Control Assoc.* 39:927–935.

Battermann, G. 1986. Decontamination of polluted aquifers by biodegradation, pp. 711–722. In: J. W. Assink and W. J. van den Brink (eds.), *1985 International TNO Conference on Contaminated Soil*. Martinus Nijhoff, Dordrecht, The Netherlands.

Bauer, J. E., and D. G. Capone. 1985. Degradation and mineralization of the polycyclic aromatic hydrocarbons anthracene and naphthalene in intertidal marine sediments. *Appl. Environ. Microbiol.* 50:81–90.

Bedard, D. L., R. Untermann, L. H. Bopp, M. J. Brennan, M. L. Haberl, and C. Johnson. 1986. Rapid assay for screening and characterizing microorganisms for the ability to degrade polychlorinated biphenyls. *Appl. Environ. Microbiol.* 51:761–768.

Bedard, D. L., R. E. Wagner, M. J. Brennan, M. L. Haberl, and J. F. Brown, Jr. 1987. Extensive degradation of Aroclors and environmentally transformed polychlorinated biphenyls by *Alcaligenes eutrophus* H850. *Appl. Environ. Microbiol.* 53:1094–1102.

Bej, A. K., R. J. Steffan, J. DiCesare, L. Haff, and R. M. Atlas. 1990. Detection of coliform bacteria in water by polymerase chain reaction and gene probes. *Appl. Environ. Microbiol.* 56:307–314.

Bell, R. A., and A. H. Hoffman. 1991. Gasoline spill in fractured bedrock addressed with in situ bioremediation, pp. 437–443. In: R. E. Hinchee and R. F. Olfenbuttel (eds.), *In Situ Bioreclamation: Applications and Investigations for Hydrocarbon and Contaminated Site Remediation*. Butterworth-Heinemann, Boston.

Beloin, R. M., J. L. Sinclair, and W. C. Ghiorse. 1988. Distribution and activity of microorganisms in subsurface sediments of a pristine study site in Oklahoma. *Microb. Ecol.* 16:85–97.

Bennedsen, M. B., J. P. Scott, and J. D. Hartley. 1987. Use of vapor extraction systems for in situ removal of volatile organic compounds from soil. *Proceedings of the Conference on Hazardous Wastes and Hazardous Materials*, Washington, DC, March 1987, pp. 92–95.

Berry, D. F., A. J. Francis, and J.-M. Bollag. 1987. Microbial metabolism of homocyclic and heterocyclic aromatic compounds under anaerobic conditions. *Microbiol. Rev.* 51:43–59.

Berry-Spark, K. L., J. F. Barker, D. Major, and C. I. Mayfield. 1986. Remediation of gasoline-contaminated ground-waters: a controlled experiment. *Proceedings of Petroleum Hydrocarbons and Organic Chemicals in Ground Water: Prevention, Detection, and Restoration*, NWWA/API. Water Well Journal Publishing, Dublin, OH.

Beveridge, T. J. 1989. Role of cellular design in bacterial metal accumulation and mineralization. *Annu. Rev. Microbiol.* 43:147–171.

Blum, D. J., and R. E. Speece. 1991. A database of chemical toxicity to environmental bacteria and its use in interspecies comparisons and correlations. *Res. J. Water Pollut. Control Fed.* 63:198–207.

Blumer, M. 1967. Equilibria and nonequilibria in organic geochemistry. *Adv. Chem. Ser.* 67:312–318.

Bollag, J.-M. 1983. Cross-coupling of humus constituents and xenobiotic substances, pp. 127–141. In: R. F. Christman and E. T. Gjessing (eds.), *Aquatic and Terrestrial Humic Materials*. Ann Arbor Science, Ann Arbor, MI.

Bone, T. L., and D. L. Balkwill. 1988. Morphological and cultural comparison of microorganisms in surface soil and subsurface sediments at a pristine study site in Oklahoma. *Microb. Ecol.* 16:49–64.

Bopp, L. H. and H. L. Ehrlich. 1988. Chromate resistance and reduction in *Pseudomonas fluorescens* strain LB300. *Arch. Microbiol.* 150:426–431.

Borden, R. C., and P. B. Bedient. 1986. Transport of dissolved hydrocarbons influenced by reaeration and oxygen limited biodegradation. I. Theoretical development. *Water Resour. Res.* 22:1973–1982.

Bosma, T. N. P., J. R. van der Meer, G. Scraa, M. E. Tros, and A. J. B. Zehnder. 1988. Reductive dechlorination of all trichloro- and dichlorobenzene isomers. *FEMS Microbiol. Ecol.* 53:223–229.

Bossert, I., and R. Bartha. 1984. The fate of petroleum in soil ecosystems, pp. 434–476. In: R. M. Atlas (ed.), *Petroleum Microbiology*. Macmillan, New York.

Bouwer, E. J., and P. L. McCarty. 1983. Transformations of halogenated organic compounds under denitrification conditions. *Appl. Environ. Microbiol.* 45:1295–1299.

Bouwer, E. J., and P. L. McCarty. 1984. Modeling of trace organics biotransformation in the subsurface. *Groundwater* 22:433–440.

Bouwer, E. J., P. L. McCarty. 1985. Utilization rates of trace halogenated organic compounds in acetate-grown biofilms. *Biotechnol. Bioeng.* 27:1564–1571.

Bouwer, E. J., and J. P. Wright. 1988. Transformations of trace halogenated aliphatics in anoxic biofilm columns. *J. Contam. Hydrol.* 2:155–169.

Bouwer, E. J., B. E. Rittmann, and P. L. McCarty. 1981. Anaerobic degradation of halogenated 1- and 2-carbon organic compounds. *Environ. Sci. Technol.* 15:596–599.

Boyd, S. A., and D. R. Shelton. 1984. Anaerobic degradation of chlorophenols in fresh and acclimated sludge. *Appl. Environ. Microbiol.* 47:272–277.

Britton, L. N. 1984. Microbial degradation of aliphatic hydrocarbons, pp. 89–129. In: D. T. Gibson (ed.), *Microbial Degradation of Organic Compounds*. Marcel Dekker, New York.

Britton, L. N. 1985. *Feasibility Studies on the Use of Hydrogen Peroxide to Enhance Microbial Degradation of Gasoline*. API Publication 4389. Texas Research Institute, Washington, DC.

Britton, L. N. 1989. *Aerobic Denitrification as an Innovative Method for In-Situ Biological Remediation of Contaminated Subsurface Sites*. Report ESL-TR-88-40. Air Force Engineering Services Center, Tyndall, Air Force Base, Panama City, FL.

Broholm, K., B. K. Jensen, T. J. Christensen, and L. Olsen. 1990. Toxicity of 1,1,1-trichloroethane and trichloroethene on a mixed culture of methane-oxidizing bacteria. *Appl. Environ. Microbiol.* 56:2488–2493.

Brown, R. A., and K. Sullivan. 1991. Integrating technologies enhances remediation. *Pollut. Eng.* 23(5):63–68.

Brown, J. F., Jr., R. E. Wagner, H. Feng, D. L. Bedard, M. J. Brennan, J. C. Carnahan, and R. J. May. 1987. Environmental dechlorination of PCBs. *Environ. Toxicol. Chem.* 6:579–593.

Brown, R. A., J. C. Dey, and W. E. McFarland. 1991. Integrated site remediation combining groundwater treatment, soil vapor extraction, and bioremediation, pp. 444–449. In: R. E. Hinchee and R. F. Olfenbuttel (eds.). *In Situ Bioreclamation: Applications and Investigations for Hydrocarbon and Contaminated Site Remediation*. Butterworth-Heinemann, Boston.

Brunner, W., F. H. Sutherland, and D. D. Focht. 1985. Enhanced biodegradation of polychlorinated biphenyls in soil by analog enrichment and bacterial inoculation. *J. Environ. Qual.* 14:324–328.

Brusseau, G. A., H.-C. Tsien, R. S. Hanson, and L. P. Wackett. 1991. Optimization of trichloroethylene oxidation by methanotrophs and the use of a colorimetric assay to detect soluble methane monooxygenase activity. *Biodegradation* 1:19–30.

Bruynesteyn, A. 1989. Mineral biotechnology. *J. Biotechnol.* 11:1–10.

Bryant, F. O., D. D. Hale, and J. E. Rogers. 1991. Regiospecific dechlorination of pentachlorophenol by dichlorophenol-adapted microorganisms in freshwater, anaerobic sediment slurries. *Appl. Environ. Microbiol.* 57:2293–2301.

Bumpus, J. A. 1989. Biodegradation of polycyclic aromatic hydrocarbons by *Phanerochaete chrysosporium*. *Appl. Environ. Microbiol.* 55:154–158.

Bumpus, J. A., and S. D. Aust. 1987. Biodegradation of environmental pollutants by the white rot fungus *Phanerochaete chrysosporium*: involvement of the lignin degrading system. *BioEassays* 6:166–170.

Bumpus, J. A., M. Tien, D. Wright, and S. D. Aust. 1985. Oxidation of persistent environmental pollutants by a white rot fungus. *Science* 228: 1434–1436.

Carlson, R. E. 1981. *The Biological Degradation of Spilled Jet Fuels: A Literature Review*. Report to USAF/AFESC, DTIC/AD-A110758. Tyndall Air Force Base, Panama City, FL.

Carpenter, D. F., N. G. McCormick, J. H. Cornell, and A. M. Kaplan. 1978. Microbial transformation of ^{14}C-labeled 2,4,6-trinitrotoluene in an activated sludge system. *Appl. Environ. Microbiol.* 35:949–954.

Castro, C E., and N. O. Belser. 1990. Biodehalogenation: oxidative and reductive metabolism of 1,1,2-trichloroethane by *Pseudomonas putida*—biogeneration of vinyl chloride. *Environ. Toxicol. Chem.* 9:707–714.

Chambers, et al. 1990. *Handbook on In Situ Treatment of Hazardous Waste-Contaminated Soils*. EPA-540-2-90-002. U.S. Environmental Protection Agency, Cincinnati, Ohio.

Chapelle, F. H., and D. Lovley. 1990. Microbial metabolism in deep coastal plain aquifers. *Appl. Environ. Microbiol.* 56:1865–1874.

Chaudhry, G. R., and S. Chapalamadugu. 1991. Biodegradation of halogenated organic compounds. *Microbiol. Rev.* 55:59–79.

Chemical Week. 1988. Bacteria show promise in soil cleanup. May 4, 1988, p. 43.

Cheng, C. N., and D. D. Focht. 1979. Production of arsine and methylarsine in soil and in culture. *Appl. Environ. Microbiol.* 38:494–498.

Cheng, H. H., K. Haider, and S. S. Harper. 1983. Catechol and chlorocatechols in soil: degradation and extractability. *Soil Biol. Biochem.* 15:311–317.

Chiang, C. Y., J. P. Salanitro, E. Y. Chai, J. D. Colthart, and C. L. Klein. 1989. Aerobic biodegradation of benzene, toluene, and xylene in a sandy aquifer: data analysis and computer modeling. *Ground Water* 27:823–834.

Chiou, C. T., and T. D. Shoup. 1985. Soil sorption of organic vapors and effects of humidity on sorptive mechanism and capacity. *Environ. Sci. Technol.* 19:1196–1200.

Chu, J. P., and E. J. Kirsch. 1972. Metabolism of pentachlorophenol by an axenic bacterial culture. *Appl. Microbiol.* 23:1033–1035.

Clark, R. R., E. S. K. Chian, and R. A. Griffin. 1979. Degradation of polychlorinated biphenyls by mixed microbial cultures. *Appl. Environ. Microbiol.* 37:680–685.

Cline, P. V., J. F. Delfino, and P. S. C. Rao. 1991. Partitioning of aromatic constituents into water from gasoline and other complex solvent mixtures. *Environ. Sci. Technol.* 25:914–920.

Cobb, G. D., and E. J. Bouwer. 1991. Effects of electron acceptors on halogenated organic compound biotransformations in a biofilm column. *Environ. Sci. Technol.* 25:1068–1074.

Collins, M. L. P., L. A. Buchholz, and C. C. Remsen. 1991. Effect of copper on *Methylomonas albus* BG8. *Appl. Environ. Microbiol.* 57:1261–1264.

Compeau, G. C., and R. Bartha. 1985. Sulfate-reducing bacteria: principal methylators of mercury in anoxic estuarine sediment. *Appl. Environ. Microbiol.* 50:498–502.

Compeau, G. C., and R. Bartha. 1987. Effect of salinity on mercury-methylating activity of sulfate-reducing bacteria in estuarine sediments. *Appl. Environ. Microbiol.* 53:261–265.

Connor, J. R. 1988. Case study of soil venting. *Pollut. Eng.* 20:74–78.

Contreras, A., S. Molin, and J.-L. Ramos. 1991. Conditional-suicide containment system for bacteria which mineralize aromatics. *Appl. Environ. Microbiol.* 57:1504–1508.

Cooksey, K. E. 1991. Biotechnology-immobilized cell research. *ESN Information Bulletin.* Office of Naval Research, European Office, No. 91-01, pp. 5–14.

Cooney, J. J., S. A. Silver, and E. A. Beck. 1985. Factors influencing hydrocarbon degradation in three freshwater lakes. *Microb. Ecol.* 11:127–137.

Craig, M. A. 1991. *Biotransformation of Ordnance Wastes Using Unique Consortia of Anaerobic Bacteria: A Proposal to the U.S. Navy.* Oregon State University, Corvallis, OR.

Crawford, R. L., and W. W. Mohn. 1985. Microbiological removal of pentachlorophenol from soil using a *Flavobacterium. Enzyme Microb. Technol.* 7:617–620.

Criddle, C. S., J. T. DeWitt, D. Grbic-Galic, and P. L. McCarty. 1990a. Transformation of carbon tetrachloride by *Pseudomonas* sp. strain KC under denitrification conditions. *Appl. Environ. Microbiol.* 56:3240–3246.

Criddle, C. S., J. T. DeWitt, and P. L. McCarty. 1990b. Reductive dehalogenation of carbon tetrachloride by *Escherichia coli* K-12. *Appl. Environ. Microbiol.* 56:3247–3254.

Cserjesi, A. J., and E. L. Johnson. 1972. Methylation of pentachlorophenol by *Trichoderma virgatum. Can. J. Microbiol.* 18:45–49.

Dalton, H., S. D. Prior, D. J. Leak, and S. H. Stanley. 1984. Regulation and control of methane monooxygenase, pp. 75–82. In: R. N. Patel and R. S. Hanson (eds.), *Microbial Growth on C_1 Compounds.* American Society for Microbiology, Washington, DC.

Daniels, S. L. 1972. The adsorption of microorganisms onto solid surfaces: a review. *Dev. Ind. Microbiol.* 13:211–253.

Davis, J. W., and C. L. Carpenter. 1990. Aerobic biodegradation of vinyl chloride in groundwater samples. *Appl. Environ. Microbiol.* 56:3878–3880.

Davis, J. W., and S. S. Madsen. 1991. The biodegradation of methylene chloride in soils. *Environ. Toxicol. Chem.* 10:463–474.

Davis-Hoover, W. J., L. C. Murdoch, S. J. Vesper, H. R. Pahren, O. L. Sprockel, C. L. Chang, A. Hussain, and W. A. Ritschel. 1991. Hydraulic fracturing to improve nutrient and oxygen delivery for in situ bioreclamation, pp. 67–82. In: R. E. Hinchee and R. F. Olfenbuttel (eds.), *In Situ Bioreclamation: Applications and Investigations for Hydrocarbon and Contaminated Site Remediation.* Butterworth-Heinemann, Boston.

de Bont, J. A. M., M. J. A. W. Vorage, S. Hartmans, and W. J. J. van den Tweel. 1986. Microbial degradation of 1,2-dichlorobenzene. *Appl. Environ. Microbiol.* 52:677–680.

Delfino, J. J., and C. J. Miles, 1985. Aerobic and anaerobic degradation of organic contaminants in Florida groundwater. *Proc. Soil. Crop Sci. Soc. Fla.* 44:9–14.

Desai, S. M., R. Govind, and H. H. Tabak. 1990. Development of quantitative structure–activity relationships for predicting biodegradation kinetics. *Environ. Toxicol. Chem.* 9:473–477.

DEVO Enterprises Inc. 1991a. *Biotreatment News*, 1(3):1–2.

DEVO Enterprises Inc. 1991b. *Biotreatment News*, 1(7):8.

de Weerd, K. A., L. Mandelco, R. S. Tanner, C. R. Woese, and J. M. Suflita. 1990. *Desulfomonile tiedjei* gen. nov. and sp. nov., a novel anaerobic, dehalogenating, sulfate-reducing bacterium. *Arch. Microbiol.* 154:23–30.

Diks, R. M. M., and S. P. P. Ottengraf. 1991. A biological treatment system for the purification of waste gases containing xenobiotic compounds, pp. 452–463. In: R. E. Hinchee and R. F. Olfenbuttel (eds.), *On-Site Bioreclamation: Processes for Xenobiotic and Hydrocarbon Treatment*. Butterworth-Heinemann, Boston.

DiStefano, T. D., J. M. Gossett, and S. H. Zinder. 1991. Reductive dechlorination of high concentrations of tetrachloroethene to ethene by an anaerobic enrichment culture in the absence of methanogenesis. *Appl. Environ. Microbiol.* 57:2287–2292.

Dmochewitz, S., and K. Ballschmiter. 1988. Microbial transformation of technical mixtures of polychlorinated biphenyls (PCB) by the fungus *Aspergillus niger. Chemosphere* 17:111–121.

Dobbins, D. C., and F. K. Pfaender. 1988. Methodology for assessing respiration and cellular incorporation of radiolabeled substrates by soil microbial communities. *Microb. Ecol.* 15:257–273.

Donnelly, K. C., K. W. Brown, C. S. Anderson, and J. C. Thomas. 1991. Bacterial mutagenicity and acute toxicity of solvent and aqueous extracts of soil samples from an abandoned chemical manufacturing site. *Environ. Toxicol. Chem.* 10:1123–1131.

Doran, J. W. 1982. Microorganisms and the biological cycling of selenium. *Adv. Microb. Ecol.* 6:1–32.

Douglass, R. H., J. M. Armstrong, and W. M. Korreck. 1991. Design of a packed column bioreactor for on-site treatment of air stripper off gas, pp. 209–225. In: R. E. Hinchee and R. F. Olfenbuttel (eds.), *On-Site Bioreclamation: Processes for Xenobiotic and Hydrocarbon Treatment*. Butterworth-Heinemann, Boston.

Doyle, R. C., J. D. Isbister, G. L. Anspach, and J. F. Kitchens. 1986. *Composting of Explosives*. Final Technical Report Contract DAAK11-84-C-0057, May.

Drobner, E., H. Huber, and K. O. Stetter. 1990. *Thiobacillus ferrooxidans*, a facultative hydrogen oxidizer. *Appl. Environ. Microbiol.* 56:2922–2923.

Dupont, R. R., W. J. Doucette, and R. E. Hinchee. 1991. Assessment of in situ bioremediation potential and the application of bioventing at a fuel-contaminated site, pp. 262–282. In: R. E. Hinchee and R. F. Olfenbuttel (eds.), *In Situ Bioreclamation: Applications and Investigations for Hydrocarbon and Contaminated Site Remediation*. Butterworth-Heinemann, Boston.

Eaton, D. C. 1985. Mineralization of polychlorinated biphenyls by *Phanerochaete chrysosporium*: a ligninolytic fungus. *Enzyme Microb. Technol.* 7:194–196.

Egdehill, R. V., and R. K. Finn. 1982. Isolation, characterization, and growth kinetics of bacteria metabolizing pentachlorophenol. *Eur. J. Appl. Microbiol. Biotechnol.* 16:179–184.

Edwards, E. A., L. E. Wills, D. Grbic-Galic, and M. Reinhard. 1991. Anaerobic degradation of toluene and xylene: evidence for sulfate as the terminal electron acceptor, pp. 463–471. In: R. E. Hinchee and R. F. Olfenbuttel (eds.), *In Situ Bioreclamation: Applications and Investigations for Hydrocarbon and Contaminated Site Remediation*. Butterworth-Heinemann, Boston.

Efroymson, R. A., and M. Alexander. 1991. Biodegradation by an *Arthrobacter* species of hydrocarbons partitioned into an organic solvent. *Appl. Environ. Microbiol.* 57:1441–1447.

Egli, C., T. Tschan, R. Scholtz, A. M. Cook, and T. Leisinger. 1988. Tranformation of tetrachloromethane to dichloromethane and carbon dioxide by *Acetobacterium woodii. Appl. Environ. Microbiol.* 54:2819–2824.

Ehrlich, H. L. 1971. Biogeochemistry of the minor elements in soil, pp. 361–384. In: A. D. McLaren and J. J. Skujins (eds.), *Soil Biochemistry*, Vol. 2. Marcel Dekker, New York.

Ehrlich, G. G., R. A. Schroeder, and P. Martin. 1985. *Microbial Populations in a Jet Fuel-Contaminated Shallow Aquifer at Tustin, California*. U.S. Geological Survey, Open-file Report 85-335. Prepared in cooperation with the U.S. Marine Corps.

Ensley, B. D. 1991. Biochemical diversity of trichloroethylene metabolism. *Annu. Rev. Microbiol.* 45:283–299.

Environment Today. 1990. Bugs took big bite of Gulf oil spills. 1(6):1, 15–16.

Environment Today. 1991a. Site remediation: 10 'can do' technologies. 2(5):23–34.

Environment Today. 1991b. New EPA database tracks innovative technologies. 2(7):46.

Environment Today. 1991c. Oak Ridge Labs' microbes tackle heavy metals. 2(3):8.

Environmental Solutions Inc. 1990. *On-Site Treatment of Hydrocarbon Contaminated Soils Manual*. ESI, Irvine, CA.

Enzinger, R. M. 1971. *Special Study of the Effect of Alpha TNT on Microbiological Systems and the Determination of the Biodegradability of Alpha TNT*. U.S. Army Project 24-017-70/71, DTIC/AD-728487.

Evans, W. C., and G. Fuchs. 1988. Anaerobic degradation of aromatic compounds. *Annu. Rev. Microbiol.* 42:289–317.

Evans, P. J., D. T. Mang, and L. Y. Young. 1991a. Degradation of toluene and *m*-xylene and transformation of *o*-xylene by denitrifying enrichment cultures. *Appl. Environ. Microbiol.* 57:450–454.

Evans, P. J., D. T. Mang, K. S. Kim, and L. Y. Young. 1991b. Anaerobic degradation of toluene by a denitrifying bacterium. *Appl. Environ. Microbiol.* 57:1139–1145.

Fallon, R. D., D. A. Cooper, R. Speece, and M. Henson. 1991. Anaerobic biodegradation of cyanide under methanogenic conditions. *Appl. Environ. Microbiol.* 57:1656–1662.

Fathepure, B. Z., and S. A. Boyd. 1988. Dependence of tetrachloroethylene dechlorination on methanogenic substrate consumption by *Methanosarsina* sp. strain DSM. *Appl. Environ. Microbiol.* 54:2976–2980.

Fathepure, B. Z., J. M. Tiedje, and S. A. Boyd. 1988. Reductive dechlorination of hexachlorobenzene to tri- and dichlorobenzenes in anaerobic sewage sludge. *Appl. Environ. Microbiol.* 54:327–330.

Fedorak, P. M., and S. E. Hrudey. 1989. Cyanide transformation in anaerobic phenol-degrading methanogenic cultures. *Water Sci. Technol.* 21:67–76.

Fedorak, P. M., D. J. Roberts, and S. E. Hrudey. 1986. The effects of cyanide on the methanogenic degradation of phenolic compounds. *Water Res.* 10:1315–1320.

Fernando, T., J. A. Bumpus, and S. D. Aust. 1990. Biodegradation of TNT (2,4,6-trinitrotoluene) by *Phanerochaete chrysosporium*. *Appl. Environ. Microbiol.* 56(6):1666–1671.

Fetzner, S., R. Muller, and F. Lingens. 1989. A novel metabolite in the microbial degradation of 2-chlorobenzoate. *Biochem. Biophys. Res. Commun.* 161:700–705.

Findlay, R. H., D. J. Morarity, and D. C. White. 1983. Improved method of determining muramic acid from environmental samples. *Geomicrobiol. J.* 3:135–150.

Flemming, C. A., F. G. Ferris, T. J. Beveridge, and G. W. Bailey. 1990. Remobilization of toxic heavy metals adsorbed to bacterial wall–clay composites. *Appl. Environ. Microbiol.* 56:3191–3203.

Fliermans, C. B., T. J. Phelps, D. Ringelberg, A. T. Mikell, and D. C. White. 1988. Mineralization of trichloroethylene by heterotrophic enrichment cultures. *Appl. Environ. Microbiol.* 54:1709–1714.

Focht, D. D., and D. Shelton. 1987. Growth kinetics of *Pseudomonas alcaligenes* C-0 relative to inoculation and 3-chlorobenzoate metabolism in soil. *Appl. Environ. Microbiol.* 53:1846–1849.

Fogel, S., R. Lancione, A. Sewall, and R. S. Boethling. 1985. Application of biodegradability screening tests to insoluble chemicals: hexadecane. *Chemosphere* 14:375–382.

Fogel, M. M., A. R. Taddeo, and S. Fogel. 1986. Biodegradation of chlorinated ethenes by a methane-utilizing mixed culture. *Appl. Environ. Microbiol.* 51:720–724.

Foght, J. M., and D. W. S. Westlake. 1988. Degradation of polycyclic aromatic hydrocarbons and aromatic heterocycles by a *Pseudomonas* species. *Can. J. Microbiol.* 34:1135–1141.

Foght, J. M., D. L. Gutnick, and D. W. S. Westlake. 1989. Effect of emulsan on biodegradation of crude oil by pure and mixed bacterial cultures. *Appl. Environ. Microbiol.* 55:36–42.

Folsom, B. R., P. J. Chapman, and P. H. Pritchard. 1990. Phenol and trichloroethylene degradation by *Pseudomonas cepacia* G4: kinetics and interactions between substrates. *Appl. Environ. Microbiol.* 56:1279–1285.

Fortnagel, P., H. Harms, R.-M. Wittich, S. Krohn, H. Meyer, V. Sinnwell, H. Wilkes, and W. Francke. 1990. Metabolism of dibenzofuran by *Pseudomonas* sp. strain HH69 and the mixed culture HH27. *Appl. Environ. Microbiol.* 56:1148–1156.

Fox, B. G., J. G. Borneman, L. P. Wackett, and J. D. Lipscomb. 1990. Halolkene oxidation by the soluble methane monooxygenase from *Methylosinus trichosporium* OB3b: mechanistic and environmental implications. 29: 6419–6427.

Francis, A., and C. J. Dodge. 1990. Anaerobic microbial remobilization of toxic metals coprecipitated with iron oxide. *Environ. Sci. Technol.* 24:373–378.

Francis, A. J., J. M. Duxbury, and M. Alexander. 1974. Evolution of dimethylselenide from soils. *Appl. Microbiol.* 28:248–250.

Frankenberger, W. T., and U. Karlson. 1991. Bioremediation of seleniferous soils, pp. 239–254. In: R. E. Hinchee and R. F. Olfenbuttel (eds.), *On-Site Bioreclamation: Processes for Xenobiotic and Hydrocarbon Treatment*. Butterworth-Heinemann, Boston.

Fredrickson, J. K., D. L. Balkwill, J. M. Zachara, S.-M. W. Li, F. J. Brockman, and M. A. Simmons. 1991a. Physiological diversity and distributions of heterotrophic bacteria in deep cretaceous sediments of the Atlantic coastal plain. *Appl. Environ. Microbiol.* 57:402–411.

Fredrickson, J. K., F. J. Brockman, D. J. Workman, S. W. Li, and T. O. Stevens. 1991b. Isolation and characterization of a subsurface bacterium capable of growth on toluene, naphthalene, and other aromatic compounds. *Appl. Environ. Microbiol.* 57:796–803.

Freedman, D. L., and J. M. Gossett. 1989. Biological reductive dechlorination of tetrachloroethylene and trichloroethylene to ethylene under methanogenic conditions. *Appl. Environ. Microbiol.* 55:2144–2151.

Freedman, D. L., and J. M. Gossett. 1991. Biodegradation of dichloromethane and its utilization as a growth substrate under methanogenic conditions. *Appl. Environ. Microbiol.* 57:2847–2857.

Freeman, H. M., and P. R. Sferra (eds.). 1990. *Innovative Hazardous Waste Treatment Technology Series,* Vol. 3: *Biological Processes*. Technomic, Lancaster, PA.

Fry, W. E., and R. L. Millar. 1972. Cyanide degradation by an enzyme from *Stemphylium loti*. *Arch. Biochem. Biophys.* 151:468–474.

Fu, J.-K., and R. G. Luthy. 1986. Effect of organic solvent on sorption of aromatic solutes onto soils. *J. Environ. Eng.* 112:346–366.

Furukawa, K. 1982. Microbial degradation of polychlorinated biphenyls (PCBs), pp. 33–57. In: A. M. Chakrabarty (ed.), *Biodegradation and Detoxification of Environmental Pollutants*. CRC Press, Boca Raton, FL.

Furukawa, K., F. Matsumura, and K. Tonomura. 1978a. *Alcaligenes* and *Acinetobacter* strains capable of degrading polychlorinated biphenyls. *Agric. Biol. Chem.* 42:543–548.

Furukawa, K., K. Tonomura, and A. Kamibayashi. 1978b. Effect of chlorine substitution on the biodegradability of polychlorinated biphenyls. *Appl. Environ. Microbiol.* 35:223–227.

Furukawa, K., N. Tomizuka, and A. Kamibayashi. 1979. Effect of chlorine substitution on the bacterial metabolism of various polychlorinated biphenyls. *Appl. Environ. Microbiol.* 38:301–310.

Galfre, G., and C. Milstein. 1981. Preparation of monoclonal antibodies: strategies and procedures. *Methods Enzymol.* 73:3–46.

Gannon, J. T., V. B. Manilal, and M. Alexander. 1991. Relationship between cell surface properties and transport of bacteria through soil. *Appl. Environ. Microbiol.* 57:190–193.

Genthner, B. R. S., W. A. Price II, and P. H. Pritchard. 1989a. Characterization of anaerobic dechlorinating consortia derived from aquatic sediments. *Appl. Environ. Microbiol.* 55:1472–1476.

Genthner, B. R. S., W. A. Price II, and P. H. Pritchard. 1989b. Anaerobic degradation of chloroaromatic compounds in aquatic sediments under a variety of enrichment conditions. *Appl. Environ. Microbiol.* 55:1466–1471.

Gerhardt, M. B., F. B. Green, R. D. Newman, T. J. Lundquist, R. B. Tresan, and W. J. Oswald. 1991. Removal of selenium using a novel algal-bacterial process. *Res. J. Water Pollut. Control Fed.* 63:799–805.

Ghiorse, W. C., and D. L. Balkwill. 1983. Enumeration and morphological characterization of bacteria indigenous to sub-surface environments. *Dev. Ind. Microbiol.* 24:213–224.

Ghiorse, W. C., and D. L. Balkwill. 1985. Microbiological characterization of subsurface environments, pp. 536–556. In: C. H. Ward, W. Gieger, and P. W. McCarty (eds.), *Ground Water Quality*. Wiley, New York.

Gibson, Subramanian. 1984.

Gibson, S. A., and J. M. Suflita. 1986. Extrapolation of biodegradation results to groundwater aquifers: reductive dehalogenation of aromatic compounds. *Appl. Environ. Microbiol.* 52:681–688.

Gibson, S. A., and J. M. Suflita. 1990. Anaerobic biodegradation of 2,4,5-trichlorophenoxyacetic acid in samples from a methanogenic aquifer: stimulation by short-chain organic acids and alcohols. *Appl. Environ. Microbiol.* 56:1825–1832.

Gibson, D. T., J. R. Koch, C. L. Schuld, and R. E. Kallio. 1968. Oxidative degradation of aromatic hydrocarbons by microorganisms. II. Metabolism of halogenated aromatic hydrocarbons. *Biochemistry* 7:3795–3802.

Godsy, E. M., D. F. Goerlits, D. Grbic-Galic, and J. Rogers. 1991. Anaerobic biodegradation of creosote contaminants in natural and simulated ground-water ecosystems. *Biotreatment News*, DEVO, Washington, DC, April, pp. 6–8.

Goldstein, R. M., L. M. Mallory, and M. Alexander. 1985. Reasons for possible failure of inoculation to enhance biodegradation. *Appl. Environ. Microbiol.* 50:977–983.

Gossett, J. M. 1987. Measurement of Henry's law constants for C1 and C2 chlorinated hydrocarbons. *Environ. Sci. Technol.* 21:202–208.

Grbic-Galic, D., and T. M. Vogel. 1987. Transformation of toluene and benzene by mixed methanogenic cultures. *Appl. Environ. Microbiol.* 53:254–260.

Greene, J. M. 1985. *Biodegradation of selected nitramines and related pollutants*. M.S. thesis, East Tennessee State University.

Grishchenkov, V. G., I. E. Fedechkina, B. P. Gaskunov, L. A. Anisimova, A. M. Boronin, and L. A. Golonleva. 1984. Degradation of 3-benzoic acid by *Pseudomonas putida* strain. *Mikrobiologiya* 52:602–606.

Grishin, S. I., and O. H. Tuovinen. 1988. Fast kinetics of Fe^{2+} oxidation in packed-bed reactors. *Appl. Environ. Microbiol.* 54:3092–3100.

Groves, F. R. 1988. Effect of cosolvents on the solubility of hydrocarbons in water. *Environ. Sci. Technol.* 22:282–286.

Gvozdyak, P. I., N. F. Mogilevich, A. F. Rylskii,and N. I. Grishchenko. 1986. Reduction of hexavalent chromium by collection strains of bacteria. *Mikrobiologiya* 55:962–965.

Hagedorn, C., E. L. McCoy, and T. M. Rahe. 1981. The potential for groundwater contamination from septic effluents. *J. Environ. Qual.* 10:1–8.

Haggblom. M. M., and L. Y. Young. 1990. Chlorophenol degradation coupled to sulfate reduction. *Appl. Environ. Microbiol.* 56:3255–3260.

Haggblom, M. M., L. J. Nohynek, and M. Salkinoja-Salonen. 1988. Degradation and O-methylation of chlorinated phenolic compounds by *Rhodococcus* and *Mycobacterium* strains. *Appl. Environ. Microbiol.* 54:3043–3052.

Haigler, B. E., S. F. Nishino, and J. C. Spain. 1988. Degradation of 1,2-dichlorobenzene by a *Pseudomonas* sp. *Appl. Environ. Microbiol.* 54: 294–301.

Hale, D. D., J. E. Rogers, and J. Wiegel. 1991. Environmental factors correlated to dichlorophenol dechlorination in anoxic freshwater sediments. *Environ. Toxicol. Chem.* 10:1255–1265.

Harker, A. R., and Y. Kim. 1990. Trichloroethylene degradation by two independent aromatic-degrading pathways in *Alcaligenes eutrophus* JMP134. *Appl. Environ. Microbiol.* 56:1179–1181.

Harms, H., R.-M. Wittich, V. Sinnwell, H. Meyer, P. Fortnagel, and W. Francke. 1990. Transformation of dibenzo-*p*-dioxin by *Pseudomonas* sp. strain HH69. *Appl. Environ. Microbiol.* 56:1157–1159.

Harris, R., and C. J. Knowles. 1983. Isolation and growth of a *Pseudomonas* species that utilizes cyanide as a source of nitrogen. *J. Gen. Microbiol.* 129:1005–1011.

Harris, R. E., A. W. Bunch, and C. J. Knowles. 1987. Microbial cyanide and nitrile metabolism. *Sci. Prog. (Oxford)* 71:293–304.

Harrison, A. P., Jr. 1984. The acidophilic thiobacilli and other acidophilic bacteria that share their habitat. *Annu. Rev. Microbiol.* 38:265–292.

Hartmann, J., W. Reineke, and H.-J. Knackmuss. 1979. Metabolism of 3-chloro-, 4-chloro-, and 3,5-dichlorobenzoate by a pseudomonad. *Appl. Environ. Microbiol.* 37:421–425.

Hartmann, J., K. Engelberts, B. Nordhaus, E. Schmidt, and W. Reineke. 1989. Degradation of 2-chlorobenzoate by in vivo constructed hybrid pseudomonads. *FEMS Microbiol. Lett.* 61:17–22.

Hartmans, S., J. A. M. de Bont, J. Tramper, and K. C. A. M. Luyben. 1985. Bacterial degradation of vinyl chloride. *Biotechnol. Lett.* 7:383–388.

Harvey, R. W., and S. P. Garabedian. 1991. Use of colloid filtration theory in modeling movement of bacteria through a contaminated sandy aquifer. *Environ. Sci. Technol.* 25:178–185.

Hazmat World. 1989. In situ TCE bioremediation field trial is "successful." 2(11):10–12.

Hazmat World. 1991. Bioremediation (Special report). 4(1):44–53.

Heitkamp, M. A., J. P. Freeman, and C. E. Cerniglia. 1987. Naphthalene biodegradation in environmental microcosms: estimates of degradation rates and characterization of metabolites. *Appl. Environ. Microbiol.* 53:129–136.

Heitkamp, M. A., J. P. Freeman, and C. E. Cerniglia. 1988. Pyrene degradation by a *Mycobacterium* sp.: Identification of ring oxidation and ring fission products. *Appl. Environ.* Microbiol. 54:2556–2565.

Hendriksen, H. V., S. Larsen, and B. K. Ahring. 1991. Anaerobic degradation of PCP and phenol in fixed-film reactors: the influence of an additional substrate. *Water Sci. Technol.* 24:431–436.

Henry, S. M., and D. Grbic-Galic. 1989. TCE transformation by mixed and pure ground water cultures, pp. 109–125. In: P. V. Roberts, L. Semprini, G. D. Hopkins, D. Grbic-Galic, P. L. McCarty, and M. Reinhard (eds.), *In-Situ Aquifer Restoration of Chlorinated Aliphatics by Methanotrophic Bacteria*. Technical Report EPA 600/2-89/033. U.S. Environmental Protection Agency, Robert S. Kerr Environmental Research Laboratory, Ada, OK.

Henry, S. M., and D. Grbic-Galic. 1991a. Influence of endogenous and exogenous electron donors and trichloroethylene oxidation toxicity on trichloroethylene oxidation by methanotrophic cultures from a groundwater aquifer. *Appl. Environ. Microbiol.* 57:236–244.

Henry, S. M., and D. Grbic-Galic. 1991b. Inhibition of trichloroethylene oxidation by the transformation intermediate carbon monoxide. *Appl. Environ. Microbiol.* 57:1770–1776.

Henschler, D., W. R. Hoos, H. Fetz, E. Dallmeier, and M. Metzler. 1979. Reactions of trichloroethylene epoxide in aqueous systems. *Biochem. Pharmacol.* 28:543–548.

Herbes, S. E., and L. R. Schwall. 1978. Microbial transformation of polycyclic aromatic hydrocarbons in pristine and petroleum-contaminated sediments. *Appl. Environ. Microbiol.* 35:306–316.

Hernandez, B. S., K. F. Higson, R. Kondrat, and D. D. Focht. 1991. Metabolism of and inhibition by chlorobenzoates in *Pseudomonas putida* P111. *Appl. Environ. Microbiol.* 57:3361–3366.

Hickey, W. J., and D. D. Focht. 1990. Degradation of mono-, di-, and trihalogenated benzoic acids by *Pseudomonas aeruginosa* JB2, *Appl. Environ. Microbiol.* 56:3842–3850.

Higgins, I. J., D. Scott, and R. C. Hammond. 1984. Transformation of C_1 compounds by microorganisms, pp. 43–87. In: D. T. Gibson (ed.), *Microbial Degradation of Organic Compounds*. Marcel Dekker, New York.

Hinchee, R. E., and M. Arthur. 1991. Bench-scale studies of the soil aeration process for bioremediation of petroleum hydrocarbons. *J. Appl. Biochem. Biotechnol.* 28/29:901–906.

Hinchee, R. E., D. C. Downey, and E. J. Coleman. 1987. Enhanced bioreclamation, soil venting and ground-water extraction: a cost-effectiveness and feasibility comparison. In: *Petroleum Hydrocarbons and Organic Chemicals in Ground Water Prevention, Detection and Restoration: A Conference and Exposition*, Houston, November 1987.

Hinchee, R. E., D. C. Downey, and P. K. Aggarwal. 1990. Use of hydrogen peroxide as an oxygen source for insitu biodegradation. Part I. Field Studies. *J. Hazard. Mater.* 27:287–299.

Hinchee, R. E., D. C. Downey, R. R. Dupont, P. Aggarwal, and R. N. Miller. 1991. Enhancing biodegradation of petroleum hydrocarbons through soil venting. *J. Hazard. Mater.* 27:315–325.

Hirsch, P., and E. Rades-Rohkohl. 1988. Some special problems in the determination of viable counts of groundwater microoganisms. *Microb. Ecol.* 16:99–113.

Hirsch, P., and E. Rades-Rohkohl. 1990. Microbial colonization of aquifer sediment exposed in a groundwater well in northern Germany. *Appl. Environ. Microbiol.* 56:2963–2966.

Hoag, G. E., and B. Cliff. 1988. The use of soil venting technique for the remediation of petroleum-contaminated soils, pp. 301–316. In: E. J. Calabrese and P. T. Kostecki (eds.), *Soils Contaminated by Petroleum: Environmental and Public Health Effects*. Wiley, New York.

Hoeppel, R. E. 1989. Biodegradation for on-site remediation of contaminated soils and groundwater at Navy sites. *Military Eng.* 81(530):31–36.

Hoeppel, R. E., R. G. Rhett, and C. R. Lee. 1980. *Fate and Enumeration Problems of Fecal Coliform Bacteria in Runoff Waters from Terrestrial Ecosystems*. TR-EL-80-9. U.S. Atomic Energy Waterways Experiment Station, Vicksburg, MS.

Hoeppel, R. E., R. E. Hinchee, and M. F. Arthur. 1991. Bioventing soils contaminated with petroleum hydrocarbons. *J. Ind. Microbiol.* 8:141–146.

Hoffsomer, J. C., L. A. Kaplan, D. J. Glover, D. A. Kubose, C. Dickinson, H. Goya, E. G. Kayser, C. L. Groves, and M. E. Sitzmann, 1978. *Biodegradability of TNT: A Three-Year Pilot Plant Study*. DTIC/AD-A061144.

Holben, W. E., J. K. Jansson, B. K. Chelm, and J. M. Tiedje. 1988. DNA probe method for the detection of specific microorganisms in the soil bacterial community. *Appl. Environ. Microbiol.* 54:703–711.

Hooper, A. B., and K. R. Terry. 1973. Specific inhibitors of ammonia oxidation in *Nitrosomonas*. *J. Bacteriol.* 115:480–485.

Horowitz, A., and R. M. Atlas. 1980. Microbial seeding to enhance petroleum hydrocarbon biodegradation in aquatic Arctic ecosystems, pp. 15–20. In: T. A. Oxley, G. Becker, and D. Allsopp (eds.), *Proceedings of the 4th International Biodeterioration Symposium*. Pitman, London.

Horowitz, A., D. R. Shelton, C. P. Cornell, and J. M. Tiedje. 1982. Anaerobic degradation of aromatic compounds in sediments and digested sludge. *Dev. Ind. Microbiol.* 23:435–444.

Horowitz, A., J. M. Suflita, and J. M. Tiedje. 1983. Reductive dehalogenation of halobenzoates by anaerobic lake sediment microorganisms. *Appl. Environ. Microbiol.* 45:1459–1465.

Huang, C. M. 1989. Ph.D. dissertation. University of California, Riverside.

Huber, H., and K. O. Stetter. 1989. *Thiobacillus prosperus* sp. nov., represents a new group of halotolerant metal-mobilizing bacteria isolated from a marine geothermal field. *Arch. Microbiol.* 151:479–485.

Huber, H., and K. O. Stetter. 1990. *Thiobacillus cuprinus* sp. nov., a novel facultatively organotrophic metal-mobilizing bacterium. *Appl. Environ. Microbiol.* 56:315–322.

Huling, S. G., B. E. Bledsoe, and M. V. White. 1990. Enhanced bioremediation utilizing hydrogen peroxide as a supplemental source of oxygen: a laboratory and field study. Unpublished report. U.S. Environmental Protection Agency. Ada, OK, 48 pp.

Huling, S. G., B. E. Bledsoe, and M. V. White. 1991. The feasibility of utilizing hydrogen peroxide as a source of oxygen in bioremediation, pp. 83–102. In: R. E. Hinchee and R. F. Olfenbuttel (eds.), *In Situ Bioreclamation: Applications and Investigations for Hydrocarbon and Contaminated Site Remediation.* Butterworth-Heinemann, Boston.

Hutchins, S. 1991. Biodegradation of monoaromatic hydrocarbons by aquifer microorganisms using oxygen, nitrate, or nitrous oxide as the terminal electron acceptor. *Appl. Environ. Microbiol.* 57:2403–2407.

Hutchins, S. R., and J. T. Wilson. 1991. Laboratory and field studies on BTEX biodegradation in a fuel-contaminated aquifer under denitrifying conditions, pp. 157–172. In: R. E. Hinchee and R. F. Olfenbuttel (eds.), *In Situ Bioreclamation: Applications and Investigations for Hydrocarbon and Contaminated Site Remediation*. Butterworth-Heinemann, Boston.

Hyman, M. R., and P. M. Wood. 1985. Suicidal inactivation and labeling of ammonia mono-oxygenase by acetylene. Biochem. J. 227:719–725.

Ide, A., Y. Niki, F. Sakamoto, I. Watanabe, and H. Watanabe. 1972. Decomposition of pentachlorophenol in paddy soil. *Agric. Biol. Chem.* 36: 1937–1944.

IIT Research Institute. 1990. *Installation Restoration and Hazardous Waste Control Technologies*. Report CETHA-TS-CR-90067. Bartlesville, OK, and Aberdeen Proving Ground, MD. August.

Inoue, A., M. Yamamoto, and K. Horikoshi. 1991. *Pseudomonas putida* which can grow in the presence of toluene. *Appl. Environ. Microbiol.* 57:1560–1562.

Isbister, J. D., R. C. Doily, and K. K. Kitchens. 1982. *Composting of Explosives*. U.S. Army Report DRXTH-TE.

Ishibashi, Y., C. Cervantes, and S. Silver. 1990. Chromium reduction in *Pseudomonas putida*. *Appl. Environ. Microbiol.* 56:2268–2270.

Jain, R. K., and G. S. Sayler. 1987. Problems and potential for in situ treatment of environmental pollutants by engineered microorganisms. *Microbiol. Sci.* 4:59–63.

Jain, R. K., G. S. Sayler, J. T. Wilson, L. Houston, and D. Pacia. 1987. Maintenance and stability of introduced genotypes in groundwater aquifer material. *Appl. Environ. Microbiol.* 53:996–1002.

Janssen, D. B., A. J. van den Wijngaard, J. J. van der Waarde, and R. Oldenhuis. 1991. Biochemistry and kinetics of aerobic degradation of chlorinated aliphatic hydrocarbons, pp. 92–112. In: R. E. Hinchee and R. F. Olfenbuttel (eds.), *On-Site Bioreclamation: Processes for Xenobiotic and Hydrocarbon Treatment.* Butterworth-Heinemann, Boston.

Jeffers, T. H. 1991. Using microorganisms to recover metals. *Calif. Geol.* 44:154–158.

Johnston, J. B., and S. G. Robinson. 1982. Opportunities for development of new detoxification processes through genetic engineering, pp. 301–314. In: J. H. Exner (ed.), *Detoxification of Hazardous Waste*. Ann Arbor Science, Ann Arbor, MI.

Jones, D. M., A. G. Douglas, R. J. Parkes, J. Taylor, W. Giger, and C. Schaffner. 1983. The recognition of biodegraded petroleum-derived aromatic hydrocarbons in recent marine sediments. *Mar. Pollut. Bull.* 14:103–108.

Kaplan, D. L., and A. M. Kaplan. 1982. Thermophilic biotransformations of 2,4,6-trinitrotoluene under simulated composting conditions. *Appl. Environ. Microbiol.* 44:757–760. Natick/TR-82-015.

Kastner, M. 1991. Reductive dechlorination of tri- and tetrachloroethylene by nonmethanogenic enrichment cultures, pp. 134–146. In: R. E. Hinchee and R. F. Olfenbuttel (eds.), *On-Site Bioreclamation: Processes for Xenobiotic and Hydrocarbon Treatment*. Butterworth-Heinemann, Boston.

Keck, H., R. C. Sims, M. Coover, K. Park, and B. Symons. 1989. Evidence for cooxidation of polynuclear aromatic hydrocarbons in soil. *Water Research* 23:1467–1476.

Keeney, D. R. 1973. The nitrogen cycle in sediment–water systems. *J. Environ. Qual.* 2:15–29.

Kido, T., T. Yamamoto, and K. Soda. 1975. Microbial assimilation of alkyl nitro compounds and formation of nitrate. *Arch. Microbiol.* 106:165–169.

Kindred, J. S., and M. A. Celia. 1989. Contaminant transport and biodegradation. 2. Conceptual model and test simulations. *Water Resour. Res.* 25:1149–1159.

King, E. F. 1984. A comparative study of methods for assessing the toxicity to bacteria of single chemicals and mixtures, p. 175. In: D. Liu and B. J. Dutka (eds.), *Toxicity Screening Procedures Using Bacteria Systems*. Marcel Dekker, New York.

King, G. M. 1988. Dehalogenation in marine sediments containing natural sources of halophenols. *Appl. Environ. Microbiol.* 54:3079–3085.

King, L. K., and B. C. Parker. 1988. A simple, rapid method for enumerating total viable and metabolically active bacteria in groundwater. *Appl. Environ. Microbiol.* 54:1630–1631.

Kirk, T. K., E. Schultz, W. J. Connors, L. F. Lorenz, and J. G. Zeikus. 1978. Influence of culture parameters on lignin metabolism in *Phanerochaete chrysosporium*. *Arch. Microbiol.* 117:277–285.

Klecka, G. M. and D. T. Gibson. 1980. Metabolism of dibenzo-*p*-dioxin and chlorinated dibenzo-*p*-dioxins by a *Beijerinckia* species. *Appl. Environ. Microbiol.* 39:288–295.

Klecka, G. M., S. J. Gonsior, and D. A. Markham. 1990. Biological transformations of 1,1,1-trichloroethane in subsurface soils and ground water. *Environ. Toxicol. Chem.* 9:1437–1451.

Knowles, C. J., and A. W. Bunch. 1986. Microbial cyanide metabolism. *Adv. Microb. Physiol.* 27:73–111.

Kohler, H. P. E., D. Kohler-Staub, and D. D. Focht. 1988. Cometabolism of polychlorinated biphenyls: enhanced transformation of Aroclor 1254 by growing bacterial cells. *Appl. Environ. Microbiol.* 54:1940–1945.

Kohring, G. W., X. Zhang, and J. Wiegel. 1989. Anaerobic dechlorination of 2,4-dichlorophenol in freshwater sediments in the presence of sulfate. *Appl. Environ. Microbiol.* 55:2735–2737.

Kovalenko, T. V., G. I. Karavaiko, and V. P. Piskunov. 1982. Effect of Fe^{3+} ions in the oxidation of ferrous iron by *Thiobacillus ferrooxidans* at various temperatures. *Mikrobiologiya* 51:156–160.

Krockel, L., and D. D. Focht. 1987. Construction of chlorobenzene-utilizing recombinants by progenitive manifestation of a rare event. *Appl. Environ. Microbiol.* 53:2470–2475.

Kuhn, E. P., and J. M. Suflita. 1989. Dehalogenation of pesticides by anaerobic microorganisms in soils and groundwater: a review, pp. 111–180. In: B. L. Sawhney and K. Brown (eds.), *Reactions and Movement of Organic Chemicals in Soils*. Special Publication 22. Soil Science Society of America, Madison, WI.

Kuhn, E. P., J. Zeyer, P. Eicher, and R. P. Schwarzenbach. 1988. Anaerobic degradation of alkylated benzenes in denitrifying laboratory aquifer columns. *Appl. Environ. Microbiol.* 54:490–496.

Kuhn, E. P., G. T. Townsend, and J. M. Suflita. 1990. Effect of sulfate and organic carbon supplements on reductive dehalogenation of chloroanilines in anaerobic aquifer slurries. *Appl. Environ. Microbiol.* 56:2630–2637.

Kurek, E., J. Czaban, and J.-M. Bollag. 1982. Sorption of cadmium by microorganisms in competition with other soil constituents. *Appl. Environ. Microbiol.* 43:1011–1015.

Lamar, R. T., and D. M. Dietrich. 1990. In situ depletion of pentachlorophenol from contaminated soil by *Phanerochaete* spp. *Appl. Environ. Microbiol.* 56:3093–3100.

Lamar, R. T., J. A. Glaser, and T. K. Kirk. 1990a. Fate of pentachlorophenol (PCP) in sterile soils inoculated with *Phanerochaete chrysosporium*: mineralization, volatilization and depletion of PCP. *Soil Biol. Biochem.* 22:433–440.

Lamar, R. T., M. J. Larsen, and T. K. Kirk. 1990b. Sensitivity to the degradation of pentachlorophenol by *Phanerochaete* spp. *Appl. Environ. Microbiol.* 56:3519–3526.

Larsen, S., H. V. Hendriksen, and B. K. Ahring. 1991. Potential for thermophilic (50°C) anaerobic dechlorination of pentachlorophenol in different ecosystems. *Appl. Environ. Microbiol.* 57:2085–2090.

Laube, V., S. Ramamoorthy, and D. J. Kushner. 1979. Mobilization and accumulation of sediment bound heavy metals by algae. *Bull. Environ. Contam. Toxicol.* 21:763–770.

Lawes, B. C., 1991. Soil-induced decomposition of hydrogen peroxide, pp. 143–156. In: R. E. Hinchee and R. F. Olfenbuttel (eds.), *In Situ Bioreclamation: Applications and Investigations for Hydrocarbon and Contaminated Site Remediation*. Butterworth-Heinemann, Boston.

Lawrence, A. W., and P. L. McCarty. 1970. Unified basis for biological treatment design and operation. *J. Sanit. Eng. Div.* 96(SA3):757–778.

Leahy, J. G., and R. R. Colwell. 1990. Microbial degradation of hydrocarbons in the environment. *Microbiol. Rev.* 54:305–315.

Leahy, M. C., M. Findlay, and S. Fogel. 1989. Biodegradation of chlorinated aliphatics by a methanotrophic consortium in a biological reactor, pp. 3–9. In: *Biotreatment: The Use of Microorganisms in the Treatment of Hazardous Materials and Hazardous Wastes.* The 2nd National Conference.

Lebron, C. A. 1989. *Fate of Ordnance Compounds in the Environment: A Literature Review.* Report TM 71-89-4. Report to Naval Civil Engineering Laboratory. Port Hueneme, CA, May.

Lee, R. F., and R. Hoeppel. 1991. Hydrocarbon degradation potential in reference soils and soils contaminated with jet fuel, pp. 570–580. In R. E. Hinchee and R. F. Olfenbuttel (eds.), *In Situ Bioreclamation: Applications and Investigations for Hydrocarbon and Contaminated Site Remediation.* Butterworth-Heinemann, Boston.

Lee, M. D., J. M. Thomas, R. C. Borden, P. B. Bedient, J. T. Wilson, and C. H. Ward. 1988. Biorestoration of aquifers contaminated with organic compounds. *Crit. Rev. Environ. Control* 18:29–89.

Lerman, A., and G. J. Brunskill. 1971. Migration of major constituents from lake sediments into lake water and its bearing on lake water composition. *Limnol. Oceanogr.* 16:880–890.

Lewis, D. L., and D. K. Gattie. 1991. The ecology of quiescent microbes. *ASM News* 57:27–32.

Linkfield, T. G., J. M. Suflita, and J. M. Tiedje. 1989. Characterization of the acclimation period before anaerobic dehalogenation of halobenzoates. *Appl. Environ. Microbiol.* 55:2773–2778.

Little, C. D., A. V. Palumbo, S. E. Herbes, M. E. Lidstrom, R. L. Tyndall, and P. J. Gilmer. 1988. Trichloroethylene biodegradation by a methane-oxidizing bacterium. *Appl. Environ. Microbiol.* 54:951–956.

Little, C. D., C. D. Fraley, M. P. McCann, and A. Matin. 1991. Use of bacterial stress promoters to induce biodegradation under conditions of environmental stress, pp. 493–498. In: R. E. Hinchee and R. F. Olfenbuttel (eds.), *On-Site Bioreclamation: Processes for Xenobiotic and Hydrocarbon Treatment.* Butterworth-Heinemann, Boston.

Lizama, H. M., and I. Suzuki. 1988. Bacterial leaching of a sulfide ore by *Thiobacillus ferrooxidans* and *Thiobacillus thiooxidans.* I. Shake flask studies. *Biotechnol. Bioeng.* 32:110–116.

Lizama, H. M., and I. Suzuki. 1989. Synergistic competitive inhibition of ferrous iron oxidation by *Thiobacillus ferrooxidans* by increasing concentrations of ferric iron and cells. *Appl. Environ. Microbiol.* 55:2588–2591.

Lovley, D. R., and D. J. Lonergan. 1990. Anaerobic oxidation of toluene, phenol, and *p*-cresol by the dissimilatory iron-reducing organism, GS-15. *Appl. Environ. Microbiol.* 56:1858–1864.

Lovley, D. R., E. J. P. Phillips, and D. J. Lonergan. 1991a. Enzymatic versus nonenzymatic mechanisms for Fe(III) reduction in aquatic sediments. *Environ. Sci. Technol.* 25:1062–1067.

Lovley, D. R., E. J. P. Phillips, Y. A. Gorby, and E. R. Landa. 1991b. Microbial reduction of uranium. *Nature (London)* 350:413–416.

Lu, J. C. S., and K. Y. Chen. 1977. Migration of trace metals in interfaces of seawater and polluted surficial sediments. *Environ. Sci. Technol.* 11:174–182.

Lund, N. C., J. Swinianski, G. Gudehus, and D. Maier. 1991. Laboratory and field tests for a biological in situ remediation of a coke oven plant, pp. 396–412. In: R. E. Hinchee and R. F. Olfenbuttel (eds.), *In Situ Bioreclamation: Applications and Investigations for Hydrocarbon and Contaminated Site Remediation.* Butterworth-Heinemann, Boston.

Lundgren, D. G., and M. Silver. 1980. Ore leaching by bacteria. *Annu. Rev. Microbiol.* 34:263–283.

Lyman, W. J., W. F. Reehl, and D. H. Rosenblatt. 1982. *Handbook of Chemical Property Estimation Methods: Environmental Behavior of Organic Compounds.* McGraw-Hill, New York.

Mackay, D. M., and J. A. Cherry. 1989. Groundwater contamination: pump-and-treat remediation. *Environ. Sci. Technol.* 23:630–636.

Mahaffey, W. R., D. T. Gibson, and C. E. Cerniglia. 1988. Bacterial oxidation of bacterial carcinogens: Formation of Polycyclic Aromatic Acids from Benz[a]-anthracene. *Appl. Environ. Microbiol.* 54:2415–2423.

Maiers, D. T., P. L. Wichlacz, D. L. Thompson, and D. F. Bruhn. 1988. Selenate reduction by bacteria from a selenium-rich environment. *Appl. Environ. Microbiol.* 54:2591–2593.

Major, D. W., C. I. Mayfield, and J. F. Barker. 1988. Biotransformation of benzene by denitrification in aquifer sand. *Ground Water* 26:8–14.

Marinucci, A. C., and R. Bartha. 1979. Biodegradation of 1,2,3- and 1,2,4-trichlorobenzene in soil and in liquid enrichment culture. *Appl. Environ. Microbiol.* 38:811–817.

Markovetz, A. J. 1971. Subterminal oxidation of aliphatic hydrocarbons by microorganisms. *Crit. Rev. Microbiol.* 1:225–237.

Marks, T. S., A. R. W. Smith, and A. V. Quirk. 1984. Degradation of 4-chlorobenzoic acid by *Arthrobacter* sp. *Appl. Environ. Microbiol.* 48: 1020–1025.

Mathewson, J. R., and R. B. Grubbs. 1989. Commercial microorganisms. *Hazmat World* 2(6):48–51.

Matthews, J. E. 1989. *Fate of PAH Compounds in Two Soil Types: Influence of Volatilization, Abiotic Loss and Biological Activity.* U.S. Environmental Protection Agency, Robert S. Kerr Environmental Research Laboratory, Ada, OK.

McCarty, P. L. 1975. Stoichiometry of biological reactions. *Prog. Water Technol.* 7:157–172.

McCarty, P. L. 1988. Bioengineering issues related to in situ remediation of contaminated soils and groundwater, pp. 143–162. In: G. S. Omenn (ed.). *Basic Life Sciences,* Vol. 45: *Environmental Biotechnology (Reducing Risks from Environmental Chemicals through Biotechnology).* Plenum Press, New York.

McCarty, P. L., L. Semprini, M. E. Dolan, T. C. Harmon, C. Tiedeman and S. M. Gorelick. 1991. In situ methanotrophic bioremediation for contaminated groundwater at St. Joseph, Michigan. In: R. E. Hinchee and R. F. Olfenbuttel (eds.), *On-Site Bioreclamation: Processes for Xenobiotic and Hydrocarbon Treatment*. Butterworth-Heinemann, Boston.

McCormick, N. G., J. H. Cornell, and A. M. Kaplan. 1984. *The Anaerobic Biotransformation of RDX, HMX and Their Acetylated Derivatives*. NATICK/TR-85-007. U.S. Army Natick R&D Center.

McKeague, J. A., M. V. Cheschire, F. Andreux, and J. Berthlin. 1986. Organo-mineral complexes in relation to pedogenesis, pp. 549–592. In: P. M. Huang and M. Schnitzer (eds.), *Interactions of Soil Minerals with Natural Organics and Microbes*. Soil Science Society of America. Madison, WI.

Mendoza, C. A., and T. A. McAlary. 1990. Modeling of ground-water contamination caused by organic solvent vapors. *Ground Water* 28:199–206.

Miguez, C. B., C. W. Greer, and J. M. Ingram. 1990. Degradation of mono- and dichlorobenzoic acid isomers by two natural isolates of *Alcaligenes denitrificans*. *Arch. Microbiol.* 154:139–143.

Mihelcic, J. R., and R. G. Luthy. 1988a. Degradation of polycyclic aromatic hydrocarbon compounds under various redox conditions in soil–water systems. *Appl. Environ. Microbiol.* 54(5):1182–1187.

Mihelcic, J. R., and R. G. Luthy. 1988b. Microbial degradation of acenaphthene and naphthalene under denitrification conditions in soil–water systems. *Appl. Environ. Microbiol.* 54(5):1188–1198.

Mihelcic, J., and R. G. Luthy. 1991. Sorption and microbial degradation of naphthalene in soil–water suspensions under denitrification conditions. *Environ. Sci. Technol.* 25:169–177.

Mikesell, M. D., and S. A. Boyd. 1985. Reductive dechlorination of the pesticides 2,4-D, 2,4,5-T and pentachlorophenol in anaerobic sludges. *J. Environ. Qual.* 14:337–340.

Mikesell, M. D., and S. A. Boyd. 1986. Complete reductive dechlorination and mineralization of pentachlorophenol by anaerobic microorganisms. *Appl. Environ. Microbiol.* 53:861–865.

Milde, G., M. Nerger, and R. Mergler. 1988. Biological degradation of volatile chlorinated hydrocarbons in groundwater. *Water Sci. Technol.* 20(3):67–73.

Miller, R. M., and R. Bartha. 1989. Evidence for liposome encapsulation for transport-limited microbial metabolism of solid alkanes. *Appl. Environ. Microbiol.* 55:269–274.

Miller, R. N., and R. E. Hinchee. 1990. *Enhanced Biodegradation through Soil Venting*. Final Report to Engineering and Service Laboratory, Tyndall Air Force Base, Panama City, FL.

Miller, R. N., C. C. Vogel, and R. E. Hinchee. 1991. A field-scale investigation of petroleum hydrocarbon biodegradation in the vadose zone enhanced by soil venting at Tyndall AFB, Florida, pp. 283–302. In: R. E. Hinchee and R. F. Olfenbuttel (eds.), *In Situ Bioreclamation: Applications and Investigations for Hydrocarbon and Contaminated Site Remediation*. Butterworth-Heinemann, Boston.

Miller, W. P., D. C. Martens, and L. W. Eelazny. 1986. Effect of sequence in extraction of trace metals from soils. *Soil Sci. Soc. Am. J.* 50:598–601.

Mokross, H., E. Schmidt, and W. Reineke. 1990. Degradation of 3-chlorobiphenyl by in vivo constructed hybrid pseudomonads. *FEMS Microbiol. Lett.* 71:179–186.

Molz, F. J., M. A. Widdowson, and L. D. Benefield. 1986. Simulation of microbial growth dynamics coupled to nutrient and oxygen transport in porous media. *Water Resour. Res.* 22:1207–1216.

Monticello, D. J., and W. R. Finnerty. 1985. Microbial desulfurization of fossil fuels. *Annu. Rev. Microbiol.* 39:371–389.

Morgan, P., and R. J. Watkinson. 1992. Factors limiting the supply and efficiency of nutrient and oxygen supplements for the in situ biotreatment of contaminated soil and groundwater. *Water Res.* 26(1):73–78.

Mozes, N., F. Marchal, M. P. Hermesse, J. L. van Haecht, L. Reuliaux, A. J. Leonard, and P. G. Rouxhet. 1987. Immobilization of microorganisms by adhesion: interplay of electrostatic and nonelectrostatic interactions. *Biotechnol. Bioeng.* 30:439–450.

Mullen, M. D., D. C. Wolf, F. G. Ferris, T. J. Beveridge, C. A. Flemming, and G. W. Bailey. 1989. Bacterial sorption of heavy metals. *Appl. Environ. Microbiol.* 55:3143–3149.

Murthy, N. B. K., D. D. Kaufman, and G. F. Fries. 1979. Degradation of PCP in aerobic and anaerobic soil. *J. Environ. Sci. Health Part B* 14:1–14.

Myers, M. E., Jr., J. Stollsteiner, and A. M. Wims. 1975. Determination of hydrocarbon-type distribution and hydrogen/carbon ratio of gasolines by nuclear magnetic resonance spectrometry. *Anal. Chem.* 47:2010–2015.

Nakahara, T., L. E. Erickson, and J. R. Gutierrez. 1977. Characteristics of hydrocarbon uptake in cultures with two liquid phases. *Biotechnol. Bioeng.* 19:9–25.

Nakamura, K., M. Sakamoto, H. Uchiyama, and O. Yagi. 1990. Organomercurial-volatilizing bacteria in the mercury-polluted sediment of Minamata Bay, Japan. *Appl. Environ. Microbiol.* 56:304–305.

National Research Council. 1981. *The Alkyl Benzenes.* National Academy Press, Washington, DC.

Neilson, A. H., A. S. Allard, P. A. Hynning, M. Remberger, and L. Lander. 1983. Bacterial methylation of chlorinated phenols and guaiacols: formation of veratroles from guaiacols and high-molecular-weight chlorinated lignin. *Appl. Environ. Microbiol.* 45:774–783.

Nelson, M. J. K., S. O. Montgomery, and P. H. Pritchard. 1988. Trichloroethylene metabolism by microorganisms that degrade aromatic compounds. *Appl. Environ. Microbiol.* 54:604–606.

Nicholas, R. B., and D. E. Giamporcaro. 1989. Nature's prescription. *Hazmat World* 2(6):30–36.

Nies, L., and T. M. Vogel. 1990. Effects of organic substrates on dechlorination of Aroclor 1242 in anaerobic sediments. *Appl. Environ. Microbiol.* 56:2612–2617.

Nirmalakhandan, N., and R. E. Speece. 1988. Structure–activity relationships. *Environ. Sci. Technol.* 7:607–615.

Norris, P. R. 1983. Iron and mineral oxidation with *Leptospirillum*-like bacteria, pp. 83–96. In: G. Rossi and A. E. Torma (eds.), *Recent Progress in Biohydrometallurgy.* Associazione Mineraria Sarda, Iglesias, Italy.

Ogunseitan, O. A., I. L. Delgado, Y.-L. Tsai, and B. H. Olson. 1991. Effect of 2-hydroxybenzoate on the maintenance of naphthalene-degrading pseudo-monads in seeded and unseeded soil. *Appl. Environ. Microbiol.* 57:2873–2879.

Okereke, A., and S. E. Stevens, Jr. 1991. Kinetics of iron oxidation by *Thiobacillus ferrooxidans*. *Appl. Environ. Microbiol.* 57:1052–1056.

Oldenhuis, R., R. L. J. M. Vink, D. B. Janssen, and B. Witholt. 1989. Degradation of chlorinated aliphatic hydrocarbons by *Methylosinus trichosporium* OB3b expressing soluble methane monooxygenase. *Appl. Environ. Microbiol.* 55:2819–2826.

Oldenhuis, R., J. Y. Oedzes, J. J. van der Waarde, and D. B. Janssen. 1991. Kinetics of chlorinated hydrocarbon degradation by *Methylosinus trichosporium* OB3b and toxicity of trichloroethylene. *Appl. Environ. Microbiol.* 57:7–14.

Olson, G. J., and F. E. Brinckman. 1986. Bioprocessing of coal. *Fuel* 65:1638–1646.

Oltmanns, R. H., H. G. Rast, and W. Reineke. 1988. Degradation of 1,4-dichlorobenzene by enriched and constructed bacteria. *Appl. Microbiol. Biotechnol.* 28:609–616.

Ong, S. K., R. E. Hinchee, R. Hoeppel, and R. Scholze. 1991. In situ respirometry for determining aerobic degradation rates, pp. 541–545. In: R. E. Hinchee and R. F. Olfenbuttel (eds.), *In Situ Bioreclamation: Applications and Investigations for Hydrocarbon and Contaminated Site Remediation*. Butterworth-Heinemann, Boston.

O'Reilly, K. T., and R. L. Crawford. 1989. Degradation of pentachlorophenol by polyurethane-immobilized *Flavobacterium* cells. *Appl. Environ. Microbiol.* 55:2113–2118.

Oremland, R. S., C. W. Culbertson, and M. R. Winfrey. 1991a. Methylmercury decomposition in sediments and bacterial cultures: involvement of methanogens and sulfate reducers in oxidative demethylation. *Appl. Environ. Microbiol.* 57:130–137.

Oremland, R. S., N. A. Steinberg, T. S. Presser, and L. G. Miller. 1991b. In situ bacterial selenate reduction in the agricultural drainage systems of Western Nevada. *Appl. Environ. Microbiol.* 57:615–617.

Osmon, J. L., and C. C. Andrews. 1978. *The Biodegradation of TNT in Enhanced Soil and Compost Systems*. DTIC/AD-E400073.

Painter, H. A. 1970. A review of the literature on inorganic nitrogen metabolism in microorganisms. *Water Res.* 4:393–450.

Palumbo, A. V., W. Eng, P. A. Boerman, G. W. Strandberg, T. L. Donaldson, and S. E. Herbes. 1991. Effects of diverse organic contaminants on trichloroethylene degradation by methanotrophic bacteria and methane-utilizing consortia, pp. 77–91. In: R. E. Hinchee and R. F. Olfenbuttel (eds.), *On-Site Bioreclamation: Processes for Xenobiotic and Hydrocarbon Treatment*. Butterworth-Heinemann, Boston.

Park, K. P., R. C. Sims, R. R. Dupont, W. J. Doucette, and J. E. Matthews. 1990. Fate of PAH compounds in two soil types: influence of volatilization, abiotic loss and biological activity. *Environ. Toxicol. Chem.* 9:187–195.

Parsons, F., G. B. Lage, and R. Rice. 1985. Biotransformation of chlorinated organic solvents in static microcosms. *Environ. Toxicol. Chem.* 4:739–742.

Pekdeger, A., and G. Matthess. 1983. Factors of bacteria and virus transport in groundwater. *Environ. Geol.* 5:49–52.

Pendrys, J. P. 1989. Biodegradation of asphalt cement-20 by aerobic bacteria. *Appl. Environ. Microbiol.* 55:1357–1362.

Pertsova, R. N., F. Kung, and L. A. Golovleva. 1984. Degradation of 3-chlorobenzoic acid in soil by pseudomonads carrying biodegradative plasmids. *Folia Microbiol.* 29:242–247.

Pettigrew, C. A., A. Breen, C. Corcoran, and G. S. Sayler. 1990. Chlorinated biphenyl mineralization by individual populations and consortia of freshwater bacteria. *Appl. Environ. Microbiol.* 56:2036–2045.

Pettigrew, C. A., B. E. Haigler, and J. C. Spain. 1991. Simultaneous biodegradation of chlorobenzene and toluene by a *Pseudomonas* strain. *Appl. Environ. Microbiol.* 57:157–162.

Pheiffer, T. H., T. J. Nunno, and J. S. Walters. 1990. EPA's assessment of European contaminated soil treatment techniques. *Environ. Prog.* 9:79–86.

Phelps, T. J., J. J. Niedzielski, R. M. Schram, S. E. Herbes, and D. C. White. 1990. Biodegradation of trichloroethylene in continuous-recycle expanded-bed bioreactors. *Appl. Environ. Microbiol.* 56:1702–1709.

Phelps, T. J., K. Malachowsky, R. M. Schram, and D. C. White. 1991. Aerobic mineralization of vinyl chloride by a bacterium of the order *Actinomycetales*. *Appl. Environ. Microbiol.* 57:1252–1254.

Phung, H. T., and J. G. Fiskell. 1972. A review of redox reactions in soils. *Soil Crop Sci. Soc. Fla. Proc.* 32:141–145.

Pignatello, J. J., M. M. Martinson, J. G. Steiert, R. E. Carlson, and R. L. Crawford. 1983. Biodegradation and photolysis of pentachlorophenol in artificial freshwater streams. *Appl. Environ. Microbiol.* 46:1024–1031.

Piotrowski, M. R. 1991. Full-scale in-situ bioremediation at a superfund site: a progress report. Presented at *EPACH's conference on Hydrocarbon Contaminated Soils and Groundwater*, Newport Beach, CA.

Poindexter, J. S. 1981. Oligotrophy: feast and famine existence. *Adv. Microb. Ecol.* 5:63–89.

Porta, A. 1991. A review of European bioreclamation practice, pp. 1–13. In: R. E. Hinchee and R. F. Olfenbuttel (eds.), *In Situ Bioreclamation: Applications and Investigations for Hydrocarbon and Contaminated Site Remediation*. Butterworth-Heinemann, Boston.

Porter, K. G., and Y. S. Feig. 1980. The use of DAPI for identifying and counting aquatic microflora. *Limnol. Oceanogr.* 25:943–948.

Pothuluri, J. V., J. P. Freeman, F. E. Evans, and C. E. Cerniglia. 1990. Fungal transformation of fluoranthene. *Appl. Environ. Microbiol.* 56:2974–2983.

Price, N. B. 1973. *Chemical Diagenesis in Sediments*. NTIS/PB-226-882.0. National Science Foundation, June.

Prior, S. D., and H. Dalton. 1985. The effect of copper ions on membrane content and methane monooxygenase activity in methanol-grown cells of *Methylococcus capsulatus* (Bath). *J. Gen. Microbiol.* 131:155–163.

Quensen, J. F., III, J. M. Tiedje, and S. A. Boyd. 1988. Reductive dechlorination of polychlorinated biphenyls by anaerobic microorganisms from sediments. *Science* 242:752–754.

Quensen, J. F., III, S. A. Boyd, and J. M. Tiedje. 1990. Dechlorination of four commercial polychlorinated biphenyl mixtures (Aroclors) by anaerobic microorganisms from sediments. *Appl. Environ. Microbiol.* 56:2360–2369.

Rahe, T. M., C. Hagedorn, E. L. McCoy, and G. F. Kling. 1978. Transport of antibiotic-resistant *Escherichia coli* through western Oregon hillslope soils under conditions of saturated flow. *J. Environ. Qual.* 7:487–494.

Ramadan, M. A., O. M. El-Tayeb, and M. Alexander. 1990. Inoculum size as a factor limiting success of inoculation for biodegradation. *Appl. Environ. Microbiol.* 56:1392–1396.

Rappe, C. 1984. Analysis of polychlorinated dioxins and furans. *Environ. Sci. Technol.* 18:78A–90A.

Rasche, M. E., R. E. Hicks, M. R. Hyman, and D. J. Arp. 1990. Oxidation of monohalogenated ethanes and *n*-chlorinated alkanes by whole cells of *Nitrosomonas europaea*. *J. Bacteriol.* 172:5368–5373.

Rasche, M. E., M. R. Hyman, and D. J. Arp. 1991. Factors limiting aliphatic chlorocarbon degradation by *Nitrosomonas europaea*: cometabolic inactivation of ammonia monooxygenase and substrate specificity. *Appl. Environ. Microbiol.* 57:2986–2994.

Reamer, D. C., and W. H. Zoller. 1980. Selenium biomethylation products from soil and sewage sludge. *Science* 208:500–502.

Regnell, O., and A. Tunlid. 1991. Laboratory study of chemical speciation of mercury in lake sediment and water under aerobic and anaerobic conditions. *Appl. Environ. Microbiol.* 57:789–795.

Reid, I. D. 1991. Intermediates and products of synthetic lignin (dehydrogenative polymerizate) degradation by *Plebia tremellosa*. *Appl. Environ. Microbiol.* 57:2834–2840.

Reineke, W., and H.-J. Knackmuss. 1984. Microbial metabolism of haloaromatics: isolation and properties of a chlorobenzene degrading bacterium. *Appl. Environ. Microbiol.* 47:395–402.

Reineke, W., and H.-J. Knackmuss. 1988. Microbial degradation of haloaromatics. *Annu. Rev. Microbiol.* 42:263–287.

Reineke, W., D. J. Jeenes, P. A. Williams, and H.-J. Knackmuss. 1982. TOL plasmid pWWO in constructed halobenzoate-degrading *Pseudomonas* strains: prevention of *meta* pathway. *J. Bacteriol.* 150:195–201.

Reinhard, M., N. L. Goodman, and J. F. Barker. 1984. Occurrence and distribution of organic chemicals in two landfill leachate plumes. *Environ. Sci. Technol.* 18:953–961.

Reinhard, M., L. E. Willis, H. A. Ball, T. Harmon, D. W. Phipps, H. F. Ridgway, and M. P. Eisman. 1991. A field experiment for the anaerobic biotransformation of aromatic hydrocarbon compounds at Seal Beach, California, pp. 487–496. In: R. E. Hinchee and R. F. Olfenbuttel (eds.), *In Situ Bioreclamation: Applications and Investigations for Hydrocarbon and Contaminated Site Remediation*. Butterworth-Heinemann, Boston.

Reganathan, V. 1989. Possible involvement of toluene-2,3-dioxygenase in defluorination of 3-fluoro-substituted benzenes by toluene-degrading *Pseudomonas* sp. strain T-12. *Appl. Environ. Microbiol.* 55:330–334.

Rifai, H. S., and P. B. Bedient. 1990. Comparison of biodegradation kinetics with an instantaneous reaction model for groundwater. *Water Resour. Res.* 26:637–645.

Rifai, H. S., P. B. Bedient, R. C. Borden, and J. F. Haasbeek. 1987. *BIOPLUME II: Computer Model of Two-Dimensional Transport under the Influence of Oxygen Limited Biodegradation in Groundwater: User's Manual.* Dept. Environmental Science and Engineering, Rice University, Houston, TX.

Rifai, H. S., P. B. Bedient, J. T. Wilson, K. M. Miller, and J. M. Armstrong. 1988. Biodegradation modeling at an aviation fuel spill site. *ASCE J. Environ. Eng.* 114:1007–1029.

Rifai, H. S., G. P. Long, and P. B. Bedient. 1991. Modeling bioremediation: theory and field application, pp. 535–541. In: R. E. Hinchee and R. F. Olfenbuttel (eds.), *In Situ Bioreclamation: Applications and Investigations for Hydrocarbon and Contaminated Site Remediation.* Butterworth-Heinemann, Boston.

Riser-Roberts, E. 1992. Bioremediation of Petroleum Contaminated Sites. C. K. Smoley, CRC Press, Inc., Boca Raton, FL.

Roberts, P. V., L. Semprini, G. D. Hopkins, D. Grbic-Galic, P. L. McCarty, and M. Reinhard. 1989. *In Situ Aquifer Restoration of Chlorinated Aliphatics by Methanotrophic Bacteria.* EPA/600/2-89/033. U.S. Environmental Protection Agency, Center for Research Information, Cincinnati, OH.

Rochkind, M. L., J. W. Blackburn, and G. S. Sayler. 1986. *Microbial Decomposition of Chlorinated Aromatic Compounds.* EPA/600/2-86/090. U.S. Environmental Protection Agency, Hazardous Office of Research and Development, Cincinnati, OH.

Rosenberg, M., and S. Kjelleberg. 1986. Hydrophobic interactions: role in bacterial adhesion. *Adv. Microb. Ecol.* 9:353–393.

Rosenberg, E., E. Englander, A. Horowitz, and D. Gutnick. 1975. Bacterial growth and dispersion of crude oil in an oil tanker during its ballast voyage, pp. 157–167. In: A. W. Borquin, D. G. Ahern, and S. P. Meyers (eds.), *Proceedings on the Impact of the Use of Microorganisms on the Aquatic Environment.* EPA 660-3-75-001. U.S. Environmental Protection Agency, Corvallis, OR.

Rosenberg, E., A. Gottlieb, and M. Rosenberg. 1983. Inhibition of bacterial adherence to hydrocarbons and epithelial cells by emulsan. *Infect. Immun.* 39:1024–1028.

Rosenberg, M., E. A. Bayer, J. Delarea, and E. Rosenberg. 1982. Role of thin fimbriae in adherence and growth of *Acinetobacter calcoaceticus* RAG-1 on hexadecane. *Appl. Environ. Microbiol.* 44:929–937.

Rossin, A. C., R. Perry, and J. N. Lester. 1982. The removal of nitrilotriacetic acid and its effect on metal removal during biological sewage treatment. 1. Adsorption and acclimatisation. *Environ. Pollut. Ser. A* 29:271–302.

Saber, D. L., and R. L. Crawford. 1985. Isolation and characterization of *Flavobacterium* strains that degrade pentachlorophenol. *Appl. Environ. Microbiol.* 50:1512–1518.

Sahu, S. K., K. K. Patnaik, M. Sharmila, and N. Sethunathan. 1990. Degradation of alpha-, beta-, and gamma-hexachlorocyclohexane by a soil bacterium under aerobic conditions. *Appl. Environ. Microbiol.* 56:3620–3622.

Saiki, R. K., D. H. Gelfand, S. Stoffel, S. J. Scharf, R. Higuchi, G. T. Horn, K. B. Mullis, and H. A. Erlich. 1988. Primer-directed enzymatic amplification of DNA with a thermostable DNA polymerase. *Science* 238:487–491.

Sakurai, I., Y. Kawamura, H. Koike, Y. Inoue, Y. Kosako, T. Nakase, Y. Kondou, and S. Sakurai. 1990. Bacterial accumulation of metallic compounds. *Appl. Environ. Microbiol.* 56:2580–2583.

Sander, P., R.-M. Wittich, P. Fortnagel, H. Wilkes, and W. Francke. 1991. Degradation of 1,2,4-trichloro- and 1,2,4,5-tetrachlorobenzene by *Pseudomonas* strains. *Appl. Environ. Microbiol.* 57:1430–1440.

Sanders, W. N., and J. B. Maynard. 1968. Capillary gas chromatographic method for determining the Ce-C12 hydrocarbons in full-range motor gasolines. *Anal. Chem.* 40:527–535.

Schaub, S. A., and C. A. Sorber. 1977. Virus and bacterial removal from wastewater by rapid infiltration through soil. *Appl. Environ. Microbiol.* 33:609–619.

Schennen, U., K. Braun, and H.-J. Knackmuss. 1983. Anaerobic degradation of 2-fluorobenzoate by benzoate-degrading, denitrifying bacteria. *J. Bacteriol.* 161:321–325.

Schieman, D. A. 1981. Advances in membrane filter applications for microbiology. In: B. J. Dutka (ed.), *Membrane Filtration*. Marcel Dekker, New York.

Schocken, M. J., and D. T. Gibson. 1984. Bacterial oxidation of the polycyclic aromatic hydrocarbons acenaphthene and acenaphthylene. *Appl. Environ. Microbiol.* 48:10–16.

Schraa, G., M. L. Boone, M. S. M. Jetten, A. R. W. van Neerven, P. J. Colberg, and A. J. B. Zehnder. 1986. Degradation of 1,4-dichlorobenzene by *Alcaligenes* sp. strain A175. *Appl. Environ. Microbiol.* 52:1374–1381.

Semprini, L., P. V. Roberts, G. D. Hopkins, and P. L. McCarty. 1990. In situ biodegradation of chlorinated ethenes. Part 2. Results of biostimulation and biotransformation experiments. *Ground Water* 28:715–727.

Semprini, L., G. D. Hopkins, P. V. Roberts, and P. L. McCarty. 1991. In situ biotransformation of carbon tetrachloride, Freon-113, Freon-11, and 1,1,1-TCA under anoxic conditions, pp. 41–58. In: R. E. Hinchee and R. F. Olfenbuttel (eds.), *On-Site Bioreclamation: Processes for Xenobiotic and Hydrocarbon Treatment*. Butterworth-Heinemann, Boston.

Sewell, G. W., and S. A. Gibson. 1991. Stimulation of the reductive dechlorination of tetrachloroethene in anaerobic aquifer microcosms by the addition of toluene. *Environ. Sci. Technol.* 25:982–984.

Shabtai, Y. 1991. Isolation and characterization of a lipolytic bacterium capable of growing in a low-water-content oil–water emulsion. *Appl. Environ. Microbiol.* 57:1740–1745.

Shiaris, M. P. 1989. Seasonal biotransformation of naphthalene, phenanthrene, and benzo(*a*)pyrene in surficial estuarine sediments. *Appl. Environ. Microbiol.* 55:1391–1399.

Shiaris, M. P., and G. S. Sayler. 1982. Biotransformation of PCB by natural assemblages of freshwater microorganisms. *Environ. Sci. Technol.* 16:367–369.

Shields, M. S. 1991. Construction of a *Pseudomonas cepacia* strain constitutive for the degradation of trichloroethylene and its evaluation for field and bioreactor studies, p. 215. *Abstracts, Annual Meeting of the American Society for Microbiology.*

Shields, M. S., S. O. Montgomery, P. J. Chapman, S. M. Cuskey, and P. H. Pritchard. 1989. Novel pathway of toluene catabolism in the trichloroethylene-degrading bacterium G4. *Appl. Environ. Microbiol.* 55:1624–1629.

Sims, R. C. 1990. Soil remediation techniques at uncontrolled hazardous waste sites: a critical review. *J. Air Waste Manage. Assoc.* 40:704.

Sims, R., and J. Bass. 1984. *Review of In-Place Treatment Techniques for Contaminated Surface Soils,* Vol. 1, *Technical Evaluation.* EPA-540/2-84/003a. U.S. Environmental Protection Agency.

Sims, R. C., and M. R. Overcash. 1983. Fate of polynuclear aromatic compounds (PNAs) in soil–plant systems. *Residue Rev.* 88:1–68.

Singer, P. C., and W. Stumm. 1970. Acidic mine drainage: the rate determining step. *Science* 167:1121–1123.

Smith, J. H., J. C. Harper, and H. Jaber. 1981. *Analysis and Environmental Fate of Air Force Distillate and High Density Fuels.* SRI International, Menlo Park, CA. Report to Engineering and Services Laboratory, Tyndall AFB, Panama City, FL. ESL-TR-81-54.

Smith, M. S., G. W. Thomas, R. E. White, and D. Ritonga. 1985. Transport of *Escherichia coli* through intact and disturbed soil columns. *J. Environ. Qual.* 14:87–91.

Sneddon, J., and C. P. Pappas. 1991. Binding and removal of metal ions in solution by an algae biomass. *Am. Environ. Lab.* 3(4):9–13.

Soczo, B., J. J. M. Staps, and K. Visscher. 1986. *Biotechnologische Bodemsanering.* RIVM report 851105002. Bilthoven, The Netherlands.

Soviero, M. M. 1989. Bacteria that eat TNT. *Popular Science,* November 1989.

Spain, J. C., and D. T. Gibson. 1991. Pathway for biodegradation of *p*-nitrophenol in a *Moraxella* sp. *Appl. Environ. Microbiol.* 57:812–819.

Spain, J. C., and S. F. Nishino. 1987. Degradation of 1,4-dichlorobenzene by a *Pseudomonas* sp. *Appl. Environ. Microbiol.* 53:1010–1019.

Spain, J. C., J. D. Milligan, D. C. Downey, and J. K. Slaughter. 1989. Excessive bacterial decomposition of H_2O_2 during enhanced biodegradation. *Ground Water* 27:163–167.

Spanggord, R. J., T. Mill, C. Tsong-Wang, W. R. Mabey, J. H. Smith, and S. Lee. March 1980. *Environmental Fate Studies on Certain Munition Wastewater Constituents Final Report, Phase I: Literature Review.* DTIC/AD-A082372.

Spanggord, R. J., J. C. Spain, S. F. Nishino, and K. E. Mortelmans. 1991. Biodegradation of 2,4-dinitrotoluene by a *Pseudomonas* sp. *Appl. Environ. Microbiol.* 57:3200–3205.

Stanlake, G. J., and R. K. Finn. 1982. Isolation and characterization of a pentachlorophenol-degrading bacterium. *Appl. Environ. Microbiol.* 44:1412–1427.

Stanley, S. H., S. D. Prior, D. J. Leak, and H. Dalton. 1983. Copper stress underlies the fundamental change in intracellular location of methane monooxygenase in methane-oxidizing organisms. *Biotechnol. Lett.* 5:487–492.

Steffan, R. J., and R. M. Atlas. 1988. DNA amplification to enhance detection of genetically engineered bacteria in environmental samples. *Appl. Environ. Microbiol.* 54:2185–2191.

Steffan, R. J., E. T. Korthals, and M. R. Winfrey. 1988. Effects of acidification on mercury methylation, demethylation, and volatilization in sediments from an acid-susceptible lake. *Appl. Environ. Microbiol.* 54:2003–2009.

Steinberg, N. A., and R. S. Oremland. 1990. Dissimilatory selenate reduction potentials in a diversity of sediment types. *Appl. Environ. Microbiol.* 56:3550–3557.

Stirling, D. L., and H. Dalton. 1977. Effect of metal-binding agents and other compounds on methane oxidation by two strains of *Methylococcus capsulates*. *Arch. Microbiol.* 114:71–76.

Strand, S. E., J. V. Wodrich, and H. D. Stensel. 1991. Biodegradation of chlorinated solvents in a sparged, methanotrophic biofilm reactor. *Res. J. Water Pollut. Control. Fed.* 63:859–867.

Strandberg, G. W., T. L. Donaldson, and L. L. Farr. 1989. Degradation of trichloroethylene and *trans*-1,2-dichloro-ethylene by a methanotrophic consortium in a fixed-film, packed-bed bioreactor. *Environ. Sci. Technol.* 23:1422–1425.

Stransky, R., and M. S. Blanchard. 1989. An in situ venting program for petroleum hydrocarbons in soil and water. *Proceedings of the 3rd NWWA National Outdoor Action Conference,* Orlando, FL, p. 607.

Stumm, W. 1984. Interpretation and measurement of redox intensity in natural waters. *Schweiz. Z. Hydrol.* 46:291–296.

Suflita, J. M. 1989. Microbial ecology and pollutant biodegradation in subsurface ecosystems, pp. 67–84. In: *Transport and Fate of Contaminants in the Subsurface*. EPA/625/4-89/019. U.S. Environmental Protection Agency.

Suflita, J. M., and G. D. Miller. 1985. Microbial metabolism of chlorophenolic compounds in ground water aquifers. *Environ. Toxicol. Chem.* 4:751–758.

Suflita, J. M., A. Horowitz, D. R. Shelton, and J. M. Tiedje. 1982. Dehalogenation: a novel pathway for the anaerobic biodegradation of haloaromatic compounds. *Science* 218:1115–1116.

Suflita, J. M., S. A. Gibson, and R. E. Demon. 1988. Anaerobic biotransformations of pollutant chemicals in aquifers. *J. Ind. Microbiol.* 3:179–194.

Sutherland, I. W. 1983. Microbial exopolysaccharides: their role in microbial adhesion in aqueous systems. *Crit. Rev. Microbiol.* 10:173–201.

Sutherland, J. B., A. L. Selby, J. P. Freeman, F. E. Evans, and C. E. Cerniglia. 1991. Metabolism of phenanthrene by *Phanerochaete chrysosporium*. *Appl. Environ. Microbiol.* 57:3310–3316.

Suzuki, T. 1977. Metabolism of pentachlorophenol by a soil microbe. *J. Environ. Sci. Health Part B* 12:113–127.

Suzuki, I., T. L. Takeuchi, T. D. Yuthasastrakosol, and J. K. Oh. 1990. Ferrous iron and sulfur oxidation and ferric iron reduction activities of *Thiobacillus ferrooxidans* are affected by growth on ferrous iron, sulfur, or a sulfide ore. *Appl. Environ. Microbiol.* 56:1620–1626.

Swain, H. M., H. J. Sommerville, and J. A. Cole. 1978. Denitrification during growth of *Pseudomonas aeruginosa* on octane. *J. Microbiol.* 107:103–112.

Taylor, B. F., W. L. Hearn, and S. Pincus. 1979. Metabolism of monofluor- and monochlorobenzoates by a denitrifying bacterium. *Arch. Microbiol.* 122:301–306.

The Bioremediation Report. 1991. Biotreatment of metals: a smorgasbord of options. December, pp. 2–3.

Thomas, J. M., and C. H. Ward. 1989. In situ biorestoration of organic contaminants in the subsurface. *Environ. Sci. Technol.* 23:760–766.

Thompson-Eagle, E. T., and W. T. Frankenberger, Jr. 1990a. Volatilization of selenium from agricultural evaporation pond water. *J. Environ. Qual.* 19:125–131.

Thompson-Eagle, E. T., and W. T. Frankenberger, Jr. 1990b. Protein-mediated selenium biomethylation in evaporation pond water. *Environ. Toxicol. Chem.* 9:1453–1462.

Tiedje, J. M., S. A. Boyd, and B. Z. Fathepure. 1987. Anaerobic degradation of chlorinated aromatic hydrocarbons. *Dev. Ind. Microbiol.* 27:117–127.

Tiedje, J. M., R. K. Colwell, Y. L. Grossman, R. E. Hodson, R. E. Lenski, R. N. Mack, and P. J. Regal. 1989. The planned introduction of genetically engineered organisms: ecological considerations and recommendations. *Ecology* 70:298–315.

Topp, E., and R. S. Hanson. 1990. Degradation of pentachlorophenol by a *Flavobacterium* species grown in continuous culture under various nutrient limitations. *Appl. Environ. Microbiol.* 56:541–544.

Torpy, M. F., H. F. Stroo, and G. Bruebaker. 1989. Biological treatment of hazardous waste. *Pollut. Eng.* 21(5):80–86.

Trevors, J. T., J. D. van Elsas, L. S. van Overbeck, and M.-E. Starodub. 1990. Transport of a genetically engineered *Pseudomonas fluorescens* strain through a soil microcosm. *Appl. Environ. Microbiol.* 56:401–408.

Trudgill, P. W. 1984. Microbial degradation of the alicyclic ring, pp. 131–180. In: D. T. Gibson (ed.), *Microbial Degradation of Organic Compounds.* Marcel Dekker, New York.

Urlings, L. G. C. M., F. Spuy, S. Coffa, and H. B. R. J. van Vree. 1991. Soil vapour extraction of hydrocarbons: in situ and on-site biological treatment, pp. 321–336. In: R. E. Hinchee and R. F. Olfenbuttel (eds.), *In Situ Bioreclamation: Applications and Investigations for Hydrocarbon and Contaminated Site Remediation.* Butterworth-Heinemann, Boston.

U.S. Environmental Protection Agency. 1991. *Innovative Treatment Technologies: Semi-annual Status Report.* EPA/540/2-91/001, No. 2. U.S. Environmental Protection Agency, Office of Solid Waste and Emergency Response. Washington, DC, September.

van Beelen, P., A. K. Fleuren-Kemila, M. P. A. Huys, A. C. P. van Montfort, and P. L. A. van Vlaardingen. 1991. The toxic effects of pollutants on the mineralization of acetate in subsoil microcosms. *Environ. Toxicol. Chem.* 10:775–789.

van den Tweel, W. J. H., N. ter Berg, J. B. Kok, and J. A. M. de Bont. 1986. Biotransformation of 4-hydroxybenzoate from 4-chlorobenzoate by *Alcaligenes denitrificans* NTB-1. *Appl. Microbiol. Biotechnol.* 25:289–294.

van den Tweel, W. J. H., J. B. Kok, and J. A. M. de Bont. 1987. Reductive dehalogenation of 2,4-dichlorobenzoate to 4-chlorobenzoate and hydrolytic dehalogenation of 4-chloro-, 4-bromo-, and 4-iodobenzoate by *Alcaligenes denitrificans* NTB-1. *Appl. Environ. Microbiol.* 53:810–815.

van der Meer, J. R., W. Roelofsen, G. Schraa, and A. J. B. Zehnder. 1987. Degradation of low concentrations of dichlorobenzenes and 1,2,4-trichlorobenzene by *Pseudomonas* sp. strain P51 in nonsterile soil columns. *FEMS Microbiol. Ecol.* 45:333–341.

van der Meer, J. R., A. R. W. van Neerven, E. J. de Vries, W. M. de Vos, and A. J. B. Zehnder. 1991. Cloning and characterization of plasmid-encoded genes for the degradation of 1,2-dichloro-, 1,4-dichloro-, and 1,2,4-trichlorobenzene of *Pseudomonas* sp. strain P51. *J. Bacteriol.* 173:6–15.

van Dort, H. M., and D. L. Bedard. 1991. Reductive *ortho* and *meta* dechlorination of a polychlorinated biphenyl congener by anaerobic microorganisms. *Appl. Environ. Microbiol.* 57:1576–1578.

van Eyk, J., and C. Vreeken. 1989. Venting-mediated removal of diesel oil from subsurface soil strata as a result of stimulated evaporation and enhanced biodegradation, pp. 475–485. In: *Hazardous Waste and Contaminated Sites, Envirotech Vienna,* Vol. 2, Session 3. ISBN 389432-009-5. Westarp Wiss., Essen, Germany.

van Eyk, J., and C. Vreeken. 1991. In situ and on-site subsoil and aquifer restoration at a retail gasoline station, pp. 303–320. In: R. E. Hinchee and R. F. Olfenbuttel (eds.), *In Situ Bioreclamation: Applications and Investigations for Hydrocarbon and Contaminated Site Remediation.* Butterworth-Heinemann, Boston.

Vannelli, T. M., M. Logan, D. M. Arciero, and A. B. Hooper. 1990. Degradation of halogenated aliphatic compounds by the ammonia-oxidizing bacterium *Nitrosomonas europae. Appl. Environ. Microbiol.* 56:1169–1171.

Venterea, R., and M. Findlay. 1991. Biofiltration for the treatment of airstreams contaminated with jet fuel vapors. *Proceedings of the New England Environmental Exposition,* May 21–23, Boston.

Vira, A., and S. Fogel. 1991. Bioremediation: the treatment for tough chlorinated hydrocarbons. *Biotreatment News* 1(11):8.

Visser, W., A. Scheffers, W. H. Batenburg-van der Vegte, and J. P. van Dijken. 1990. Oxygen requirements of yeasts. *Appl. Environ. Microbiol.* 56:3785–3792.

Vogel, T. M., and D. Grbic-Galic. 1986. Incorporation of oxygen from water into toluene and benzene during anaerobic fermentative transformation. *Appl. Environ. Microbiol.* 52:200–202.

Vogel, T. M., and P. L. McCarty. 1985. Biotransformation of tetrachloroethylene to trichloroethylene, dichloroethylene, vinyl chloride and carbon dioxide under methanogenic conditions. *Appl. Environ. Microbiol.* 49:1080–1083.

Vogel, T. M., C. S. Criddle, and P. L. McCarty. 1987. Transformations of halogenated aliphatic compounds. *Environ. Sci. Technol.* 21:722–736.

Voice, T. C., and W. J. Weber. 1983. Sorption of hydrophobic compounds by sediments, soils and suspended solids. I. Theory and background. *Water Res.* 17:1433–1441.

Voice, T. C., C. P. Rice, and W. J. Weber, Jr. 1983. Effect of solids concentration on the sorptive partitioning of hydrophobic pollutants in aquatic systems. *Environ. Sci. Technol.* 17:513–518.

Wackett, L. P., and D. T. Gibson. 1988. Degradation of trichloroethylene by toluene dioxygenase in whole-cell studies with *Pseudomonas putida* F1. *Appl. Environ. Microbiol.* 54:1703–1708.

Wackett, L. P., and S. R. Householder. 1989. Toxicity of trichloroethylene to *Pseudomonas putida* F1 is mediated by toluene dioxygenase. *Appl. Environ. Microbiol.* 55:2723–2725.

Wackett, L. P., G. A. Brusseau, S. R. Householder, and R. S. Hanson. 1989. Survey of microbial oxygenases: trichloroethylene degradation by propane-oxidizing bacteria. *Appl. Environ. Microbiol.* 55:2960–2964.

Walia, S., A. Khan, and N. Rosenthal. 1990. Construction and applications of DNA probes for detection of polychlorinated biphenyl-degrading genotypes in toxic organic-contaminated soil environments. *Appl. Environ. Microbiol.* 56:254–259.

Walker, J. D. 1988. Effects of chemicals on microorganisms. *J. Water Pollut. Control. Fed.* 60:1106–1121.

Walker, J. D., and R. R. Colwell. 1976. Enumeration of petroleum-degrading microorganisms. *Appl. Environ. Microbiol.* 31:198–207.

Walker, S. G., C. A. Flemming, F. G. Ferris, T. J. Beveridge, and G. W. Bailey. 1989. Physicochemical interaction of *Escherichia coli* cell envelopes and *Bacillus subtilis* cell walls with two clays and the ability of the composite to immobilize heavy metals from solution. *Appl. Environ. Microbiol.* 55:2976–2984.

Wang, X., X. Yu, and R. Bartha. 1990. Effect of bioremediation on polycyclic aromatic hydrocarbon residues in soil. *Environ. Sci. Technol.* 24:1086–1089.

Ward, C. T., and F. Matsumura. 1978. Fate of 2,3,7,8-tetra-chlorodibenzo-*p*-dioxin (TCDD) in a model aquatic environment. *Arch. Environ. Contam. Toxicol.* 7:349–357.

Watanabe, I. 1973. Isolation of pentachlorophenol-decomposing bacteria from soil. *Soil Sci. Plant Nutr.* 19:109–116.

Watts, R. J., P. N. McGuire, H. Lee, and R. E. Hoeppel. 1989. Effect of concentration on the biological degradation of petroleum hydrocarbons associated with in situ soil-water treatment. *1989 ASCE National Conference on Environmental Engineering,* Austin, TX, July 10–12.

Webster, J. J., G. T. Hampton, J. T. Wilson, W. C. Ghiorse, and F. R. Leach. 1985. Determination of microbial cell numbers in subsurface samples. *Ground Water* 23:17–25.

Werner, P. 1991. German experiences in the biodegradation of creosote and gaswork-specific substances, pp. 496–517. In: R. E. Hinchee and R. F. Olfenbuttel (eds.), *In Situ Bioreclamation: Applications and Investigations for Hydrocarbon and Contaminated Site Remediation.* Butterworth-Heinemann, Boston.

Wetzel, R. S., C. M. Durst, D. H. Davidson, and D. J. Sarno. 1987. *In Situ Biological Treatment Test at Kelly Air Force Base,* Vol. II: *Field Test Results and Cost Model*. Science Applications International Corp., McLean, VA. Prepared for Engineering and Services Laboratory, Tyndall Air Force Base, Panama City, FL, July. ESL-TR-85-52, Vol. II.

White, D. C., R. J. Bobbie, J. D. King, J. S. Nickels, and P. Amoe. 1979a. Lipid analysis of sediments for microbial biomass and community structure, pp. 87–103. In: C. D. Litchfield and P. L. Sayfried (eds.), *Methodology for Biomass Determination and Microbial Activities in Sediments*. ASTM STP 673. American Society for Testing and Materials, Philadelphia.

White, D. C., W. M. Davis, J. S. Nickels, J. D. King, and R. J. Bobbie. 1979b. Determination of the sedimentary microbial biomass by extractable lipid phosphate. *Oecologia (Berlin)* 40:51–62.

White, J. M., D. D. Jones, D. Huang, and J. J. Gauthier. 1988. Conversion of cyanide to formate and ammonia by a pseudomonad from industrial wastewater. *J. Ind. Microbiol.* 3:263–272.

Widdowson, M. A., and C. M. Aelion. 1991. Application of a numerical model to the performance and analysis of an in situ bioremediation project, pp. 227–244. In: R. E. Hinchee and R. F. Olfenbuttel (eds.), *In Situ Bioreclamation: Applications and Investigations for Hydrocarbon and Contaminated Site Remediation*. Butterworth-Heinemann, Boston.

Wiggins, B. A., and M. Alexander. 1986. Role of protozoa in microbial acclimation to low concentrations of organic chemicals. *Abstracts Annual Meeting of the American Society for Microbiology,* p. 300.

Wiggins, B. A., and M. Alexander. 1988. Role of chemical concentration and second carbon sources in acclimation of microbial communities for biodegradation. *Appl. Environ. Microbiol.* 54:2803–2807.

Wiggins, B. A., S. H. Jones, and M. Alexander. 1987. Explanations for the acclimation period preceding the mineralization of organic chemicals in aquatic environments. *Appl. Environ. Microbiol.* 53:791–796.

Wiley, B. J. (ed.). 1985. Thirty-third Conference in Microbiological Deterioration of Military Material. NATICK/TR-85/057L U.S. Army Natick R&D Center, ATTN: STRNC-YEP, Natick, MA 01760–5020.

Wilkinson, T. G., J. J. Topiwalal, and G. Hamer. 1974. Interactions in a mixed bacterial population growing on methane in continuous culture. *Biotechnol. Bioeng.* 16:41–59.

Wilson, J. T., and B. H. Wilson. 1985. Biotransformation of trichloroethylene in soil. *Appl. Environ. Microbiol.* 49:242–243.

Wilson, J. T., J. F. McNabb, D. L. Balkwill, W. C. Ghiorse. 1983. Enumeration and characterization of bacteria indigenous to a shallow water-table aquifer. *Ground Water* 23:17–25.

Wilson, J. T., L. E. Leach, M. Henson, and J. N. Jones. 1986b. *In Situ Biorestoration as a Groundwater Remediation Technique*. GWMR 56–64.

Wilson, B. H., G. B. Smith, and J. F. Rees. 1986a. Biotransformations of selected alkylbenzenes and halogenated aliphatic hydrocarbons in methanogenic aquifer material: a microcosm study. *Environ. Sci. Technol.* 20:997–1002.

Wimpee, C. F., T.-L. Nadeau, and K. H. Nealson. 1991. Development of species-specific hybridization probes for marine luminous bacteria by using in vitro DNA amplification. *Appl. Environ. Microbiol.* 57:1319–1324.

Winter, R. B., K. M. Yen, and B. D. Ensley. 1989. Efficient degradation of trichloroethylene by a recombinant *Escherichia coli. Bio/Technology* 7:282–285.

Won, W. D., L. H. DiSalvo, and J. Ng. 1976. Toxicity and mutagenicity of 2,4,6-trinitrotoluene and its microbial metabolites. *Appl. Environ. Microbiol.* 31:576–580.

Wong, P. T. S., Y. K. Chau, and P. L. Luxon. 1975. Methylation of lead in the environment. *Nature (London)* 253:263–264.

Wood, J. M. 1974. Biological cycles for toxic elements in the environment. *Science* 183:1049–1052.

Woods, S. L., J. F. Ferguson, and M. M. Benjamin. 1989. Characterization of chlorophenol and chloromethoxybenzene biodegradation during anaerobic treatment. *Environ. Sci. Technol.* 23:62–68.

Woolson, E. A. 1977. Generation of alkylarsines from soils. *Weed Sci.* 25:412–416.

Yang, J., and R. E. Speece. 1986. The response, acclimation, and remedial treatment of an enriched methanogenic culture to cyanide. *Toxicol. Assess. Int. Q.* 1:431–454.

Yaws, C. L., and H.-C. Yang. 1990. Water solubility data for organic compounds. *Pollut. Eng.* 22:70–75.

Yaws, C. L., H.-C. Yang, J. R. Hopper, and K. C. Hansen. 1990. 232 hydrocarbons: water solubility data. *Chem. Eng.* 97:177–181.

Yong, R. N., L. P. Tousignant, R. Leduc, and E. C. S. Chan. 1991. Disappearance of PAHs in a contaminated soil from Mascouche, Quebec, pp. 377–395. In: R. E. Hinchee and R. F. Olfenbuttel, (eds.), *In Situ Bioreclamation: Applications and Investigations for Hydrocarbon and Contaminated Site Remediation.* Butterworth-Heinemann. Boston.

Zaidi, B. R., Y. Murakami, and M. Alexander. 1989. Predation and inhibitors in lake water affect the success of inoculation to enhance biodegradation of organic chemicals. *Environ. Sci. Technol.* 23:859–863.

Zajic, J. E., and C. J. Panchel. 1976. Bio-emulsifiers. *Crit. Rev. Microbiol.* 5:39–66.

Zhang, X., and J. Wiegel. 1990. Sequential anaerobic degradation of 2,4-dichlorophenol in freshwater sediments. *Appl. Environ. Microbiol.* 56:1119–1127.

Zieve, R., and P. J. Peterson. 1981. Factors influencing the volatilization of selenium from soil. *Sci. Total Environ.* 19:277–284.

Zieve, R., and P. J. Peterson. 1985. Sorption of dimethylselenide by soils. *Soil Biol. Biochem.* 17:105–107.

Zimmerman, R., R. Iturriaga, and J. Becker-Birck. 1978. Simultaneous determination of the total number of aquatic bacteria and the number thereof involved in respiration. *Appl. Environ. Microbiol.* 36:926–935.

Zitrides, T. G. 1990. Bioremediation comes of age. *Pollut. Eng.* 22(5): 57–62.

8

Saturated Zone Remediation of VOCs Through Sparging

Ann N. Clarke and Robert D. Norris
ECKENFELDER INC.
Nashville, Tennessee

David J. Wilson
Vanderbilt University
Nashville, Tennessee

INTRODUCTION

The unique properties of chlorinated solvents have caused them to be widely used at industrial sites and governmental installations in connection with manufacturing operations, research, maintenance, and so on. The chloride content of these organic compounds causes them to be denser than water. Such dense nonaqueous-phase liquids (DNAPLs) move under the force of gravity through the unsaturated zone and continue to sink through the saturated zone. DNAPLs are impeded by whatever impermeable layer underlies the aquifer. When this occurs, pools of DNAPLs can accumulate. Additionally, as the solvent settles through the saturated zone, ganglia (irregular globules of DNAPL in the interstices of the soil) remain in the saturated soil. Work by Schwille (1988) has indicated that as much as 10 gallons of DNAPL per cubic yard of soil can remain as ganglia distributed throughout the saturated zone.

Not only does DNAPL cause contamination of groundwater in dissolved and ganglia forms, but the pooled DNAPL at the bottom of the saturated zone provides a source of long-term recontamination after the overlying groundwater and/or soil is remediated. Additionally, the pooled DNAPL can penetrate the underlying impermeable boundary, as is quite frequently seen in the contamination of fractured bedrock. This contamination is difficult to remove, and remediation, if possible at all, is very slow. For the length of time the DNAPL has had to diffuse into a porous bedrock, a comparable amount of time is required for it to diffuse back out of the bedrock. This time is extended further by the fact that some of the DNAPL continues its inward migration even after the source of contamination has been removed and diffusion outward has begun. Therefore, any technologies that can facilitate the remediation of both the free-phase DNAPL pool and the ganglia in the saturated zone as well as the solvent dissolved in the groundwater itself would be of great importance. Sparging may well be one such technology, although it is still under development and our understanding of it in this regard needs improvement.

There are currently two broad approaches to sparging volatile organic compounds (VOCs) from the saturated zone using air. One involves the use of individual sparging wells placed throughout the contaminated zone to remove volatile components from across a wide area. Wells are screened over a narrow interval located at the bottom of an aquifer or below the deepest contamination within the aquifer, as shown in Figure 1. Compressed air is forced from the well screen and flows radically outward and upward through the soils. As the air bubbles move through the soils, volatile compounds dissolved in the groundwater or sorbed to the surface of soil particles are volatilized and swept

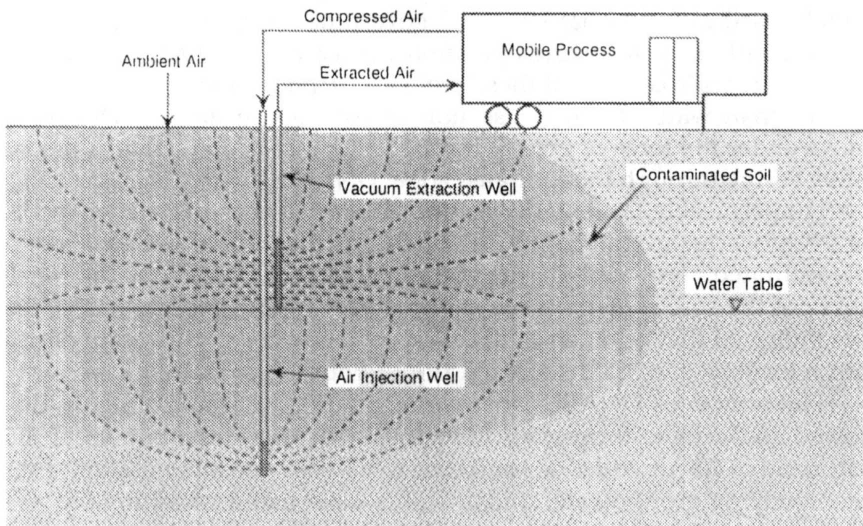

Figure 1 Generalized diagram of saturated zone sparging/in situ vapor stripping of the vadose zone.

to the unsaturated zone with the air bubbles. The process of air injection increases dissolved oxygen concentrations, which can lead to increased rates of biooxidation of biodegradable substances and can act to disturb the soil particles, thus increasing the opportunity for DNAPLs to be dissolved or volatilize. If wells are judiciously (fortunately) placed, pools and ganglia of DNAPL can be aggressively reduced.

The transfer of volatiles to the unsaturated zone requires that an in situ vapor stripping system be used to capture the volatile constituents and transfer them to the surface for concentration using either activated carbon or a condensation unit or for destruction using either a thermal treatment unit or a vapor-phase bioreactor. For biodegradable volatile constituents it may be possible to design a system to use the unsaturated zone as a bioreactor.

The injection of air into groundwater can also cause increased lateral movement of the groundwater. This raises the potential for accelerated migration of dissolved contaminants. The propensity for migration can be reduced by operating the aeration system on an off/on cycle. In many cases, it is advisable to incorporate a groundwater recovery system with air sparging.

The other approach uses one or more aeration curtains, oriented at right angles to the flow of the groundwater plume to remove volatiles from groundwater as it flows from the contaminated area, as shown in Figure 2. Aeration curtains can be created in trenches backfilled with porous media. The trenches have a horizontal slotted pipe near the bottom of the trench to supply air. As the groundwater flows through the trench, the rising air bubbles strip the volatiles to the top of the trench. Typically, a vapor recovery system with some form of off-gas treatment would be incorporated. As with the aeration wells, biodegradation may occur within the aeration curtain, thus reducing the need for off-gas treatment.

BACKGROUND

Many sites of concern are contaminated with VOCs of low water solubility from leaking underground storage tanks (Lyman and Noonan, 1990), spills, improper waste disposal, and so on. In the vadose zone one can take advantage of the volatility of these compounds to remove them by soil vapor extraction, but this technique cannot be used below the water table (Pedersen and Curtis, 1991). The traditional approach to dealing with VOC contaminated aquifers has been to pump the affected groundwater to the surface and treat it by means of air stripping, granular activated carbon absorption, or both.

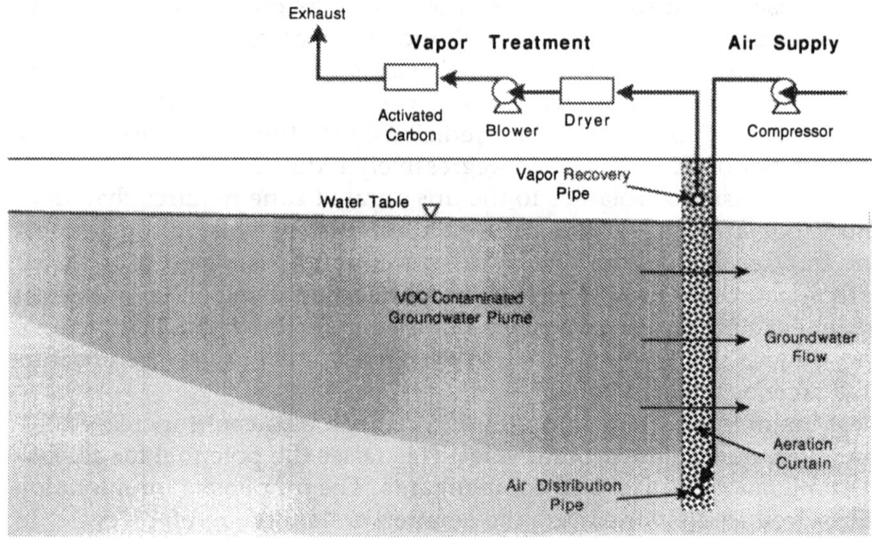

Figure 2 Aeration curtain concept for in situ groundwater sparging.

The pump-and-treat method is the oldest and probably the most widely selected for remediating techniques for sites involving contaminated groundwater. The last decade has seen an unprecedented surge in the implementation of groundwater pump-and-treat systems throughout the country. In fiscal year 1989 alone, over 60 records of decision selected groundwater pump and treat systems as part of the remedial action. The vast majority of these systems are aimed at purging groundwater of VOCs. The capital and operational costs of such systems can be high. The total operational costs can become extremely high since many of these systems will have to continue operating for quite long periods of time. This is particularly true when the aquifer is contaminated by DNAPLs or where cleanup kinetics are controlled by the diffusion-limited release of contaminants (see, e.g., Mutch et al., 1992).

The pump-and-treat method is satisfactory if the contaminant is relatively soluble in water and if the aquifer medium has a relatively constant permeability. The pump-and-treat method is less satisfactory if the contaminant is of low solubility in water and is more dense than water, so that free-phase DNAPL may be present, and/or if the permeability of the medium is highly variable (i.e., fractured porous rock, clay lenses in sand or gravel, etc.). In the first case, the rate of equilibrium between the droplets/ganglia of DNAPL and the mobile aqueous phase may be quite slow, and in the second, diffusion of contaminant from stagnant pore water in the porous domains of low permeability into the mobile water may be severely rate limiting. The situation may be even more unfavorable if there are pools of DNAPL at the bottom of the saturated zone. In any case, cleanup will be slow, with prolonged tailing. Rebound of contaminant concentration in the groundwater may also be observed after pump-and-treat operations are stopped. A major problem in the remediation of contaminated groundwater is the question of how to deal with the very severe mass transfer rate limitations that arise if DNAPL is present and/or if fractured porous bedrock is contaminated. DNAPLs that are potentially amenable to removal by sparging include trichloroethylene (TCE), perchloroethylene (PCE), 1,1,1-trichloroethane (1,1,1-TCA), carbon tetrachloride, and other reasonably volatile chlorinated solvents. DNAPLs having low volatility and small Henry's constants, such as polychlorinated biphenyls (PCBs), are not good prospects for removal by sparging.

The literature contains several detailed discussions concerning the factors controlling the removal of organic substance from soils and groundwater. Naymik (1987) included a discussion of fractured porous media in a review of the dynamics of groundwater. A common approach has been to use a dual-porosity model, with high porosity with generally low permeability for the porous blocks, and low porosity with high permeability for the fracture system (Huyakorn et al., 1983; Bibby, 1981; Rasmuson and Neretneiks, 1980, 1981). All results indicate that diffusion kinetics limitations on contaminant removal rates are severe.

Weber and his associates (Powers et al., 1991) have recently published a study of nonequilibrium dissolution of NAPLs in subsurface systems. They noted the importance of nonequilibrium effects in the solution of "blobs" of NAPL in advecting groundwater, and ascribed these to rate-limited mass transport between the nonaqueous and aqueous phase, bypassing of advecting aqueous phase around contaminated regions having a low aqueous phase permeability, and nonuniform flow due to aquifer heterogeneities.

Wilson et al. have examined the implications of matrix diffusion effects for the cleanup of fractured porous rock and other heterogeneous aquifers (Wilson and Mutch, 1990; Wilson 1992; Mutch et al., 1992). Under quite reasonable circumstances the remediation of such aquifers by conventional pump-and-treat methods was estimated by means of a mathematical model to take decades to centuries.

As it has become more and more apparent that the very low diffusion processes involved in pump-and threat operations must result in extremely long remediation times at many sites in which the zone of saturation has become contaminated, workers have searched for alternatives that would permit more rapid cleanups to the target levels. One of the more promising of these alternatives appears to be sparging in the zone of saturation.

The USEPA conducted a symposium on soil venting (soil vapor extraction) in Houston, Texas, from April 29 to May 1, 1991, at which approaches to sparging were described and results were presented. However, sparging, or in situ groundwater stripping, has been tested on a very limited basis in the United States. At the Houston meeting, Herrling and Stamm (1991) described the use of vacuum-vaporizer wells (a type of sparging well) for in situ removal of strippable VOCs in the vadose and saturated zones. This technology is now established in Germany, and these workers presented both experimental results and a mathematical model for design purposes. At the same meeting, Brown and Fraxedas (1991) described the use of a simpler configuration (simple air injection or sparging wells) which Ground Water Technology, Inc., has used successfully in the United States for the in situ removal of VOCs from contaminated groundwater. Hiller (1991) also noted the use in Germany of vadose zone soil vapor extraction in conjunction with groundwater aeration by means of air injection wells for the in situ removal of VOCs in groundwater. The modeling work that has been done to date appears to be restricted to the removal of dissolved VOCs, and therefore does not address the very serious mass transport problems associated with the removal of DNAPLs.

In these sparging techniques, one uses air injection to generate convective currents in the aquifer in the vicinity of the well. This will increase local turbulence, thereby enhancing mass transport of VOCs between the aqueous phase and the organic liquid phase, and will also provide a vapor-phase route for the efficient removal of dissolved VOCs from the aquifer. The contaminated vapor phase can then be captured for treatment by any of the methods used in soil vapor extraction; activated carbon and catalytic oxidation are the methods most commonly used for chlorinated VOCs. If the VOCs are biodegradable, they might be destroyed along with other biodegradable constituents in both the saturated and vadose zones by aerobic soil microorganisms as the oxygen-rich injected gas moves up through the vadose zone.

VERTICAL WELL SPARGING

This technology is relatively new in the United States but has been used for several years in Europe. An overview of the technique was recently provided by Middleton and Hiller (1990). Sparging systems make use of the volatility and limited water solubility of some organic compounds. Air is introduced at a point below the groundwater surface. As the air moves radically outward and upward to the unsaturated zone, it carries with it the VOCs that have partitioned into the gas phase. As the air containing the VOC vapors reaches the unsaturated zone, it is captured by an in situ vapor stripping system (see Figure 1). Because there exists some potential for increasing the migration of dissolved constituents during sparging, sparging systems should generally be operated with a groundwater recovery system in place.

As with all in situ technologies, successful implementation is very site specific. Therefore, on-site pilot scale testing of saturated zone sparging is needed to support optimum design of a full-scale facility. Information is needed on (1) the effective radius of influence of wells; (2) the effects of gas flow rate and pressure; (3) the nature and quantity of the site constituents; (4) the hydrogeology of the saturated zone, including water level; (5) the characterization of the vadose zone; and (6) if biodegradable constituents are present, the CO_2 and O_2 concentrations in the groundwater.

A sparging well treatability study was conducted by ECKENFELDER INC. at an active industrial site in California (Confident Client, 1991). Sparging resulted in an average 49% decrease in the total VOC concentration in the groundwater at two test locations. The TCE present in the groundwater exhibited a 48% reduction (23 mg/L to 12 mg/L) at one location and a 25% reduction (10 mg/L to 7.5 mg/L) at the other after 3 h of sparging.

AERATION CURTAIN TECHNOLOGY

Horizontal wells have recently been installed and tested for in situ remediation of groundwater and soils at the Department of Energy (DOE) Savannah River Site (Kaback et al., 1989; Kaback and Looney, 1989). One deep horizontal well, installed below the water table, was used for air injection to strip VOCs from the groundwater, while a shallow horizontal well in the vadose zone recovered the vapor-phase VOCs. This concept is based on directional drilling technology developed for the oil industry (Langseth, 1990).

The aeration curtain approach to in situ groundwater stripping has several advantages over the horizontal well approach. Because the aeration curtain would be installed using trenching technology, the curtain placement would be much more precise than horizontal well technology can achieve. The deep well at the Savannah River Site missed the target azimuth by 53°. This is significant since the well lateral reach was approximately 300 ft. In addition, the aeration curtain would provide much more uniform air stripping than would the horizontal well since the curtain would be constructed of a gravel packing of uniform width and uniform characteristics. Nonuniform air dispersion due to clay lenses without a sandy aquifer would be a common occurrence with horizontal well air injection. Air injected into an undisturbed formation can be expected to tend to channel in a nonuniform manner since capillary pressure within the saturated soil resists the intruding air. A much lower capillary pressure is anticipated within a gravel or crushed crock aeration curtain.

Advances in subsurface construction techniques permit construction of lateral curtains in both shallow and deep aquifers at a small fraction of the cost of older conventional technologies. One of these new techniques is the use of specially adapted trenching machines that can, simultaneously, excavate a narrow trench, lay in a perforated pipe, and embed it in gravel. A second technological advance is the use of polymer slurry trenches which permit excavation of a deep trench filled with a biodegradable polymer slurry into which a perforated pipe and gravel can be placed (Day, 1990). The slurry biodegrades to a liquid in a matter of hours, leaving the gravel in place for use as the aeration curtain. The aeration curtain is also expected to enhance in situ biodegradation by supplying oxygen to the aquifer. Future development activities could include injection of nutrients with the air to promote in situ biodegradation of organic compounds having lower Henry's constants. The essential components of the technology are illustrated in Figure 2. The lateral air distribution pipe is placed across the leading edge of the VOC plume at a depth below the bottom of the plume. If the full vertical thickness of the aquifer is contaminated, the lateral air distribution pipe would be placed at the base of the aquifer. If the plume is confined to the upper part of the aquifer, the distribution pipe would be placed at a depth just below the bottom of the plume, as shown in Figure 2.

Air is then injected into the lateral air distribution pipe from a surface compressor. The rising air bubbles would strip VOCs from the groundwater as the groundwater passes horizontally across the rising airstream. The off gas can then be collected by a parallel vapor recovery pipe in the overlying unsaturated zone or by using conventional vertical well vadose zone vapor stripping technology. These options provide flexibility for implementation of this groundwater stripping technique at a variety of sites. The collected off gas would then be treated by carbon adsorption, catalytic oxidation, regenerative thermal oxidation, biological treatment, or other techniques. Figure 2 illustrates the use of carbon adsorption. Depending on the nature and concentration of groundwater contaminants, as well as upon regulatory requirements, there may also be cases where off-gas collection and treatment would not be necessary. In such cases the cost of the technology would be significantly lessened, since neither the vapor recovery pipe nor the soil gas treatment system would be required. It may also be desirable in some situations to construct a multiple-stage in situ stripping system by installing two or more parallel aeration curtains. Higher degrees of groundwater treatment would be expected with such a system, since the second curtain would provide stripping of those contaminants evading removal by the first curtain. This has been examined theoretically by Wilson (1991) and is discussed later in this chapter. One would expect that this technology could be used successfully and cost-effectively at both large and small sites.

THEORY AND MATHEMATICAL MODELING

Simple steady-state and nonsteady-state models for the removal of dissolved volatile organic compounds from aquifers by sparging techniques have been developed. A method was developed for estimating the streamlines and transit times of water in a stagnant or nearly stagnant aquifer in the vicinity of a sparging well. The resulting flow velocities are then used in a model for the sparging of a VOC obeying Henry's law; this model is discussed later in the chapter (Wilson, 1992). These flow velocities can also be used in the development of a model for the removal of droplets/ganglia of DNAPL distributed in an aquifer in the vicinity of a sparging well; here solution and diffusion kinetics can be extremely important. Also, a method was developed for using vadose zone soil gas pressure measurements in the vicinity of a single sparging well to estimate the radius of cone around the well through which the injected gas rises to the surface of the zone of saturation (Wilson et al., 1993a,b).

Aeration Curtain Model

A local equilibrium model has also been developed to describe the removal of dissolved VOCs by in situ sparging by means of one or more horizontal lateral slotted pipes at the bottom of trenches filled with gravel or crushed rock and oriented at right angles to the direction of flow of the groundwater. The effects of air injection rate, groundwater flow rate, aquifer thickness, number of theoretical transfer units (related to vertical dispersion in the curtain), Henry's constant for the VOC, and initial VOC concentrations were explored. The geometry of the model is shown in Figure 3. The symbols used are as follows.

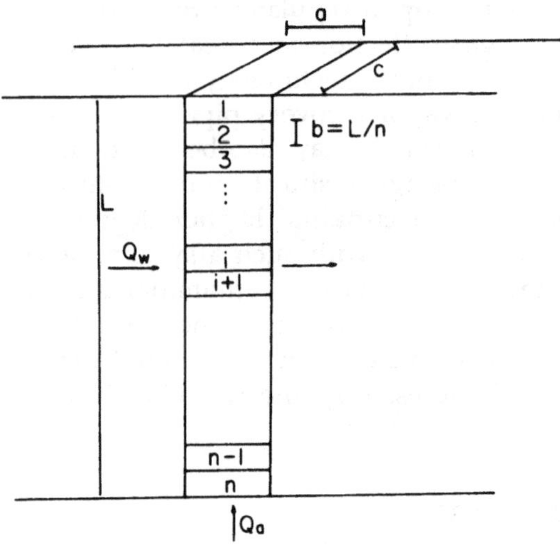

Figure 3 Geometry and notation for the aeration curtain.

L = thickness of aquifer, m
a = width of aeration curtain, m
c = length of aeration curtain, m
n = number of equivalent theoretical transfer units into which the aeration curtain is partitioned for analysis
b = thickness of an equivalent theoretical transfer unit, m
Q_w = total water flow rate through the curtain, m^3/s
Q_a = total airflow rate through the curtain, m^3/s
c_o = incoming aqueous contaminant concentration, kg/m^3
K_H = Henry's constant of the contaminant, dimensionless
ν = voids fraction in the crushed rock/gravel curtain, dimensionless

The hydrostatic pressure at the bottom of the ith cell (transfer unit) is given by

$$P_1 = \left(1 + \frac{ib}{10.336}\right) \text{atm} \tag{1}$$

so the volumetric airflow rate at the bottom of the jth cell is given by

$$Q_{a_j} = \frac{Q_a}{1 + jb/10.336} \tag{2}$$

We shall neglect the volume of air compared to the volume of water in a cell. Let m_i be the mass of contaminant in the ith cell (kg). Then

$$\frac{dm_i}{dt} = \frac{Q_w}{n}(c_o - c_{w_i}) + Q_{a_i} c_{a_{i+1}} - Q_{a_{i-1}} c_{a_i} \tag{3}$$

The steady-state assumption for the system then permits us to set equation (3) equal to zero.

We assume that Henry's law applies to the VOC in the aquifer, which gives

$$c_{a_j} = K_H c_{w_j} \qquad j = 1, 2, \ldots, n \tag{4}$$

On substituting equations (2) and (4) into equation (3), we obtain

$$0 = \frac{Q_w}{n}(c_o - c_{w_i}) + \frac{Q_a K_H c_{w_{i+1}}}{1 + ib/10.336} - \frac{Q_a K_H c_{w_i}}{1 + (i-1)b/10.336} \tag{5}$$

Solving equation (5) for c_{w_i} gives

$$c_{w_i} = \left(\frac{Q_w c_o}{n} + \frac{Q_a K_H c_{w_{i+1}}}{1 + ib/10.336}\right) \Bigg/ \left[\frac{Q_w}{n} + \frac{Q_a K_H}{1 + (i-1)\,b/10.336}\right] \qquad i = 1, 2, \ldots, n-1 \tag{6}$$

and

$$c_{w_n} = \frac{Q_w c_o}{n} \Bigg/ \left[\frac{Q_w}{n} + \frac{Q_a K_H}{1 + (n-1)\,b/10.336}\right] \tag{7}$$

The mean contaminant concentration in the water that has passed through the curtain is given by

$$c_{\text{out}} = \sum_{i=1}^{n} \frac{c_{w_i}}{n} \tag{8}$$

and the percent removal of VOC from the passing groundwater by the curtain is

$$R = 100\left(1 - \frac{c_{\text{out}}}{c_o}\right) \tag{9}$$

Let us next investigate the dependence of R, the percent removal of VOC by the aeration curtain, on the model parameters. Default parameters used in the calculations are listed in Table 1. The effect of airflow rate is given in Table 2. Percent VOC removal increases with increasing

Table 1 Aeration Curtain Default Parameters

Depth of aquifer	5 m
Thickness of aeration curtain	0.5 m
Length of aeration curtain	100 m
Number of theoretical transfer units	10
Voids fraction of the crushed rock curtain	0.4
Contaminant	Trichloroethylene
Henry's constant of contaminant (15°C)	0.2821
Contaminant concentration in groundwater incident on the aeration curtain	100 mg/L
Total water flow across the aeration curtain	0.1 m^3/s
Total airflow generating the curtain	2 m^3/s

airflow rate, but there is no abrupt transition as is the case with countercurrent aerators. Evidently, if the VOC removal is not sufficient at an airflow rate that is near the maximum feasible, a minor increase in flow rate will not be adequate to solve the problem.

The groundwater flow rate affects VOC removal as shown in Table 3. Percent VOC removal decreases with increasing groundwater flow rate, as expected, and again there is no relatively sharp transition to an overloaded condition such as one sees with countercurrent separations. The aeration curtain is a cross-current separation technique, so this is not surprising.

Variation in the number of theoretical transfer units representing the aeration curtain has only a minor effect on the percent VOC removal. Thus there is little advantage to be gained by reducing longitudinal dispersion in the air in the curtain.

Table 2 Effect of Airflow Rate on Percent VOC Removal

Airflow rate (m^3/s)	Percent VOC removal
0.5	67.02
1.0	80.96
1.5	86.64
2.0	89.71
3.0	92.96
5.0	95.68

Table 3 Effect of Groundwater Flow Rate on Percent VOC Removal

Groundwater flow rate (m^3/s)	Percent VOC removal
0.02	97.80
0.025	97.27
0.05	94.65
0.10	89.71
0.15	85.16
0.20	80.96
0.50	61.43

The model equations are linear in the VOC concentrations, so one expects that percent VOC removal will be exactly independent of the VOC concentration of the entering groundwater. This was found to be the case. A very weak dependence of percent removal on aquifer thickness was found; the difference in R for aquifers of 3 and 10 m thickness was less than 1%.

The size of the Henry's constant of the VOC has a very marked effect on R, as indicated in Table 4. This is such that the utility of the single aeration curtain technique for VOCs having Henry's constants of less than about 0.1 (dimensionless) is questionable. Such VOCs include 1,2-dichlorobenzene (0.060), 1,2-dichloroethane (0.055), 1,1,2-trichloroethane (0.027), tetralin (0.044), ethylene dibromide (0.020), 1,2-dichloropropane (0.053), dibromochloromethane (0.019), 1,2,4-trichlorobenzone (0.044), methyl ethyl ketone (0.016), and methyl isobutyl ketone (0.016).

The aeration curtain is a cross-current technique, so that it is essentially a single-stage method. One can improve its performance

Table 4 Effect of Henry's Constant, K_H, on Percent VOC Removal

K_H (dimensionless)	Percent VOC removal
0.30	90.28
0.2821 (TCE)	89.71
0.20	85.95
0.10	74.73
0.05	58.23

Table 5 Percent VOC Removals, R, from Multiple-Curtain Systems

n	R			
1	90.00	75.00	50.00	25.00
2	99.00	93.75	75.00	43.75
3	99.90	98.44	87.50	57.81
4	99.99	99.61	93.75	68.36

significantly by constructing a system having a number of aeration curtains in series. A simple analysis gives equation (10) for R_n, the percent VOC removal resulting from n aeration curtains in series.

$$R_n = 100\left[1 - \left(1 - \frac{R}{100}\right)^n\right] \tag{10}$$

Table 5 shows R_n values for systems with multiple aeration curtains. The useful range of the aeration curtain technique evidently can be extended significantly by employing multiple curtains. These can be separated only sufficiently to ensure that there is no movement of sparging air from one curtain to another.

Sparging by a Single Vertical Well

Here we develop a relatively simple model for sparging by means of a single vertical well that is injecting air near the bottom of the aquifer. As with the aeration curtain model above, we shall concern ourselves here only with the removal of dissolved VOCs. The model breaks down into two components: the calculation of the groundwater flow pattern in the vicinity of a sparging well; and the modeling of the movement and removal of the dissolved VOC.

Groundwater Flow in the Vicinity of a Sparging Well

Consider the situation illustrated in Figure 4, where we have a sink $-Q$ at the bottom of the aquifer and a source of water $+Q$ at the top of the aquifer, representing (1) the intake of water from the bottom of the aquifer into the bottom of the sparging zone and (2) the discharge of water from the top of the sparging zone in the near vicinity of the air pipe back into the top of the aquifer. Let h be the thickness of the aquifer. We have no-normal-flow boundary conditions at the bottom and top of the aquifer. The method of images can be used to construct a solution to Laplace's equation satisfying these boundary conditions and having the desired sources and sinks; this velocity potential function is

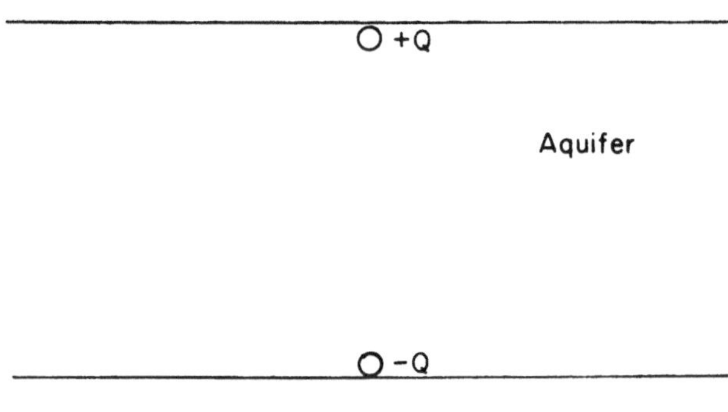

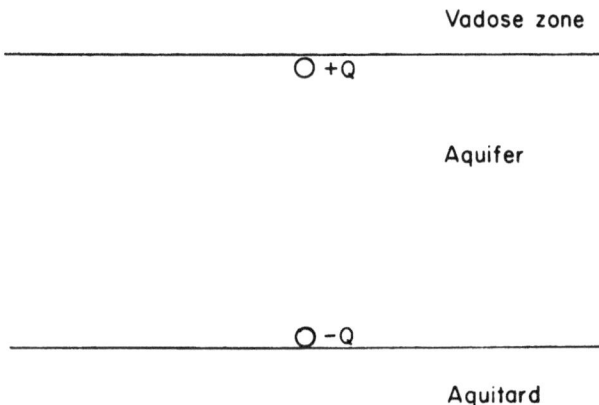

Figure 4 Locations of source and sink representing the water intake and discharge of a sparging well.

$$W = \sum_{n=-\infty}^{\infty} A \left\{ \frac{2Q}{\sqrt{r^2 + [z - (2n + 1)\,h]^2}} - \frac{2Q}{\sqrt{r^2 + (z - 2nh)^2}} \right\} \qquad (11)$$

The constant A is determined by the requirement that

$$2Q = \int_0^{2\pi} \int_0^{\pi} A \frac{2Q}{\rho^2} \rho^2 \sin\theta \, d\theta \, d\phi \qquad (12)$$

which yields $A = 1/(4\pi)$. Then

$$W = \sum_{n=-\infty}^{\infty} \frac{Q}{2\pi} \left\{ \frac{2Q}{\sqrt{r^2 + [z - (2n + 1)\,h]^2}} - \frac{2Q}{\sqrt{r^2 + (z - 2nh)^2}} \right\} \qquad (13)$$

The linear velocity of the fluid is given by

$$v = -\frac{1}{\nu} \nabla W \qquad (14)$$

which gives for the velocity components v_r and v_z the following:

$$v_r = \frac{Q}{2\pi\nu} \sum_{n=-\infty}^{\infty} \left(\frac{r}{\{r^2 + [z - (2n + 1)\,h]^2\}^{3/2}} - \frac{r}{[r^2 + (z - 2nh)^2]^{3/2}} \right) \qquad (15)$$

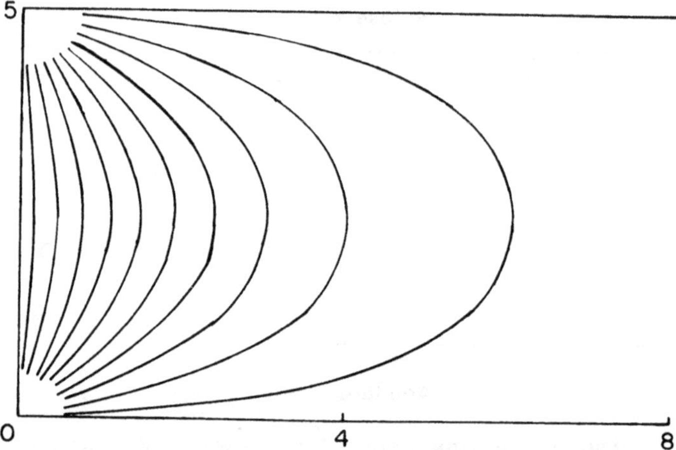

Figure 5 Streamlines in an aquifer in the vicinity of a single sparging well. The parameters used and the transit times along each of the streamlines are given in Table 6.

and

$$v_z = \frac{Q}{2\pi\nu} \sum_{n=-\infty}^{\infty} \left(\frac{z - (2n + 1)h}{\{r^2 + [z - (2n + 1)h]^2\}^{3/2}} - \frac{z - 2nh}{[r^2 + (z - 2nh)^2]^{3/2}} \right) \tag{16}$$

The flow paths and transit times are then calculated by integrating the parametric equations

$$\frac{dr}{dt} = v_r(r, z) \tag{17}$$

and

$$\frac{dz}{dt} = v_z(r, z) \tag{18}$$

Flow paths for such a system are shown in Figure 5. The parameters used in the calculations, the transit time of each of the streamlines shown, the maximum radii of the streamline, and the volumes enclosed by the figures of revolution generated by rotating the streamlines around the axis between the sink and the source are shown in Table 6. A dimensionless time $\tau = h^3t/Q$ and a dimensionless distance $x' = x/h$ can be used to calculate from Figure 5 and Table 6 the transit times for

Table 6 Sparging Well Parameters, Transit Times, Streamline Maximum Radii, and Volumes Enclosed

Thickness of aquifer: 5m
Aquifer porosity: 0.3
Induced water flow rate: 0.01 m^3/s

Transit times (s)	Maximum radius of streamline (m)	Volume enclosed by streamline (m^3)
1,490	0.147	0.156
1,580	0.445	1.441
1,770	0.755	4.198
2,110	1.086	8.867
2,680	1.452	16.254
3,690	1.872	27.995
5,610	2.379	47.354
9,780	3.036	81.812
21,930	4.003	154.294
104,650	6.003	394.710

geometrically similar streamlines with different values of the aquifer thickness h and the flow rate Q. Table 6 shows that a volume around the sparging well of approximately $1.2h^3$ is flushed out fairly rapidly (within about 6 h with these parameters), but that flushing times increased up to 29 h if the volume of influence of the well is increased to a value of $3.2h^3$. Evidently, the practical radius of influence of a sparging well in an isotropic medium of constant permeability is somewhere around $0.8h$.

Movement of VOC by a Sparging Well

We use the velocity field calculated in the preceding section to develop an n-compartment model for the operation of a sparging well operating in a stagnant or nearly stagnant aquifer. The setup is shown in Figure 6.

The flow of fluid between the ith and the $(i + 1)$th annular volume elements in the top half of the domain is given by

$$Q_i^{out} = \int_0^{2\pi} \int_{h/2}^{h} \nu v_r \, d\theta \, dz \tag{19}$$

$$= Qr_i^2 \left(\int_{h/2}^{h} \sum_{n=-\infty}^{\infty} \{r_i^2 + [z - (2n + 1)h^2\}^{-3/2} - [r_i^2 + (z - 2nh)^2]^{-3/2} dz \right) \tag{20}$$

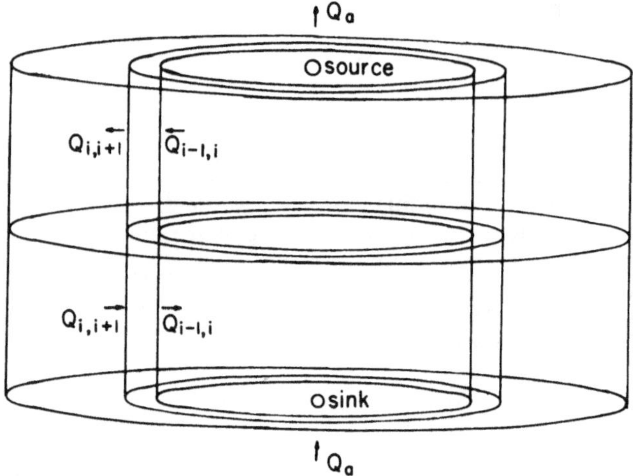

Figure 6 An n-compartment sparging well model, showing the partitioning of the domain in the vicinity of the sparging well into $2n$ annular volume elements.

Integration and insertion of the limits of integration then give

$$Q_i^{\text{out}} = Q \sum_{n=-\infty}^{\infty} \left[\frac{-1}{[1 + (r_i/2nh)^2]^{1/2}} + \frac{1}{\{1 + [r_i/(2n + 1/2)h]^2\}^{1/2}} + \frac{1}{\{1 + [r_i/(2n - 1)h]^2\}^{1/2}} - \frac{1}{\{1 + [r_i/(2n - 1/2)h]^2\}^{1/2}} \right] \tag{21}$$

as the flow rate outward from the ith to the $(i + 1)$th annular volume element in the top half of the domain. Henceforth we denote this as $Q_{i,i+1}$.

In the bottom half of the domain the flow rates are the same in magnitude but in the opposite direction (from the periphery of the domain toward the well). We use these flow rates to get an estimate of the movement of dissolved VOC between the volume elements by advection. This, together with the assumption that the central volume element is being sparged with air at a flow rate Q_a, that this volume element is well mixed, and that Henry's law applies, yields the following model equations for the sparging system:

$$V_i \frac{dc_i}{dt} = Q_{i-1,i}(c_{i-1} - c_i) + Q_{i,i+1}(c_{i+1} - c_i) \qquad i = 2,3, \ldots \tag{22}$$

Table 7 Default Parameters for the Runs Plotted in Figures 7 to 10[a]

Domain radius = 5 m
Aquifer thickness = 5 m
Radius of zone of contamination = 5 m
Number of annular compartments used in model = 20
Airflow rate Q_a = 0.01 m^3/s
Induced water circulation rate Q_w = 0.01 m^3/s
Henry's constant = 0.1 (dimensionless)
Porosity of aquifer medium = 0.3
Initial VOC concentration in groundwater = 25 mg/L

[a]Departures from these values are indicated in the figure captions.

and

$$V_1 \frac{dc_1}{dt} = Q_{1,2}(c_2 - c_1) - Q_a K_H c_1 \tag{23}$$

Here V_j is the volume of the jth annual volume element,

$$V_j = [j^2 - (j - 1)^2]\pi(\Delta r)^2 h \tag{24}$$

Some results obtained with this model are shown in Figures 7 to 10. Default parameters used in these runs are listed in Table 7; variations from these values are indicated in the captions to the figures. In Figure 7 the airflow rate through the sparging well Q_a is varied and the induced water circulation Q_w is held constant. In Figure 8 the water circulation induced by the sparging well is varied and the airflow rate is held constant. We see in both figures that increased flow rates result in increased VOC removal rates, but the removal rates increase less than linearly with Q_a and with Q_w. In Figure 9, Q_a and Q_w are increased together, with Q_w proportional to Q_a. One would expect that the actual situation would be rather similar to this (Q_w increasing with increasing Q_a), except that the functional dependence of Q_w on Q_a is uncertain. In the runs plotted in Figure 9, the removal rate is proportional to Q_a. Removal rates increase with increasing Henry's constant, as shown in Figure 10; removal rate increases sublinearly with K_H for the parameter values used in these runs.

Both this model and the aeration curtain model are for handling the removal of dissolved VOC. They are not applicable if nonaqueous-phase liquid (NAPL) is present, since in that situation the rate-limiting step is almost certain to be the rate of dissolution of the nonaqueous-phase liquid into the aqueous phase, which is not included in these models. One would expect that the flow velocities calculated above would be helpful

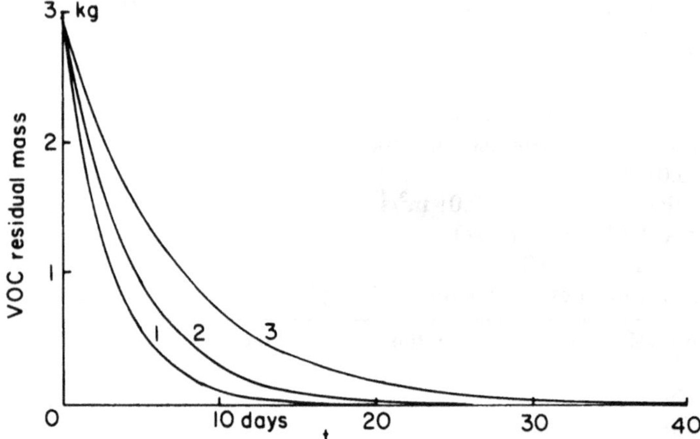

Figure 7 VOC sparging simulation using the n-compartment model. Effect of airflow rate on removal rate of VOC. See Table 7 for default run parameters. Q_a = 0.01, 0.005, and 0.0025 m^3/s for runs 1, 2, and 3, respectively.

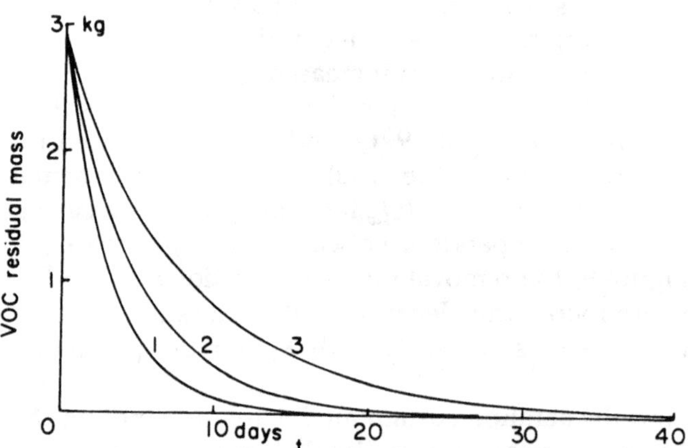

Figure 8 VOC sparging simulation using the n-compartment model. Effect of induced water flow rate on removal rate of VOC. See Table 7 for default run parameters. Q_w = 0.01, 0.005, and 0.0025 m^3/s for runs 1, 2, and 3, respectively.

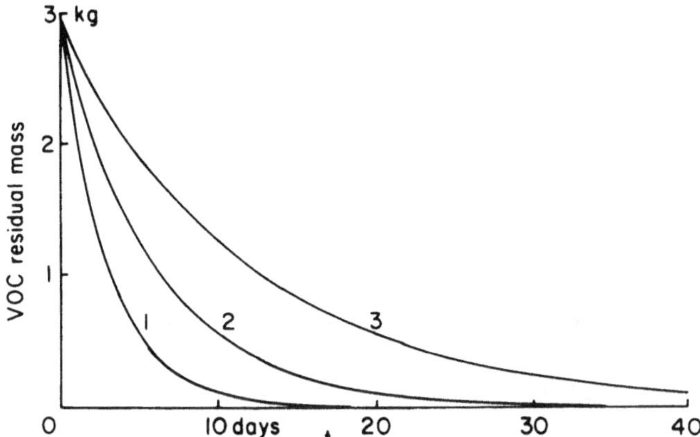

Figure 9 VOC sparging simulation using the n-compartment model. Effect of linked water and airflow rates on removal rate of VOC. See Table 7 for default run parameters. In these runs $Q_a = Q_w = 0.01$ (run 1), 0.005 (run 2), and 0.0025 m^3/s (run 3).

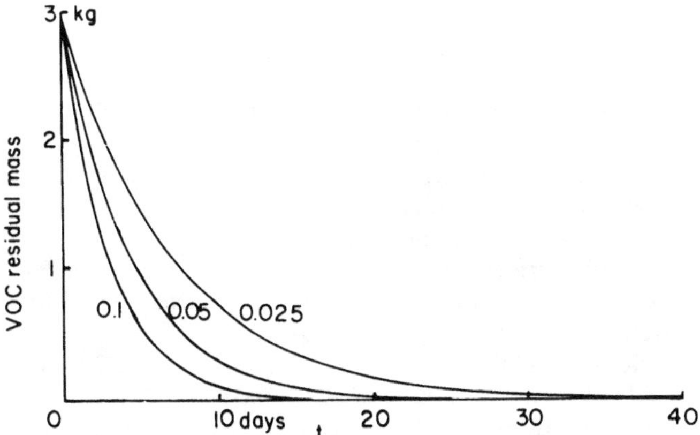

Figure 10 VOC sparging simulation using the n-compartment model. Effect of Henry's constant. Default parameters are given in Table 7. Henry's constants are 0.1, 0.05, and 0.025 as indicated.

in the development of a model for NAPL removal, since mass transfer from the NAPL phase is surely affected by the velocity of the water streaming past the stationary NAPL surfaces.

REFERENCES

Bibby, R. 1981. Mass transport of solutes in dual-porosity media. *Water Resour. Res.* 17:1075.

Brown, R. A., and R. Fraxedas. 1991. Air sparging: extending volatilization to contaminated aquifers. *Symposium on Soil Venting,* Houston, April 29–May 1; Robert S. Kerr Environmental Research Laboratory and National Center for Ground Water Research, sponsors.

Day, S. R. 1990. Deep groundwater collection trenches by the biopolymer drain method. *Superfund '90, Proceedings of the 11th National Conference of the Hazardous Materials Control Research Institute,* Washington, DC, November.

Herrling, B., and J. Stamm. 1991. Vacuum-vaporizer-wells (UVB) for in situ remediation of volatile and strippable contaminants in the unsaturated and saturated zones. *Symposium on Soil Venting,* Houston, April 29–May 1; Robert S. Kerr Environmental Research Laboratory and National Center for Ground Water Research, sponsors.

Hiller, D. H. 1991. Performance characteristics of vapor extraction system operated in Europe. *Symposium on Soil Venting,* Houston, April 29—May 1; Robert S. Kerr Environmental Research Laboratory and National Center for Ground Water Research, sponsors.

Huyakorn, P. S., B. H. Lester, and C. R. Faust. 1983. Finite element techniques for modeling groundwater flow in fractured aquifers. *Water Resour. Res.* 19:1019.

Kaback, D. S., and B. B. Looney. 1989. *Status of In Situ Air Stripping Tests and Proposed Modifications: Horizontal Wells AMH-1 and AMH-2 Savannah River Site.* WSRC-RP-89-0544. Westinghouse Savannah River Company, Savannah River Laboratory, Aiken, SC.

Kaback, D. S., B. B. Looney, J. C. Corey, L. M. Write, and J. L. Steele. 1989. Horizontal wells for in situ remediation of groundwater and soils. *NWWA 3rd Outdoor Action Conference Proceedings,* Orlando, FL, May.

Langseth, D. E. 1990. Hydraulic performance of horizontal wells. *Superfund '90, Proceedings of the 11th National Conference of the Hazardous Materials Control Research Institute,* Washington, DC, November, pp. 398–408.

Lyman, W. J., and D. C. Noonan. 1990. *Assessing UST Corrective Action Technologies: Site Assessment and Selection of Unsaturated Zone Treatment Technologies.* Camp, Dresser & McKee, Inc., EPA/600/2-90/001. U.S. Environmental Protection Agency, Risk Reduction Engineering Laboratory, Cincinnati, OH.

Middleton, A. C., and D. H. Hiller. 1990. In situ aeration of groundwater, a technology overview. *Conference on Prevention and Treatment of Soil and Groundwater Contamination in the Petroleum Refining and Distribution Industry,* October 16–17, Montreal, Quebec, Canada.

Mutch, R. D., Jr., J. I. Scott, and D. J. Wilson. 1992. Cleanup of fractured rock aquifers: implementations of matrix diffusion. *Environ. Monit. Assess.*, 19, 1–26, 1992.

Naymik, T. G. 1987. Mathematical modeling of solute transport in the subsurface. *CRC Crit. Rev. Environ. Control* 17:229.

Osejo, R. E., and D. J. Wilson. 1991. Soil cleanup by in situ aeration. IX. Diffusion constants of volatile organics and removal of underlying liquid. *Sep. Sci. Technol.*, *26*, 1991.

Pedersen, T. A., and J. T. Curtis. 1991. *Soil Vapor Extraction Technology Reference Handbook*. Camp, Dresser & McKee, Inc., EPA/540/2-91/003. U.S. Envivironmental Protection Agency, Risk Reduction Engineering Laboratory, Cincinnati, OH.

Powers, S. E., C. O. Louriero, L. M. Abriola, and W. J. Weber, Jr. 1991. Theoretical study of the significance of nonequilibrium dissolution of nonaqueous phase liquids in subsurface systems. *Water Resour. Res.* 27:463.

Rasmuson, A., and I. Neretneiks. 1980. Exact solution for diffusion in particles and longitudinal dispersion in packed beds. *AIChE J.* 26:686.

Rasmuson, A., and I. Neretneiks. 1981. Migration of radionuclides in fissured rock: the influence of micropore diffusion and longitudinal dispersion. *J. Geophys. Res.* 86(B5):3749.

Schwille, F. 1988. *Dense Chlorinated Solvents in Porous and Fractured Media: Model Experiments* (English translation). Lewis Publishers, Chelsea, MI.

Wilson, D. J. 1992. Groundwater cleanup by in situ sparging. II. Modeling of dissolved VOC removal. *Sep. Sci. Technol.*, 27, 1675–1690, 1992.

Wilson, D. J., and R. D. Mutch, Jr. 1990. Migration of pollutants in groundwater. IV. Modeling of the pumping of contaminants from fractured bedrock. *Environ. Monit. Assess.* 15:183.

Wilson, D. J., R. D. Mutch, and J. I. Scott. 1993. Matrix diffusion effects in the cleanup of heterogeneous aquifers. *Environ. Monit. Assess.*, 26, 49–64, 1993.

Wilson, D. J., S. Kayano, R. D. Mutch, Jr., and A. N. Clarke. 1992b. Groundwater cleanup by in situ sparging. I. Mathematical modeling. *Sep. Sci. Technol.* 27:1023.

9

In Situ Vitrification

Kenton H. Oma
ECKENFELDER INC.
Nashville, Tennessee

INTRODUCTION AND BACKGROUND

In situ vitrification (ISV) is the process of melting contaminated soil, buried wastes, or sludges in place to render the material nonhazardous. The ISV process produces a glassy and/or crystalline solid matrix that is resistant to leaching and more durable than natural granite or marble. ISV technology is based on the concept of joule heating to melt the soil or sludge electrically. The molten soil is heated as high as 1600 to 2000°C which acts to destroy organic pollutants by pyrolysis.

The ISV process was initially developed by U.S. Department of Energy (USDOE) to provide enhanced isolation of previously disposed radioactive wastes (Oma et al., 1982). The technology is a spinoff of glass melter research that has been conducted by the USDOE since the early 1970s for treatment of high-level radioactive wastes. The ISV technology was first tested in 1980, and since that time the potential for the technology to immobilize many organic and inorganic hazardous chemical wastes became obvious. There are several general areas where the ISV process might be applied to hazardous wastes: contaminated soil sites, burial grounds, tanks that contain hazardous material in the form of either sludge or salt cake, process sludges, and others.

DESCRIPTION AND STATUS OF TECHNOLOGY

ISV Process Description

ISV is a process that melts soil and sediment electrically for the purpose of treating free and/or containerized contaminants present within the treatment volume. Most ISV applications involve melting of natural soils. However, other materials, such as process sludge, mill tailings, sediments, process chemicals, or other naturally occurring inorganics, may be treated effectively.

To accomplish ISV, four electrodes are inserted into the soil to the desired treatment depth. To initiate the process, a conductive mixture of flaked graphite and glass frit is placed among the electrodes to act as the starter path for the electric circuit. Soil at ambient temperature does not have sufficient electric conductivity to allow initiation of the process without the startup path. An electric current passed between the electrodes through the graphite and frit path initiates the melting process. The graphite starter path is eventually consumed by oxidation and the current is transferred to the surrounding molten soil, which is then electrically conductive. Figure 1 illustrates the operating sequence of the ISV process. As the melt grows downward and outward, the nonvolatile elements become part of the melt matrix and the organic compounds are destroyed by pyrolysis. The pyrolyzed by-products migrate to the surface of the vitrified zone, where they combust in the presence of air. Inorganic materials are dissolved into or are encapsulated in the vitrified mass. Convective currents within the melt uniformly mix materials that are present in the soil. When the desired melt depth and volume have been achieved, the electric current is discontinued and the molten volume is allowed to cool and solidify. During the process, a hood is placed over the affected area to collect any combustion gases and entrained particles for off-gas treatment.

Individual ISV "settings" may grow to encompass a total melt mass of up to 1000 tons and a maximum width of about 30 ft. Single ISV setting melt depths as great as 30 ft are considered possible with the ex-

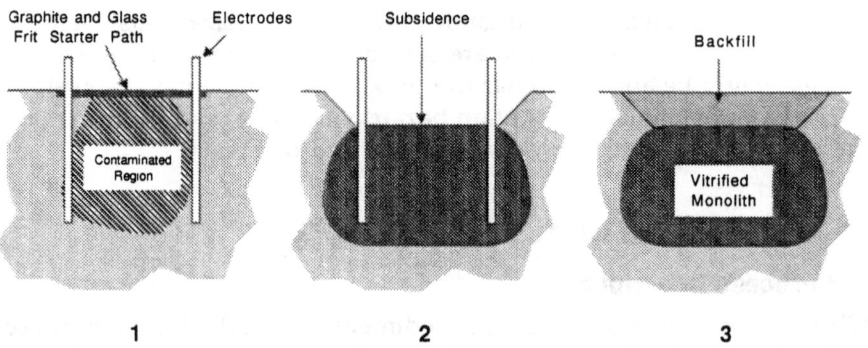

Figure 1 In situ vitrification operating sequence.

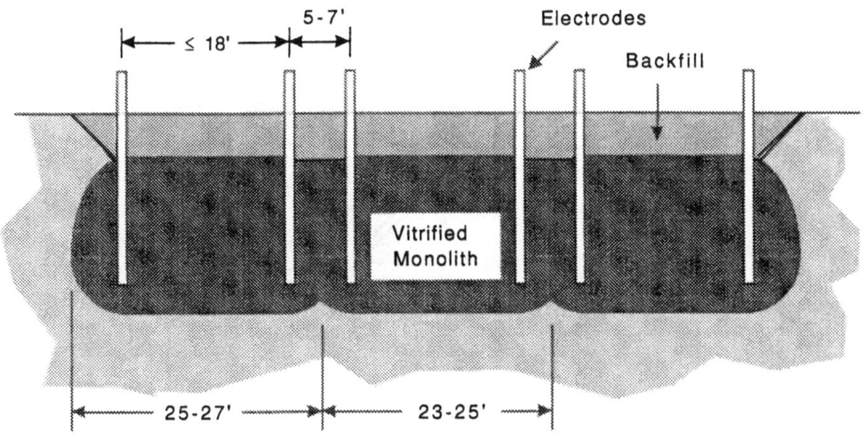

Figure 2 Adjacent ISV melt configuration.

isting large-scale ISV equipment, although depths of up to 19 ft have been demonstrated. Figure 2 illustrates how adjacent ISV settings are positioned to fuse together and to process the desired volume completely at a waste site.

Specific site characteristics must be considered in determining the applicability of ISV. In the event that feasibility tests indicate problems in soil conductance or vitrification, sand, soda ash, and/or glass frit can be mixed with the soil to improve the process. A combination of high soil permeability and the presence of groundwater can create economic limitations to the process. The process can work with fully saturated soils; however, the soil moisture must be evaporated before the soil will begin to melt. If the soil moisture is being recharged by an aquifer, there is an additional economic impact. Soils with a permeability higher than 10^{-4} cm/s are considered difficult to vitrify in the presence of groundwater and would probably require temporary groundwater diversion (if practical) during processing.

The basic components of a large-scale ISV system are illustrated in Figure 3. Three-phase electric power is usually taken from the local utility at transmission voltages of 12.5 or 13.8 kV. Alternatively, the power may be generated on site by a portable diesel generator. The standard three-phase AC power is supplied to a special multitapped transformer (Scott-Tee connected) that converts the power into two-phase AC for melting the soil. A typical large-scale ISV power supply has 16 voltage taps with an available voltage turn-down ratio in excess of 10:1. This

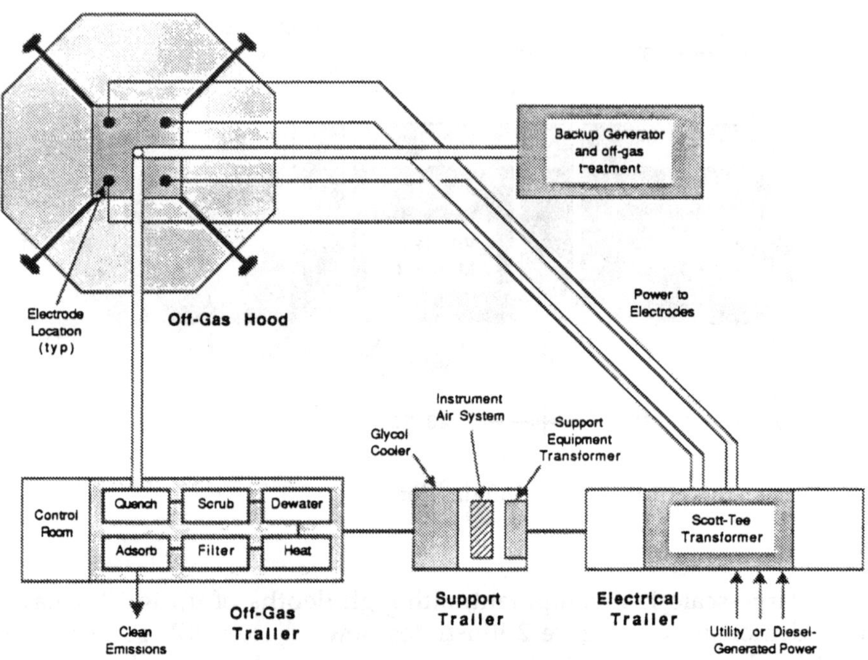

Figure 3 Basic components of large-scale ISV.

wide range of available voltages permits a high power input to the vitrification zone throughout the ISV operation. This is important since initially high voltages are required during startup when the conductive path is very small and the electrical resistance is relatively high, while later in the ISV operations, electric resistance is many times lower because of the increased mass of molten soil between the electrodes. Electrical power is transferred to the electrodes through flexible insulated conductors.

Electrodes are either graphite or of a molybdenum and graphite design and are described in more detail in the section "Electrode Configuration." The graphite electrodes are fed into the ground as the ISV melting proceeds. The molybdenum and graphite electrodes are installed in the ground utilizing casings that are vibrated or driven into place, followed by vibratory extraction of the casings after the electrodes are in place. The electrodes are typically left in place once the ISV operation is completed; however, they may be removed for recycling or for partial reuse.

An off-gas collection hood is positioned over the vitrification area. The hood serves to contain the off gases during processing and to support the electrode connectors. Flow of air through the hood is controlled to maintain a negative pressure of 0.5 to 1.0 in. water column. An ample air sweep through the hood provides excess oxygen for combustion when pyrolysis products and organic vapors are present. The excess air, off gases, and combustion products are drawn from the hood by an induced-draft blower into the off-gas treatment system, which utilizes a quencher, controlled pH scrubber, mist eliminator, heater, particulate filtration, and activated carbon adsorption. A self-contained glycol cooling system is used to cool the quenching and scrubbing liquid. This eliminates the need for a constant cooling-water supply on site. The moisture content of the exhaust airstream is controlled by controlling the scrub solution temperature to accommodate the moisture that is removed from the treatment volume during processing. In this way, the volume of scrub solution generated can be minimized.

Typically, the volume of gases evolved from the ISV melt represent less than 1% by volume of the total volume of air passed through the off-gas treatment system. Also, experience has shown that very little, if any, hazardous material is emitted from the melt during processing. After processing for a time, the scrub solution, particle filters, and activated carbon may contain sufficient contaminants to warrant treatment or disposal themselves. The scrub water can be passed through diatomaceous earth and activated carbon, followed by reuse of the water or discharge to a publicly owned treatment works (POTW). The particulate filters and activated carbon can be placed in subsequent ISV settings for reprocessing. As an alternative, the activated carbon can be sent off site for treatment and reactivation of the carbon and ultimate recycle.

Status of Technology and Scales of Development

The ISV technology was originally developed by the USDOE for possible application to soils previously contaminated with radioactive transuranic materials, related refuse, and related process chemicals that exist at the Hanford Nuclear Reservation in Washington state. In addition to the USDOE, other government agencies and organizations have contributed to the development of the technology, including the U.S. Environmental Protection Agency (USEPA), U.S. Department of Defense (USDOD), the Electric Power Research Institute (EPRI), Battelle Pacific Northwest Laboratory (PNL), Geosafe Corporation, and numerous other private companies. The USDOE mission of ISV, while initially focusing on the Hanford Nuclear Reservation, has expanded to include remediation of waste at other sites, including Oak Ridge National Laboratory (ORNL), Savannah River Plant (SRP), Idaho National Engineering Laboratory (INEL), and others.

Table 1 ISV Development Scales

Scale	Power supply	Electrode separation		Mass vitrified
Bench	10 kW	10–20 cm	(4–8 in)	5–10 kg
Engineering	30 kW	20–30 cm	(8–12 in)	0.05–1 T
Pilot	500 kW	0.9–1.2 m	(3–4 ft)	10–50 T
Large	3,750 kW	3.5–5.5 m	(11.5–18 ft)	400–1000 T

The original development work by the USDOE in this previously unexplored area of technology resulted in a basic patent (Brouns et al., 1983). The USDOE subsequently granted Battelle PNL exclusive rights to application of the technology in the field of nonradioactive and non-government hazardous waste remediation. Battelle PNL, in turn, has sublicensed the Geosafe Corporation for commercial deployment of the ISV technology.

Numerous ISV experimental tests under a variety of conditions and with a variety of waste types have been conducted (Oma et al., 1982, 1983; Timmerman et al. 1983; Timmerman and Oma 1985; Buelt et al. 1987). Table 1 presents the various scales of ISV testing units that have been used to develop and adapt the technology. Over 160 bench-, engineering-, pilot-, and large-scale ISV tests have been conducted and have demonstrated the general feasibility and widespread application of the process. A brief description of the ISV developmental units and the two existing large-scale operational units follows.

Bench-Scale ISV

Bench-scale ISV tests are the smallest and most economical tests to perform when initially determining the feasibility of ISV for processing a new waste/soil type. Bench-scale ISV tests are performed utilizing either two or four electrodes. Electrode diameters have ranged from 1.3 to 3.8 cm ($\frac{1}{2}$ to $1\frac{1}{2}$ in), depending on the test configuration and objectives. Electrode separations of 10 to 15 cm (4 to 6 in) are typical of two electrode bench-scale tests, while electrode separations of 20 cm (8 in) are typical for the four electrode tests. A two-electrode bench-scale test can typically be performed within a 5-gal container, while the four-electrode test requires a 55- or 85 gal-drum for containment. Thermocouples are typically placed in the soil at incremental depths between the electrodes and to the sides perpendicular to the electrodes in order to monitor the depth and width of the vitrification zone. Analysis of the vitrified block and surrounding soil after bench-scale ISV testing provides data as to the product quality and migration of contaminants adjacent to the melt

Figure 4 Vitrified block produced by two-electrode bench-scale ISV system.

zone. Collection and analysis of off-gas samples provide a qualitative indication as to the off gas that might be expected during large-scale ISV. Figure 4 shows a typical vitrified block produced with a two-electrode bench-scale system.

Engineering-Scale ISV

The ISV concept was first demonstrated in 1980 using an engineering-scale system. The engineering-scale system utilizes four electrodes, ranging from 2.5 to 5.1 cm (1 to 2 in) in diameter and has been used to develop and test new concepts for the process. The multiple-tap Scott-Tee-connected transformers with saturable core reactors for power control (same as those used for large-scale ISV) were first developed on the engineering-scale system. In addition, the concept of electrode feeding was first demonstrated at this scale.

Current-generation engineering-scale ISV systems utilize a 30-kW Scott-Tee transformer. The transformer has 16 voltage taps, similar to the transformers used for the large-scale systems. The engineering-scale tests are conducted in an outer container that houses the test soil

Figure 5 Engineering-scale ISV system.

drum and electrodes. Most of the data on process limitations were gathered using engineering-scale ISV systems. Engineering-scale systems have been used to vitrify from approximately 50 kg to 1 metric ton of soil. Figure 5 shows the power control console and the soil container for an engineering-scale ISV system.

Pilot-Scale ISV

Two pilot-scale systems have been utilized for the development of ISV. The first pilot-scale system (Oma et al., 1983) was used to determine scale-up feasibility from the engineering scale. This initial pilot-scale unit was tested successfully on four different occasions prior to design and fabrication of a second pilot-scale ISV system. Both pilot-scale systems used four electrodes with a separation of 0.9 to 1.2 m (3 to 4 ft), an off-gas treatment system, and an off-gas containment hood over the waste site.

The second pilot-scale ISV system had two important design enhancements. First, it was designed to be mobile, with the off-gas system and power supply unit permanently mounted in a mobile semitrailer; and second, it was designed for processing radioactively contaminated soils. The mobile pilot-scale power system is a Scott-Tee transformer rated at 500 kW. The transformer has four voltage tap settings (1000, 650, 430, and 250 V) to accommodate changing resistances to the molten soil during vitrification. The mobile pilot-scale ISV system is shown in Figure 6. (Timmerman and Oma, 1985).

During operations, the off gases pass from the containment hood through a venturi-ejector scrubber, Hydro-Sonic scrubber, water separator, condenser, a second water separator, heater, two stages of high-efficiency particulate air (HEPA) filters, and a blower (Oma and Timmerman, 1985). A schematic of the pilot-scale off-gas system is shown in Figure 7. With the exception of the second-stage HEPA filter and blower, all off-gas components are housed in a removable containment module that has glove port access for remote radioactive operations. The containment module is maintained under a slight vacuum. The off gases are cooled using a closed-loop glycol cooling system that consists of an air/liquid heat exchanger, coolant storage tank, and pump. The cooling liquid is recycled through heat exchangers that cool the scrub liquids going to the venturi-ejector and the Hydro-Sonic scrubbers. In this way the dew point and water balance in the system can be maintained by controlling the off-gas exit temperature. Particle filtration consists of a first stage with two parallel HEPA filters and a single second-stage (backup) HEPA filter. The parallel first-stage

Figure 6 Mobile pilot-scale ISV system.

design permits change out of filters without interrupting ISV operations. During pilot-scale tests that involve soils contaminated with organics, the second-stage HEPA filter can be replaced with an activated carbon absorber to remove any volatile organic compounds (VOCs) that are emitted.

Large-Scale ISV

Two large-scale ISV systems have been designed and fabricated. The initial system was designed for testing on USDOE radioactive and mixed waste sites (Buelt and Carter, 1986), while the second unit was designed for remediating industrial waste sites (Geosafe, 1989).

Large-Scale USDOE System. This system is designed to accommodate electrode separations of 3.5 to 5.5 m (11.5 to 18 ft) and vitrification depths up to approximately 9 m (30 ft), this maximum depth being contingent on the soil properties and depth to groundwater. The large-scale USDOE system is shown in Figure 8. The system consists of a 12.2 × 12.2 m (40 × 40 ft) off-gas hood and three semitrailers that contain the equipment. The first trailer is an open flatbed design that holds a 3750-kVA Scott-Tee transformer and the closed-loop glycol cooling unit. The second trailer is an oversized trailer that houses the off-gas treatment system. The third trailer houses the process control system, motor starters, and all other equipment essential to controlling the ISV

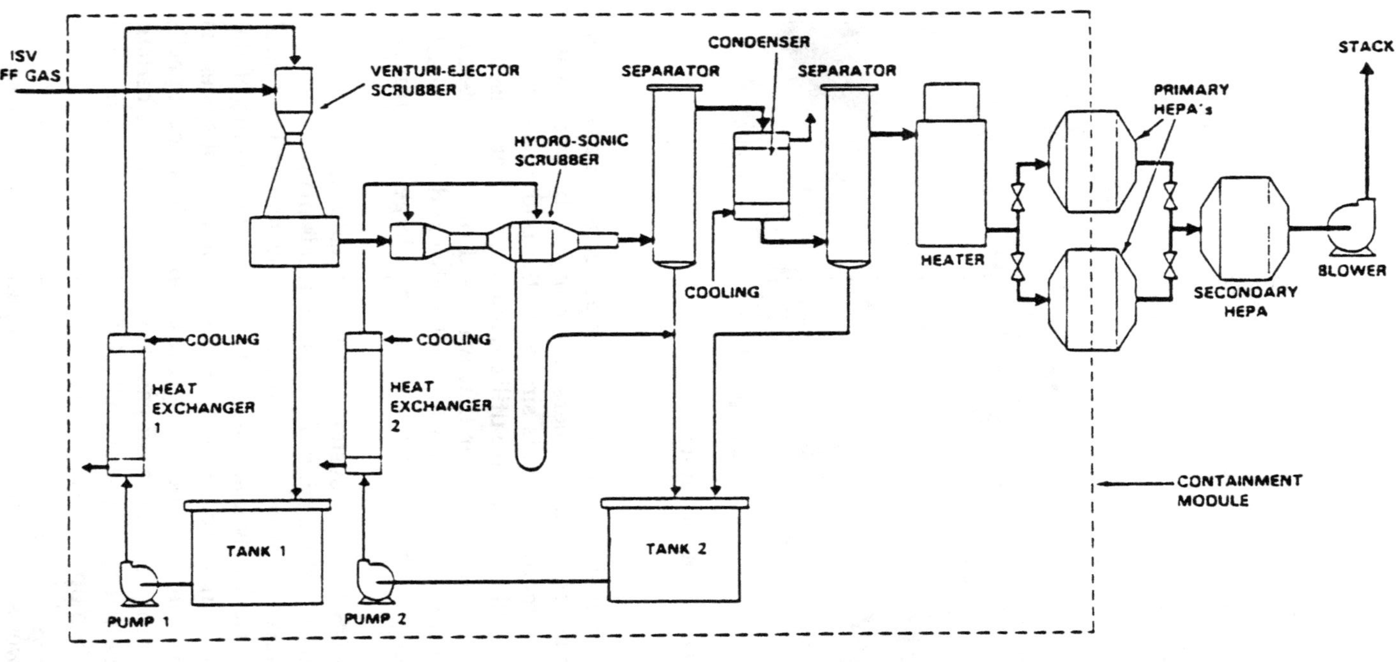

Figure 7 Pilot-scale ISV off-gas system schematic. (Courtesy of Pacific Northwest Laboratory.)

Figure 8 Large-scale USDOE ISV system.

process. The process trailers are designed to be pulled either as one unit of three trailers or as individual trailers for transport between waste sites. The off-gas treatment system is similar to the pilot-scale system except that parallel scrubber units are utilized and a quencher is used in place of the venturi-ejector scrubber that was used on the pilot-scale system. Since this unit was designed for radioactive and mixed waste site testing, the off-gas equipment is housed in a large containment module that has glove port access.

The large-scale USDOE ISV system has safety features and redundant features throughout the design. Process control is maintained by a distributed microprocessor monitoring and control system and a control console for the electrical power supply. All critical components of the process control system have redundant backups. The control system is used to monitor and control important process parameters and to activate backup equipment or reroute off-gas flow automatically if certain equipment fails. The entire large-scale USDOE ISV system can be routinely operated by two operators.

Large-Scale Geosafe ISV System. This large-scale system has essentially the same power supply and electrode separations as those of the large-scale USDOE system. The overall physical size is somewhat smaller than the USDOE system; however, the equipment is still mounted on three process trailers (refer to Figure 3). The first trailer contains a control room and the off-gas treatment system. The second trailer contains the closed-loop glycol cooling system, an instrument air system, and a stepdown transformer. The third trailer contains the Scott-Tee transformer.

The Geosafe ISV off-gas system is similar to the USDOE system except that the containment module is not required since the system is not intended for use with radioactive wastes. Off gas is passed from the off-gas hood to a quencher, where it is cooled to the dew point, then scrubbed, using a tandom-nozzle Hydro-Sonic scrubber and then dewatered prior to reheating and filtration. Off-gas filtration consists of a first stage of particulate filters followed by a second stage of activated carbon adsorbers for cleanup of any residual VOCs that might remain in the off gas. The Geosafe system also includes a backup generator and a cool-down off-gas system that provides emergency off-gas treatment in the event of system power failure.

The off-gas hood for the large-scale Geosafe ISV system is designed to cover a larger area than the USDOE system hood. The hood covers an area that is 16.8 m (55 ft) across and is in the shape of an octagon (see Figure 3). The hood is designed for easy setup and breakdown for transport between job sites.

The large-scale ISV process and equipment possess the general capabilities presented in Table 2. For typical applications, large-scale ISV equipment is capable of processing:

- A mass in the range of 800 to 1000 metric tons (T) of soil in a single setting
- A maximum melt area of about 84 m^2 (900 ft^2)
- A maximum depth of 9 m (30 ft)

Table 2 General Large-Scale ISV System Physical Capabilities

Surface area dimensions	3 × 3 m (10 × 10 ft) minimum 9 × 9 m (30 × 30 ft) maximum
Depth	1.5 to 2.1 m (5 to 7 ft) economic minimum 9 m (30 ft) estimated single setting maximum
Processing rate	4 to 6 T/h
Melt mass	800 to 1000 T (maximum)
Melt temperature	1600 to 2000°C (soils)
Power level	1.9 MW/phase (3.8 MW total)
Electrical consumption	0.9 to 1.1 kWh/kg (0.4 to 0.5 kWh/1b) soil

For most soil applications, it is not possible to reach all three of these limits simultaneously. That is, the maximum tonnage would occur in a 9 × 9 m (30 × 30 ft) setting before reaching the 9 m (30 ft) depth. Thus depth and setting areal dimensions are trade-offs to be considered during remedial design.

Depth Determination

During the engineering- and pilot-scale ISV development phase, growth of the vitrification zone was monitored using thermocouples placed in the soil at predesignated positions. These in-ground thermocouples are not considered feasible for monitoring vitrification depth at actual waste sites with the large-scale ISV equipment; therefore, a remote depth monitoring concept was developed and tested. The depth monitoring system is shown in Figure 9 (Oma et al., 1989a). The system consists of a depth transmitter assembly, which is mounted below one or more of the electrodes (or which is placed in a separate hole), and a surface receiving unit, which receives radio transmissions from the

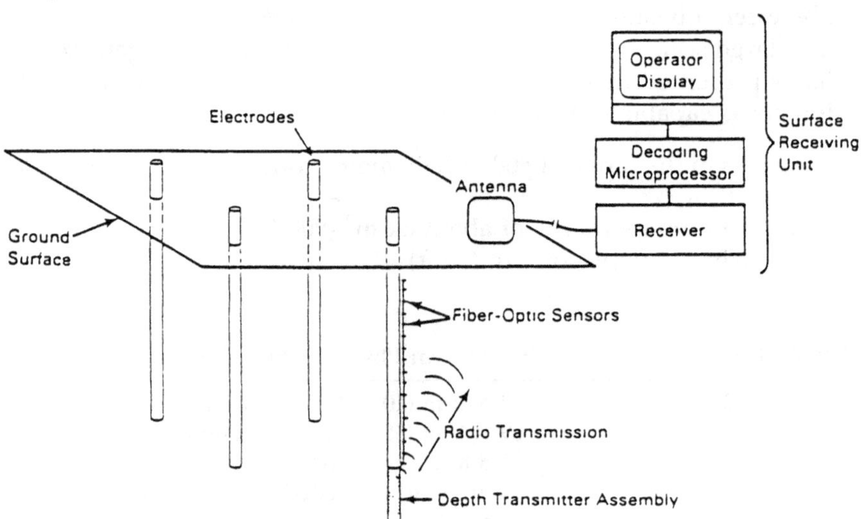

Figure 9 Depth monitoring system. (Courtesy of Pacific Northwest Laboratory.)

depth transmitter and converts the transmissions to useful depth data. The transmitter assembly is connected to a series of fiber-optic sensors that extend from the transmitter up the electrode and terminate at different levels along the electrode. As the ISV process melts downward, the sensors that come in contact with the molten soil transmit a self-generated light to the signal processor within the transmitter assembly. The depth data are then passed to a radio transmitter, which then transmits the data to the receiver located at the ground surface. The receiver passes the information to a second processor which decodes the data and displays them for the ISV operations personnel. The depth transmitter assembly portion of the monitoring system is expended during the ISV operations and, as such, is designed with economics in mind. The depth monitoring system can provide accurate depth data during the ISV process to ensure that the remediation melt depth criteria are achieved.

Electrode Configuration

Development of an electrode that will withstand the severe ISV processing environment for prolonged periods of time has been one of the more challenging aspects of the ISV development. Electrodes inserted into the ground must withstand several different hostile environments. The vitrified soil mass reaches temperatures as high as 2000°C during ISV and is very reducing in nature. In contrast, the portion of the electrode protruding above the molten soil is in an air (oxidizing) atmosphere. The portion of each electrode immediately above the molten soil is hot due to thermal conduction from the melt zone below and must withstand the high-temperature oxidizing atmosphere. The portion of the electrode most susceptible to the effects of the hostile environment is the portion located in the subsidence region. In this region the electrodes are initially exposed to the high-temperature molten soil (reducing environment), and after subsidence, they are exposed to the air within the hood plenum.

A composite electrode design has been developed and applied successfully which uses a multibarrier concept to inhibit corrosion and oxidation to acceptably low rates. The electrode design (Figure 10) consists of a 5.1-cm (2-in)-diameter molybdenum (Mo) rod inserted into the contaminated soil that is to be vitrified. The upper region of the Mo electrode, which is exposed to air when the molten surface subsides during processing, is encased in a graphite collar that has a 6.4 cm (2.5 in.) ID and a 30.5 cm (12 in.) OD (typical). To further protect the Mo rod, the

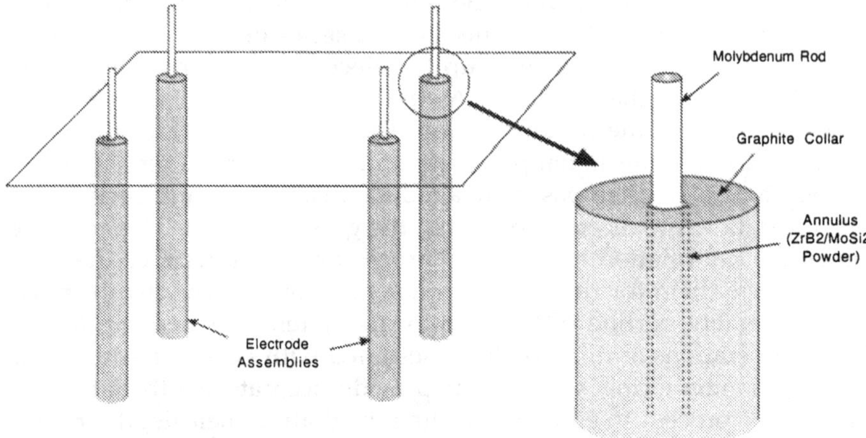

Figure 10 Large-scale ISV electrode configuration.

annulus between the collar and the rod is filled with a mixture of zirconium diboride (ZrB_2) and molybdenum disilicide ($MoSi_2$) powder. In addition, a coating of like material is applied directly to the Mo rods. This powder was selected because of its resistance to oxidation and its relatively high electrical conductivity. For some ISV operations, the 30.5-cm (12-in)-diameter graphite collar has been extended to the bottom of the vitrification zone to provide additional structural integrity to the Mo rod. For this configuration, the annulus below the subsidence region is typically filled with Mo powder, which is less costly than ZrB_2/$MoSi_2$ powder and which is compatible with the Mo rod at processing temperatures.

While the composite electrode design has proven effective during large-scale ISV operations, the cost of electrode materials constitutes about one-fourth to one-third of the total cost of ISV. For this reason, the less costly configuration of electrode feeding has been developed.

Electrode Feeding

The concept of electrode feeding was first demonstrated using the engineering-scale ISV system (Farnsworth et al., 1990a). The concept improves the performance of ISV and reduces the electrode materials cost. Electrode feeding was first developed for the INEL for application to waste sites that contain buried solid transuranic and mixed waste.

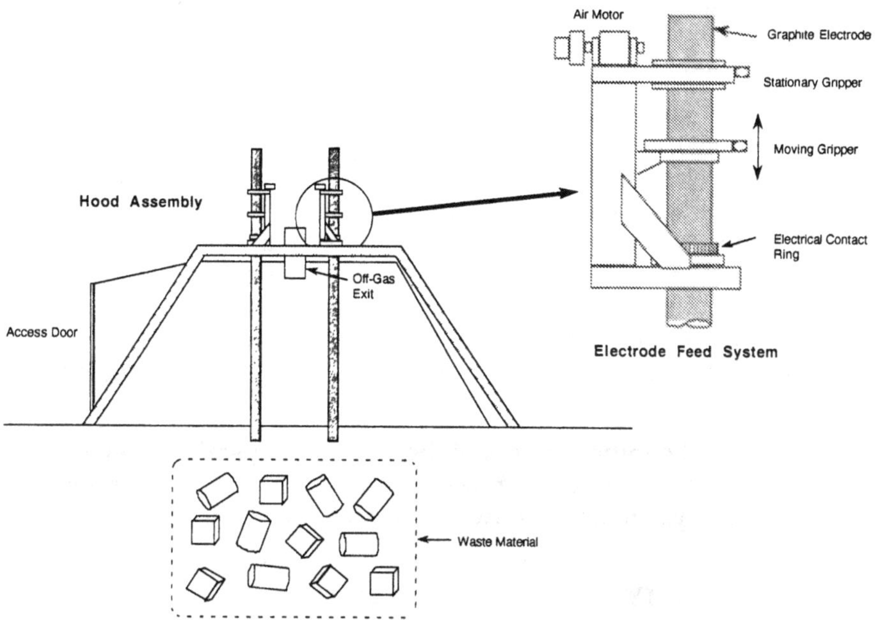

Figure 11 Electrode feeding concept.

Electrode feeding is accomplished using solid graphite electrodes which are considerably less costly than the composite design described in the preceding subsection. Figure 11 shows the electrode feeding concept. Each of the four graphite electrodes is gripped by an electrode feeder that is mounted on top of the off-gas containment hood. Each electrode feeder consists of two electrode grippers, one stationary and one moving, and an electrical contact ring. For safety reasons, the electrode grippers are powered by air-operated motors. For ISV startup, the electrodes are fed through the grippers and are inserted in the soil at a shallow depth judged sufficient to establish a stable vitrified soil zone. The ISV process is started using conventional startup paths of graphite powder and glass frit. Once the melt depth has reached the bottom of the electrodes, the electrode feeders are used to lower the electrodes in pace with the advancing melt zone. If desired, the electrode can be gravity fed by releasing the grippers. If the soil waste area contains a substantial quantity of metal inclusions, a pool of metal will form at the bottom of the melt zone. If signs of electrical shorting across a metal pool occur, the electrodes are raised slightly using the grippers to prevent the full short-circuit condition. The electrode adjustment process is continued until the desired melt depth is achieved. Some of the inherent advantages of electrode feeding are listed below.

1. The need for penetrating the waste zone to install electrodes is eliminated.
2. Improved electrical control is obtained that enables higher average power input and more efficient ISV operations.
3. Electrode feeding permits recovery from a metal shorting condition.
4. The use of graphite electrodes in place of the composite electrode design results in a substantial cost savings for ISV.
5. Electrode feeding permits easy reinsertion of a new electrode in the event that electrode failure (i.e., breakage) occurs during operations.

Because of the initial success of the engineering-scale ISV electrode feeding tests, electrode feeding has been incorporated into the mobile pilot-scale ISV system and the two large-scale ISV systems.

PRODUCT QUALITY

The vitrified product that is produced by ISV technology has superior long-term characteristics and permanence compared to other technologies because (1) hazardous organic and some inorganic compounds are irreversibly destroyed or removed from the treatment volume, and (2) remaining hazardous inorganic compounds are chemically incorporated into a residual product that is capable of withstanding long-term environmental exposure without effect. Because of these demonstrated behaviors, the ISV process is considered a permanent remediation solution.

The ISV process is capable of immobilizing hazardous inorganic compounds in the residual glass and crystalline product (Timmerman and Lokken, 1984). The vitrified waste form has been subjected to a variety of leaching tests including: a 24-h Soxhlet test, a 28-day materials characterization center test (MCC-1), EPA's EP Tox Test, and EPA's TCLP test. A comparison of the resulting corrosion rate during Soxhelt extraction with published data is presented in Figure 12 and shows that the bulk leaching rate of vitrified soil is significantly less than that of a marble or bottle glass and is comparable to Pyrex and granite (Oma et al., 1983). This is due primarily to the high silica and alumina contents and low alkali metal content of the typical ISV glasses. The Soxhlet extraction test results also indicated that ISV glass that contains a crystalline

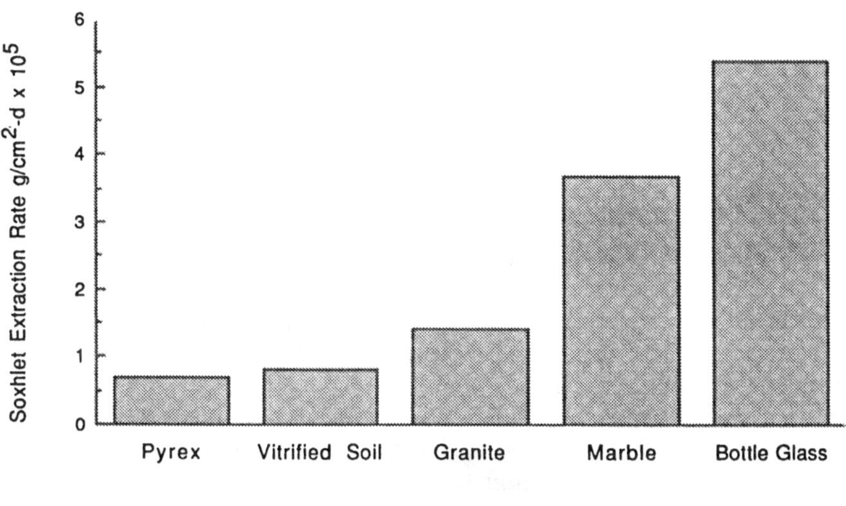

Figure 12 Leach resistances of selected materials.

phase exhibits slightly improved leach resistance. In the 28-day MCC-1 leach test, glasses with and without a crystalline phase were examined. Glasses with an observable crystalline phase exhibited equal or better leach resistance than that of the single-phase glass. This is consistent with Soxhlet weight loss measurements. The EP Tox and TCLP tests show a uniformly low leaching rate for heavy metals of about 5×10^{-7} g/cm^2 per day or lower (Geosafe, 1989). Based on extensive testing, the vitrified material appears to qualify for delisting since the treated material can be shown to pass both the EP Tox and TCLP tests.

Another indication of the durability of the ISV waste form is found in a study of the weathering of obsidian, a naturally occurring (volcanic origin) glasslike material physically and chemically similar to the ISV waste form (Geosafe, 1989). In the natural environment, obsidian has a hydration rate constant of about 1 to 20 μm^2 per 1000 years. A linear hydration rate constant of 10 μm^2 per 1000 years produces a highly conservative estimate of less than 1 mm hydrated depth for the ISV waste form over a 10,000 year time span. In addition, the formation of hydrated products will encase the monolith in a thin hydration film that will inhibit further hydration. The ISV glass hydration rate constant is predicted to be 5 μm^2 per 1000 years at 77°F (e.g., for glass exposed to the air) and 1 μm^2 per 1000 years at 50°F (e.g., for glass buried underground). The values are comparable to those found for obsidian hydration rate constants in the field for similar average weathering temperatures.

The long-term stability of obsidian in nature is controlled by three mechanisms: (1) alteration (weathering), (2) devitrification (recrystallation), and (3) hydration (water absorption). A review of the literature indicates that the usual controlling mechanism is devitrification. Studies of the mean age of natural glasses indicate that obsidian has a mean life of about 18 million years. Considering the similarity of the ISV waste form to obsidian, one can reasonably postulate that the mean life of the vitrified material would be in the realm of geological time (i.e., thousands to millions of years).

Based on USEPA tests of ISV product incorporating waste from a Superfund site, the ISV product has been shown to have excellent structural strength, averaging 10 times the strength of unreinforced concrete in tension and compression. It has also been demonstrated to be unaffected by freeze–thaw and wet–dry cycling.

The ISV process has been demonstrated on a wide variety of soils. A total of 18 different soils from Washington, Idaho, Tennessee, Nevada, Ohio, Wisconsin, and South Carolina, as well as two industrial sludges, have been vitrified successfully during ISV tests. In rare instances it may be necessary to add fluxes to the soil to improve its melting properties during ISV. Early ISV tests showed that inorganic materials were uniformly distributed throughout the vitrified glass area due to convective mixing currents within the molten soil (Timmerman and Oma, 1985). Because of convective mixing, the final vitrified product is homogeneous in nature, with uniform waste content throughout.

EXPERIENCE WITH HAZARDOUS MATERIALS

The ISV process is applicable to hazardous organics, hazardous inorganics, radioactive components, and mixed waste types within certain limits. The ISV process simultaneously treats the various waste types present within the treatment volume. This simultaneous capability is an advantage of ISV relative to other technologies. Figure 13 shows the potential combination of waste types applicable to ISV.

The ability of the ISV equipment to process the various waste types effectively relates primarily to the design of the off-gas collection and treatment system. The system has specific volumetric throughput and heat removal capabilities. Limits must be placed on the concentration of combustible organic compounds and other materials that may vaporize during processing, to ensure an adequate safety factor relative to volu-

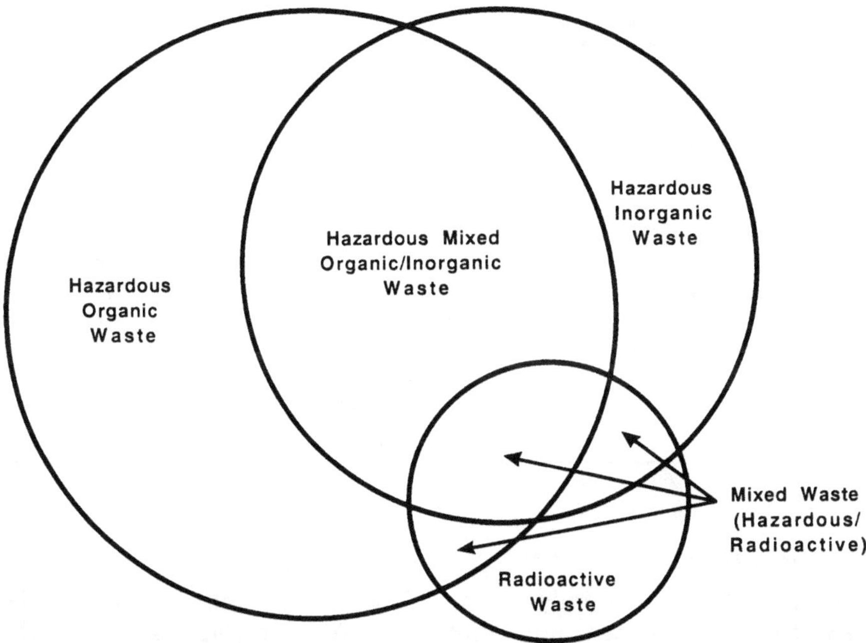

Figure 13 Combinations of wastes applicable to ISV.

metric throughput and heat removal capacities. For immobilization of inorganic wastes, it is necessary to analyze the contaminants in conjunction with the soil matrix to determine if the final vitrified product will be satisfactory. It is necessary to evaluate the concentration of specific inorganic wastes relative to their solubility levels in glass, the expected retention percentage in the melt of the inorganic constituents, and the desired leaching characteristics from the residual product.

As a general rule, organic contaminants at concentrations as high as 5 to 10% by weight and inorganic concentrations as high as 5 to 15% by weight in soil may be acceptable for ISV treatment (Geosafe, 1989). There are waste specific exceptions to this generalization; therefore, consideration of ISV as an alternative should include an applicability evaluation for the specific site based on site characterization information.

The ISV test database includes experience with a large number of waste types. The waste materials that have been involved in previous ISV tests at significant concentration levels are listed in Table 3. The following subsections present ISV performance data during selected tests involving hazardous materials.

Table 3 Waste Materials Processed in ISV Tests

Heavy metals	Liquid organics	Solid materials	Radioactive materials
Lead	PCBs	Asbestos	Plutonium
Nickel	Dioxin	Wood	Amerecium
Cadmium	Trichloroethylene	Buna rubber	Radium
Arsenic	Carbon tetrachloride	PVC	Uranium
Barium	Dichlorobenzene	Polyethylene	Radon
Zinc	Benzene	Neoprene	Cesium
Mercury	Methylene chloride	Ion-exchange resin	Ruthenium
Copper	Toluene	Teflon	Cobalt
Aluminum	Ethylene glycol	Paper	Strontium
Iron	Methyl ethyl ketone	Cotton	
		Polypropylene	
		DDT, DDD, DDE	

Industrial Sludge Test

An industrial sludge heavily laden with zirconium lime was solidified effectively during bench- and pilot-scale ISV tests (Buelt and Freim, 1986). The basis for conducting the pilot-scale demonstration was established by two successful bench-scale feasibility tests. During the bench-scale tests, the mass and volume of the sludge were reduced to 30 and 15%, respectively, of their original values. The encouraging bench-scale test results led to the decision to test the process using the mobile pilot-scale ISV system.

To conduct the pilot-scale tests, the sludge was placed in a 3-m-diameter by 3½-m-tall culvert buried vertically in the ground. The sludge was filled to grade level and electrodes were inserted inside the culvert through the sludge material. Because of the extensive consolidation that the sludge exhibited during melting, startup was not achieved until soil and soda ash were added to the sludge surface. A total of 3.8 m^3 of soil was added to the surface of the sludge and provided sufficient molten material to maintain a continuous electrical path among the four electrodes during and after startup. A total of 100 kg of soda ash (Na_2CO_3) were added to the sludge surface to improve the electrical conductivity of the sludge once molten. The soda ash decomposed to Na_2O and supplemented the NaCl and KCl in the sludge and the natural Na_2O associated with the soil cover. The pilot-scale test was completed successfully in 3 days, processing a total of 8.9 m^3 of sludge and reaching a vitrification depth of 3.0 m, significantly reducing the volume of the sludge and driving moisture and associated process effluents from the sludge. The sludge originally contained 55 to 70% moisture. However, this did not pose a problem for the ISV process. Analysis of the off gas showed the retention of the fluorides, chlorides, and sulfates in the melt; destruction of organics and nitrates was extremely high, which minimized the requirements placed on the off-gas treatment system.

Pilot-Scale Radioactive Test

A pilot-scale ISV radioactive test was performed to demonstrate the ISV processing containment and confinement of radioactivity contaminated soil and to evaluate the off-gas behavior of volatile or entrained radionuclides (Timmerman and Oma, 1985). The pilot-scale radioactive test was conducted using a makeup site in which known quantities of radionuclides were introduced into the soil. Soil containing radioactive isotopes of americium, plutonium, cesium, ruthenium, strontium, and cobalt were placed in a 5-gallon container and centrally position within the zone to be vitrified.

The test successfully demonstrated the processing containment of radionuclides both within the vitrified mass and in the off-gas system. No environmental release of radioactive material was measured during test operations. The vitrified soil showed greater than 99% retention of all radioactive isotopes. Loss to the off-gas system varied from less than or equal to 0.03% for particulate materials (plutonium and strontium) to 0.8% for cesium, which is a more volatile element. The off-gas system effectively contained both volatile and entrained radioactive materials. Analysis of the vitrified soil revealed that all radionuclides were distributed throughout the vitrified zone, some more uniformly than others. No migration of radionuclides outside the vitrified zone occurred, as indicated by analysis of soil samples adjacent to the block. Waste leaching studies (Oma et al., 1983; Timmerman and Lokken, 1984) indicate an exceptable durability of the ISV product. A summary of radionuclide melt retention data for the pilot-scale radioactive test is presented in Table 4. These data show that most species are retained quite well by the molten soil. Nonvolatile radionuclides (cobalt, strontium, plutonium, and americium) have very high retentionnates, ranging from approximately 99.9% to greater than 99.99%. Retention of semivolatile radionuclides (cesium and ruthenium) was also quite good (with retentions greater than 99.2%).

ISV of PCB-Containing Harbor Sediments

A bench-scale process feasibility study using ISV was performed successfully on PCB-containing sediments from New Bedford harbor (Reimus, 1988). The destruction efficiency (DE) for the ISV system without the benefit of off-gas treatment was measured to be greater than

Table 4 Melt Retention Data for Pilot-Scale Radioactive Test

Radionuclide	Melt retention (%)
Transuranics	
^{241}Am	99.992
$^{239/240}Pu$	99.977
^{238}Pu	99.978
Total Pu	99.978
Fission products	
^{137}Cs	99.23
^{106}Ru	99.82
^{90}Sr	99.968
^{60}Co	99.945

99.9985%. This removal efficiency represents the amount of PCBs that were effectively treated by ISV and were not released to the off-gas system. Based on a single-stage activated carbon filter having a 99.9% organic removal efficiency, the ISV system soil-to-stack destruction and removal efficiency (DRE) with a single-stage carbon filter is estimated to be greater than 99.99999% (seven nines). This is greater than the six-nines efficiency required by 40 CFR 761.70 for PCB incinerators. The vitrified product produced was subjected to leach testing using the USEPA Toxic Characteristic Leaching Procedure (TCLP). The TCLP was used because it is applicable to ash product from Resource Conservation and Recovery Act (RCRA), type B, polychlorinated biphenyl (PCB) incinerators. The concentrations of regulated metals in the leach extract were two orders of magnitude or more below the regulatory limits. The TCLP extract was also analyzed for organics, all of which were below the detection limit. Other ISV testing of PCB containing soils has been done with similar positive results (Timmerman, 1985, 1989a,b).

ISV of Dioxin-Contaminated Soils

A bench-scale ISV test was performed to determine the applicability of the process for treating dioxin-contaminated soils (Mitchell, 1987). Soil contaminated to a level of 100 ppb 2,3,7,8-tetrachlorodibenzo-*p*-dioxin was vitrified during a 4-h test producing about 3 kg of vitrified soil. A DE of greater than 99.995% based on analytical detection limits was calculated for destruction of dioxins by the ISV process alone. This value does not include the removal efficiency of the off-gas system. An activated carbon filter typically has 99.9% efficiency; therefore, an off-gas treatment system containing a carbon filter would result in an overall DRE of greater than 99.99999% (seven nines) for dioxin. This efficiency exceeds the USEPA requirement of six nines by an order of magnitude.

ISV of Soil and Limestone at Oak Ridge National Laboratory

A pilot-scale field demonstration of ISV was carried out on ORNL on a $\frac{3}{8}$-scale model of an ORNL radioactive liquid waste disposal trench. Nonradioactive isotopes of Cs_2CO_3 and $SrCO_3$ were used to simulate ^{137}Cs and ^{90}Sr that are contained in several abandoned ORNL seepage disposal trenches (Carter et al., 1988; Spalding and Jacobs 1989). The pilot-scale test was conducted after successful laboratory- and engineering-scale tests in which ORNL soil and limestone were melted. During the pilot-scale test, a 20-metric ton mass of vitrified product was produced during a 110-h test operations period. After the ISV block was sufficiently cool, core samples were taken and later one side of the trench was excavated to expose the vitrified product. The melt tended to proceed along the long axis of the trench to a greater degree than into the undistributed host soil. This effect can be considered an advantage for application of the ORNL liquid waste disposal trenches because fewer electrode placements would be required to vitrify the given length of trench than if a symmetrical block were produced. The retention of Sr in the vitrified zone was excellent; >99.999% of the Sr was incorporated into the ISV product and did not appear in the off-gas process system. The retention of Cs was encouraging at 99.88%.

The ISV product was analyzed and found to be a mixture of microcrystalline and glass phases with the microcrystalline phase dominating in the more slowly cooling regions of the trench. The bulk chemical compositions of the crystalline and glass phases were essentially identical. Both the Cs and Sr were quite uniformly distributed in all vitrified samples from the trench, indicating that thermal convection in the molten state had been extensive.

Both the crystalline and glass phases were tested for leach resistance and durability by using standardized test methods for nuclear waste glasses. Although the crystalline phase was slightly less leach resistant than the glass phase, both phases were as durable as several vitreous forms considered for high-level radioactive waste. The leachability of Sr was greatly reduced over its ambient leachability from contaminated soil and waste. The leachability of Cs from vitrified phases increased marginally over that in soil, where its fixation by illitic minerals results in a low leachability. However, the increase in Cs leachability from the ISV product was more than offset by a decrease in surface area after vitrification (by up to five orders of magnitude), which resulted in a gain in durability. Overall, much of the improvement in the waste form durability after vitrification results from a large decrease in the waste material surface area when soil is transformed to glass.

ISV of Fuel Oil and Heavy Metals at Arnold Air Force Base

Bench- and pilot-scale ISV tests were conducted on soils from Arnold Engineering and Development Center, Arnold Air Force Base, Tennessee. Soils from a fire training pit that is contaminated with fuel oils and heavy metals from fire training exercises was successfully vitrified during both bench- and pilot-scale tests (Timmerman, 1989a,b; Lominac et al., 1989). Initial testing and analysis of the soil indicated that a lower-melting electrically conductive fluxing additive (such as sodium carbonate) was required as an addition to the soil to make ISV processing work effectively. The soil composition contained a high alumina and silica content, which is suitable for making a good glass product, but ISV also requires a sufficient quantity of alkali metal elements (Na and K) to lower the melting temperature and provide electrical conduction. The bench-scale test soil contained approximately 1% of these alkali elements; however, typically 5% is required for effective ISV processing. Therefore, additives were used during testing. However, additives were placed differently in the pilot-scale test than they were in the initial bench-scale tests. The initial tests were of a smaller scale and allowed total blending of the flux (Na_2CO_3) with entire soil volume. To prevent disturbing the contaminated soil during pilot-scale testing, a cover layer of flux was placed over the site prior to ISV. The depth of the cover layer was 0.9 m (3 ft) and the flux concentration was 27%.

The pilot-scale test was performed on the edge of the fire training pit area. This location was chosen to allow application of ISV to an actual portion of the contaminated site and to allow monitoring thermal transport to the clean surrounding soil. Pretest and post test soil core sampling were performed to obtain before and after soil profiles.

Organic contaminants were effectively destroyed to the 89% level for the fuel oil contaminated site, solely by the ISV melt, exclusive of any off-gas treatment. The overall ISV system DRE was 99.85%, which includes the off-gas treatment system. Samples of vitrified soil passed both the Extraction Procedure Toxicity (EP Tox) and TCLP leach tests and showed that all metals of concern are below leach release limits. The results indicate that inorganic contaminants are immobilized to a level that should allow the site to be listed as nonhazardous according to regulatory criteria. This test also indicated that fluxing agents could be used successfully to modify the melting properties of the soil during ISV. It was found during the pilot-scale test with the 0.9 m (3 ft) of fluxing material cover that the process operations efficiency was reduced due to extreme lowering of the electrical conductivity of the molten material. Because of this, to ensure achieving the desired melting depth, it is recommended that the fluxing additives be added to the entire vitrification volume (if they are needed) by using soil mixing or injection techniques instead of concentrating them on the cover soil layer.

ISV of Soils Containing High Metals Loading

Bench- and engineering-scale ISV tests have been performed to determine if ISV is a feasible long-term confinement technology for previously buried solid transuranic and mix wastes located at INEL. Some of the waste areas contain significant quantities of metals, and for this reason, several tests were conducted to determine the feasibility of ISV when high metals loadings are present (Oma et al., 1989b; Farnsworth et al., 1990a,b). Two tests were conducted using the conventional ISV process with electrodes inserted into the soil. The ISV process successfully vitrified soils containing localized metal concentrations as high as 42% by weight without requiring special methods to prevent electrical shorting within the melt zone. Vitrification of this localized concentration resulted in a 15.9% by weight metals content in the resulting ISV vitrified block. The upper ISV metals limit is related to the quantity of metal that accumulates at the bottom of the molten zone. Even though ISV of soil with these high metal loadings was successful, electrical imbalances and electric shorting did occur during the tests, resulting in a decreased operating efficiency.

To improve operational efficiency when high metal concentrations are present, an engineering-scale test was conducted to demonstrate the effectiveness of electrode feeding (Farnsworth et al., 1990a). This engineering scale test used actual soils from INEL that were mixed with representative concentrations of carbon steel and stainless steel. The test successfully demonstrated the feasibility and benefits of electrode feeding during ISV processing. The electrode feed test also demonstrated gravity feeding of electrodes that was continued until the electrodes encountered an electrical short. This implies that for a contaminated soil site where metal pooling is not a consideration, electrode feeding may be possible by gravitational force without a mechanized feed system. When the electrode encountered an electrical short, the operations were continued successfully simply by lifting the electrodes slightly, avoiding the shorting condition. In addition to the other benefits, the use of electrode feeding provides a contingency for recovering from electrode failure simply by reinserting the unbroken portion of the electrode into the vitrified melt. Successful results of these tests provide the basis for continued scaleup of the electrode feeding system as described earlier.

ISV APPLICATION ALTERNATIVES

Waste materials may be arranged in a variety of ways for ISV processing, depending on the specific characteristics and needs of the site (Geosafe, 1989). Figure 14 illustrates the classical ISV application. Such application arrangements are applicable to a landfill or to general soil contamination conditions where the contaminants are deep enough to allow economic application of ISV (as a rule of thumb, waste depths of greater than 5 to 7 ft are preferred). Figure 15 illustrates how the ISV process would be applied to contaminated soils under and around a surface impoundment. In such cases it is common to level the impoundment either by removing the berm, by filling the depressed volume with soil, or by a combination of the two.

Figures 16 and 17 illustrate two variations of waste material staging arrangements. Figure 16 represents the case where a trench may be provided for staging of the contaminated materials completely below grade. Figure 17 illustrates how materials may be staged totally or partially above grade if desired. Staging of contaminated materials

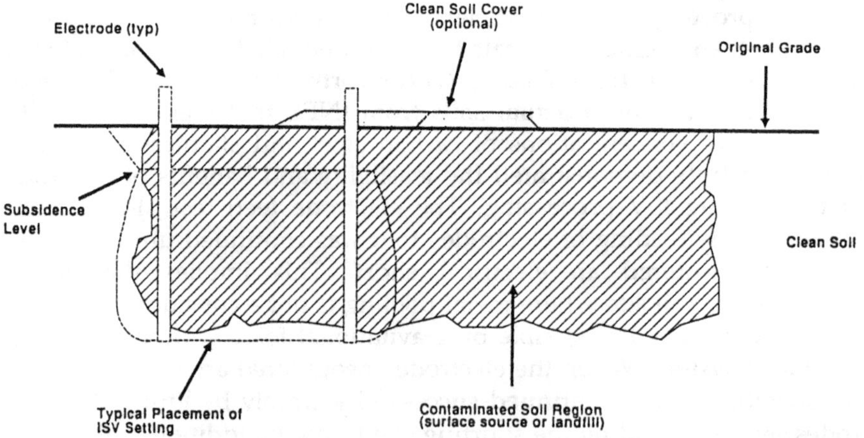

Figure 14 In situ processing arrangement. (Courtesy of Geosafe Corporation.)

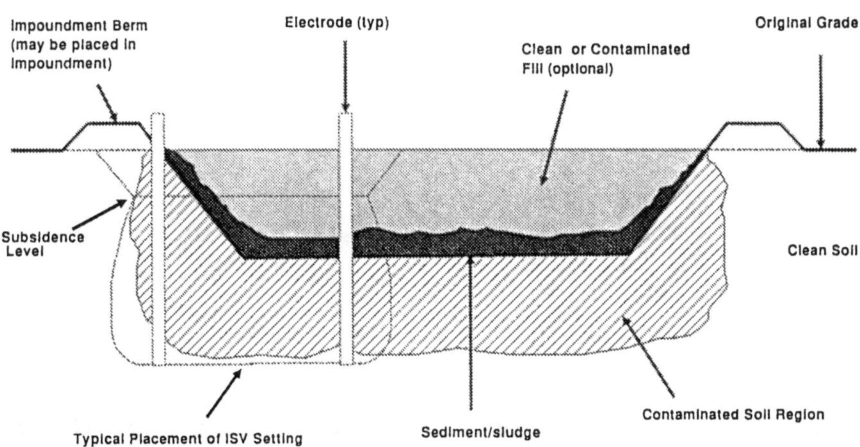

Figure 15 Modified in situ processing arrangement. (Courtesy of Geosafe Corporation.)

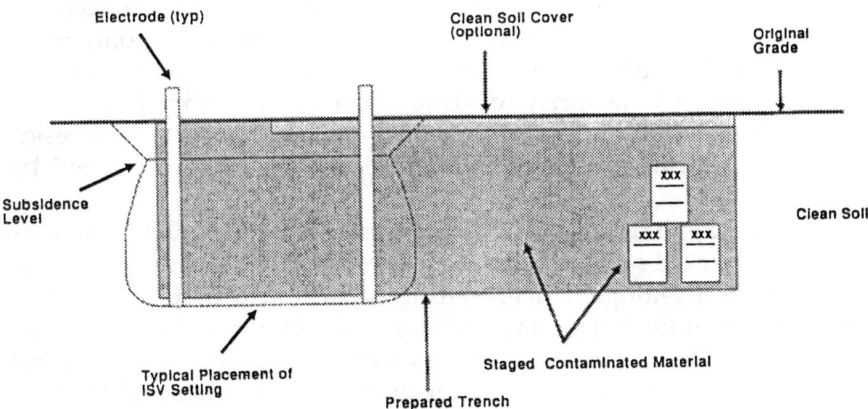

Figure 16 Below-grade staging arrangement. (Courtesy of Geosafe Corporation.)

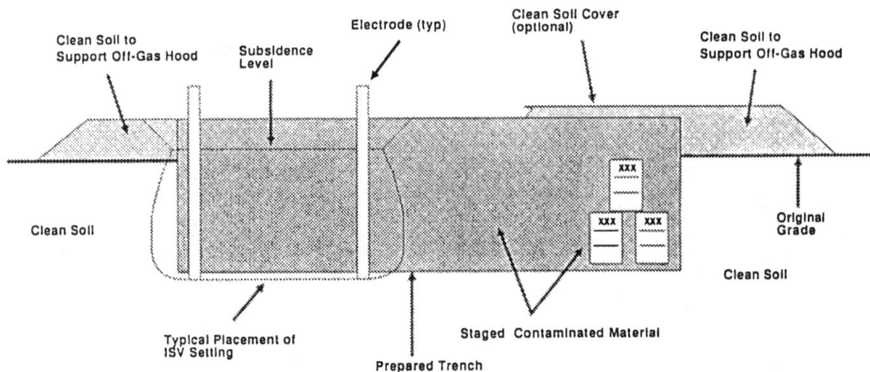

Figure 17 Above-grade staging arrangement. (Courtesy of Geosafe Corporation.)

should be considered when (1) the contamination is very shallow, (2) the contamination cannot be processed where it presently is located, (3) it is considered necessary to add soil to or above the contaminated materials, and/or (4) it is desired to minimize the area of treatment by consolidation.

Figure 18 illustrates a special variation of the staging concept as may be applied to remediation of contaminated tanks or containers. In this case the tank or other container is filled with soil or suitable glass-forming material prior to vitrification of the complete volume. Metal tank materials are expected to melt and become part of the residual product during treatment. Nonmetal tanks will be destroyed by the processing.

For deep applications of ISV [e.g., greater than 9 m (30 ft)], the special case of stacked settings is illustrated in Figure 19 as an option. In the case of stacked settings, a deep setting is performed first, followed by placement of additional material on top of the first monolith for subsequent processing. If the stack settings were to be performed on in-situ soils, it would first be necessary to excavate the upper material to allow vitrification of the lower setting. This would not be the case if the stack setting were performed on staged contaminated material. Figure 20 is a variation of the stack setting concept for the special case of staging high-volume-reduction materials (e.g., sludge or tailings undergoing a significant percentage of conversion to gaseous products during

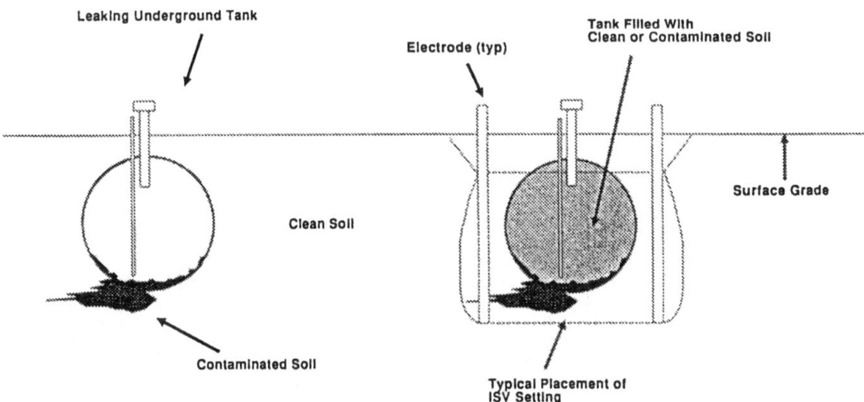

Figure 18 Tank/container setting arrangement. (Courtesy of Geosafe Corporation.)

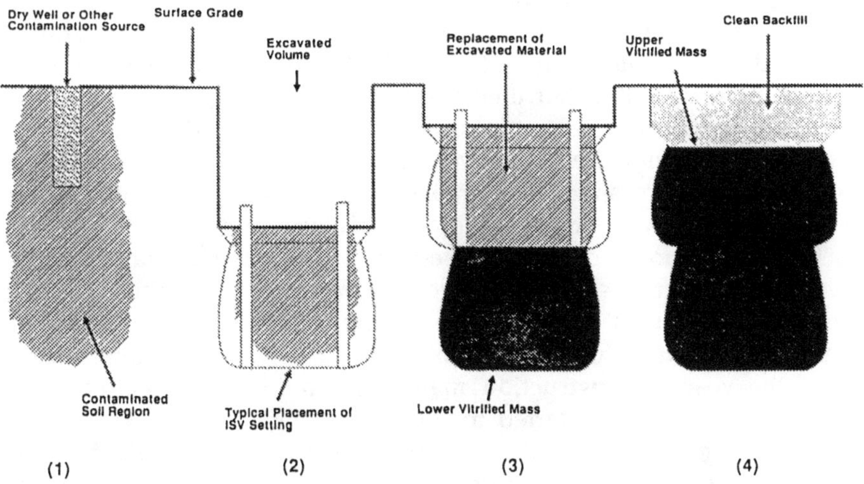

Figure 19 Stacked setting processing arrangement. (Courtesy of Geosafe Corporation.)

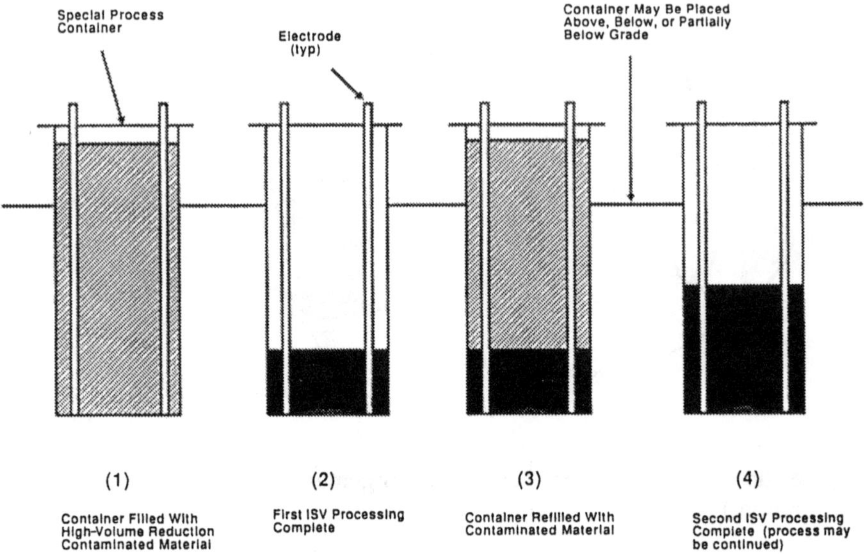

Figure 20 Process-container repetitive setting arrangement. (Courtesy of Geosafe Corporation.)

processing). In this case a process container may be utilized to hold the staged material. After completing the first setting, the container would be refilled for a second setting. Such activities would be repeated until the container was full of vitrified product.

Figure 21 illustrates a continuous feeding process wherein contaminated material would be fed between the electrodes, allowing a period of continuous processing. Such an arrangement may be another possibility for cases of high-volume-reduction materials.

The ISV process could be used to produce structural footings since the product has extremely good load-bearing capabilities suitable for foundations (compressive strengths typically an order of magnitude greater than unreinforced concrete). In remote areas, where the cost of construction materials would be prohibitive, the native soil could be converted to footings using an ISV system with a portable generator. In addition, the vitrified product would be well suited to provide erosion protection for either coastal or stream applications.

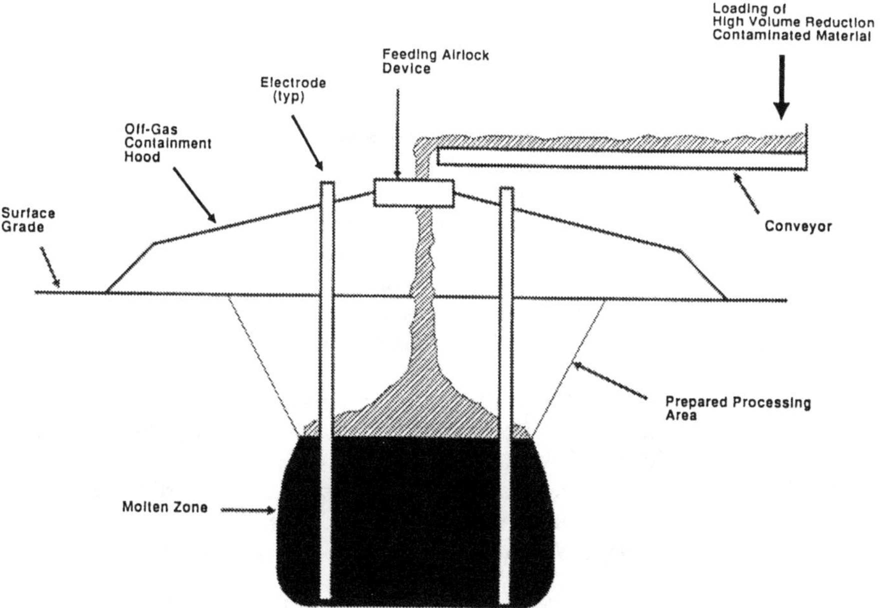

Figure 21 Continuous-feed processing arrangement. (Courtesy of Geosafe Corporation.)

REFERENCES

Brouns, R. A., J. L. Buelt, and W. F. Bonner. 1983. In situ vitrification of soil. U.S. patent 4,376,598 (March).

Buelt, J. L., and J. G. Carter. 1986. *Description and Capabilities of the Large-Scale In Situ Vitrification Process*. PNL-5738. Pacific Northwest Laboratory, Richland, WA.

Buelt, J. L., and S. T. Freim. 1986. *Demonstration of In Situ Vitrification for Volume Reduction of Zirconia/Lime Sludges*. Prepared for Teledyne Wah Chang by Battelle, Pacific Northwest Laboratories, Richland, WA.

Buelt, J. L., C. L. Timmerman, K. H. Oma, V. F. FitzPatrick, and J. G. Carter. 1987. *In Situ Vitrification of Transuranic Wastes: An Updated Systems Evaluation and Applications Assessment*. PNL-4800, Supplement 1. Pacific Northwest Laboratory, Richland, WA.

Carter, J. G., S. S. Koegler, and S. O. Bates. 1988. *Process Performance of the Pilot-Scale In Situ Vitrification of a Simulated Waste Disposal Site at the Oak Ridge National Laboratory*. PNL-6530. Pacific Northwest Laboratory, Richland, WA.

Farnsworth, R. K., K. H. Oma, and C. E. Bigelow. 1990a. *Initial Tests on In Situ Vitrification Using Electrode Feeding Techniques*. PNL-7355. Pacific Northwest Laboratory, Richland, WA.

Farnsworth, R. K., K. H. Oma, and M. A. H. Reimus. 1990b. "*Crucible Melts and Bench-Scale ISV Tests on Simulated Wastes in INEL Soils*. PNL-7344. Pacific Northwest Laboratory, Richland, WA.

Geosafe. 1989. "*Application and Evaluation Considerations for In Situ Vitrification Technology: A Treatment Process for Destruction and/or Permanent Immobilization of Hazardous Materials*. GSC 1901, Geosafe Corporation, Kirkland, WA.

Lominac, K. J., R. C. Edwards, and C. L. Timmerman. 1989. Pilot-scale in situ vitrification at Arnold Engineering Development Center, Arnold AFB, Tennessee. *Superfund '89, Proceedings of the 10th National Conference*, Washington, DC, November 27–29.

Mitchell, S. J. 1987. *In Situ Vitrification of Dioxin-Contaminated Soils*. Prepared for American Fuel and Power Corporation, Panama City, FL, by Pacific Northwest Laboratory, Richland, WA.

Oma, K. H., and C. L. Timmerman. 1985. Off-gas treatment and characterization for a radioactive in situ vitrification test. CONF-840806, *Proceedings of the 18th Department of Energy Nuclear Airborne Waste Management and Air Cleaning Conference*, pp. 683–701.

Oma, K. H., R. K. Farnsworth, and J. M. Rusin. 1982. *In Situ Vitrification: Application Analysis for Stabilization of Transuranic Waste*. PNL-4442. Pacific Northwest Laboratory, Richland, WA.

Oma, K. H., D. R. Brown, J. L. Buelt, V. F. FitzPatrick, K. A. Hawley, G. B. Mellinger, B. A. Napier, D. J. Silviera, S. L. Stein, and C. L. Timmerman. 1983. *In Situ Vitrification of Transuranic Wastes: Systems Evaluation and Applications Assessment*. PNL-4800, Pacific Northwest Laboratory, Richland, WA.

Oma, K. H., K. C. Davis, and C. L. Timmerman. 1989a. A telemetry system for monitoring melt depth during in situ vitrification. *Meeting of the American Nuclear Society*, Dallas, TX, June 1988.

Oma, K. H., M. A. H. Reimus, and C. L. Timmerman. 1989b. *Support for the In Situ Vitrification Treatability Study at the Idaho National Engineering Laboratory: FY 1988 Summary*. PNL-6787. Pacific Northwest Laboratory, Richland, WA.

Reimus, M. A. H. 1988. *Feasibility Testing of In Situ Vitrification of New Bedford Harbor Sediments*. Prepared for Ebasco Services, Inc., Arlington, VA, and REM III Program, issued by Ebasco Services, Inc., Pacific Northwest Laboratory, Richland, WA.

Spalding, B. P., and G. K. Jacobs. 1989. *Evaluation of an In Situ Vitrification Field Demonstration of a Simulated Radioactive Liquid Waste Disposal Trench*. ORNL/TM-10992, Oak Ridge National Laboratory, Oak Ridge, TN.

Timmerman, C. L. 1985. In situ vitrification of PCB-contaminated soils. In *Proceedings: 1985 EPRI PCB Seminar*. CS/EA/EL-4480. Electric Power Research Institute, Palo Alto, CA, and Pacific Northwest Laboratory, Richland, WA.

Timmerman, C. L. 1989a. *Feasibility Testing of In Situ Vitrification of Arnold Engineering Development Center Contaminated Soils*. PNL-6780. Pacific Northwest Laboratory, Richland, WA.

Timmerman, C. L. 1989b. *Feasibility Testing of In Situ Vitrification of PCB-Contaminated Soil from a Spokane, WA Site*. Prepared for Geosafe Corporation, Kirkland, WA, by Pacific Northwest Laboratory, Richland, WA.

Timmerman, C. L., and R. O. Lokken. 1984. Characterization of vitrified soil produced by in situ vitrification. In: *Nuclear Waste Management Advances in Ceramics*. G. G. Wicks and W. A. Ross (eds.), Vol. 8, pp. 619–626. The American Ceramic Society, Columbus, OH, and Pacific Northwest Laboratory, Richland, WA.

Timmerman, C. L., and K. H. Oma. 1985. A pilot-scale radioactive test using in situ vitrification. *Nucl. Technol.* 71(2):471–481.

Timmerman, C. L., R. A. Brouns, J. L. Buelt, and K. H. Oma. 1983. In situ vitrification: pilot-scale development. *Nucl. Chem. Waste Manage.* 267:4.

10

Soil Surfactant Flushing/Washing

David J. Wilson
Vanderbilt University
Nashville, Tennessee

Ann N. Clarke
ECKENFELDER INC.
Nashville, Tennessee

INTRODUCTION

In this chapter our object is to provide an overview of two related technologies that are still very much in the development stage: soil surfactant flushing and soil surfactant washing. The term *soil surfactant flushing* is used to refer to in situ treatment of the contaminated soil or other matrix with surfactant solution. Soil surfactant washing denotes processes in which the soil is excavated for treatment aboveground with surfactant solution.

At present, these technologies are not as well developed as a number of others discussed in this book. They do look sufficiently promising for some applications that their further development is warranted, especially since simple water flushing often requires unacceptably long times for remediation or yields residual contaminant concentrations that are above acceptable limits.

We hope here to provide information on these techniques to managers and regulators who may have to participate in decisions as to whether the use of such innovative and relatively untested technologies is warranted in any particular case. It is also hoped that this chapter will be of use to engineers who must read reports and recommendations involving the use of surfactant flushing/washing with some degree of critical understanding.

Surfactant washing is an ex situ process—the soil is excavated. Possible configurations for surfactant washing include heaping the contaminated material on plastic liners or other impermeable barriers and irrigating the piles with surfactant solution (which is then recovered), batch washing of the soil in tanks or lined pits, or continuous-flow washing in countercurrent or normal modes. In situ surfactant flushing involves the delivery of surfactant solution to the contaminated medium in place by irrigation and/or injection wells; the contaminant-laden surfactant solution is then pumped up for treatment (and possible recycle) by recovery wells. Obviously, this approach must be limited to sites where the hydrogeology is such that one can guarantee that all of the contaminated zone will be reached by surfactant solution and all of the contaminated surfactant solution will be withdrawn by the recovery wells. The in situ technique is illustrated schematically in Figure 1.

The range of applicability of surfactant flushing/washing is by no means universal. The contaminants must be hydrophobic, and can but need not be volatile. Suitable compounds include chlorinated pesticides, PCBs, semivolatiles (such as chlorinated benzenes, naphthalene, etc.), petroleum products (ranging from gasoline through diesel and jet fuel, kerosene, oils, to greases), plasticizers such as the dialkylphthalates, chlorinated solvents such as trichloroethylene, and aromatic solvents (benzene, toluene, xylenes, ethylbenzene). These techniques are not suitable for the removal of inorganics (such as compounds containing toxic metal ions), and they are not necessary for the removal of water-

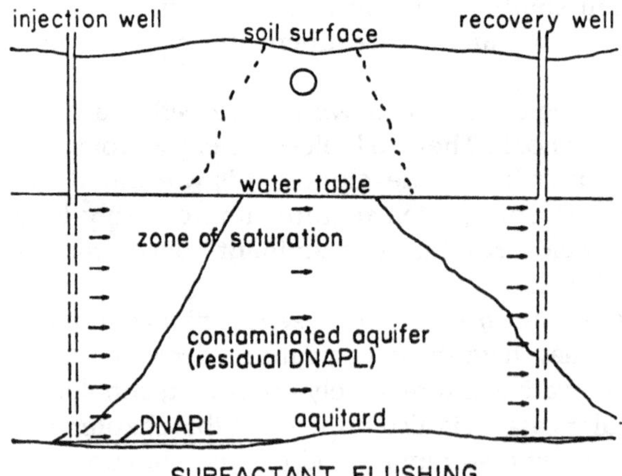

Figure 1 Schematic diagram of in situ surfactant flushing.

soluble compounds such as the lower alcohols, acetone, and methyl ethyl ketone. The latter will be removed during the course of surfactant flushing of hydrophobic compounds without adding to the cost of the operation. Incidentally, often much of the equipment used for surfactant flushing/washing can be used for metals removal simply by using other reagents, such as chelating agents or acids.

A second constraint is that one must be able to deliver the surfactant solution to the contaminant. If the soil permeability is very low, surfactant flushing may be unable to do this and an ex situ washing technique may be needed. Third, we reiterate the importance of complete recovery of contaminant-laden surfactant solution. At some sites geological factors may make this an uncertain or expensive matter, in which case one must either use the ex situ technique or employ a different technology altogether.

PRINCIPLES, HISTORICAL BACKGROUND, AND LITERATURE REVIEW

To understand the workings of these techniques, one must understand something of the principles of surfactant solution behavior. Surfactants are amphipathic molecules or ions—that is, one portion of the molecule is hydrophobic (water-hating), while the other portion is hydrophilic (water-loving). Hydrophobic portions (tails) are usually hydrocarbon chains typically containing 12 or more carbon atoms. Hydrophilic portions (heads) are usually ionic (—COO^-, —SO_4^-, —$N(CH_3)_3^+$, for example) or polar (oxyethylene chains, —NH_2, etc.). In surfactant solutions the surfactants tend to concentrate at air/water interfaces, where the hydrophilic heads can be hydrated in the water yet the hydrophobic tails need not disrupt the hydrogen-bonded structure of the water by being immersed in the aqueous phase. By concentrating at the air/water interface of the solution, the surfactant species are able to reduce the free energy of the system, thereby increasing its stability.

As the concentration of surfactant in an aqueous solution is increased, one eventually reaches a point where the surfactant molecules/ions can employ a second mechanism to reduce the contact between their hydrophobic tails and the aqueous phase. At and above the *critical micelle concentration* species tend to clump together in clustered aggregates; in these structures (micelles) the hydrophobic tails are in the interior of the micelle, away from the water, while the hydrophilic heads are on the surface of the roughly spherical micelle. Each micelle therefore has as its interior a very tiny droplet of nonpolar hydrocarbon phase. The surfactant concentrations at which these micelles start to

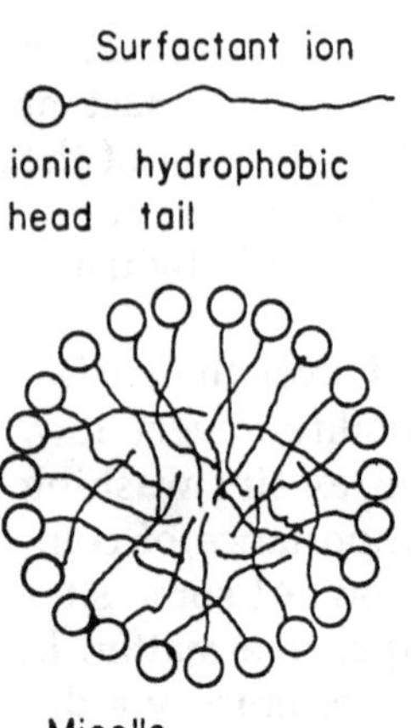

Figure 2 Surfactant molecule/ion, and a representation of a surfactant micelle in a surfactant solution somewhat above the critical micelle concentration.

form varies with the type (ionic or nonionic) and identity of the surfactant and with the total salts concentration of the aqueous phase. Micellar structure is illustrated in Figure 2. Vold and Vold (1983) give an extensive discussion of the behavior of surfactant solutions and of micelle formation; this is also treated by Rosen (1989), Scamehorn (1986), Lucassen-Reynders (1981), and McBain (1950).

Aqueous surfactant solutions with surfactant concentrations substantially above the CMC contain a significant volume of nonpolar phase in the interiors of the micelles. This nonpolar phase is able to dissolve relatively nonpolar solutes—hydrophobic compounds such as PCBs, organic solvents, chlorinated pesticides, and the like. This phenomenon, known as *micellar solubilization*, is illustrated in Figure 3. The solubilities of these hydrophobic compounds in water are typically quite small, on the order of 1 mg/L for PCBs, for instance. In surfactant solutions, on the other hand, their solubilities are increased manyfold by micellar solubilization. Cleanups that might require hundreds of years if simple water flushing were used may be complete within a year or two if surfactants are employed, due to the ability of surfactant solutions to dissolve perhaps 100-fold to 1000-fold more contaminant per unit volume than can be dissolved in water. Two early but very useful references on solubilization are McBain and Hutchinson's book (1955) and a review article by Klevens (1950). The thermodynamics of solubilization (or mixed micelle formation) has been discussed by Hall and Pethica (1967), Mukerjee (1971a,b), and Wayt and Wilson (1989), among others.

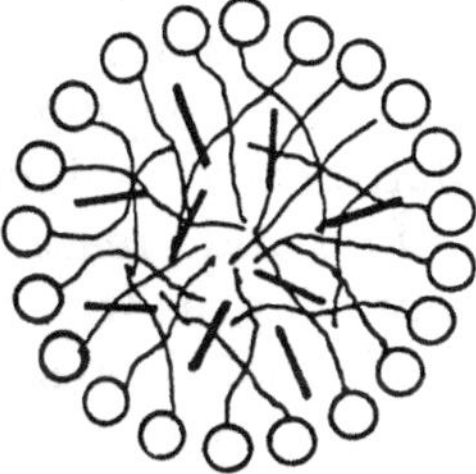

Figure 3 Schematic representation of micellar solubilization of a hydrophobic organic compound.

The use of aqueous surfactant systems for the recovery of oil has been common for some years and certainly suggests the employment of surfactants for the remediation of soils and aquifers contaminated with hydrophobic organics (see, e.g., Hill et al., 1973; Klins et al., 1976). Use of surfactants for remediation purposes apparently began with the Texas Research Institute (1979, 1985), which carried out successfully a study of the use of surfactants for the recovery of gasoline by in situ flushing with a mixture of an anionic and a nonionic surfactant.

One of the more extensive studies of surfactant flushing of contaminated soils is that of Science Applications International Corp. (SAIC), which prepared a report on the use of aqueous nonionic surfactants for flushing PCBs, chlorinated phenols, and petroleum hydrocarbons from contaminated soils (Ellis et al., 1984; Ellis et al., 1985). They found removal efficiencies of over 90% with surfactant solutions containing 1.5% surfactants, and noted that these removals were orders of magnitude greater than those obtained by flushing with water alone. These workers used a 1:1 blend of two nonionic surfactants.

The SAIC group also carried out treatability studies of the aqueous surfactant–contaminant solutions resulting from surfactant flushing, and noted that a serious problem arose here. They were unable to find a process that would allow recovery of the surfactant for reuse, essential for cost-effective application of the technique. They tried a number of separation methods unsuccessfully, and indicated that the development of a scheme in which the surfactant solution could be separated from the contaminants and recycled was a crucial next step in the development of surfactant flushing.

Nash et al. (1987) carried out a small-scale field trial of surfactant flushing at Volk Air National Guard Base, Wisconsin. Soil heavily contaminated with oil and other hydrocarbons at a fire pit used in training was treated in laboratory columns and in situ. The lab studies were quite encouraging. The field studies were plagued by clogging and heavy rains, and hydrocarbon removal was not statistically significant. Nash commented that the extremely high levels of oil and grease that were present in the soil may have led to severe channeling, as well as to the clogging of several of the test holes. We conclude that Nash's negative findings demonstrate that surfactant solutions are not effective in cleaning soil if they cannot penetrate it. His encouraging lab column results suggest that a soil washing technique might prove effective even with such heavily contaminated material.

Castle and his co-workers (1985) developed a mobile soil washing system for use at Superfund sites which includes the possible use of surfactants, as well as chelating agents, acids, and so on. There are some problems with the use of this apparatus with surfactants. Tabak and Traver (1990) and Traver, et al. (1989) reported on the U.S. Environmental Protection Agency's (USEPA's) studies of washing technologies for the removal of petroleum products such as gasoline and home heating oil from a synthetic soil matrix. Two additives were used, CitriKleen and an unidentified surfactant. This bench-scale work simulated the mobile soils washing unit that the USEPA has developed. It was found that removals from actual site soils were less efficient than removals from the synthetic mix, that there were losses of soil as large as 45%, that the wash water contained 20% suspended solids, and that foaming was excessive. Evidently, the mobile soils washing unit in its present form is not well adapted for use with surfactants, and additional work will be required on the treatment of the spent washwaters and fines. The USEPA (1986) published an extensive review of early surfactant flushing work and the surface chemistry on which it is based; this report includes a good deal of useful data.

Roy and Griffin (1988) have reviewed the decontamination of soils with both surfactants and chelating agents, and McDermott and his co-workers (1988) have reported on the removal of PCBs from soils by both biodegradation and surfactant extraction. Another study reports on the surfactant flushing of creosote (Hazardous Materials Control Research Institute, 1987). Amdurer et al. (1986) have compiled information on a number of commercially available surfactants and related their properties to the possibility of improvements in surfactant-assisted flushing. Their report also notes the need for continued research on this technology.

Valsaraj and Thibodeaux (1989) have shown that there is a useful correlation between octanol/water partition coefficients and the micellar phase/water partition coefficients that arise in surfactant flushing and surfactant washing. Since a very large number of octanol/water partition coefficients are available in the literature (Montgomery and Welkom, 1990; Verschueren, 1983; Mercer et al., 1990) such a correlation permits the easy estimation of micellar phase/water partition coefficients, of which relatively few have been determined.

Our group has carried out both laboratory-scale studies and mathematical modeling of in situ surfactant flushing and ex situ surfactant washing. The mathematical modeling included surfactant flushing in a lab column and also with an injection well and a recovery well, either unconfined in the aquifer or enclosed within a rectangular slurry wall to prevent escape of surfactant solution from the site (Wilson, 1989a). A second paper examined the nature of the solubilization process and developed a method for estimating the partition coefficient of a hydrophobic contaminant between the aqueous phase and the micellar phase present in surfactant solutions having surfactant concentrations above the CMC (Wayt and Wilson, 1989). Laboratory studies included the investigation of the effective solubilities of naphthalene, dichlorobenzene, and biphenyl in solution having various concentrations of sodium dodecylsulfate (SDS), demonstration that SDS solutions are capable of extracting these model contaminants from sand and sand/clay mixtures with high efficiency, and demonstration that the contaminant-laden surfactant solutions could be reclaimed by gentle extraction with hexane or mineral oil. Valsaraj's correlation of octanol/water partition coefficients with micellar phase/water partition coefficients was also verified and extended to five more compounds. (Gannon et al., 1989; Wilson, 1989b; Wilson and Clarke, 1988). A recent paper (Clarke et al., 1991) presents data on the bench-scale removal of PCBs from high-clay soil by surfactant washing. Countercurrent liquid–liquid extraction was used for removal of nonvolatile contaminants from the contaminated surfactant solution, and thin-film aeration was shown to remove volatiles. Models for batch–batch, batch–continuous flow, and countercurrent flow surfactant soil washing were developed, and the effects of the model parameters were explored.

Kunze and Gee (1989) investigated the use of Triton-X-100 (a nonionic surfactant), as well as acids and CitriKleen, in washing soil from a CERCLA site. Improved removal of PCBs was noted, but the use of a nonionic surfactant precluded easy separation of the surfactant solution from the contaminants for reuse. The authors concluded that the technology had very promising aspects, but that additional bench and pilot scale work was needed to demonstrate feasibility. One reason why past workers have selected nonionic surfactants is that the critical micelle concentrations of these are much lower than those of ionic surfactants, so that a larger fraction of surfactant is in micellar form and therefore effective in solubilizing hydrophobic contaminants.

Other researchers are investigating in situ flushing at the bench level, focusing on specific problems. Dworkin et al. (1988) indicated that in situ soil flushing in combination with in situ biodegradation could be a cost-effective way to remediate creosote-contaminated soils. Kuhn and Piontek (1989) also looked at this technology for a contaminated wood-preserving site. They achieved up to 98% removal from soil cores in the laboratory using combinations of agents.

Esposito et al. (1988a,b) developed four surrogate contaminated soils [synthetic analytical reference matrices (SARMs)] spiked with specified amounts of seven volatile organics, three semivolatiles, As, Cd, Cr, Cu, Pb, Ni, and Zn. The clean matrix had an average cation-exchange capacity of 133 mEq per 100 g, average total organic carbon of 3.2%, average pH of 8.5, and an average weight percent grain size distribution of gravel, sand, silt, and clay of 3, 56, 28, and 12%, respectively. Four SARMs were prepared: high organic–low metal, low organic–low metal, low organic–high metal, and high organic–high metal. Bench-scale experiments were used to assess soil washing treatments using water, surfactant, or EDTA solutions. Soil washing with chelating agent–water and detergent–water was 93% effective for removal of metals; semivolatile and volatile organics were removed with efficiencies of 87 and 99%, respectively.

PHASED EVALUATION OF SURFACTANT FLUSHING/WASHING

Surfactant remediation techniques are still in a rather exploratory stage, so their evaluation for use at any particular site involves more uncertainty than one finds with techniques in more advanced states of development. It is convenient to carry out the evaluation in phases to avoid unnecessary expense and effort.

The initial phase involves literature review and examination of data available from the remedial investigation. This is simply a first screening of the techniques to see if they can be ruled out with relatively little effort. It involves three points, as follows.

First, are the chemical contaminants present appropriate for remediation by surfactant techniques? Removal of soluble organics is readily accomplished by flushing with water alone, so that the additional expense and complication of surfactants can be avoided. Soluble metals are also more readily removable with water or other reagents (such as chelating agents or acids) than with surfactants; some surfactants form slightly soluble compounds with metals which may actually interfere with their removal. Surfactants are not effective for removing metal precipitates (hydroxides, carbonates, sulfides). Volatile, semivolatile, and slightly volatile hydrophobic organics are suitable for removal by solubilization with surfactants. Compounds that should be suitable for removal with surfactants include chlorinated solvents (tetrachloroethylene, trichloroethylene, 1,1,1-trichloroethane, etc.); aromatic solvents (benzene, toluene, xylenes, ethylbenzene); other volatile, semivolatile, and slightly volatile hydrocarbons (petroleum ether, gasoline, diesel and jet fuel, kerosene, fuel oils, etc.); greases; PCBs, chlorinated pesticides (DDT, chlordane, dieldrin, etc.); plasticizers [such as bis(2-ethylhexyl)phthalate]; and polynuclear aromatic hydrocarbons. The octanol/water partition coefficients of the contaminants can be found in such references as Mercer et al. (1990), Montgomery and Welkom (1990), or Verschueren (1983) and the micelle/water partition coefficient K_{sw} estimated from Valsaraj's (1989) formula as modified by Clarke et al. (1991),

$$\log_{10}K_{sw} = 1.12 \log_{10}K_{ow} - 0.686 \quad (1)$$

Second, is the soil reasonably homogeneous and permeable so that one will be able to move surfactant solution through all of the contaminated region at adequate flow rates? If there are strata of lower permeability (silt, till, etc.), will it be possible to maintain sufficient pressure gradients across these strata to force surfactant solution through them at an acceptable rate? In soil vapor stripping one can count on diffusion to help move VOCs from regions of low permeability. This is not true in surfactant flushing, since diffusion constants in the liquid phase are roughly 1/10,000 as large as diffusion constants in the gas phase. Therefore, all removal must be by advecting surfactant solution. Well logs are a useful source of information about soil characteristics. Ranges of hydraulic conductivities of soils of various types are shown in Figure 4.

Third, is the hydrogeology such that one is certain that no contaminant-laden surfactant solution will escape capture by the recovery well(s)? If groundwater flow is slow and there is a continuous aquitard underlying the domain to be treated, this may be accomplished merely by pumping slightly more fluid from the recovery well than is being injected. If the groundwater movement is more rapid, careful alignment of the injection and recovery wells may be necessary, or one

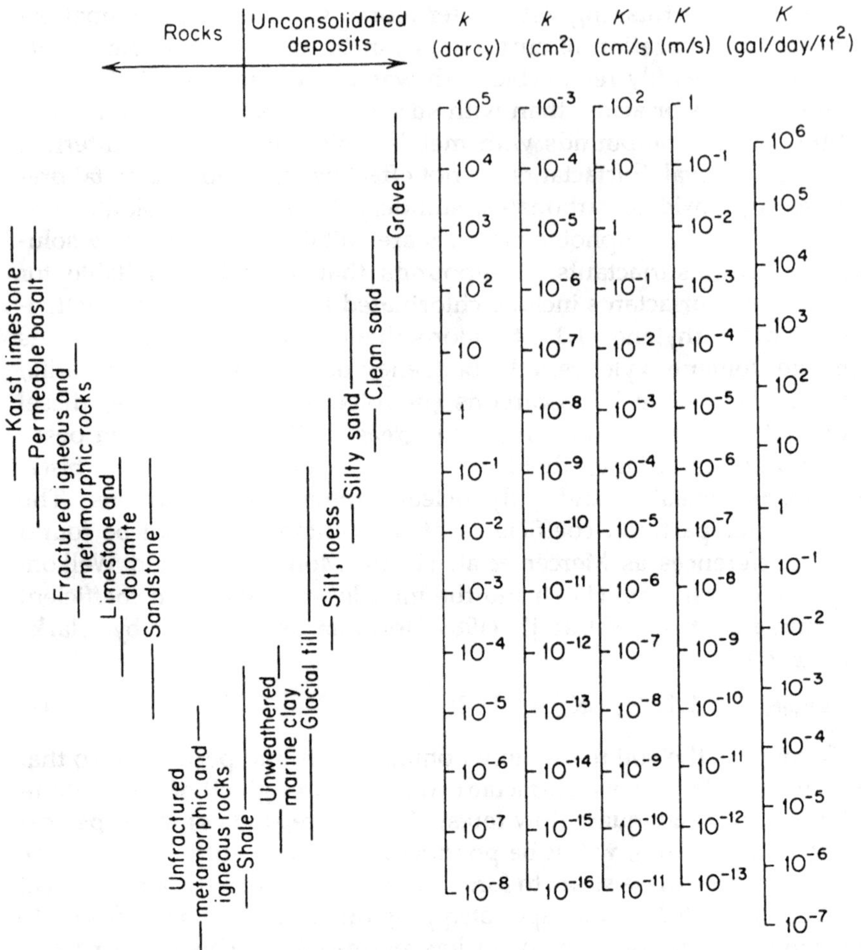

Figure 4 Range of values of hydraulic conductivity and permeability. (From R. A. Freeze and J. A. Cherry, *Groundwater*, Prentice-Hall, Englewood Cliffs, N.J., 1979.)

may need to put a slurry wall barrier around the domain of contamination. If the domain of contamination cannot be hydraulically isolated from the rest of the aquifer (and possibly other aquifers as well), it may be necessary to use ex situ surfactant washing or another technique altogether.

At this point one may be able to eliminate surfactant technologies as unsuitable for one or more reasons. If they have not been eliminated, one enters the next phase of evaluation. This involves laboratory studies to answer three crucial questions on a site-specific basis. Since these techniques are still very much in the development stage, somewhat more work may be needed at this point than is required with better established methods. The questions to be answered are as follows.

Can bench-scale surfactant washing achieve the mandated levels of cleanup when used on samples of contaminated materials from the site? Flask tests run on a shaker or column flushing tests will certainly be more efficient than full-scale surfactant washing or surfactant flushing in the field. If the lab-scale tests are unable to meet the cleanup criteria required, there is little or no chance that larger-scale operations will be successful. If column flushing tests indicate that the soil permeability is too low to make flushing feasible, one can only expect this problem to be more acute when using in situ flushing on undisturbed soils.

Can the soil fines be adequately separated from the contaminated surfactant solution? This is particularly likely to be a problem when soil washing with vigorous mixing is used and the matrix is high in clay. pH adjustment to achieve the point of zero charge of the suspended material should be investigated, as might the use of alum, ferric chloride, and ionic polymers of the sort used to facilitate coagulation in wastewater treatment.

Can the contaminants be removed from the used surfactant solution to the point where it can be reused? Use of surfactant on a one-pass basis would be quite expensive, both in terms of the cost of the surfactant itself and the cost of treating the contaminant-laden surfactant solution by conventional techniques. Volatiles and semivolatiles can probably be removed by trickling aeration in a packed column. Organics of low volatility will probably require countercurrent liquid–liquid extraction with a solvent. Use of a volatile solvent such as hexane permits further separation and recycle of constituents and/or reduced volumes of waste for final treatment or disposal. This is consistent with the requirements of the Superfund Amendments and Reauthorization Act of 1986.

If surfactant flushing and/or washing have not been eliminated by the screening process described above, one is ready to make a very preliminary estimate of costs for comparison with competing technologies. Then, if circumstances warrant, a decision must be made as to whether an in situ or an ex situ technique is to be used, following which a pilot-scale operation is designed and carried out. This provides input to the final decision as to whether the technique is to be used, data for improved estimates of cleanup times and costs, and design information for the full-scale facility.

MICELLAR SOLUBILIZATION

Micelle formation and micellar solubilization of hydrophobic compounds have been the subjects of a number of rather complex and sophisticated treatments (see, e.g., Vold and Vold, 1983; Israelachvili, 1985; McBain and Hutchinson, 1955; Wayt and Wilson, 1989; Hall and Tiddy, 1981). We present a rather simple two-phase distribution approach described by Vold and Vold (1983) and discussed in detail by Gannon et al. (1989).

Assume that the concentrations of the contaminant in the interiors of the micelles and in the aqueous phase outside the micelles are related by the simple distribution law

$$K_{sw} = \frac{C_{\text{micelle}}}{C_{\text{aqueous}}} \tag{2}$$

where C_{micelle} is the contaminant concentration in the hydrophobic phase inside the micelles, mol/L, and C_{aqueous} is the contaminant concentration in the aqueous phase outside the micelles, mol/L. This formula is a good approximation for the distribution of solutes between water and solvents immiscible in water. Since the micellar interiors are rather like microdroplets of hexane or other hydrocarbon solvent, this would appear to be a reasonable model for calculating solubilizate concentrations in the micellar interiors, with values of K_{aw} expected to be approximately those for the distribution coefficients of the solutes between water and hexane.

Let us assume that the aqueous phase is in equilibrium with solid (liquid) contaminant and that the contaminant concentration in the aqueous phase is therefore the saturation concentration, C_s (mol/L). The contaminant concentration in a micelle is then given by $C_{\text{micelle}} = K_{sw}C_s$. The total effective contaminant concentration in the solution is then given by

$$C_{\text{total}} = \frac{\text{moles solute in aqueous phase} + \text{moles solute in micelle interiors}}{\text{total volume of solution}} \tag{3}$$

If the surfactant solution is relatively dilute (<5%), the number of moles of solute in 1 L of solution is given approximately by $C_s \cdot 1$ L.

The volume of micelle interior phase in 1 L of solution is given by $1\ \text{L} \cdot (C_{\text{surf}} - \text{CMC}) \cdot V_{\text{tail}}$, where C_{surf} is the total molar surfactant concentration, CMC is the critical micelle concentration of the surfactant (about 0.008 M or 2.31 g/L for SDS), and V_{tail} is the molar volume of the surfactant hydrocarbon tail. For SDS, the molar volume of dodecane,

$C_{12}H_{26}$, should be a good approximation to V_{tail}. If one divides the molecular weight of dodecane, 170.33 g/mol, by its density, 0.766 g/mL, one obtains a molar volume of 0.222 L/mol.

The number of moles of contaminant dissolved in this volume of micellar phase is given by

$$1 \text{ L} \cdot (C_{surf} - \text{CMC})V_{tail}C_{micelle} \tag{4}$$

Use of equation (2) then yields

$$C_{total} = C_s[1 + K_{sw}V_{tail}(C_{surf} - \text{CMC})] \tag{5}$$

if $C_{surf} > \text{CMC}$, and

$$C_{total} = C_s$$

if $C_{surf} <= C_s$.

A plot of C_{total} versus C_{surf} is therefore expected to be flat at surfactant concentrations below the CMC, and to be linearly increasing with surfactant concentration above the CMC with a slope of $C_sK_{sw}V_{tail}$. Values for K_{sw} are expected to be rather similar to the distribution coefficients for the contaminants between hydrocarbons such as hexane or octane, and water. If so, the slope of the plot, which measures the surfactant's solubilizing power for the solute, can be well estimated without any measurements of solubilization. This theory also explains the good linear correlations of micelle/water distribution coefficients and octanol/water partition coefficients.

Distribution coefficients for *p*-dichlorobenzene (DCB), naphthalene, and biphenyl between water and hexane were measured by extracting 1 L of saturated aqueous solution with 1 mL of hexane, analyzing, and substituting the results into

$$K_{hw} = \frac{V_w}{V_h}\left(\frac{C_w^\circ}{C_w} - 1\right) \tag{6}$$

where V_w = volume of aqueous phase
V_n = volume of organic (hexane) phase
C_w° = initial solute concentration in aqueous phase
C_w = final solute concentration in aqueous phase

Values of the distribution coefficients and the aqueous solubilities of these compounds are listed in Table 2. These distribution coefficients and the value of V_{tail} calculated above, 0.222 L, were used to calculate theoretical values for the slopes of plots of C_{total} versus C_{surf} for these three compounds; the slopes were then converted to (mg/L)/m*M*. Experimental values for the slopes were obtained from the data plotted in

Table 1 Aqueous Solubilities and Hexane/Water Distribution Coefficients for DCB, Naphthalene, and Biphenyl

Compound	Aqueous solubility (mol/L)	Distribution coefficient, K_{hw} (dimensionless)
DCB	5.4×10^{-4}	950
Naphthalene	2.3×10^{-4}	1490
Biphenyl	5.3×10^{-6}	3870

Figures 5 to 7. The experimental and theoretical values are compared in Table 2; the results suggest that this very simple approach permits a fairly good qualitative estimate of the solubilizing power of a surfactant for any particular hydrophobic contaminant. This should be useful in preliminary screening of surfactant flushing/washing.

A plot of $\log_{10}$ SDS micelle/water partition coefficients versus $\log_{10}$ octanol/water partition coefficients is shown in Figure 8. Data for methylene chloride, chloroform, and carbon tetrachloride are from Valsaraj et al. (1989); data for DCB, naphthalene, and biphenyl are from Gannon et al. (1989); data for dieldrin and heptachlor are from Clarke et al. (1991). A least-squares fit of the eight data points gives

$$\log_{10}K_{sw} = 1.12 \log_{10}K_{ow} - 0.686 \tag{7}$$

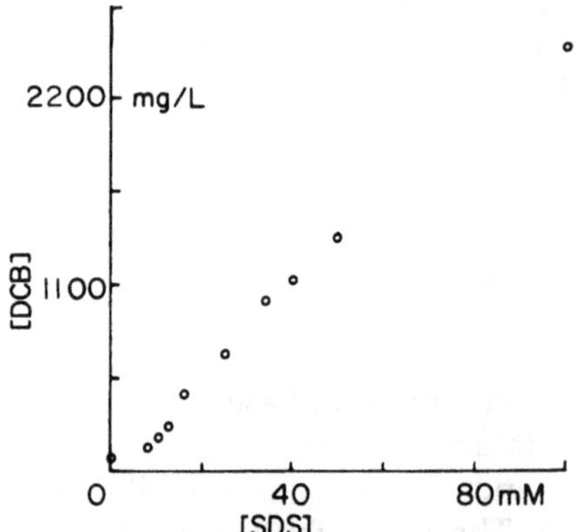

Figure 5 Plot of *p*-dichlorobenzene (DCB) solubility (mg/L) versus aqueous sodium dodecylsulfate (SDS) concentration (m*M*). (From Gannon et al., 1989.)

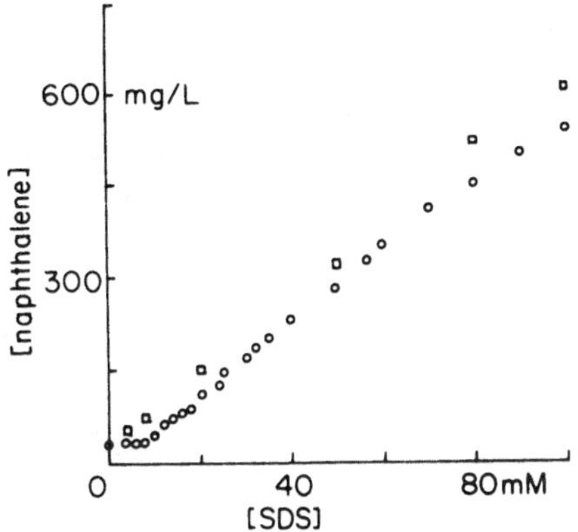

Figure 6 Naphthalene solubility (mg/L) versus aqueous SDS concentration (m*M*) (circles). Squares represent runs in which 0.1 *M* NaCl was present. (From Gannon et al., 1989.)

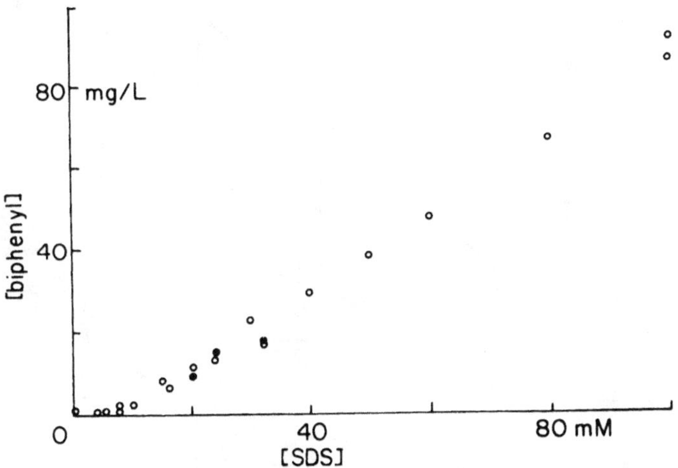

Figure 7 Biphenyl solubility (mg/L) versus aqueous SDS concentration (m*M*). Solid circles are for runs with recycled SDS solution. (From Gannon et al., 1989.)

Table 2 Solubilizing Power of SDS for DCB, Naphthalene, and Biphenyl [(mg/L)/m*M*]

Compound	Observed slope	Calculated slope
DCB	26 ± 1 mg/mmol	17 mg/mmol
Naphthalene	5.7 ± 0.1	9.7
Biphenyl	0.94 ± 0.02	0.70

with a correlation coefficient of 0.981 as the equation relating K_{sw} for SDS solutions (as described previously) to the octanol/water partition coefficient for any given hydrophobic compound.

Increases in solubility of hydrophobics in the presence of surfactants are very substantial, as seen in Table 3. These give some idea of the differences in cleanup times one might expect between simple water flushing and surfactant flushing/washing. Table 4 gives $\log_{10}K_{ow}$ values and some other properties for a number of compounds that are likely candidates for removal by surfactant flushing or washing techniques.

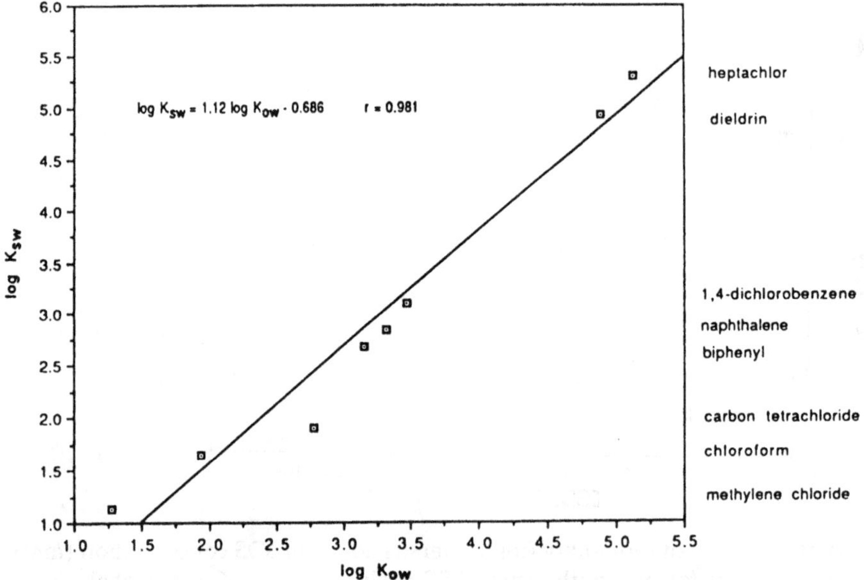

Figure 8 Correlation of octanol/water partition coefficient (K_{ow}) with SDS surfactant/water partition coefficient (K_{sw}). (From Clarke et al., 1991.)

Table 3 Solubilities of Model Compounds in Water and 100 m*M* SDS Solution

Compound	Solubility in water (mg/L)	Solubility in SDS (mg/L)	Ratio
DCB	79	2390	30
Naphthalene	29	524	18
Biphenyl	0.82	86	105

LABORATORY-SCALE WORK

Batch shaken or stirred flask experiments are useful and inexpensive for preliminary assessment of the ability of surfactant solutions to remove contaminants from the soil matrix at a particular site. Because of the simplicity of the setup, a rather broad spectrum of experimental conditions can be explored relatively quickly. Parameters that can be explored include the identity of the surfactant, the concentration of surfactant to be used, the various types of soil matrix present at the site, and the duration of the washing process. The results of such experiments are directly relevant to surfactant washing techniques involving a good deal of agitation. Since the original soil structure is utterly destroyed during the shaking or stirring, these experiments give one a most favorable estimate of the performance of surfactant flushing/washing techniques in which the structure of the soil is undisturbed (flushing) or perhaps disturbed relatively slightly.

The procedure used for this test (ECKENFELDER INC., 1990) is as follows. The soil sample is thoroughly mixed to attain as homogeneous a distribution of contaminants as possible without losing significant amounts of volatiles. Three portions of this sample are taken for analysis. The desired number of 100-g portions of the sample are then placed in 1-L amber glass jars and 80-mL portions of surfactant solutions of the composition to be studied are added. Surfactant concentrations should be above the critical micelle concentration. One also prepares one or more deionized water controls.

The jars are then shaken or mixed with magnetic stirrers for the desired period of time, and the suspensions allowed to settle. The supernatant surfactant solutions are decanted, filtered through 0.45-μm filters to remove remaining fines, and analyzed. Fresh portions of surfactant solution are added to the soil in the jars, and the process is repeated. The procedure is continued through as many steps as needed. Initially, we used periods of 24 h for each surfactant washing step; these turned out to be quite a bit longer than needed. Depending on the

Table 4 Properties of Organic Compounds Relevant to Surfactant Flushing/Washing[a]

Compound	Molecular weight (gm/mol)	Vapor pressure (torr)	Aqueous solubility (mg/L)	$\log_{10}K_{ow}$
Aroclor 1260	372 (ave.)	4.05×10^{-5}	0.0027	7.14
Aroclor 1254	327 (ave.)	7.7×10^{-5}	0.012	6.03
Benzene	78	95	1750	2.12
Biphenyl	154	0.06	7.5	3.88
bis(2-Ethylhexyl) phthalate	390.5	2×10^{-7}	0.0285	3.98
Carbon tetrachloride	153.8	90	757	2.64
Chlordane	409.8	10^{-5}	0.56	3.32
Chlorobenzene	112.5	11.7	466	2.84
Chloroform	119	151	8200	1.97
1,2-Dichlorobenzene	147	1.47	137	3.38
1,3-Dichlorobenzene	147	2.28	123	3.60
1,4-Dichlorobenzene	147	1.18	79	3.60
cis-Dichloroethylene	97	208	3500	0.70
trans-Dichloroethylene	97	324	6300	0.48
1,1-Dichloroethylene	97	600	2250	1.84
Dieldrin	381	7.78×10^{-7}	0.186	3.87
Endrin	380.9	2×10^{-7}	0.024	5.34
Ethylbenzene	106	7	152	3.15
Heptachlor	373	3×10^{-4}	0.03	3.87–5.44
Naphthalene	128	0.23	31.7	3.44
Tetrachloroethylene	165.8	17.8	150	2.60
Toluene	92	36.7	500	2.69
1,1,1-Trichloroethane	133.4	123	1500	2.50
Trichloroethylene	131	19.9	1.11	2.89
Vinyl chloride	62.5	2660	2670	1.38
o-Xylene	106	7	175	3.12
m-Xylene	106	10	130	3.16
p-Xylene	106	10	198	3.15

Sources: Hazardous Substance Data Bank, National Library of Medicine; *NIOSH Pocket Guide to Chemical Hazards* (1987); Mercer et al. (1990).
[a]At room temperature and atmospheric pressure.

nature of the contaminants and the matrix, 2 to 5 h is probably ample for each washing step. After the surfactant washing is completed, the soil portions in the jars are analyzed for contaminants.

If a few extra jars are included in the experiments, one can carry out studies on the coagulation and settling of clay fines that may be present in the surfactant solution. Removal of most of these fines from the surfactant solution is necessary if surfactant reuse is planned. We have found that adjustment of the pH of the solution with small amounts of acid greatly increased the settling rate of the fines, as did the addition of small amounts of ferric chloride. Other approaches that might profitably be used include alum and ionic polymers.

The presence of substantial quantities of surfactant interferes with the standard methods of solvent extraction used in the analytical methods commonly employed. If surfactant solutions are extracted with organic solvents by means of shaking in a separatory funnel or other mechanically vigorous procedures, extremely persistent emulsions result. To avoid this, we use a gentle extraction procedure in which the organic liquid (hexane) is layered above the aqueous phase in an Erlenmeyer flask. The flask is stoppered with an aluminum foil–wrapped stopper and stirred overnight on a magnetic stirrer at a speed that does not disrupt the interface between the two phases (i.e., no droplets are formed). After extraction, an aliquot of the organic phase is taken for analysis by gas chromatography.

While requiring more space and time than the jar tests just described, column experiments usually more closely represent what will be done in the field. Ellis et al. (1985), Nash (1987), Gannon et al. (1989), and ECKENFELDER INC. (1990) have given descriptions of lab-scale surfactant flushing columns. Ellis and his co-workers studied the removal of PCBs, pentachlorophenol, and hydrocarbon oils and greases; Nash investigated oils and greases. Gannon et al. (1989) removed *p*-dichloro-benzene (DCB), naphthalene, and biphenyl; they used relatively small Pyrex columns (4.3 cm ID by 120 cm length) as shown in Figure 9. The column is assembled and about 250 g of washed sand is placed in the column to prevent clogging of the glass frit at the bottom of the column with fines. The soil sample (generally 250 g of material) is then put in the column; this is then topped with another 50 g of washed sand. The samples were clay–sand mixtures of various proportions that had been spiked with contaminant. Spiking is done by dissolving 1 g of the contaminant (*p*-dichlorobenzene, naphthalene, or biphenyl) in 20 mL of hexane, then adding this solution slowly, with vigorous shaking, to the soil sample. The mixture is then vigorously shaken for about 2 min, after which the hexane (along with some of the contaminant) is allowed to evaporate. This requires about 30 min. SDS solution (50 to 100 m*M*, roughly) or water is added to the column until the liquid level is about 1 cm above the top of the overlying sand layer, and the column is allowed to stand for 24 h to permit trapped air to escape. SDS solution is then added to the column and allowed to flow through the soil sample at the desired rate. Effluent samples are taken for analysis as needed;

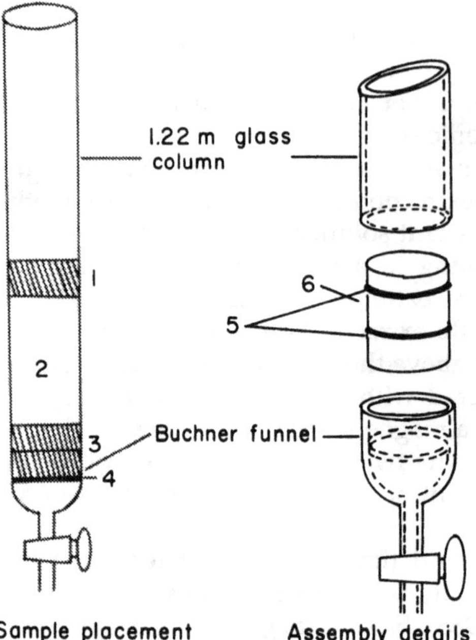

Figure 9 Diagram of a laboratory soil surfactant flushing column. 1, 50-g sand layer; 2, spiked soil or soil/sand sample; 3, 250-g sand filter layer; 4, coarse fritted glass disk in Buchner funnel; 5, O-rings; 6, aluminum collar.

we took 20-mL samples at 100-mL intervals and analyzed these by ultraviolet absorption. Samples that are excessively concentrated in the aromatic contaminant are diluted with SDS solution, rather than water, to avoid precipitation of the contaminant.

Naphthalene and DCB concentrations in the column effluents are shown in Figure 10; effluent biphenyl concentrations are shown in Figure 11. Figure 12 shows the fraction of DCB removed as a function of the volume of SDS solution passed through the column for two clay–sand mixtures. The process is kinetically controlled, as seen in Figure 13, which shows fraction DCB removed versus SDS effluent volume for two different flow rates.

Pilot- and field-scale studies are, as yet, very few in number. Nash's small-scale field test at Volk Air National Guard Base, Wisconsin, was unsuccessful, despite his very encouraging laboratory results. The site was an old fire pit, used for training and heavily saturated with oil and

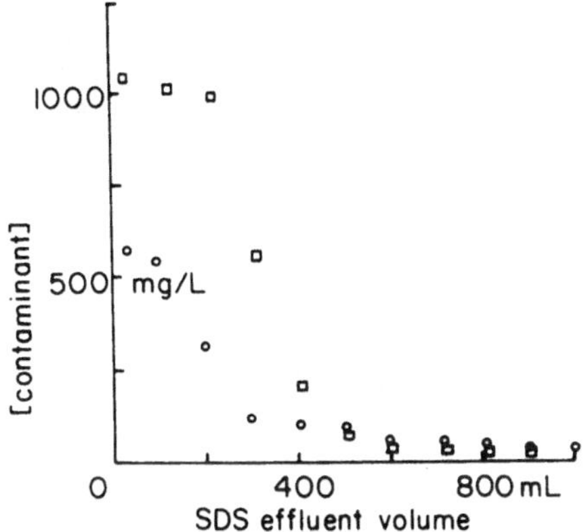

Figure 10 Effluent naphthalene (circles) and DCB (squares) concentrations versus volume of SDS solution (100 m*M*) passed through spiked sea sand in surfactant flushing columns. (From Gannon et al., 1989.)

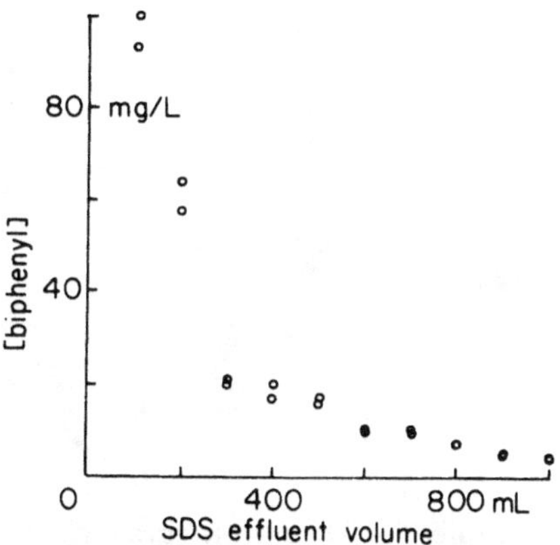

Figure 11 Effluent biphenyl concentration versus volume of 100 m*M* SDS passed through spiked sea sand in a surfactant flushing column. (From Gannon et al., 1989.)

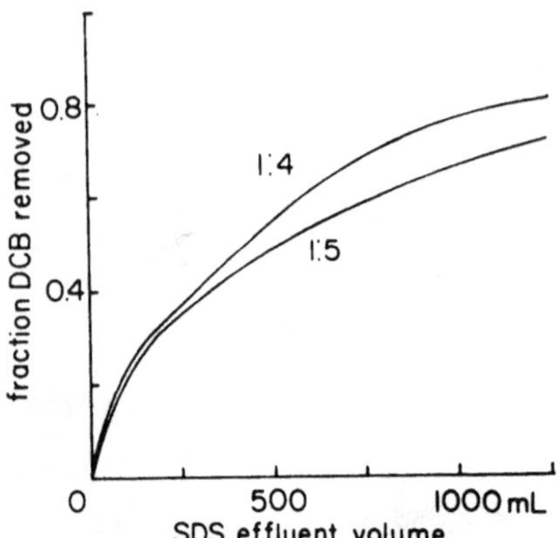

Figure 12 Removal efficiencies of 0.3-g samples of DCB versus volume of 100 m*M* SDS passed through a surfactant flushing column. Soil samples were red clay: sand in the proportions 1:4 and 1:5. (From Gannon et al., 1989.)

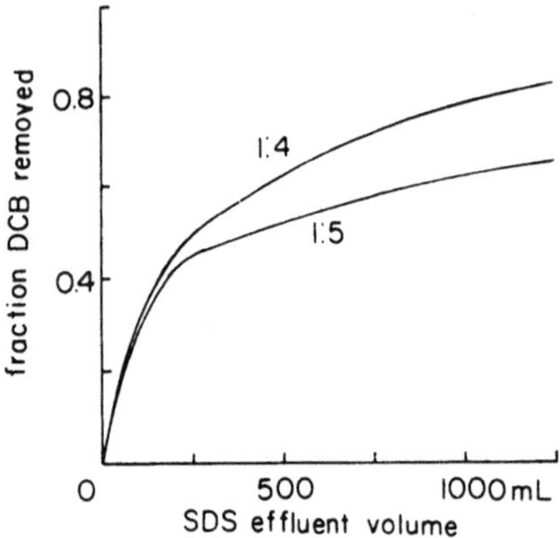

Figure 13 Removal efficiencies of 0.3-g samples of DCB from gray clay: sand (1:5) mixtures versus volume of 100 m*M* SDS passed through the columns. Curve 1 shows results for a column through which the SDS was gravity fed continuously as rapidly as possible. Curve 2 shows results for a column through which the SDS solution moved at a rate of 100 mL/h. (From Gannon et al., 1989.)

grease (1000 to 13,500 mg/kg). Several of the holes used in the field test clogged during the course of the work, and the results were complicated by a heavy rain that may have washed material into the holes.

SURFACTANT RECLAMATION

The importance of surfactant solution recovery, treatment, and reuse is dictated by the relatively high concentrations of surfactant that must be used (1% or more, typically), the high cost of surfactants, and the load on conventional waste treatment facilities that one-pass use of surfactant would impose. Ellis et al. (1985) have shown that use of nonionic surfactants precludes the use of a substantial number of potential surfactant recovery techniques. Their work encouraged us to focus on such ionic surfactants as SDS.

Volatiles and semivolatiles can be removed by air stripping. Clarke et al. (1991) air stripped *o*-xylene, toluene, trichloroethylene, and 1,2-dichlorobenzene from 50 m*M* SDS solution in countercurrent columns packed with Raschig rings (see Figure 14). Liquid residence times in the columns were approximately 4 min, which was sufficient to remove 98% of the more volatile compounds; dichlorobenzene (vapor pressure 1.2 torr at 20°C) required a longer stripping time. This was achieved by using two columns in series. One column was 76 cm long by 7.5 cm ID; the other, 90 × 7.5 cm. The results are summarized in Table 5.

Underwood (see Gannon et al., 1989; Clarke et al., 1991) removed contaminants of low volatility from SDS solutions by means of batch or countercurrent extraction with an organic solvent. Her apparatus is sketched in Figure 15. The column is 90 cm long by 4 cm ID. A plastic scouring pad was placed at the interface between the surfactant solution and the overlying hexane layer to facilitate hexane droplet coalescence. A dispersion head at the bottom of the column produced hexane droplets about 0.5 cm in diameter. The hexane was recirculated through the column and was replaced about every 30 min. In the countercurrent mode the hexane flow rate was 90 mL/min; that of the surfactant solution was 10 mL/min. The use of such large droplets of hexane is to facilitate droplet coalesce and prevent emulsion formation. Data on the removal of *p*-dichlorobenzene (DCB), naphthalene, biphenyl, PCBs (Askarel, an insulating fluid containing 41.8% Aroclor 1260), and dieldrin from 50 m*M* SDS solution were obtained. Results on a batch run with DCB are shown in Figure 16. Table 6 gives results for the removal of dieldrin and PCBs by countercurrent extraction with hexane. Reclaimed surfactant solution was successfully used to

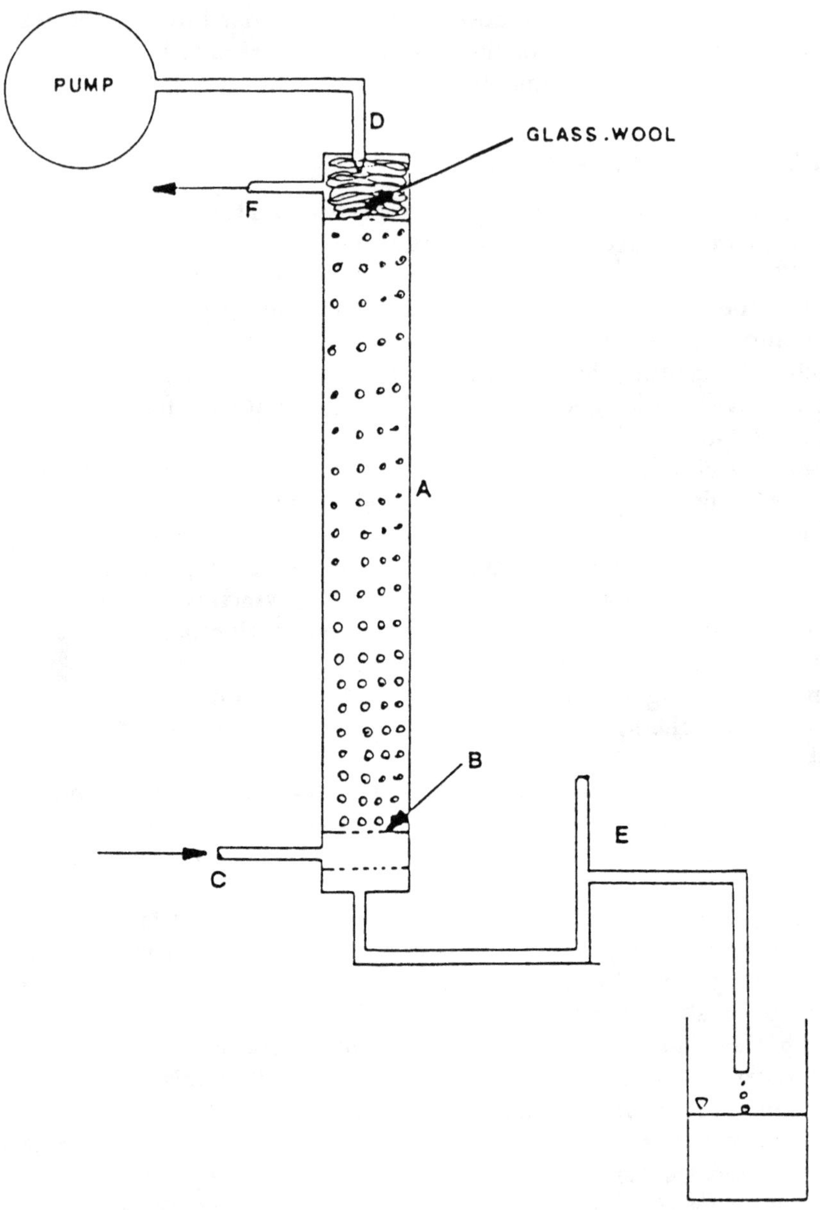

Figure 14 Diagram of a packed column for air stripping of volatile organics from surfactant solution. (From Clarke et al., 1991.)

Table 5 Results of Thin-Film Aeration of Surfactant in Packed Columns

Compound	Initial conc. (mg/L)	Final conc. (mg/L)	Percent removal	Liquid flow rate (mL/min)	Air flow rate (mL/min)	Treatment time (min)
o-Xylene	50	0.6	99	30	500	4
Toluene	500	11	98	28	500	4
Trichloroethylene	30,000	4100	86	26	500	4
1,2-Dichlorobenzene	500	95	81	11	1500	8

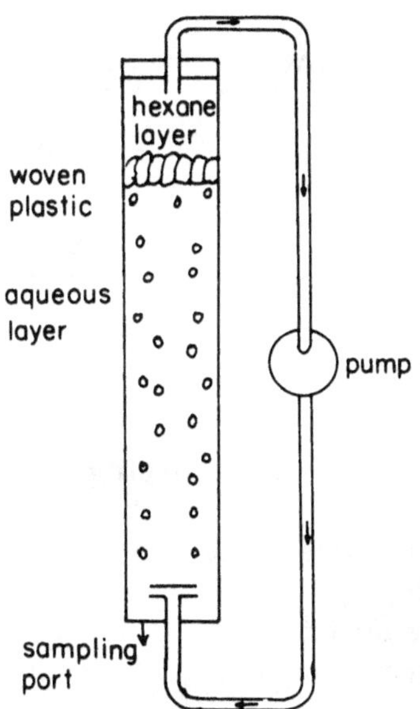

Figure 15 Batch mode column for carrying out gentle extraction of surfactant solution contaminated with nonvolatile organics. (From Gannon et al., 1989.)

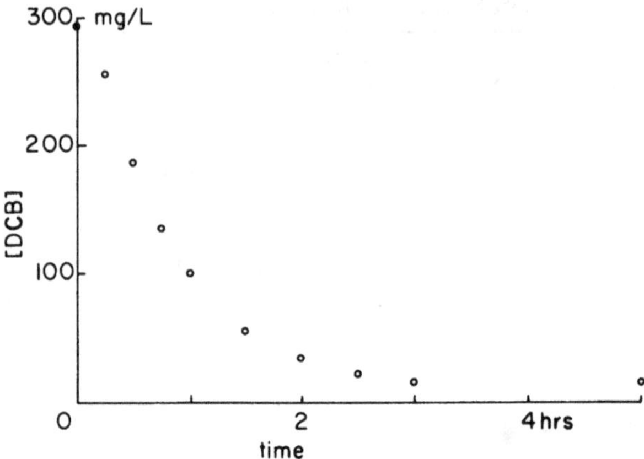

Figure 16 DCB concentration in the aqueous phase versus time for a run in the batch mode column surfactant recovery apparatus. The hexane flow rate was 19 mL/min, and the SDS concentration in the aqueous phase was 50 m*M*. (From Gannon et al., 1989.)

Table 6 SDS Reclamation by Countercurrent Extraction with Hexane

Compound	Initial conc.[a] (mg/L)	Final conc. (mg/L)	Percent removal	Treatment time (h)
Dieldrin	87	2.4	97	2.0
PCBs (in saturated[b] oil)	—	—	82[c]	1.5

[a]In 50 m*M* SDS solution.
[b]Oil sample containing 41.8% by weight Aroclor 1260.
[c]Based on linear response calibration curve of spectrophotometer.

remove biphenyl from a clay–sand mixture in a surfactant flushing column experiment.

In dealing with soils that contain appreciable quantities of clay, one can anticipate having to remove fines from the contaminant-laden surfactant solution before the contaminants are removed. Methods that appear promising for this include pH adjustment and treatment with ferric chloride, alum, or polymer.

MATHEMATICAL MODELING

In this section we discuss a number of models for modeling various modes of ex situ surfactant washing, surfactant flushing in laboratory columns, and in situ surfactant flushing. Some of these make the assumption of local equilibrium between contaminant in the stationary phase(s) and contaminant in the mobile micellar solution. Others take into account the possibility of diffusion and desorption kinetics being limiting. If contaminant must diffuse any appreciable distance before it reaches advecting surfactant solution, one can expect these techniques to be in serious difficulty because of the very small values of diffusion constants in the liquid phase. The models included are the following:

1. Batch process surfactant soil washing, local equilibrium model
2. Batch process surfactant soil washing, diffusion limited
3. Batch column with continuous-flow surfactant, diffusion limited
4. Countercurrent surfactant soil washing
5. In situ surfactant flushing

We first consider local equilibrium and diffusion-controlled models for soil surfactant washing in batch and continuous-flow modes.

Local Equilibrium Model for Batch Process Surfactant Soil Washing

See Figure 17 for a diagram of the apparatus. We follow Clarke et al. (1991). The container [of volume V (m^3)] is loaded with pulverized contaminated soil of porosity ν and is then filled with surfactant solution of concentration C (kg/m^3). The initial contaminant concentration in the soil is m_o (kg/m^3), and the contaminant concentration in the soil after the surfactant solution has been allowed to equilibrate with the soil and is then drained away is m (kg/m^3). We assume that the concentration of contaminant in the surfactant solution after equilibration, c (kg/m^3) is given by

$$c(m, C) = [c_o + K_D(C - \mathrm{CMC})] \frac{m}{m_{1/2} + m} \tag{8}$$

where c_o = contaminant solubility in water, kg/m^3
K_D = distribution coefficient for contaminant in the surfactant being used, dimensionless; $K_D = K_{sw} V_{\mathrm{tail}} C_o$
C = surfactant concentration, kg/m^3
CMC = surfactant critical micelle concentration, kg/m^3
$m_{1/2}$ = isotherm parameter to take into account the strength of the contaminant–soil binding, kg/m^3

Contaminant conservation for a 1-m^3 portion of soil and the surfactant solution it contains then yields

$$m_o = \nu c + m \tag{9}$$

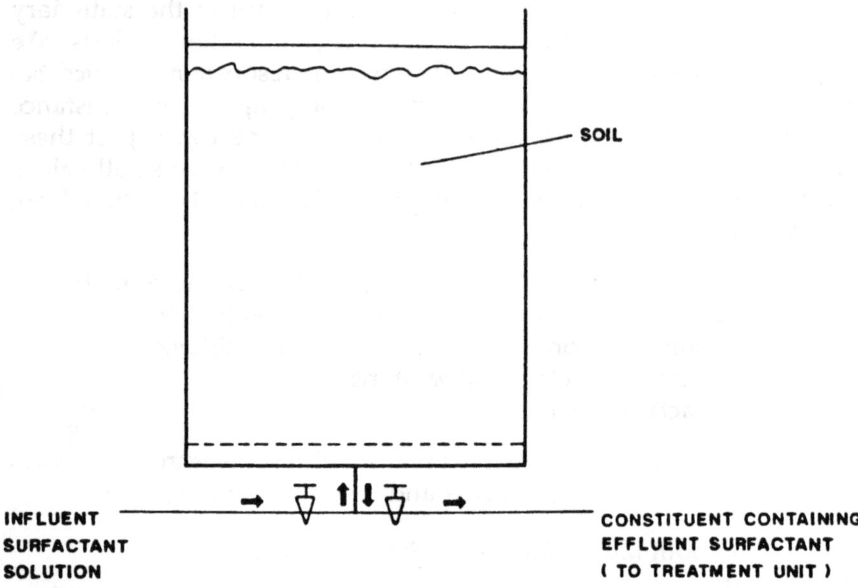

Figure 17 Schematic of a batch soil washing apparatus. (From Clarke et al., 1991.)

Define $K_s = c_o + K_D(C - \text{CMC})$ and substitute equation (8) into equation (9) to get

$$m_o = \nu K_s \frac{m}{m_{1/2} + m} + m \tag{10}$$

This is readily solved for m. If several washings are needed to reach the desired level of removal, one proceeds recursively; after $n + 1$ washings the contaminant concentration in the soil is given by

$$m_{n+1} = \frac{(m_n - \nu K_s - m_{1/2}) + \sqrt{(m_n - \nu K_s - m_{1/2})^2 + 4m_n m_{1/2}}}{2} \tag{11}$$

The sign in equation (11) is selected so that m_{n+1} lies in the range 0, m_n, where m_n is the soil contaminant concentration after n washings. This model was used to fit data on the removal of weathered-in Aroclor 1260 from a rather high-clay soil by means of jar tests.

Diffusion-Limited Model for Batch Process Surfactant Soil Washing

If the soil to be treated is lumpy (i.e., contains pieces of porous rock, lumps of porous clay of low aqueous permeability, etc.), the diffusion of contaminant from the interiors of these blocks may be rate limiting in the soil washing process. Also, desorption kinetics may be rate limiting. In this section the model above is modified to deal with such rate-limited systems.

Assume that the blocks of porous low-permeability medium can be approximated by spheres of radius a (m). Solution of the diffusion equation in spherical coordinates for a spherically symmetric porous solid is done by separation of variables, with a zero concentration boundary condition at the surface of the sphere. The smallest eigenvalue of the system is found to be

$$\lambda_1 = D_{\text{eff}}\left(\frac{\pi}{a}\right)^2 \tag{12}$$

where D_{eff}, the effective diffusion constant of the contaminant in the porous medium (m^2/s), is given by

$$D_{\text{eff}} = D\nu^{4/3} \tag{13}$$

Here D is the diffusion constant of the contaminant in bulk water, ν is the porosity, and equation (13) results from assuming complete saturation in Millington and Quirk's (1961) formula.

A conservative estimate of the rate of diffusion transport from the block of porous medium is to assume that during the course of one washing we have an exponential decay of the soil contaminant concentration from its initial value at the start of the washing toward the equilibrium value of the soil contaminant concentration as calculated in the preceding section. The rate constant for this exponential decay is taken as λ_1. The recursion formula derived in the preceding section is then modified as follows. First, one calculates m^*_{n+1}, the equilibrium value of m_{n+1}, as above. Then

$$\Delta m^* = m_n - m^*_{n+1} \tag{14}$$

gives the change in soil contaminant concentration that would result if equilibrium were achieved during this washing. Since the process is diffusion limited, the actual change in soil contaminant concentration that takes place during this washing period (of duration Δt) is given by

$$\Delta m = [1 - \exp(-\lambda_1 \Delta t)]\, \Delta m^* \tag{15}$$

Then the actual soil contaminant concentration after $n + 1$ washes is given by

$$m_{n+1} = m_n - \Delta m \tag{16}$$

Diffusion constants of ethanol, *n*-butanol, and sucrose in water are 1.24, 0.56, and 0.52 $\times$ 10^{-9} m^2/s (Levine, 1988). If we take 1 $\times$ 10^{-9} m^2/s as a representative diffusion constant for the contaminants, a soil porosity of 0.3, and a soil lump radius of 0.5 cm (0.005 m), the value of λ_1 is 7.928 $\times$ $10^{-9}s^{-1}$, and the characteristic time associated with the diffusion process is 1460 days, the washing period required to approach within about 37% of equilibrium in one stage of washing. Such long times are obviously unacceptable and indicate that fairly fine pulverization of the contaminated soil is necessary if its permeability is sufficiently low that the surfactant solution flows around the soil lumps rather than through them. If the quantity

$$D\nu^{4/3}\left(\frac{\pi}{a}\right)^2 \Delta t$$

is greater than unity, one can expect a washing to remove close to the amount of contaminant predicted by equilibrium calculations. (Here Δt is the duration of a washing.) If this quantity is significantly less than unity, removal will be quite inefficient.

Diffusion-Limited Model for Batch Column with Continuous-Flow Surfactant

Figure 18 diagrams an apparatus in which the soil is placed in a column and surfactant solution is passed upward through the column continuously. The notation is as above, with the following additions.

Q = surfactant flow rate, m^3/s

m_i = soil contaminant concentration in the ith compartment into which the column is formally partitioned, kg/m^3

c_i = aqueous contaminant concentration in the *i*th compartment, kg/m^3

c_i^e = equilibrium aqueous contaminant concentration in the *i*th compartment, kg/m^3

m_i^e = equilibrium soil contaminant concentration, *i*th compartment, kg/m^3

M_i = mass of contaminant in the *i*th compartment, kg

R = column radius, m

h = height of column, m

n = number of compartments into which column is formally partitioned

dz = h/n, thickness of one compartment

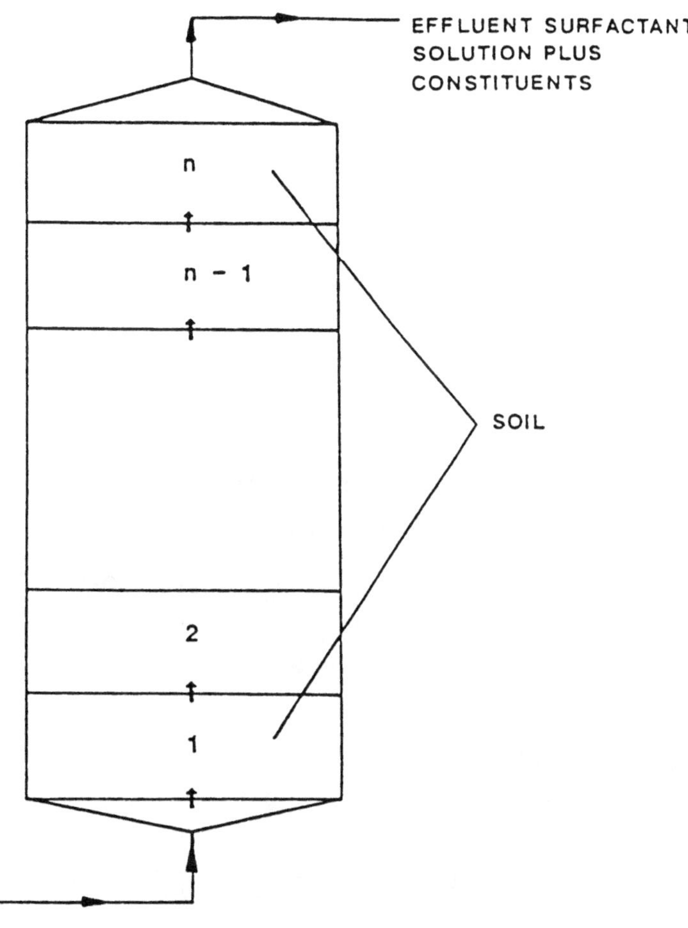

Figure 18 Schematic of a batch-continuous flow soil washing apparatus. (From Clarke et al., 1991.)

Then

$$M_i = [\nu c_i + (1 - \nu)m_i]\,\Delta V \tag{17}$$

where $\Delta V = \pi R^2\,dz$. The adsorption isotherm, equation (10), yields

$$c_i^e = K_s \frac{m_i^e}{m_{1/2} + m_i^e} \tag{18}$$

The rate of change of contaminant mass in the ith compartment is given by

$$\frac{dM_i}{dt} = Q(c_{i-1} - c_i) \tag{19}$$

Diffusion transport of contaminant is represented by means of the lumped-parameter approach used in the preceding section. We assume that

$$\left(\frac{dc_i}{dt}\right)_{\text{diff}} = \lambda_1(c_i^e - c_i) \tag{20}$$

We consider a short time interval Δt such that the effect of advection on c_i can be neglected. Then it is easily shown that

$$c_i(\Delta t) - c_i(0) = [c_i^e - c_i(0)]\,[1 - \exp(-\lambda_1\,\Delta t)] \tag{21}$$

Combining advective and diffusive terms finally gives

$$\frac{dc_i}{dt} = \frac{Q}{\Delta V}(c_{i-1} - c_i) + (c_i^e) - c_i)\frac{1 - \exp(-\lambda_1\,\Delta t)}{\Delta t} \tag{22}$$

The c_i^e are calculated from equation (18) and the mass balance equation

$$M_i = \Delta V[\nu c_i^e + (1 - \nu)m_i^e] \tag{23}$$

This yields

$$M_i = \Delta V\left[\nu c_i^e + (1 - \nu)\frac{m_{1/2}c_i^e}{K_s - c_i^e}\right] \tag{24}$$

rearrangement of which gives a quadratic equation in c_i^e which is readily solved. With a formula for c_i^e, it is possible to integrate equations (19) and (22) forward in time to describe the behavior of the column. A standard predictor–corrector formula works well for this. This model was implemented on a microcomputer and its behavior was explored. The standard parameter set is given in Table 7; departures from these values are indicated in the figures.

Figure 19 shows the effect on removal rate of variations in the surfactant flow rate through the column. A significant increase in removal rate is seen as the flow rate is increased from 0.5 to 1.0 L/s, but relatively little further increase in removal rate is observed as the flow rate is increased from 1 to 2 L/s. At this point the diffusion of contaminant from the interiors of the soil lumps is becoming the rate-controlling factor,

Table 7 Standard Parameter Set for Continuous-Flow Surfactant Soil Washing Column

Column height	2 m
Column diameter	1 m
Column flow rate	1 L/s
Number of compartments into which the column is partitioned	10
Soil porosity	0.3
Soil density	1.7 g/cm^3
Initial contaminant concentration	10^4 mg/kg
Contaminant solubility in water	1 mg/L
Distribution coefficient of contaminant in surfactant, K_D	2
Surfactant concentration	10 g/L
Surfactant critical micelle concentration	1 g/L
Isotherm parameter, $m_{1/2}$	1 kg/m^3
Time increment in numerical integration	10 s
Diffusion constant of contaminant in water	10^{-9} m^2/s
Effective diameter of soil lumps	0.1 cm

and further increases in surfactant flow rate will produce relatively small increases in contaminant removal rate.

The effects of increasing the surfactant concentration from 5 g/L to 10 to 20 g/L are shown in Figure 20. The increase from 5 g/L to 10 g/L results in a near-doubling of the rate of removal, but the increase from 10 g/L to 20 g/L results in an increase in contaminant removal rate that is significantly less than a doubling. At a surfactant concentration of 20 g/L the contaminant removal is apparently limited by diffusion of contaminant from the interiors of the soil lumps at this flow rate, so little is to be gained by using still higher surfactant concentrations.

Diffusion rates are proportional to the square of the effective diameter of the soil lumps, so one can expect enormous decreases in contaminant removal rates as the lump diameter is increased. This is borne out by the plots in Figure 21. In these the lump diameters are 0.2, 0.5, 1, 2, and 4 mm. Diffusion limitation is unimportant for the first two runs, becomes significant at a lump diameter of 1 mm, and is overwhelmingly the controlling factor for lump diameters of 2 and 4 mm. These results indicate the importance of adequately comminuting the soil to be treated if it contains porous lumps of low permeability and diameter greater than 1 to 2 mm.

Variation of the number of compartments into which the column is partitioned for mathematical representation was explored; increasing

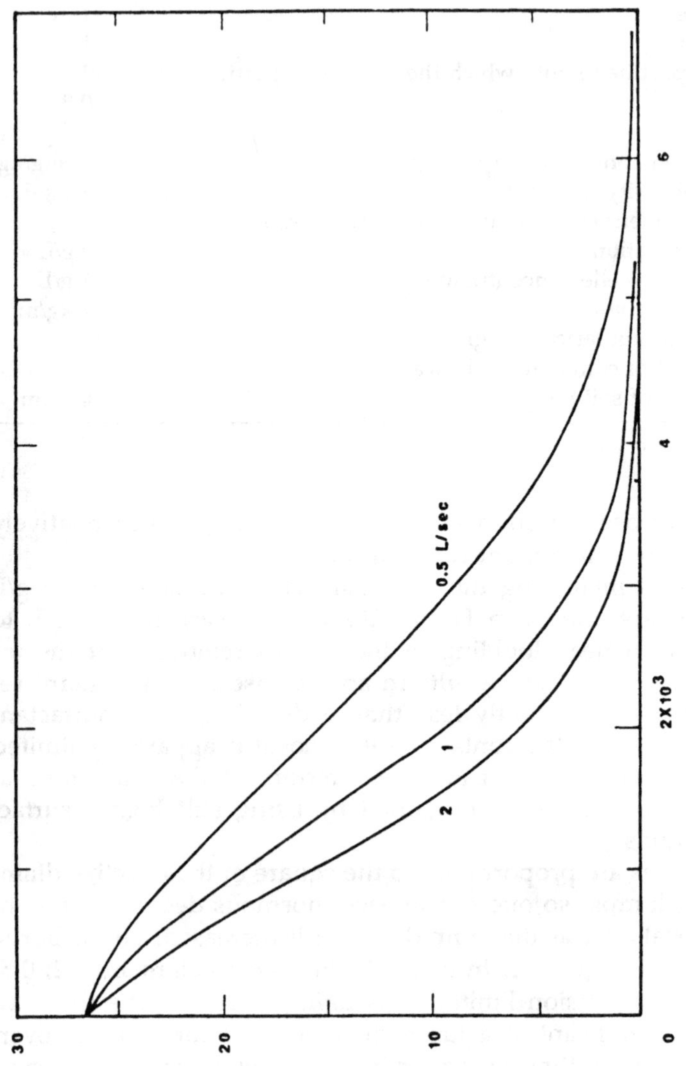

Figure 19 Batch-continuous flow operation; plots of total remaining contaminant mass versus time. Dependence on surfactant solution flow rate. Parameters as in Table 7 except as indicated. (From Clarke et al., 1991.)

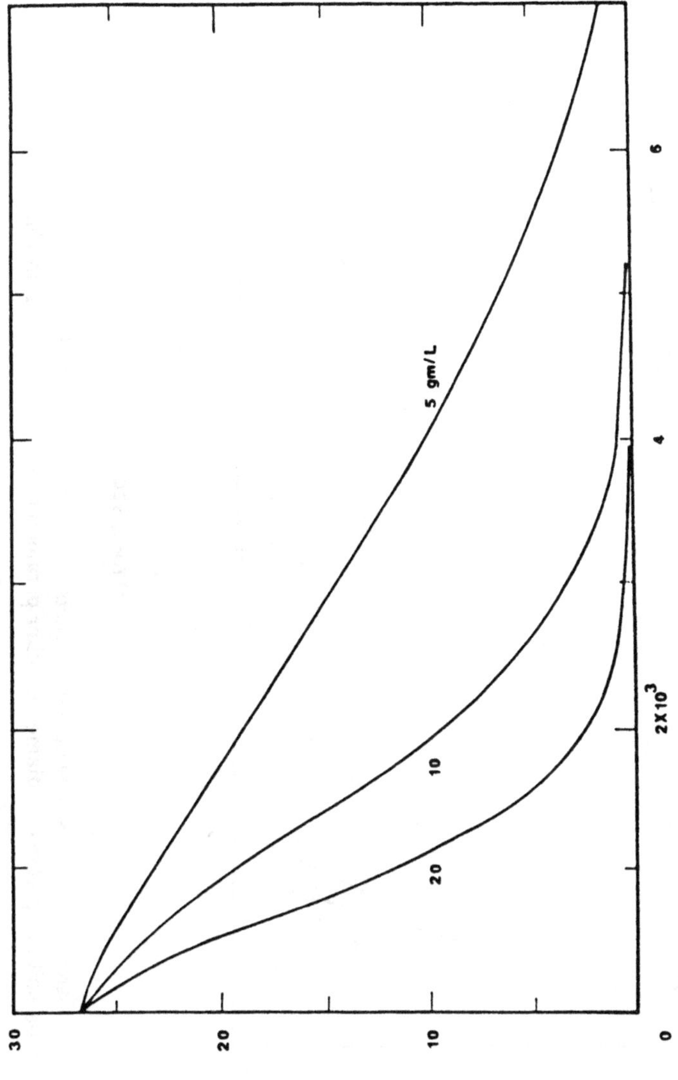

Figure 20 Batch-continuous flow operation; plots of total remaining contaminant mass versus time. Dependence on surfactant concentration. Parameters as in Table 7 except as indicated. (From Clarke et al., 1991.)

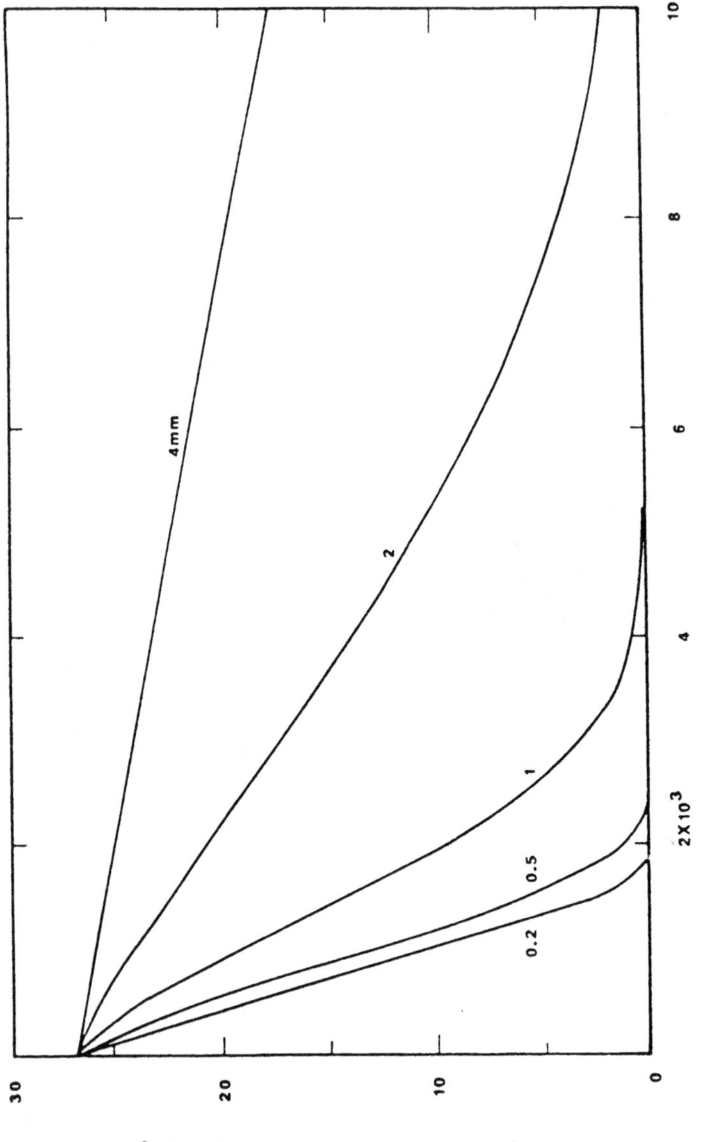

Figure 21 Batch-continuous flow operation; plots of total remaining contaminant mass versus time. Dependence on soil lump effective diameter. Other parameters as in Table 7. (From Clarke et al., 1991.)

the number of compartments from 5 to 40 resulted in very minor change in the appearance of the removal curve. Since this parameter can be used to represent axial dispersion, we conclude that little effort need be made to increase the number of equivalent theoretical plates in the column, and that minor variations in soil characteristics are not expected to interfere seriously with the process.

Countercurrent Surfactant Soil Washing Model

The inconvenience and high labor costs of batch operation are avoided if the process is run in a completely continuous-flow mode. A sketch of the setup is given in Figure 22. The countercurrent flow column is partitioned into n compartments, as before. Each contains a mobile aqueous phase (moving upward) and a mobile soil phase (moving downward). We carry out a mass balance on each phase, including advection of both soil and aqueous phase and diffusion transport (lumped-parameter approach) between the two. Notation is as before, with the following additions and modifications.

Q_s = rate of soil loading, m^3/s
Q_w = flow rate of surfactant solution, m^3/s
v_s = downward velocity of soil relative to the lab, m/s
ν = soil porosity
C = surfactant concentration, kg/m^3
c_i^w = contaminant concentration in the aqueous phase, ith compartment, kg/m^3
c_i^s = concentration of contaminant held in the soil, ith compartment, kg/m^3
$\Delta V_s = (1 - \nu)Q_s h/v_s n$ = volume of soil (excluding pores) in one compartment, m^3
$\Delta V_w = (\pi R^2 h/n) - \Delta V_s$ = volume of mobile water in one compartment, m^3

Advective transport alone in the aqueous phase gives

$$\left(\Delta V_w \frac{dc_i^w}{dt}\right)_{\text{adv}} = Q_w(c_{i-1}^w - c_i^w) \tag{25}$$

Similarly, advective transport alone in the soil phase gives

$$\left(\Delta V_s \frac{dc_i^s}{dt}\right)_{\text{adv}} = Q_s(c_{i+1}^s - c_i^s) \tag{26}$$

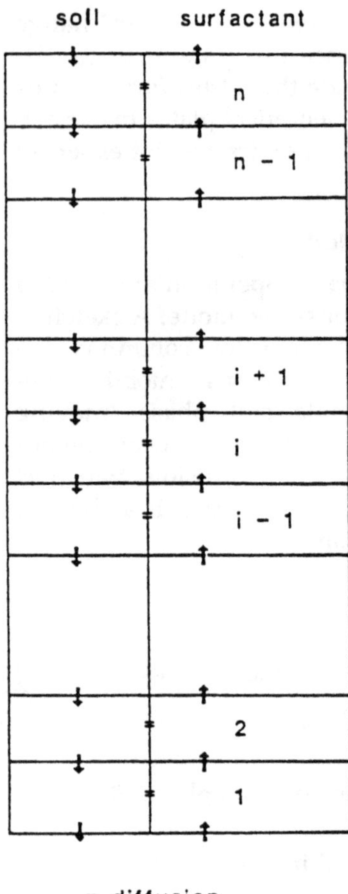

Figure 22 Mathematical partitioning of a countercurrent flow soil washing column.

We next examine diffusion transport between the soil phase and the aqueous phase by means of the lumped-parameter approach. The total mass of contaminant in the *i*th compartment is assumed to remain constant during the very short time interval during which we focus on diffusion transport. Thus

$$M_i = \Delta V_w c_i^w + \Delta V_s c_i^s = \text{constant} \tag{27}$$

We use the same adsorption isotherm as before, so at local equilibrium we have

$$c_i^{we} = K_s \frac{c_i^{se}}{m_{1/2} + c_i^{se}} \tag{28}$$

Equation (27) is also valid at local equilibrium, which yields

$$M_i = \Delta Vwc_i^{we} + \Delta Vsc_i^{we} \tag{29}$$

Substitution of equation (28) into equation (29) yields a quadratic equation in c_i^{se}, which is readily solved.

Next, we make the lumped-parameter assumption, that the diffusive mass transport of contaminant between the soil phase and the aqueous phase is governed by

$$\left(\frac{dc^s}{dt}\right)_{\text{diff}} = \lambda_1(c^{se} - c^s) \tag{30}$$

This is integrated as before, and we then use for the diffusion transport term

$$\left(\frac{dc_i^s}{dt}\right)_{\text{diff}} = (c_i^{se} - c_i^s)\frac{1 - \exp(-\lambda_1 \Delta t)}{\Delta t} \tag{31}$$

The expression for $(dc_i^w/dt)_{\text{diff}}$ is obtained by noting that for diffusion only, M_i in equation (27) is constant, so its rate of change with respect to time vanishes. This yields, after rearrangement,

$$\left(\frac{dc_i^w}{dt}\right)_{\text{diff}} = \frac{\Delta V_s}{\Delta V_w}\left(\frac{dc_i^s}{dt}\right)_{\text{diff}} \tag{32}$$

so that

$$\left(\frac{dc_i^w}{dt}\right)_{\text{diff}} = \frac{\Delta V_s}{\Delta V_w}(c_i^s - c_i^{se})\frac{1 - \exp(-\lambda_1 \Delta t)}{\Delta t} \tag{33}$$

Finally, the diffusive and advective terms are combined to obtain the differential equations modeling the countercurrent flow soil washing column. These are

$$\frac{dc_i^w}{dt} = \frac{Q_w}{\Delta V_w}(c_{i-1}^w - c_i^w) + \frac{\Delta V_s[1 - \exp(-\lambda_1 \Delta t)]}{\Delta V_w \qquad \Delta t}(c_i^s - c_i^{se}) \tag{34}$$

Table 8 Standard Parameter Set for CounterCurrent Flow Model

Column height	2 m
Column diameter	1 m
Rate of soil loading	0.001 m^3/s
Flow rate of surfactant solution	0.002 m^3/s
Downward linear velocity of soil in column	0.01 fm/s
Soil porosity	0.3
Soil density	1700 kg/m^3
Surfactant concentration	10 g/L
Surfactant critical micelle concentration	1 g/L
Initial contaminant soil concentration	10,000 mg/kg
Contaminant solubility in water	1 mg/L
Isotherm parameter, $m_{1/2}$	1 kg/m^3
Contaminant distribution coefficient, K_D	2
Effective soil lump diameter	0.1 cm
Contaminant diffusion constant in water	$10^{-9}m^2/s$
Number of compartments into which column is formally partitioned	10
dt	10

and

$$\frac{dc_i^s}{dt} = \frac{Q_s}{\Delta V_s}(c_{i+1}^s - c_i^s) - \frac{1 - \exp(-\lambda_1 \Delta t)}{\Delta t}(c_i^s - c_{ei}^s) \tag{35}$$

The model was implemented on a microcomputer and a number of runs were made to ascertain the dependence of countercurrent flow column performance on the model parameters. The standard parameter set is given in Table 8; variations from this are noted in Tables 9 to 12.

Table 9 shows the extremely large effect of soil particle size on column performance. With the other model parameters having the values given in Table 8, soil cleanup is essentially complete for particles of 1 mm diameter or less, but only about 70% of the contaminant is removed if the particles have a diameter of 2.5 mm. As before, adequate comminution of the material being washed is essential to success if porous lumps of low permeability are present. Scholze and Milanowski (1983) found this to be the case experimentally in an earlier field study of soil washing.

The effect of the linear velocity with which the soil is carried through the apparatus is shown in Table 10. A rather abrupt deterioration of separation efficiency is observed. A linear velocity of 4 cm/s yields an effluent soil contaminant concentration of less than 1 mg/kg,

Table 9 Effect of Soil Lump Diameter on Effluent Soil Contaminant Concentration

Lump diameter (cm)	Effluent soil contaminant conc. (mg/kg)
0.10	3.83×10^{-3}
0.15	6.61
0.20	654
0.25	2980

while a velocity of 8 cm/s yields an effluent soil contaminant concentration of almost 15 mg/kg. In designing a system one needs to ensure that the linear soil velocity is sufficiently small (i.e., the contact time is sufficiently large) to avoid this loss of efficiency.

The surfactant solution flow rate turns out to be another rather critical variable, as shown in Table 11. A 25% decrease in the flow rate below 10 L/s results in an increase in effluent soil contaminant concentration by five orders of magnitude. A decrease in flow rate to 5 L/s results in only roughly 50% removal of the contaminant. One should select operating parameters such that the surfactant flow rate is comfortably in excess of the critical value below which effluent soil contaminant concentrations increase so drastically.

In Situ Surfactant Flushing Model

If the nature of the site permits, in situ surfactant flushing can be used. This should result in significant cost reductions, since no soil handling

Table 10 Effect of Linear Velocity of Soil Through the Column on Effluent Soil Contaminant Concentration

v_s (cm/s)	Effluent soil contaminant concentration (mg/kg)
1	3.83×10^{-3}
2	4.19×10^{-2}
4	0.796
8	14.7
16	169
24	519

Table 11 Effect of Surfactant Solution Flow Rate on Effluent Soil Contaminant Concentration

Q_w (L/s)	Effluent soil contaminant concentration (mg/kg)
5	5370
7.5	97.4
10	3.83×10^{-3}

is required. If the domain of contamination is well below the water table, one may be forced to use this technique, since excavation of the contaminated material may be out of the question. In this section we present local equilibrium models for in situ surfactant flushing (Wilson, 1989a). These are two-dimensional, to permit use of the models on microcomputers.

Construction of a model for surfactant flushing breaks down into two tasks: calculation of the flow field in the vicinity of the wells, and use of this flow field and a suitable isotherm to model the movement of contaminant under the influence of the moving surfactant solution. We address the calculation of the flow field first.

Let us calculate the velocity field resulting from an array of injection and recovery wells in an aquifer of constant thickness in which the natural unperturbed flow of water is of constant direction and magnitude. We shall consider steady-state flow only. We define a velocity potential W such that

$$v = \nabla W \tag{36}$$

where

$$W = \sum_{i=1}^{N} c_i \log_e[(x - a_i)^2 + (y - b_i)^2] + v_x^o x + v_y^o y \tag{37}$$

Wells are located at the points $(a_i, b)_i$. It can be shown that the c_i are related to the well flow rates Q_i by

$$c_i = \frac{Q_i}{4\pi h \nu} \tag{38}$$

where h = thickness of aquifer, m
ν = porosity of aquifer

Q_i = flow rate of ith well, m^3/s, positive for injection wells, negative for recovery wells

The natural, unperturbed flow of the groundwater is assumed to be constant and uniform, with velocity components v_x^o *and* v_y^o. The velocity components of the flow field in the presence of the wells are then given by

$$v_x = \frac{1}{2\pi h \nu} \sum_{i=1}^{N} \frac{Q_i(x - a_i)}{(x - a_i)^2 + (y - b_i)^2} + v_x^o \tag{39}$$

$$v_y = \frac{1}{2\pi h \nu} \sum_{i=1}^{N} \frac{Q_i(y - b_i)}{(x - a_i)^2 + (y - b_i)^2} + v_y^o \tag{40}$$

The results above are suitable for use when the domain of interest is not bounded by any impermeable barriers. One of the problems with in situ surfactant flushing, however, is the possibility that surfactant-solubilized toxics may escape capture by a recovery well and become widely disseminated through the aquifer. This could be avoided by enclosing the zone of contamination within a slurry wall extending down to the aquitard beneath the contaminated aquifer. We therefore next address the calculation of the velocity field of an array of injection and recovery wells in a domain surrounded by an impermeable boundary.

The velocity potential function W satisfies Laplace's equation as before:

$$\nabla^2 W = 0 \tag{41}$$

The effects of the injection and recovery wells are included by writing

$$\begin{aligned} W &= \sum_{p=1}^{N} \frac{Q_p}{4\pi h \nu} \log_e[(x - a_p)^2 + (y - b_p)^2] + U \\ &= S + U \end{aligned} \tag{42}$$

where the logarithmic terms generate the sources and sinks corresponding to the injection and recovery wells. The function S is a solution to Laplace's equation, as is the function U. U is to be constructed so that W satisfies the desired boundary conditions. Let us choose the domain $0 < x < a$; $0 < y < b$ as the region to be enclosed. Then the boundary conditions on W are

$$\frac{\partial W}{\partial x}(0, y) = 0 \tag{43}$$

$$\frac{\partial W}{\partial x}(a, y) = 0 \tag{44}$$

$$\frac{\partial W}{\partial y}(x, 0) = 0 \tag{45}$$

$$\frac{\partial W}{\partial y}(x, b) = 0 \tag{46}$$

Equation (41) is then represented by a discrete mesh approximation,

$$0 = W_{i-1,j} + W_{i+1,j} + W_{i,j-1} + W_{i,j+1} - 4W_{i,j} \tag{47}$$
$$i = 2,3, \ldots, n_x - 1; \quad j = 2,3, \ldots, n_y - 1$$

Here it is assumed that $\Delta x = \Delta y$ and that

$$W_{i,j} = W[(i - ½)\,\Delta x, (j - ½)\,\Delta y]$$

In terms of the $U_{1,j}$equation (47) yields

$$U_{i,j} = ¼(U_{i-1,j} + U_{i+1,j} + U_{i,j-1} + U_{i,j+1}) \tag{48}$$

for the interior points of the mesh. Use of the boundary conditions permits the calculation of $U_{i,j}$ on the boundaries and in the corners of the domain; for example, equation (43) allows the calculation of values of U along the boundary $s = 0$; these are

$$U_{1,j} = ⅓(U_{2,j} + U_{1,j-1} + U_{1,j+1} + S_{2,j} + S_{1,j-1} + S_{1,j+1} - 3S_{1,j}) \tag{49}$$

At the corners one obtains such equations as

$$U_{11} = ½(U_{12} + U_{21} + S_{12} + S_{21} - 2S_{11}) \tag{50}$$

The resulting system of equations is then solved by iteration, using an overrelaxation method. Convergence is quite rapid.

Plots of the streamlines of the flow can readily be obtained by numerical integration of the equations

$$\frac{dx}{dt} = v_x(x, y) \tag{51}$$

$$\frac{dy}{dt} = v_y(x, y) \tag{52}$$

The velocities are calculated from equations (39) and (40) if one is dealing with an unbounded domain. In the course of mapping the streamlines one can also obtain the time required for an element of liquid to transit the trajectory being mapped out; this permits one to identify regions in the domain from which contaminant removal will be particularly slow.

Streamlines can also be computed from equations (51) and (52) if the velocity potential has been computed numerically as described above. In the vicinity of the point (x_i, y_i) one expands $W(x, y)$ in a Taylor's series about this point, uses finite-difference representations for the derivatives (through second order), and obtains the velocity components by differentiation of the resulting expression with respect to x or y.

We are now ready to begin the second part of the task—that of modeling the movement of contaminant under the influence of the flowing surfactant. To construct a model that can be run on microcomputers, we assume that the surfactant concentrations along all streamlines originating at an injection well are all equal to the surfactant concentration of the injected surfactant solution. Contaminated regions that are traversed by streamlines originating at an injection well will be cleaned up, sooner or later. Contaminated regions that lie outside the domain flushed by the injected surfactant solution will not be cleaned up. (The extremely slow cleanup resulting from flushing hydrophobic compounds with water alone is neglected.)

We assume that

$$\frac{\partial m}{\partial t} = \nabla\cdot(D\,\nabla c) - \nu\nabla\cdot(vc) \tag{53}$$

where m = soil contaminant concentration, kg/m^3
D = dispersivity tensor, m^2/s
c = contaminant concentration in the aqueous phase, kg/m^3
ν = soil porosity
v = linear velocity of fluid, m/s

and

$$c = [c_o + K_D(C - \text{CMC})]\,\frac{m}{m + m_{1/2}} \tag{54}$$

where these terms have been defined previously. Because of the numerical dispersion associated with discrete approximation of the advection term, we drop the dispersive term in equation (53) when approximating it over a mesh of points covering the domain of interest. This yields

$$\begin{aligned}\frac{dm_{ij}}{dt} = \nu h[&\Delta y v_{ij}^{L} S(v^{L}) c_{i-1,j}^{1} + \Delta y v_{ij}^{L} S(-v^{L}) c_{ij}^{1}\\ &- \Delta y v_{ij}^{R} S(-v^{R}) c_{i+1,j}^{1} - \Delta y v_{ij}^{R} S(v^{R}) c_{ij}^{1}\\ &+ \Delta x v_{ij}^{B} S(v^{B}) c_{i,j-1}^{1} + \Delta x v_{ij}^{B} S(-v^{B}) c_{ij}^{1}\\ &- \Delta x v_{ij}^{T} S(-v^{T}) c_{i,j+1}^{1} - \Delta x v_{ij}^{T} S(v^{T}) c_{ij}^{1}]\end{aligned} \tag{55}$$

where $m_{ij}(t)$ is the mass of contaminant in the ijth volume element, kg, and

$$S(x) = \begin{cases} 0, & x < 0 \\ 1, & x > 0 \end{cases}$$

and the velocities are calculated from the velocity potential W by simple finite-difference formulas.

Equations (55) are then integrated forward in time. The overall progress of cleanup is monitored by calculating the total mass of contaminant remaining in the domain of interest, given by

$$M_{\text{total}} = \sum_{i=1}^{n_x} \sum_{j=1}^{n_y} m_{ij}(t) \tag{56}$$

More detail can be obtained about the movement of contaminant by examining the masses of contaminant in the various volume elements during the course of the run. This is most readily done by calculating the quantities

$$I_{ij} = \text{Int}\left[\frac{9m_{ij}(t)}{m_{ij}(0)}\right] \tag{57}$$

and plotting these integers as an array of the same shape as the domain of interest. Here Int(u) is the largest integer less than u. Logarithmic scales that cover a much wider range can also be constructed; one such is

$$J_{ij} = \text{Int}\left[\frac{1}{\log_e 2} \log_e \frac{m_{ij}(0)}{m_{ij}(t)}\right] \tag{58}$$

For this scale, as J_{ij} goes from 0 to 9, the contaminant concentration decreases to 1/512 of its initial value, with each integer increase in J_{ij} corresponding to a concentration decrease by a factor of $\frac{1}{2}$.

We next display a number of flow fields generated by an injection well and a recovery well in an unbounded aquifer of uniform thickness and constant porosity. Then we look at flow fields generated by an injection well and a recovery well enclosed by a slurry wall—an impermeable rectangular boundary. Finally, we examine some surfactant flushing run simulations.

In the flow fields shown in Figures 23 to 26 the aquifer is unbounded. The injection (I) and recovery (R) wells are 40 m apart, and both are located on the x-axis. The aquifer thickness is 1 m and the porosity is 0.2.

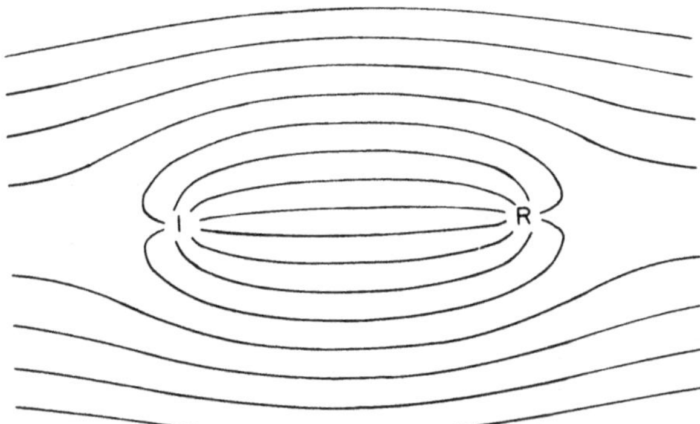

Figure 23 Velocity field around an injection well and a recovery well in an unbounded aquifer. I, injection well; R, recovery well. The distance between the two wells is 40 m, the thickness of the aquifer is 1 m, and the porosity of the aquifer is 0.2. The unperturbed flow velocity in the x-direction (v_x) is 0.001 m/s; the unperturbed flow velocity in the y-direction (v_y) is 0. The injection and recovery rates are both 0.01 m^3/s. (From Wilson, 1989a.)

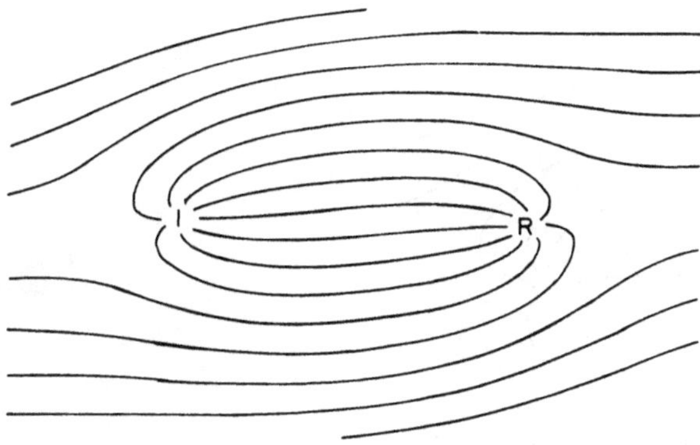

Figure 24 Velocity field around an injection well and a recovery well in an unbounded aquifer. $v_x = 0.001$, $v_y = 0.0002$ m/s; all other parameters as in Figure 23. Misalignment of the wells results in injected liquid escaping capture by the recovery well. (From Wilson, 1989a.)

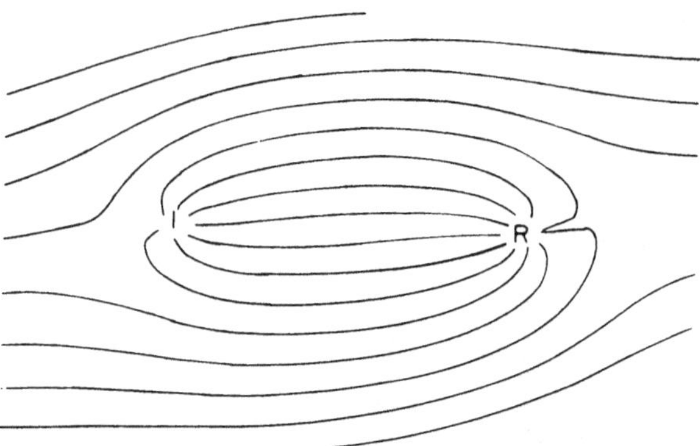

Figure 25 Velocity field around an injection well and a recovery well in an unbounded aquifer. Recovery flow rate = 0.0125 m^3/s; v_x = 0.001, v_y = 0.0002 m/s; all other parameters are as in Figure 23. Here overpumping of the recovery well compensates for misalignment. (From Wilson, 1989a.)

The importance of having the recovery well fairly precisely downstream from the injection well is illustrated in Figures 23 and 24. In Figure 23 the unperturbed, natural flow (in the absence of the wells) has an x-component of velocity equal to 0.001 m/s, and a y-component equal to zero. Both wells are being operated at a flow rate of 0.01 m^3/s. All of the streamlines originating at the injection well terminate at the recovery well, indicating that the surfactant solution injected is completely recovered. (Recall that dispersive mixing has been neglected.) In

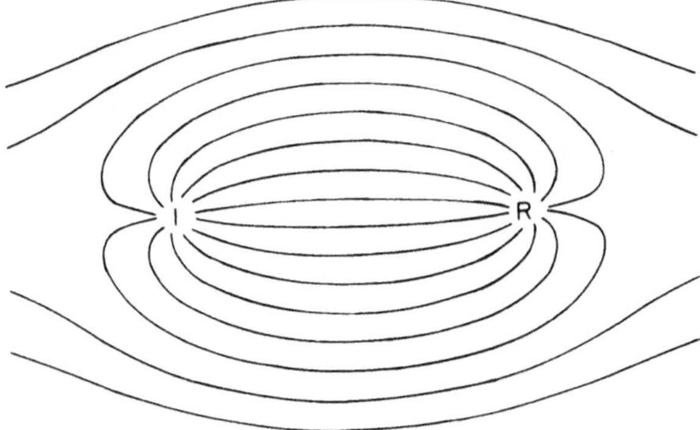

Figure 26 Velocity field around an injection well and a recovery well in an unbounded aquifer. Injection and recovery flow rates = 0.02 m^3/s; all other parameters are as in Figure 23. (From Wilson, 1989a.)

Figure 24 the unperturbed x- and y-components of the flow velocity are 0.001 and 0.0002 m/s, and both wells are operated at a flow rate of 0.01 m^3/s. Some of the streamlines originating at the injection well do not terminate at the recovery well, but pass on to the right. Evidently, for this setup some of the injected surfactant solution is escaping into the aquifer.

One can overcome this problem of sensitivity to proper alignment of the wells, but only at a price. The streamlines in Figure 25 describe the motion of water when the unperturbed x- and y-components of the flow velocity are 0.001 and 0.0002 m/s (as in Figure 24), the injection well is operating at 0.01 m^3/s, and the recovery well is pumping 0.0125 m^3/s. In this configuration all of the surfactant solution injected appears to be recovered, despite the fact that the recovery well is not directly downstream from the injection well. However, three streamlines that do not originate at the injection well terminate at the recovery well. These describe the movement of surfactant-free groundwater, which is pumped up the recovery well. This extra pumping is costly and results in undesirable dilution of the surfactant solution if it is to be recycled after treatment.

Increasing the flow rates through the injection and recovery wells broadens the area that is flushed by surfactant, as seen by comparing Figure 23 (well flow rates = 0.01 m^3/s) with Figure 26 (well flow rates = 0.02 m^3/s). A similar broadening of the domain of influence of the injection and recovery well pair is found if the unperturbed flow rate of the groundwater is reduced. This suggests that one might improve the surfactant flow pattern in the contaminated zone by the judicious placement of a slurry wall or other barrier to reduce the natural flow velocity of groundwater in and near the domain of contamination.

A major concern with in situ surfactant flushing is the possibility of pollutant-laden surfactant solution escaping capture by the recovery well(s) and moving off the site. This could result from either improper design or failure of the pump on the recovery well. A possible technique for reducing the probability of contaminant escape is to place an impermeable barrier around the zone of contamination and extending down to the aquitard. Flow fields for an injection well and a recovery well operating at the same flow rates are shown in Figures 27 and 28. In Figure 27 the domain being flushed is a 60×60 m square, with the well located 5 m in from the middle of opposite sides. (These drawings are somewhat distorted by differences in the horizontal and vertical scales.) The streamline pattern indicates fairly complete coverage of the domain, with relatively small regions of stagnation in the corners.

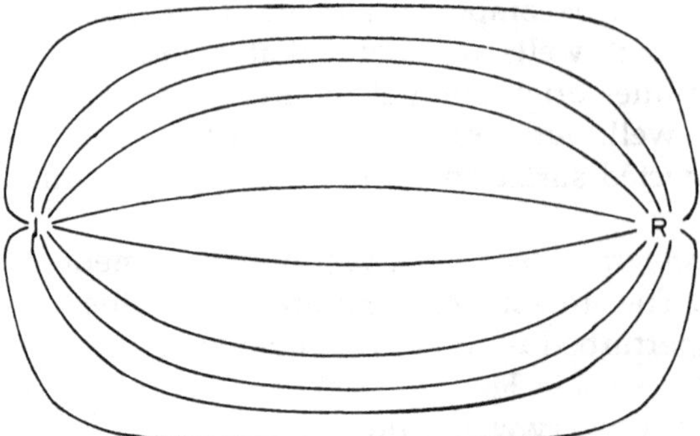

Figure 27 Velocity field around an injection well and a recovery well within a domain surrounded by an impermeable rectangular barrier. The domain is 60 × 60 m in size; the injection well is located at (5,30), and the recovery well is located at (55,30). Both wells are operated at the same flow rate. Note that the scale factors differ for the horizontal and vertical axes. (From Wilson, 1989a.)

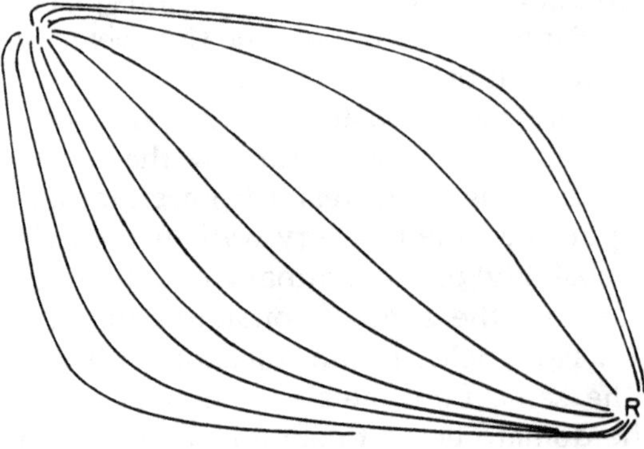

Figure 28 Velocity field around an injection well and a recovery well within a domain surrounded by an impermeable rectangular barrier. The injection well is located at (5,5), and the recovery well is located at (55,55); all other parameters are as in Figure 27. (From Wilson, 1989a.)

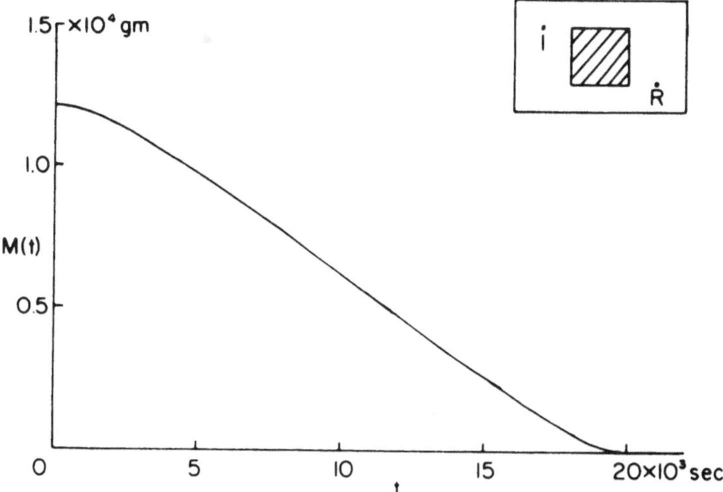

Figure 29 Plot of total mass of contaminant versus time during a surfactant flushing operation. The model parameters are given in Table 12, and the positions of the wells and the distribution of contaminant are shown in the inset to the figure. (From Wilson, 1989a.)

The wells are located in diagonally opposite corners of a 60 × 60 m domain in Figure 28; the wells are 5 m in from each of the nearby sides. This configuration appears to leave two regions of stagnation in the two corners not containing wells. Qualitatively, the amount of soil being poorly flushed appears to be somewhat larger than is the case in Figure 27. Figures 29 and 30 are plots of the total mass of contaminant in the domain of interest as a function of the flushing duration. Model parameters for Figures 29 to 31 are given in Table 12.

For the runs shown in Figures 29 and 30 removal appears to be rather linear with time after a fairly short period during which the slug of contaminant is being flushed over to the recovery well. One obtains extensive tailing, however, if a portion of the contaminated region lies in areas of the zone of interest through which movement of surfactant is quite slow, such as the corners of rectangular regions enclosed by a barrier. This is shown by the run plotted in Figure 31. The geometry of the system is exactly as in Figure 30, except that the zone of contamination fills the entire region that is enclosed within the slurry wall boundary. (Note that the scales in Figure 31 are different from those used in Figures 29 and 30.) In Figure 31 tailing is quite extensive, and it is evident

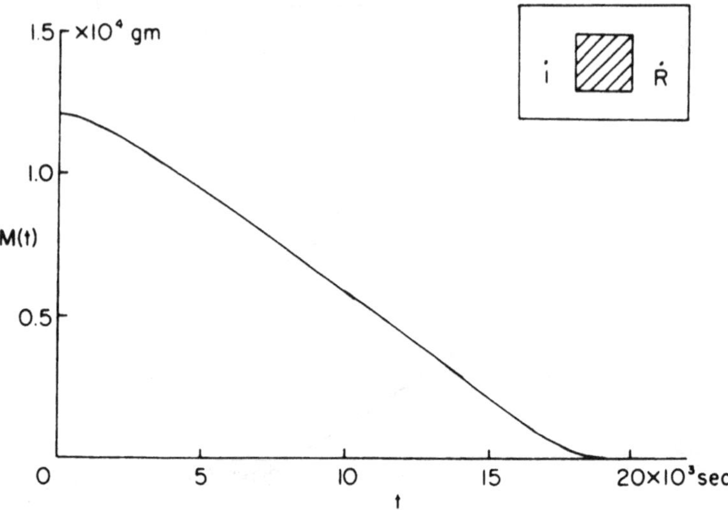

Figure 30 Plot of total mass of contaminant versus time during a surfactant flushing operation. The model parameters are given in Table 12, and the geometry of the system is shown in the inset. (From Wilson, 1989a.)

that cleanup will require far more time than is required for the runs plotted in Figures 29 and 30. One would be well advised to design slurry wall barrier systems in such a fashion that the corners of the enclosed domain do not contain significant amounts of contaminant.

The results presented above are based on a rather simple adsorption isotherm and the assumption of local equilibrium. If one has adsorption sites in the aquifer material which strongly bind the contaminant, cleanup could be much slower. This would also be true if diffusion of contaminant from lumps of porous material of low permeability were important. Preliminary laboratory studies are therefore needed to determine the ability of the surfactant solution to move the given contaminants through the aquifer material present at the site of interest. Data from such lab studies can be interpreted by means of one or more of the models described above under the heading of soil washing.

A technique suggested by R. D. Mutch, Jr. (personal communication, 1990) for investigating kinetic effects in in situ surfactant flushing consists of pumping down a slug of surfactant solution, leaving it in contact with the aquifer for a specified time, and then pumping it back out for analysis. Several experiments having different contact times should give information as to the time constant(s) associated with the

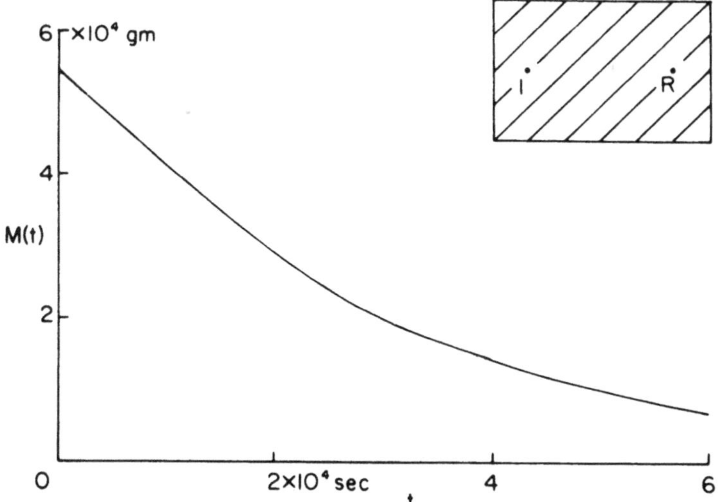

Figure 31 Plot of total mass of contaminant versus time during a surfactant flushing operation. The model parameters are given in Table 12, and the geometry of the system is shown in the inset. (From Wilson, 1989a.)

processes by which contaminant equilibrates between the stationary phase(s) and the mobile surfactant solution.

NEED FOR FUTURE RESEARCH AND DEVELOPMENT

As surfactant washing/flushing techniques are still in their infancy as compared to several other technologies, there are a number of needs for information that must be met before these techniques will achieve established status. We mention a few of these here.

Very little pilot-scale work has been done to date. Additional pilot-scale work on a range of soil matrices and contaminants is very badly needed to support the feasibility of these surfactant-based techniques. These pilot-scale data should be compared with laboratory data, the results of both jar tests and lab column work. We need to determine whether lab data can be used to predict adequately the performances of surfactant washing and flushing operations under field conditions.

At present, relatively little is known about the adsorption isotherms of contaminants that are distributed between a soil matrix and a surfactant solution. This information is needed both for preliminary qualitative assessment and for the development of improved models.

Table 12 System Parameters Used for Modeling of Surfactant Flushing

Dimensions of domain of interest
$x_{min} = 0$
$x_{max} = 30$ m
$y_{min} = 0$
$y_{max} = 20$ m
Aquifer thickness = 1.0 m
Aquifer voids fraction = 0.2
Injection rate = 0.01 m^3/s
Recovery rate = 0.01 m^3/s
Location of zone of contamination
Figures 29 and 30
$x_{min} = 10$ m
$x_{max} = 20$ m
$y_{min} = 5$ m
$y_{max} = 15$ m
Figure 31
$x_{min} = 0$
$x_{max} = 30$ m
$y_{min} = 0$
$y_{max} = 20$ m
Coordinates of injection and recovery wells
Figure 29
Injection: (5,5)
Recovery: (25,15)
Figures 30 and 31
Injection: (5,10)
Recovery: (25,10)
Initial contaminant concentration = 100 g/m^3
Contaminant isotherm parameters
Contaminant solubility in water = 1 mg/L
K_D, mg/L of contaminant per mg/L of surfactant = 0.01
Surfactant critical micelle concentration = 2300 mg/L
Influent surfactant concentration = 5000 mg/L
$m_{1/2}$, soil adsorption parameter = 10 g/m^3

The reclamation of contaminant-laden surfactant solution has been investigated in a preliminary way. Pilot-scale studies are needed, and methods for reclaiming nonionic surfactants need to be found. In connection with surfactant reclamation, more work on methods for the removal of fines is needed if surfactant-based techniques are to be successful with matrices containing clays. If in situ surfactant flushing is done in aquifers without confining boundaries, some degree of overpumping is necessary to guarantee complete recovery of the contaminant-laden solution from the aquifer. If the surfactant solution is to be reused, methods for removing the excess water from this recovered diluted surfactant solution must be developed.

Finally, all of the information above will be needed to develop credible preliminary cost figures for these technologies.

At this point one might decide to write off these technologies, since they are so undeveloped. We note, however, that in 1990 they were still in approximately the same stage of development that soil vacuum extraction was in back in the early 1980s. They present problems, but the prospects of one 100- to 1000-fold increases in cleanup rates compared to simple water flushing are very alluring and certainly warrant further investigation.

REFERENCES

Amdurer, M. 1986. *Systems to Accelerate In Situ Stabilization of Waste Deposits*. EPA/540/20-86/002. U.S. Environmental Protection Agency, Cincinnati, OH.

Castle, C., J. Bruck, D. Sappington, and M. Erbaugh. 1985. Research and development of a soil washing system for use at Superfund sites. *Proceedings of the 6th National Conference on Management of Uncontrolled Hazardous Waste Sites*, November. Hazardous Materials Control Research Institute, Silver Spring, MD, p. 452.

Clarke, A. N., P. D. Plumb, T. K. Subramanyam, and D. J. Wilson. 1991. Soil cleanup by surfactant washing. I. Laboratory results and mathematical modeling. *Sep. Sci. Technol.*, 26:301.

Dworkin, D., D. J. Messinger, and R. M. Shapote. 1988. In situ flushing and bioreclamation technologies at a creosote-based wood treatment plant. *Proceedings of the 5th National Conference on Hazardous Waste and Hazardous Materials*, Las Vegas, April.

Eckenfelder Inc. 1990. *Surfactant Flushing/Washing: An Innovative Hazardous Waste Treatment*. USEPA SBIR I 68D90125, March.

Ellis, W. D., J. R. Payne, A. N. Tatuni, and F. J. Freestone. 1984, The development of chemical countermeasures for hazardous waste contaminated soil. *Hazardous Materials Spills Conference*. U.S. Environmental Protection Agency, Cincinnati, OH, p. 116.

Ellis, W. E., J. R. Payne, and G. D. McNabb. 1985. *Treatment of Contaminated Soils with Aqueous Surfactants*. USEPA Report EPA/600/2-85/129, PB 86-122561.

Esposito, P., J. Hessling, B. B. Lock, M. Taylor, M. Szabo, R. Thurnau, C. Rogers, R. Traver, and E. Barth. 1988a. Results of evaluations of contaminated soil treatment methods in conjunction with the CERCLA BDAT program. *Proceedings of the APCA 81st Annual Meeting*, Paper 88/6B5; see *Chem. Abstr.* 110:198600b.

Esposito, M. P., B. B. Lock, J. Greber, and R. P. Traver. 1988b. *Superfund Standard Analytical Reference Matrix Preparation and Results of Physical Soils Washing Experiments*. USEPA Report EPA-600/9-88/021; see *Chem. Abstr.* 112:185211s.

Gannon, O. K., P. Bibring, K. Raney, J. A. Ward, D. J. Wilson, J. L. Underwood, and K. A. Debelak. 1989. Soil cleanup by in situ surfactant flushing. III. Laboratory results. *Sep. Sci. Technol.* 24:1073.

Hall, D. G., and B. A. Pethica. 1967. Thermodynamics of micelle formation, p. 516. In: M. Schick (ed.), *Nonionic Surfactants*. Marcel Dekker, New York.

Hall, D. G., and G. J. T. Tiddy. 1981. Surfactant solutions: dilute and concentrated, p. 55. In: E. H. Lucassen-Reynders (ed.), *Ionic Surfactants: Physical Chemistry of Surfactant Action*. Marcel Dekker, New York.

Hazardous Materials Control Research Institute. 1987. *Evaluating In Situ Surfactant for Creosote Cleanup*. HMCRI, Silver Spring, MD, November, p. 5.

Hazardous Substance Data Bank, National Library of Medicine.

Hill, H. J., J. Reisberg, and G. L. Stegemeier. 1973. Aqueous surfactant systems for oil recovery. *J. Pet. Technol.* 25:186.

Israelachvili, J. N. 1985. Thermodynamic and geometric aspects of amphiphile aggregation into micelles, vesicles and bilayers, and the interactions between them, p. 24. In: V. DeGiorgio and M. Corti (eds.), *Physics of Amphiphiles: Micelles, Vesicles, and Microemulsions*. North-Holland, Amsterdam.

Klevens, K. B. 1950. Solubilization. *Chem. Rev.* 47:1.

Klins, M. A., S. M. Farouqali, and C. D. Stahl. 1976. *Tertiary Recovery of the Bradford Crude by Micellar Slugs and Three Different Polymer Buffers*. ERDA Contract E(40-1)-5078.

Kuhn, R. C., and K. R. Piontek. 1989. A site-specific in situ treatment process development program for wood preserving site. *R. S. Kerr Technical Assistance Program: Oily Waste, Fate Transport, Site Characterization, and Remediation*. Denver, CO, May.

Kunze, M. E., and J. R. Gee. 1989. Bench- and pilot-scale case studies for metals and organics removals from CERCLA site soils. *HMCRI's 10th National Conference—Superfund*, Washington, DC, p. 207.

Levine, I. N. 1988. *Physical Chemistry*, 3rd ed. McGraw-Hill, New York, p. 483.

Lucassen-Reynders, E. H. 1981. *Anionic Surfactants: Physical Chemistry of Surfactant Action*. Marcel Dekker, New York.

McBain, J. W. 1950. *Colloid Science*. D. C. Heath, Boston.

McBain, M. E. L., and E. Hutchinson. 1955. *Solubilization*. Academic Press, New York.

McDermott, J. B., R. Unterman, M. J. Brennan, R. E. Brooks, D. P. Mobley, C. C. Schwartz, and D. K. Dietrich. 1988. Two strategies for PCB soil remediation: biodegradation and surfactant extraction. *AIChE Meeting*, New Orleans, March.

Mercer, J. W., D. C. Skipp, and D. Giffin. 1990. Basics of Pump-and-Treat Ground-Water Remediation Technology. U.S. EPA Report No. EPA-600/8-90/003.

Millington, R. J., and J. M. Quirk. 1961. Permeability of porous solids. *Trans. Faraday Soc.* 57:1200.

Montgomery, J. H., and L. M. Welkom. 1990. *Groundwater Chemicals Desk Reference*. Lewis Publishers, Chelsea, MI.

Mukerjee, P. 1971a. Solubilization of benzoid acid derivatives by nonionic surfactants. Location of solubilizates in hydrocarbon core of micelles and poly-(oxyethylene) mantle. *J. Pharm. Sci.* 60:1528.

Mukerjee, P. 1971b. Analysis of distribution model for micellar solubilization using thermodynamics of small systems. Nonideality of solubilization of benzoid acid derivatives in nonionic surfactants. *J. Pharm. Sci.* 60:1531.

Nash, J. H. 1987. *Field Studies of In Situ Soil Washing*. USEPA Report EPA/600/2-87/110, PB 88-146808.

Nash, J. H., R. P. Traver, and D. C. Downey. 1987. *Surfactant Enhanced In Situ Soil Washing*. U.S. Environmental Protection Agency Hazardous Waste Environmental Research Laboratory, Edison, NJ, September.

NIOSH Pocket Guide to Chemical Hazards. 1987. National Institute of Occupational Safety and Health, Washington, DC.

Rosen, M. J. 1989. *Surfactants and Interfacial Phenomena*, 2nd ed. Wiley-Interscience, New York.

Roy, W. R., and R. A. Griffin. 1988. *Surfactant- and Chelate-Induced Decontamination of Soil Materials: Current Status*. Open File Report 21. University of Alabama Environmental Institute for Waste Management Studies, Tuscaloosa, AL.

Scamehorn, J. F. 1986. *Phenomena in Mixed Surfactant Systems*. ACS Symposium Series 311. American Chemical Society, Washington, DC.

Scholz, R., and J. Milanowski. 1983. *Project Summary, Mobile System for Extracting Spilled Hazardous Materials from Excavated Soils*. USEPA Report EPA-600/S2-83-100. See also *Mobile System for Extracting Spilled Hazardous Materials from Excavated Soils*. NTIS PB 84-123637, the complete report.

Tabak, M. E., and R. P. Traver. 1990. Evaluation of EPA soil washing technology for remediation at UST sites. *16th Annual EPA Hazardous Waste Research Symposium*, Cincinnati, OH, April.

Texas Research Institute. 1979. *Underground Movement of Gasoline on Groundwater and Enhanced Recovery by Surfactants*. American Petroleum Institute, Washington, DC, September.

Texas Research Institute. 1985. *Test Results of Surfactant Enhanced Gasoline Recovery in Large-Scale Model Aquifer*. API Publication 4390. American Petroleum Institute, Washington, DC, April.

Traver, R. P., et al. 1989. Evaluation of USEPA soil washing technology for remediation at UST sites. *HMCRI's 10th National Conference—Superfund*, Washington, DC, p. 202.

U.S. Environmental Protection Agency. 1986. Surfactant-assisted flushing, p. 126. In: *Systems to Accelerate In Situ Stabilization of Waste Deposits*. USEPA Report EPA/540/2-86/002.

Valsaraj, K. T., and L. J. Thibodeaux. 1989. Relationships between micelle–water and octanol–water partition constants for hydrophobic organics of environmental interest. *Water Res.* 23:183.

Verschueren, K. 1983. *Handbook of Environmental Data on Organic Chemicals*, 2nd ed. Van Nostrand Reinhold, New York.

Vold, R. D., and M. J. Vold. 1983. *Colloid and Interface Chemistry*. Addison-Wesley, Reading, MA.

Wayt, H. J., and D. J. Wilson. 1989. Soil cleanup by in situ surfactant flushing. II. Theory of micellar solubilization. *Sep. Sci. Technol.* 24:905.

Wilson, D. J. 1989a. Soil cleanup by in situ surfactant flushing. I. Mathematical modeling. *Sep. Sci. Technol.* 24:863.

Wilson, D. J. 1989b. In situ surfactant flushing: a developing technology. *Proceedings of the 2nd Annual Hazardous Materials Conference/Central*, Rosemont, IL, March 14–16, p. 27.

Wilson, D. J., and A. N. Clarke. 1988. Soil cleanup by in situ surfactant flushing. I. Mathematical modeling and lab scale results. *Proceedings of the DOE Model Conference*, Oak Ridge, TN, October 3–7, Vol. 3, p. 735.

CONTRIBUTORS

Richard J. Ayen holds a Ph.D. degree in chemical engineering from the University of Illinois. His current position is Vice President and General Manager at RUST Remedial Services Inc., Clemson Technical Center, Inc., in Anderson, South Carolina. Dr. Ayen is responsible for treatability studies and proposal support for soil and sludge remediation activities. He directs activities that provide technical assistance to fixed treatment sites, and he is responsible for the piloting and start-up of new hazardous waste treatment processes. Current areas of activity include thermal desorption for remediation of contaminated soils and sludges, advanced technologies for dechlorination of PCBs and for soil washing, process simulation and modeling, stabilization formulation development, and refinery waste management. Prior to joining Clemson Technical Center, Dr. Ayen was Manager, Process Development Department, with Stauffer Chemical Company and was a member of the chemical engineering department faculty at the University of California, Berkeley. He is the author of numerous technical publications and holds two patents.

Ann N. Clarke is Director, Remedial Technologies Development Division, ECKENFELDER INC., Nashville, Tennessee, and has over twenty years' experience in environmental management. Among other activities, she has directed state-of-the-art studies for PCBs, PCDDs, and PCDFs in complex construction and environmental media. She also has experience in the study of the fate of chemicals, toxic substances assessments, environmental assessments (national and international), hazardous waste sampling program design including QA/QC, analytical aspects of trace level organic compounds, soil and groundwater contamination studies, and estuary water quality studies. Dr. Clarke is currently working on innovative remedial options at several Superfund sites. She is also a past member of the Board of Directors of the National Environmental Trainers Association; a member of the Scientific Review Panel, National Library of Medicine, and Environmental Engineering Review Panel, USEPA; and the author of several books and over 100 publications. Dr. Clarke has a Ph.D. degree from Vanderbilt University, Nashville, and holds M.A. and B.S. degrees in chemistry and the M.A. degree in geology.

James H. Clarke is Chairman and President of ECKENFELDER INC., an environmental science and engineering firm specializing in industrial waste management. Dr. Clarke has over 20 years of experience in environmental chemistry and chemical risk assessment. His primary areas of interest include the fate and transport of chemicals in the environment, the design of environmental data acquisition programs for evaluation of the risks associated with chemical releases, and innovative and emerging technologies for hazardous waste site remediation. Dr. Clarke is a member of the Remediation of Buried and Tank Wastes Committee of the National Academy of Science. He is an Adjunct Professor with the Department of Civil and Environmental Engineering of Vanderbilt University and serves on the faculty of several continuing education programs, including those of the American Institute of Chemical Engineers, the Center for Professional Advancement, and several universities. Dr. Clarke has also served as an expert witness in several cases involving the release of industrial chemicals to the environment and the remediation technologies that are feasible and appropriate. Dr. Clarke received the B.A. degree in chemistry from Rockford College, and the Ph.D. degree in theoretical physical chemistry from The Johns Hopkins University.

Jesse R. Conner is a Senior Research Scientist, in the Department of Stabilization Systems, with RUST Remedial Services Inc., and has nearly 40 years of professional experience, mostly in the environmental and waste treatment areas. He co-founded several companies in the hazardous waste stabilization field, is the author of a highly regarded book on chemical fixation and solidification, and consults extensively in this area. Currently he is involved in the evaluation and development of innovative stabilization systems for "Land Ban" and remedial action requirements, the supervision of treatability studies on stabilization, and regulatory review and interaction with the EPA on stabilization. He has been active for 20 years in research, testing, commercial development, operation, and management of hazardous waste treatment. He is widely known as an expert in waste stabilization and fixation. He was the winner of the IR-100 Award (1970) and the John C. Vaaler Award (1972) for the Chem-Fix Process and holds eight U.S. patents. He frequently contributes to professional publications in the field and has made numerous presentations on chemical fixation and stabilization at technical meetings.

Paul R. dePercin graduated from the University of Maryland in Chemical Engineering in 1971. He joined the United States Environmental Protection Agency in 1971 and worked eight years in air and water enforcement in various offices (Baton Rouge, Chicago, and Denver). In 1979, he joined the Office of Research and Development in Cincinnati, performing air emissions and control research from industrial facilities. This led to research on air emissions from RCRA and Superfund sites, and hazardous waste treatment processes.

Since 1985 Dr. dePercin has been a project engineer in the Superfund Innovative Technology Evaluation (SITE) program, performing field demonstrations and research of new and innovative hazardous waste treatment technologies. He also supports the Superfund Technical Assistance Response Team (START) by providing technical assistance to the Regional and State Regulatory Agencies.

Robert E. Hinchee is Research Leader in the Department of Environmental Technology Development at Battelle Memorial Institute in Columbus, Ohio. He has extensive experience in the development and application of technology to the assessment and remediation of contaminated sites. He has been involved in the development and application of new technologies including soil gas surveying, soil venting, in situ bioremediation, and TCE cometabolic bioremediation at more than 200 sites throughout the United States. He has had direct experience in the design and evaluation of groundwater pump and treat systems as well as soil treatment and has been responsible for design and implementation of field demonstration in situ processes such as forced-air soil venting, enhanced bioreclamation, and in-place stabilization systems. Dr. Hinchee is the senior technical leader on the North Tank Area bioremediation demonstration as part of Tinker Air Force Base's Innovative Technology Demonstration program. He is also Project Manager of joint

Air Force/EPA studies at Hill Air Force Base, Utah, and at Eielson Air Force Base, Alaska, in which bioventing is being coupled with soil heating to accelerate and optimize bioremediation. As program manager under contract to the U.S. Air Force, Dr. Hinchee oversaw the development of a pilot-scale (200 1) reactor for cometabolic treatment of TCE-contaminated groundwater at Tinker Air Force Base, Oklahoma. He has designed and supervised installation and evaluation of innovative soil venting systems for in situ removal of volatile organics from the vadose zone at numerous sites throughout the United States. Dr. Hinchee has also served as project manager and engineer in charge of a large full-scale bioreclamation demonstration program for the U.S. Air Force and as project manager for an ARCO bioventing study near Kenai, Alaska. He organized and chaired the International Symposium on In Situ and On-Site Bioreclamation in San Diego in March, 1991, and is currently organizing the 1993 symposium. His prior professional experience includes work at EA Engineering Science and Technology, Utah Water Research Laboratory, Logan, Utah, U.S. Department of Agriculture, Jackson, Wyoming, and Louisiana State University, Baton Rouge. Dr. Hinchee received the B.S. degree in zoology/chemistry from Utah State University (1974), the M.S. degree in oceanography from Louisiana State University (1977), and the Ph.D. degree in civil and environmental engineering from Utah State University (1983).

Ronald E. Hoeppel is a soil microbiologist in charge of bioremediation studies in the Environmental Restoration Division at the Naval Civil Engineering Laboratory, Port Hueneme, California. He has conducted and managed environmental research studies at Department of Defense Facilities and for Army Civil Works projects for the past 20 years. These include land treatment of wastewater, dredged material geochemical studies, aquatic herbicide development and testing, and environmental microbiology. He has been conducting fuel and ordnance bioremediation laboratory and field studies for the past seven years, with emphasis on bioventing low-volatility fuels. Field studies include large-scale bioremediation of jet fuel, diesel, and waste oil contamination at several Navy bases involving both heap pile and in situ techniques. He has over 50 professional publications in various subject areas and has served on many advisory committees concerned with bioremediation and innovative cleanup technologies. He is a member of the American Society for Microbiology and the Society of Environmental Toxicology and Chemistry. He received the B.S. degree in biology/geology from California Polytechnic State University, and the M.S. degree in soil microbiology from New Mexico State University.

Robert D. Mutch, Jr. P.Hg., P.E., is an Executive Vice President and Corporate Director for Hydrogeology and Waste Management for ECKENFELDER INC. He is also an Adjunct Professor at Manhattan College in Riverdale, New York. He has 20 years of experience in the fields of hydrogeology and land disposal engineering. His work has included the investigation and remedial design of hundreds of municipal and hazardous waste disposal sites, including dozens of Superfund sites. He has designed and, in most cases, supervised construction of 14 miles and 1,500,000 square feet of subsurface cutoff wall, 2½ square miles of low-permeability landfill caps, 15 miles of retrofitted leachate collection systems, and numerous groundwater recovery systems, ranging from 50,000–2,000,000 gallons per day. He has also provided consultation to the United Nations Environmental Programme (UNEP) in regard to international landfill problems. He holds B.S. and M.S. degrees in civil engineering from New Jersey Institute of Technology. He is certified by the American Institute of Technology and by the American Institute of Hydrology as a professional hydrogeologist, and he is a licensed professional engineer in the states of New Jersey and New York. Mr. Mutch has numerous publications in the areas of secure landfill design, landfill remedial measures, and groundwater hydrology of landfill sites. He is a member of the National Water Well Association, the American Society of Civil Engineers, and the National Environmental Trainers Association. His current activities focus on development of in situ treatment technologies for remediation of Superfund sites, including such technologies as in situ biotreatment, in situ vapor stripping, and in situ soil flushing.

Robert D. Norris is Technical Director of Bioremediation Services for ECKENFELDER INC., and has more than 11 years experience in environmental applications and consulting. He was involved for four years in the environmental applications of hydrogen peroxide and has experience in the application of chemical, physical, and biological treatments of soils and ground waters. Since 1983, he has been involved in the development and application of bioremediation technologies. He holds four U.S. patents on various aspects of bioremediation and has conducted in situ and ex situ bioremediation under a wide range of conditions including designing and/or managing of over 40 bioremediation projects. These include treatment of petroleum hydrocarbons in aquifers, in situ treatment of unsaturated soils, soil cell treatment, and land farming of contaminated soils. He was project manager for the first field demonstration of the use of hydrogen peroxide for bioremediation of aquifers. Dr. Norris's involvement in bioremediation has included laboratory research, development of laboratory protocols for treatability tests, management of pilot tests, management of commercial bioremediation projects, evaluation of bioremediation feasibility, design of bioremediation systems, regulatory interactions, and management of bioremediation groups. He has a B.S. degree in chemistry from Beloit College (Wisconsin) and a Ph.D. degree in organic chemistry from the University of Notre Dame (Indiana).

Kenton H. Oma is Assistant Technical Director for the Remedial Technologies Development Division of ECKENFELDER INC. in Nashville, Tennessee. He is currently involved in laboratory and pilot-scale treatability testing of innovative technologies. During his professional career, Mr. Oma has gained in-depth expertise in developing and testing waste immobilization processes. He is one of the developers of the In Situ Vitrification (ISV) process. ISV is a thermal treatment process that converts in-ground waste and contaminated soil into a chemically inert, stable glass and crystalline product by passing an electric current between electrodes inserted in the soil. He holds three patents on the ISV technology and a fourth patent on an in situ heating process. He and his coworkers received the Outstanding Engineering Achievement Award from the National Society of Professional Engineers in 1986 for developing the ISV process. Mr. Oma holds a B.S. degree from Montana State University and an M.S. degree from Rice University, both in chemical engineering. He has authored over 40 publications in the area of hazardous and radioactive waste processing and remediation.

Carl R. Palmer holds a B.S. degree in chemical engineering from the University of Illinois and is a Registered Professional Engineer in the State of Illinois. His current position is Project Manager, RUST Remedial Services Inc., Clemson Technical Center, Inc., in Anderson, South Carolina. Mr. Palmer performs process engineering and project management in support of Clemson Technical Center's process development program. He also performs technical services for the company's treatment, storage and disposal facilities. Since his employment with Clemson Technical Center, (formerly Chemical Waste Management) in 1987, he has actively supported the design, construction, start-up testing and operation of the X*TRAX™ thermal desorption systems, including the lab, pilot, and commercial units. He is qualified to act as an Operations Manager for both the pilot and commercial X*TRAX systems. Prior to joining Chemical Waste Management, Mr. Palmer was a Nuclear Process Engineer in the Radwaste Division of Sargent & Lundy Engineers.

Danny D. Reible is a Professor of Chemical Engineering at Louisiana State University and Shell Chair of Environmental Engineering in the Chemical Engineering Department at the University of Sydney in Australia. He received a Ph.D. degree in chemical engineering in 1982 from the California Institute of Technology. His research activities have been focused on the movement and fate of chemicals in the natural environment. He has more than 35 journal publications on topics such as dispersion in the atmosphere, contaminant adsorption and desorption from sediments, and the movement and dissolution of nonaqueous phase liquids in soils. Together with Dr. James Clarke and Dr. Louis Thibodeaux, he teaches a short course on chemodynamics for the American Institute of Chemical Engineers and consults for several companies in this same area. He is active in several professional organizations including the American Institute of Chemical Engineers, of which he is currently Chair of the Executive Board of the National Programming Committee.

Joanna I. Scott is a mathematical modeler at ECKENFELDER INC., where she has developed and worked with models of groundwater flow, contaminant transport, soil vapor stripping, and others. She has also worked with statistics and geostatistics, including kriging. Prior to joining ECKENFELDER INC., she worked for The Analytic Sciences Corporation modeling various aspects of nuclear waste disposal. She has a B.A. degree from Rice University in chemistry and mathematics and a Ph.D. degree from M.I.T. in physical chemistry.

Carl P. Swanstrom holds a B.S. degree in chemical engineering from the Illinois Institute of Technology. His current position is President of CPS Consulting in Naperville, Illinois, and he is a consultant to RUST Remedial Services Inc. Mr. Swanstrom is an experienced operational manager and start-up engineer. He has been heavily involved in the commercialization of an innovative low-temperature thermal desorption technology, X*TRAX™, which will replace incineration in many applications. This is the most important new treatment technology now evolving in Clemson Technical Center. This effort includes management of the treatability studies for thermal desorption using a small pilot version of X*TRAX. He recently participated in the successful start-up of the full-scale version of the technology. Other activities include tracking and managing activities relating to soil roaster and sludge dryer technologies. Prior activities include management of design and start-up activities for a wide range of waste treatment systems including drummed solvent reclamation, reactive solids disposal systems, development of solidification/stabilization procedures, and waste acid treatment. Prior to his tenure as Senior Project Manager at Chemical Waste Management, Mr. Swanstrom was a Senior Project Leader at Beatrice Foods Corporation. He is the author of numerous technical publications and holds two patents.

David J. Wilson is Professor of Chemistry and Environmental Engineering at Vanderbilt University. His studies in the environmental area include work on the foam flotation of toxic metals from aqueous systems, on precipitation processes and clarifier operation, and on the solvent sublation of trace organics from water. His research interests also include hazardous waste site remediation, particularly experimental work and mathematical modeling studies on soil vapor extraction, surfactant flushing, air sparging in the zone of saturation, diffusion-limited pump and treat operations, and low-temperature thermal treatment of contaminated soils. Dr. Wilson is a Senior Research Associate with ECKENFELDER INC., and does extensive consulting work on environmental problems. He is co-author (with Dr. Ann Clarke) of a book on foam flotation, and he has authored about 250 publications. He was an Alfred P. Sloan Fellow and taught at the University of Rochester and the University of Ife, Nigeria, before coming to Vanderbilt. Dr. Wilson received a B.S. degree in chemistry from Stanford University and a Ph.D. degree in physical chemistry from the California Institute of Technology.

Index

For Product Safety Concerns and Information please contact our EU representative GPSR@taylorandfrancis.com
Taylor & Francis Verlag GmbH, Kaufingerstraße 24, 80331 München, Germany

www.ingramcontent.com/pod-product-compliance
Lightning Source LLC
Chambersburg PA
CBHW060419220326
41598CB00021BA/2223

* 9 7 8 0 3 6 7 4 0 2 2 2 8 *